Advances in Remote Sensing for Forest Fire Management

AF752344

Earth Observation Series

Advances in Remote Sensing for Forest Fire Management

Technologies, Applications, Case Studies and Future Directions

Edited by

Carmen Quintano

Electronic Technology Department, IuFOR, University of Valladolid, Valladolid, Spain

Alfonso Fernández-Manso

Agrarian Science and Engineering Department, University of León, Ponferrada, Spain

José Manuel Fernández-Guisuraga

Area of Ecology, Department of Biodiversity and Environmental Management, Faculty of Biological and Environmental Sciences, University of León, León, Spain; Institute for the Environment and Global Change (IMACG), University of León, León, Spain

Leonor Calvo

Area of Ecology, Department of Biodiversity and Environmental Management, Faculty of Biological and Environmental Sciences, University of León, León, Spain; Institute for the Environment and Global Change (IMACG), University of León, León, Spain

Series Editor

George P. Petropoulos

ELSEVIER

Elsevier
Radarweg 29, PO Box 211, 1000 AE Amsterdam, Netherlands
125 London Wall, London EC2Y 5AS, United Kingdom
50 Hampshire Street, 5th Floor, Cambridge, MA 02139, United States

Copyright © 2027 Elsevier Ltd. All rights are reserved, including those for text and data mining, AI training, and similar technologies.

For accessibility purposes, images in electronic versions of this book are accompanied by alt text descriptions independently provided by Elsevier, generated with the aid of AI. For more information, see https://www.elsevier.com/about/accessibility or Radarweg 29, 1043 NX Amsterdam, Netherlands.

Books and Journals published by Elsevier comply with applicable product safety requirements. For any product safety concerns or queries, please contact our authorised representative, Elsevier B.V., at productsafety@elsevier.com.

Publisher's note: Elsevier takes a neutral position with respect to territorial disputes or jurisdictional claims in its published content, including in maps and institutional affiliations.

No part of this publication may be reproduced or transmitted in any form or by any means, electronic or mechanical, including photocopying, recording, or any information storage and retrieval system, without permission in writing from the publisher. Details on how to seek permission, further information about the Publisher's permissions policies and our arrangements with organizations such as the Copyright Clearance Center and the Copyright Licensing Agency, can be found at our website: www.elsevier.com/permissions.

This book and the individual contributions contained in it are protected under copyright by the Publisher (other than as may be noted herein).

Notices
Knowledge and best practice in this field are constantly changing. As new research and experience broaden our understanding, changes in research methods, professional practices, or medical treatment may become necessary.

Practitioners and researchers must always rely on their own experience and knowledge in evaluating and using any information, methods, compounds, or experiments described herein. In using such information or methods they should be mindful of their own safety and the safety of others, including parties for whom they have a professional responsibility.

To the fullest extent of the law, neither the Publisher nor the authors, contributors, or editors, assume any liability for any injury and/or damage to persons or property as a matter of products liability, negligence or otherwise, or from any use or operation of any methods, products, instructions, or ideas contained in the material herein.

ISBN: 978-0-443-41692-7

For Information on all Elsevier publications
visit our website at https://www.elsevier.com/books-and-journals

Publisher: Peter Llewellyn
Acquisitions Editor: Menka Mehta
Editorial Project Manager: Ashavari Bhattacharya
Production Project Manager: Gomathi Sugumar
Cover Designer: Bakyalakshmi S.

Working together
to grow libraries in
developing countries
www.elsevier.com • www.bookaid.org

Typeset by MPS Limited, Chennai, India

Contents

List of contributors

Miguel Aguilera Facultad de Ciencias Agrarias y Forestales, Centro de Desarrollo del Secano Interior, Universidad Católica del Maule, Talca, Chile

Ernesto Alvarado-Celestino School of Environmental and Forest Sciences, University of Washington, Seattle, WA, United States

Juan Gabriel Álvarez-González Department of Agroforestry Engineering, Higher Polytechnic School of Engineering, University of Santiago de Compostela, Lugo, Spain

Jose Manuel Álvarez-Martínez Area of Ecology, Department of Biodiversity and Environmental Management, Faculty of Biological and Environmental Sciences, University of León, León, Spain; Institute for the Environment and Global Change (IMACG), University of León, León, Spain

David Beltrán-Marcos Area of Ecology, Department of Biodiversity and Environmental Management, Faculty of Biological and Environmental Sciences, University of León, León, Spain

Carlos Blanco-Alegre Department of Physics, Universidad de León, León, Spain

Allison Bredder Department of Geographical Sciences, University of Maryland, College Park, MD, United States

Carlos Briones-Herrera Faculty of Forestry, University of Juarez del Estado de Durango, Durango, Mexico

Robert E. Burgan Rocky Mountain Research Station, USDA Forest Service, Missoula, MT, United States

Ana Calvo Department of Physics, Universidad de León, León, Spain

Leonor Calvo Area of Ecology, Department of Biodiversity and Environmental Management, Faculty of Biological and Environmental Sciences, University of León, León, Spain; Institute for the Environment and Global Change (IMACG), University of León, León, Spain

Jordan Caraballo-Vega Department of Geographical Sciences, University of Maryland, College Park, MD, United States; NASA Goddard Space Flight Center, Greenbelt, MD, United States

Alberto Cerrillo R&D Department, Vexiza, León, Spain

Dong Chen Department of Geographical Sciences, University of Maryland, College Park, MD, United States

José Javier Corral-Rivas Faculty of Forestry, University of Juarez del Estado de Durango, Durango, Mexico

Martín Cuahutle Cuahutle National Commission for the Understanding and Use of Biodiversity, Mexico City, Mexico

Lilia de Lourdes Manzo-Delgado National Earth Observation Laboratory, Autonomous University of Mexico, Mexico City, Mexico

Giandomenico De Luca Institute of BioEconomy (IBE), National Research Council of Italy (CNR), Sesto Fiorentino, Italy

Mesías Edwin Domínguez Faculty of Forestry, University of Juarez del Estado de Durango, Durango, Mexico

Fabiola Esquerra National Forestry Commission, Zapopan, Jalisco, Mexico

Massimiliano Farris Department of Geography, University of Chile, Santiago, Chile

Paulo M. Fernandes CITAB - Centre for the Research and Technology of Agro-Environmental and Biological Sciences, University of Trás-os-Montes e Alto Douro, Vila Real, Portugal

Víctor Fernández-García Department of Engineering and Agricultural Sciences, School of Agricultural and Forestry Engineering, Universidad de León, Ponferrada, León, Spain

José Manuel Fernández-Guisuraga Area of Ecology, Department of Biodiversity and Environmental Management, Faculty of Biological and Environmental Sciences, University of León, León, Spain; Institute for the Environment and Global Change (IMACG), University of León, León, Spain

Alfonso Fernández-Manso Agrarian Science and Engineering Department, University of León, Ponferrada, Spain

Roberto Fraile Department of Physics, Universidad de León, León, Spain

María Fuentealba Institute of Geography, Pontificia Universidad Católica de Chile, Santiago, Chile; Instituto de Ecología & Biodiversidad, Pontificia Universidad Católica de Chile, Santiago, Chile; Centro UC Desierto de Atacama, Pontificia Universidad Católica de Chile, Santiago, Chile

Mariano Garcia Universidad de Alcalá, Departamento de Geología, Geografía y Medio Ambiente, Environmental Remote Sensing Research Group, Alcalá de Henares, Madrid, Spain

Rebecca Kate Gibson NSW Department of Climate Change, Energy, the Environment and Water, Parramatta, New South Wales, Australia

Colvert Gomez National Earth Observation Laboratory, Autonomous University of Mexico, Mexico City, Mexico

João Gonçalves BIOPOLIS Program in Genomics, Biodiversity and Land Planning, CIBIO, Centro de Investigação em Biodiversidade e Recursos Genéticos, InBIO Laboratório Associado, Vairão, Portugal

Miguel Iglesias González R&D Department, Vexiza, León, Spain

César Alberto Robles Gutiérrez National Forestry Commission, Zapopan, Jalisco, Mexico

Francisca Gutiérrez-Cáceres Centro Regional Copernicus-LAC Chile, University of Chile, Santiago, Chile

Katharine Haynes NSW National Parks and Wildlife Service, Parramatta, New South Wales, Australia

Ursula Berenice García Herrera National Forestry Commission, Zapopan, Jalisco, Mexico

Enrique J. Jardel-Pelaez South Coast University Centre, University of Guadalajara Atlas de Navarro, Autlán de Navarro, Mexico

William Matthew Jolly Rocky Mountain Research Station, USDA Forest Service, Missoula, MT, United States

Amin Khairoun Universidad de Alcalá, Departamento de Geología, Geografía y Medio Ambiente, Environmental Remote Sensing Research Group, Alcalá de Henares, Madrid, Spain

Tatiana Loboda Department of Geographical Sciences, University of Maryland, College Park, MD, United States

Luis Filipe Lopes Forest Research Centre, School of Agriculture, University of Lisbon, Lisbon, Portugal

María Isabel Cruz López National Commission for the Understanding and Use of Biodiversity, Mexico City, Mexico

Pablito Marcelo López-Serrano Faculty of Forestry and Timber Industry, University of Juarez del Estado de Durango, Durango, Mexico

Elena Marcos Department of Biodiversity and Environmental Management, Faculty of Biological and Environmental Sciences, University of León, León, Spain

Oscar Enrique Balcázar Medina South Coast University Centre, University of Guadalajara Atlas de Navarro, Autlán de Navarro, Mexico

Giuseppe Modica Department of Veterinary Sciences, University of Messina, Messina, Italy

Norma Angélica Monjarás-Vega Faculty of Forestry, University of Juarez del Estado de Durango, Durango, Mexico

Graham Moore NSW National Parks and Wildlife Service, Parramatta, New South Wales, Australia

Juan Miguel Campos Muñoz National Forestry Commission, Zapopan, Jalisco, Mexico

Ana Novo Misión Biolóxica de Galicia, Consejo Superior de Investigaciones Científicas, Av. de Vigo, Santiago de Compostela, Spain

Martin Ibarra Ochoa National Meteorological Service, National Water Commission, Mexico City, Mexico

Patricia Oliva Universidad de Alcalá, Departamento de Geología, Geografía y Medio Ambiente, Environmental Remote Sensing Research Group, Alcalá de Henares, Madrid, Spain

Elijah Orland Department of Geographical Sciences, University of Maryland, College Park, MD, United States; University of Maryland - Baltimore County, Baltimore, MD, United States

Sean A. Parks Aldo Leopold Wilderness Research Institute, Rocky Mountain Research Station, USDA Forest Service, Missoula, MT, United States

José M.C. Pereira Forest Research Centre, Associate Laboratory TERRA, School of Agriculture, University of Lisbon, Lisbon, Portugal

Diego Rafael Perez-Salicrup Institute for Research into Ecosystems and Sustainability. Autonomous University of Mexico, Morelia, Mexico

Marín Pompa-García Faculty of Forestry, University of Juarez del Estado de Durango, Durango, Mexico

Tony Prior NSW National Parks and Wildlife Service, Parramatta, New South Wales, Australia

Carmen Quintano Electronic Technology Department, IuFOR, University of Valladolid, Valladolid, Spain

Adrián Regos Misión Biolóxica de Galicia, Consejo Superior de Investigaciones Científicas, Av. de Vigo, Santiago de Compostela, Spain

Víctor Resco de Dios Department of Forest and Agricultural Science and Engineering, University of Lleida, Lleida, Spain

Rainer Ressl National Commission for the Understanding and Use of Biodiversity, Mexico City, Mexico

Jaime Briseño Reyes Faculty of Forestry, University of Juarez del Estado de Durango, Durango, Mexico

Yair Ricardez National Forestry Commission, Zapopan, Jalisco, Mexico

Marcos Rodrigues Department of Geography and Land Management, University of Zaragoza, Zaragoza, Spain; GEOFOREST-IUCA group, University Institute for Environmental Sciences, University of Zaragoza, Zaragoza, Spain

Marcela Rosas-Chavoya Faculty of Forestry, University of Juarez del Estado de Durango, Durango, Mexico

Ana Daría Ruiz-González Department of Agroforestry Engineering, Higher Polytechnic School of Engineering, University of Santiago de Compostela, Lugo, Spain

Alicia Verónica Salas National Forestry Commission, Zapopan, Jalisco, Mexico

Pablo Sarricolea Center for Climate and Resilience Research (CR)2, University of Chile, Santiago, Chile; Department of Geography, University of Chile, Santiago, Chile; Centro Regional Copernicus-LAC Chile, University of Chile, Santiago, Chile

João M.N. Silva Forest Research Centre, Associate Laboratory TERRA, School of Agriculture, University of Lisbon, Lisbon, Portugal

Adrián Silva-Cardoza Faculty of Forestry, University of Juarez del Estado de Durango, Durango, Mexico

Susana Suárez-Seoane Biodiversity Research Institute (IMIB), University of Oviedo–CSIC–Principality of Asturias, Mieres, Asturias, Spain; Department of Organisms and Systems Biology, University of Oviedo, Mieres, Asturias, Spain

Greg Summerell NSW National Parks and Wildlife Service, Parramatta, New South Wales, Australia

Diana Aimé Tinoco-Orozco Faculty of Forestry, University of Juarez del Estado de Durango, Durango, Mexico

Daniel José Vega-Nieva Faculty of Forestry, University of Juarez del Estado de Durango, Durango, Mexico

Olga Viedma Department of Environmental Sciences, University of Castilla-La Mancha, Toledo, Spain

Matt Wiseman NSW National Parks and Wildlife Service, Parramatta, New South Wales, Australia

Part I

Introduction and background

The growing challenge of forest fires

1

Leonor Calvo[1,2] *and Elena Marcos*[3]
[1]Area of Ecology, Department of Biodiversity and Environmental Management, Faculty of Biological and Environmental Sciences, University of León, León, Spain, [2]Institute for the Environment and Global Change (IMACG), University of León, León, Spain, [3]Department of Biodiversity and Environmental Management, Faculty of Biological and Environmental Sciences, University of León, León, Spain

1.1 Evolution of fire regime attributes over time

Most terrestrial ecosystems have coevolved with fire over millions of years, and in some cases, fire is essential for maintaining their structure and function, such as in the case of savannah systems (Harris et al., 2016). Ecosystems are adapted to particular fire regimes; however, significant alterations in fire regime attributes can lead to substantial ecological and socio-economic impacts (Doerr and Santín, 2016; Feng et al., 2021). Among these, perhaps the most critical are the loss of human lives and properties, as well as negative effects on public health (Balch et al., 2020).

Historically, and across broad temporal and spatial scales, fire regimes have been primarily shaped by climate and its interaction with vegetation (Harris et al., 2016; Molinari et al., 2018). This interaction determined fuel availability, flammability, and the likelihood of ignition (Bowman et al., 2009). Over the last several millennia, however, human activities have increasingly influenced wildfire regimes through changes in vegetation patterns, ignition sources, and climate. Thus, wildfire regimes have become increasingly influenced by anthropogenic factors (Balch et al., 2020; Ellis et al., 2022; Hagmann et al., 2021).

More recently, it has been shown that, globally, there has been no significant increase in two major fire regime attributes, burned area and frequency, over the past two decades (Fernández-García and Alonso-González, 2023). Nevertheless, for the first time, an increase in the extent of high-severity fires has been documented, indicating more widespread ecological damage to ecosystem structure and functioning. The same study reports that the number of extreme wildfires, defined by exceptionally high energy release (Molinari et al., 2018), has doubled in the last 20 years (Cunningham et al., 2024). These global trends must be contextualized at the regional level due to substantial variability. In general, regions exhibiting increases in mean burn severity and the extent of high-severity fires outnumber those showing declines. The most pronounced changes are generally observed at higher latitudes, such as in boreal-dominated systems, but also in temperate broadleaf and mixed forests of South America, as well as in the tropical moist broadleaf forests of Australia (Fernández-García and Alonso-González, 2023). As a result, we are now

Advances in Remote Sensing for Forest Fire Management.
DOI: https://doi.org/10.1016/B978-0-443-41692-7.00018-4
© 2027 Elsevier Ltd. All rights are reserved, including those for text and data mining, AI training, and similar technologies.

facing rapid and substantial changes in fire behavior, which not only alter fire suppression strategies but also require a transformation in fire prevention and management approaches.

Overall, considering changes in fire regime patterns is a necessary step toward defining sustainable land management strategies within the context of climate change. This, in turn, supports the development of mitigation, adaptation, and planning measures aimed at building resilient socio-ecological systems in the face of increasing wildfire risk (Moritz et al., 2014).

1.2 Factors contributing to forest fires

The occurrence of wildfires in ecosystems depends on the interaction of at least four key factors: ignition sources, continuous fuel availability, drought conditions, and favorable weather (wind, high temperatures, and low relative humidity). In general, wildfires occur when thresholds associated with these drivers are crossed (Pausas and Keeley, 2021). In this context, climate change, characterized by rising global temperatures, increased climatic variability, and a higher frequency of extreme events such as heatwaves and droughts, has emerged as a major factor facilitating the crossing of these thresholds, thereby promoting fire occurrence and extending fire seasons.

Rising temperatures significantly reduce fuel moisture by increasing evaporation rates, leading to drier and more flammable vegetation that enhances both ignition likelihood and fire spread (Wasserman and Mueller, 2023). As surface temperatures increase, wildfire risk intensifies, resulting in longer and more dangerous fire seasons (Miller and Safford, 2012). This warming trend is often accompanied by declining precipitation. During dry seasons, vegetation moisture decreases markedly, and consecutive dry periods exacerbate fuel desiccation, leading to highly flammable conditions. In the long term, such dynamics affect plant productivity and water availability, modifying fuel loads and continuity, and ultimately altering fire regimes (Parks et al., 2016; Abatzoglou et al., 2025).

One of the fire regime attributes most affected by climate change is fire severity, defined as the degree of ecological change induced by fire, including impacts on both aboveground and belowground vegetation as well as soil (Keeley, 2009). Warmer and drier conditions promote higher fire severity by increasing vegetation flammability, with dry years showing a greater proportion of high-severity fire-affected areas compared to wetter years (Wasserman and Mueller, 2023).

Climate change has also altered the spatial and temporal distribution of wildfires. According to Zhang et al. (2024), accelerated global warming has intensified the frequency of extreme fire events in various regions worldwide, as exemplified by the 2023 wildfires in Canada. Furthermore, fires increasingly tend to cluster in previously affected areas, fire-prone regions such as the Mediterranean, exhibiting more aggregated spatial patterns and increasingly erratic interannual cycles (Baïle et al., 2021).

Importantly, wildfires are not only a consequence of climate change but also a driver of it. Abatzoglou et al. (2025) found that carbon emissions from extreme fire years can exceed annual industrial emissions. This is due not only to the direct release of greenhouse gases during combustion, but also to changes in surface albedo and the reduction in forests' carbon sequestration capacity. As such, a positive feedback loop between wildfire activity and climate change has been identified.

Scientific evidence strongly suggests that climate change is substantially altering fire regimes—affecting fire frequency, magnitude, and duration—and forest ecosystems now face novel scenarios dominated by more extreme and less manageable fires (Cunningham et al., 2024). In this context, understanding and modeling the influence of climate on fire severity is essential to assess the ecological impacts of fire and to anticipate future responses under warmer and drier conditions (Mueller et al., 2020). In parallel, the increase in human-caused ignitions and assets at risk demands more effective allocation of resources and funding for fire management (Parks et al., 2016).

However, climatic factors alone do not fully explain the complexity of global changes in fire regimes. Two additional drivers play a critical role: (1) altered ignition patterns, particularly related to human behavior, and (2) changes in fuel structures, shaped by land-use changes, fire suppression, drought-induced tree mortality, and landscape fragmentation.

Today, human activity is the dominant driver of wildfire frequency, intensity, size, and spatial distribution. This influence arises both directly—through intentional or accidental ignitions—and indirectly, via land-use change, fire suppression policies, urban expansion into flammable landscapes (wildland–urban interface, WUI), and broader socio-economic transformations. In fire-prone regions such as the Mediterranean and southern Europe, over 90% of wildfires are human-caused (Rodrigues et al., 2016; European Commission – Joint Research Centre, 2023). Similar proportions have been reported in Chile (de la Barrera et al., 2018) and the United States (Balch et al., 2020), with slightly lower figures in Australia (Price and Bradstock, 2013). Ignitions may be intentional (e.g., vandalism, land-use conflicts, economic interests) or accidental (e.g., recreation, poorly managed burning). Additionally, landscape transformations, especially in areas undergoing rural abandonment, promote fuel continuity through secondary succession, facilitating fire spread (Calvo et al., 2002).

Fire suppression policies, widely implemented in Europe and other fire-prone regions, have also contributed to the current fire crisis. These policies have led to a well-documented paradox: by eliminating low- and moderate-intensity fires, fuel accumulates, increasing the probability of more severe fires in the future. Kreider et al. (2024) further define this phenomenon as the "suppression bias," wherein low-intensity fires are preferentially extinguished, while high-intensity fires—often under extreme weather conditions—escape control. As a result, ecosystems, species, and human communities are increasingly exposed to high-severity fires, reducing the diversity of fire effects and amplifying the impacts of both climate change and fuel accumulation. Addressing this situation requires shifting from regressive suppression strategies to more progressive and adaptive approaches that support coexistence with fire under moderate conditions (Kreider et al., 2024). This transition may be as effective as other key interventions, such as mitigating global warming, reducing human ignitions, and modifying forest structure (e.g., eliminating ladder fuels and reducing horizontal continuity) to lower fire severity.

It is also essential to account for the socio-economic conditions of fire-prone regions. Factors such as population density, unemployment, rural depopulation, and the expansion of the WUI are closely linked to increased ignition risk (Rodrigues et al., 2016). WUI expansion plays a critical role in wildfire dynamics, as it increases human exposure and ignition potential. Although WUIs may act as physical barriers to fire spread, they also create conflict between infrastructure protection and the need to maintain natural fire regimes. Consequently, these areas are particularly vulnerable and subject to strong societal pressure for fire suppression, which may, paradoxically, increase long-term fuel loads and the likelihood of extreme events (Rodrigues et al., 2016; Fernández-García et al., 2022; Beltrán-Marcos et al., 2023).

In sum, contemporary wildfire dynamics cannot be understood without recognizing the central role of human activity. From ignition and suppression to landscape transformation and policy decisions, the human footprint is omnipresent. This influence can amplify or moderate climate impacts, disrupt natural fire cycles, and reshape ecological trajectories. Acknowledging this reality means accepting that fire management is, at its core, the management of our own actions and choices. Studies conducted in southern Europe have shown that increasing landscape heterogeneity reduces fire severity (García-Llamas et al., 2019; García-Llamas et al., 2020). Such heterogeneity can be promoted through prescribed fire and livestock grazing, fostering mosaic landscapes. At the same time, studies conducted by Fernandez-Guisuraga and Fernandes (2024) in Portugal have highlighted the importance of prescribed burning in reducing the severity of future wildfires, with these beneficial effects lasting for more than four years. Therefore, there is an urgent need to reassess current fire management strategies and integrate prescribed burning and ecological restoration as key preventive tools (Rielli et al., 2023). Moreover, the paradigm of absolute fire suppression must be re-evaluated. Experts advocate for adaptive management strategies that include prescribed fire, targeted fuel reduction in strategic zones, and land-use planning that limits unregulated WUI expansion.

1.3 Economic consequences of forest fires

Wildfires are among the most severe disturbances affecting ecosystems and human societies in the 21st century. Their frequency and intensity have increased in recent decades due to the combined effects of climate change, intensive land-use practices, and global socio-economic dynamics (Kalogiannidis et al., 2023). While their ecological consequences are well documented, the economic dimension of wildfires is equally profound. Direct and indirect costs extend far beyond firefighting expenditures, encompassing losses of forest resources, damage to infrastructure, public health impacts, reduced tourism revenues, population displacement, and the long-term degradation of ecosystem services. This section examines the multiple layers of economic consequences associated with wildfires, drawing on empirical evidence from diverse geographical contexts.

Direct losses are the most immediate and visible consequences of wildfires. They include the destruction of forest and agricultural resources, property damage, and the costs of suppression operations. The most tangible impact is the loss of timber and other forest products. Velichkova et al. (2024), in their study of Bulgaria, demonstrate how wildfires destroyed significant volumes of valuable timber, degraded plantations, and undermined the protective ecological functions of forests. These losses directly affect the forestry sector by reducing raw material availability for industries such as construction and energy. Agriculture is also severely affected. Fires destroy crops, pastures, and livestock infrastructure, leading to significant income losses for farmers and ranchers. In Greece, empirical studies show that wildfires reduce agricultural productivity, damage grazing areas, and diminish land value, thereby affecting rural economies at multiple levels (Kalogiannidis et al., 2023). Such impacts exacerbate rural vulnerability and undermine community resilience. Infrastructure losses represent another major category of direct damage. The destruction of homes, farm facilities, roads, and power networks generates enormous reconstruction costs. Depending on the scale of the event, such costs may reach billions of dollars (Hanson et al., 2013). A striking example is California in 2020, when fires affected 1.74 million hectares, caused more than USD 19 billion in economic losses, and devastated critical infrastructure (Safford et al., 2022). Suppression costs further compound these losses. Firefighting has become an increasingly expensive undertaking for governments worldwide. Soontha and Bhat (2024) report that suppression expenditures in Mediterranean countries have risen between 65% and 87% over the last decade (Posavec et al., 2023). In Europe alone, wildfire damages in 2022 were estimated at USD 2.33 billion, while in Spain, suppression costs from the 2025 fire season may exceed USD 8.84 billion. Similarly, the Missionary Ridge Fire in Colorado generated USD 90.3 million in direct costs, of which USD 37.7 million were devoted to firefighting (Mackes et al., 2007). These data highlight how suppression, though essential, represents a growing burden on public finances.

While direct damages are visible and quantifiable, indirect costs often surpass them in scope and duration. They extend into the health sector, labor markets, migration patterns, and social services. Public health systems are particularly affected. Wildfires release vast amounts of ash, carbon, and toxic particles into the atmosphere, exacerbating respiratory, dermatological, and ocular diseases (Kalogiannidis et al., 2023). Hariri-Ardebili et al. (2022) emphasize the strain placed on healthcare infrastructures, while Soontha and Bhat (2024) quantify the relationship between wildfire incidence and health expenditures: a 1% increase in wildfire activity correlates with a 0.03%–0.08% rise in public healthcare spending. The associated costs of treating cardiovascular and respiratory conditions, along with the psychological toll on affected populations, are substantial. Fatalities and injuries among civilians and firefighters add to these losses, requiring compensation payments and long-term support services. Population displacement is another significant indirect effect. The European Environment Agency reported nearly 590,000 wildfire-related displacements in the EU between 2008 and 2023, including 234,287 in 2023 alone. In the United States, McConnell et al. (2024) link highly destructive fires to migration flows, particularly when entire communities lose housing and infrastructure. In

Russia, Dmitrenko et al. (2022) document rural depopulation caused by recurrent fires, accompanied by migration toward urban centers. This process creates new costs for governments, including the provision of social services and the loss of agricultural and forestry productivity in affected regions. Tourism, a sector highly dependent on landscapes and environmental quality, is also disrupted. Cillis et al. (2023) found that wildfires in protected areas of Basilicata (Italy) reduced visitor numbers and threatened ecotourism-based livelihoods. In Mediterranean regions where rural and nature-based tourism constitutes a significant source of income, these effects are particularly damaging. Losses are not limited to the year of the fire but often extend for years due to reputational damage and long-term landscape degradation.

One of the most challenging aspects of wildfire economics lies in quantifying long-term rehabilitation costs. Watersheds and forest ecosystems may take decades to recover, and restoration efforts are both costly and uncertain (Barrett, 2018). Postfire flooding, soil erosion, and the spread of invasive species further exacerbate damages (Robichaud et al., 2009). In India, Kumar et al. (2024) document how wildfires accelerate soil degradation and desertification, undermining food security and long-term agricultural productivity. The erosion of soil fertility demands continuous investments in restoration and conservation, while also constraining rural development opportunities. Beyond tangible losses, wildfires also disrupt ecosystem services that provide immense economic value. Forests play a critical role in carbon sequestration, water regulation, and biodiversity maintenance. The destruction of these services is difficult to monetize but represents a substantial hidden cost. Hanson et al. (2013) propose the use of Habitat Equivalency Analysis (HEA) to evaluate ecosystem service losses, showing that their economic value can exceed that of timber and other marketable forest products. Posavec et al. (2023) estimate that wildfires in Croatia between 2013 and 2020 caused over EUR 326 million in damages, including reforestation costs and the loss of nontimber functions such as recreation and biodiversity.

The magnitude of wildfire-related losses underscores the need for rethinking management strategies. Current approaches often prioritize suppression, but growing evidence highlights the superior economic efficiency of prevention. Dmitrenko et al. (2022) argue that preventive measures, although costly, reduce overall expenditures compared with suppression and reconstruction. According to the European Commission and the World Bank, every euro invested in prevention saves between four and seven euros in avoided damages. Empirical research supports this claim: an analysis of the Las Conchas Fire in the United States found that preventive measures reduced suppression costs by 88% and total expenditures (prevention plus suppression) by 55% relative to a scenario without prevention (Houtman et al., 2018).

Effective wildfire management also requires institutional coordination. In Croatia, Posavec et al. (2023) observe that the lack of cooperation between forestry authorities and governmental agencies undermines sustainable management strategies. They recommend climate-adaptive forestry, which involves the use of fire-resistant species and biological restoration programs. At the international level, Kalogiannidis et al. (2023) emphasize the importance of cooperative frameworks to facilitate the exchange of financial, technological, and human resources. Given the transboundary nature of wildfires and their amplification by climate change, international collaboration is essential for building regional and global resilience.

Wildfires impose economic costs that extend far beyond immediate damages to forests and agriculture. Their impacts permeate multiple sectors, including public health, infrastructure, tourism, migration, and ecosystem services. Evidence from Europe, the United States, Asia, and Russia indicates that these costs amount to billions of euros and dollars annually, and that their long-term consequences often outweigh the visible short-term damages. The complexity and magnitude of these losses highlight the need for integrated strategies combining prevention, sustainable forest management, adaptive restoration, and international cooperation. Moreover, the economic valuation of lost ecosystem services must be incorporated into public policies to reflect the true costs of wildfires. Only through a multidimensional and forward-looking approach can societies mitigate the impacts of wildfires and enhance the resilience of ecosystems and communities in an era of intensifying climate risks.

1.4 Environmental impacts of forest fires

Historically, fire has played a fundamental ecological role in regulating and transforming ecosystems. Today, however, its frequency and intensity are amplified by climate change and human activities, leading to increasingly severe environmental consequences.

Wildfires cause substantial alterations in vegetation, influencing structure, composition, and ecosystem processes. Fire acts as an ecological disturbance that, while often exerting negative pressures on plant communities, also triggers adaptive mechanisms that have shaped the dynamics and evolution of many fire-prone ecosystems. Processes such as heat-stimulated germination, resprouting of fire-adapted species (Calvo et al., 2002; Calvo et al., 2012), and the release of nutrients through ash deposition promote vegetation recolonization and primary productivity (Gajendiran et al., 2024). The capacity for postfire regeneration, however, depends strongly on community type, soil moisture, and fire severity (Fernández-Guisuraga and Fernandes, 2024). When fires are frequent or particularly intense, regeneration pathways may be disrupted. For instance, repeated high-severity events can hinder conifer recruitment (Schüle et al., 2023) and facilitate replacement by shrub-dominated formations (Fernández-García et al., 2020).

Soil is among the most severely affected ecosystem components, undergoing profound physical, chemical, and biological modifications after fire. Fire severity is a key determinant of both the magnitude and persistence of these effects. Notable alterations include changes in aggregate stability, hydrophobicity, and soil structure, which directly influence erodibility and water retention capacity (Mataix-Solera et al., 2011). Fires can also mobilize potentially toxic elements (PTEs) and alter their bioavailability, with implications for soil quality, agricultural productivity, and food security (Terzano et al., 2021). At the geochemical level, fire modifies nutrient and trace-element cycles, promotes reactions that rarely occur under natural conditions, and alters metal speciation in ways that may increase toxicity (Roshan and Biswas, 2023).

These transformations extend to the soil microbiota. Fire typically reduces microbial biomass, enzymatic activity, and community diversity, thereby compromising key processes such as nutrient recycling and soil fertility (Barreiro and Díaz-Raviña, 2021). In Mediterranean conifer ecosystems, medium-term legacies of fire severity have been observed in soil biochemistry and microbiology, persisting for years after the disturbance (Fernández-García et al., 2019). Although resilience mechanisms may allow for partial recovery of soil functions, recurrent wildfires can produce long-lasting degradation of soil structure and functionality, ultimately limiting ecosystem regeneration capacity.

In the broader context of climate change, wildfires are a major source of greenhouse gas emissions, particularly carbon dioxide (CO_2) and methane (CH_4), which exacerbate global warming. It is estimated that 350–450 million hectares burn annually worldwide, involving the combustion of billions of tons of dry biomass and releasing approximately 1.5–3 petagrams of carbon (PgC) per year. These emissions include not only CO_2 and CH_4 but also other compounds that alter atmospheric composition and functioning, with direct consequences for human health (Leblon et al., 2016). The massive release of carbon stored in vegetation makes wildfires a critical driver within climate feedback loops. According to the Global Fire Emissions Database (GFED), between 2001 and 2024, fires emitted an annual average of 1.7 gigatons (Gt) of CO_2 equivalent, of which about 87% was CO_2, underscoring their central role in the global greenhouse gas balance.

Wildfires profoundly reshape ecosystems by altering vegetation, soil properties, and biogeochemical cycles. Although historically a natural regulator, their increased frequency and intensity due to climate change and human activities now pose escalating environmental threats. High-severity and recurrent fires disrupt vegetation regeneration, hinder conifer recovery, and promote long-term soil degradation and microbial loss. Moreover, wildfires are a major global source of greenhouse gases, releasing 1.5–3 PgC annually and reinforcing climate feedback loops. Together, these effects underscore the dual role of wildfires as both ecological drivers and accelerants of global change, demanding integrated management within climate adaptation strategies.

1.5 Management and mitigation strategies for reducing human-caused forest fires

Mitigating the impacts of wildfires requires a comprehensive approach that integrates preventive, technological, and ecological strategies adapted to diverse socio-environmental contexts. Among preventive measures, education and outreach programs targeting schools, youth, and local communities have proven effective in significantly reducing the number of human-caused ignitions. Complementary to these efforts, the implementation of burn permits provides oversight of fire use and minimizes the risk of uncontrolled burns (Abt et al., 2015). Community-based fire management, involving local populations, nongovernmental organizations, and public authorities, has also demonstrated effectiveness by fostering shared responsibility in wildfire risk management (Jurvélius, 2004).

Fuel management is another key component. Prescribed burning is an effective tool to reduce fuel accumulation and, consequently, the likelihood of high-severity fires. However, its application requires strict safety protocols and suitable environmental conditions, as improper use may generate adverse impacts on air quality and human health (Neary and Leonard, 2021). In areas adjacent to infrastructure or settlements, mechanical fuel reduction represents a viable alternative, albeit at higher economic costs (Tayari et al., 2023).

Technological advances have created new opportunities for wildfire management using predictive models based on machine learning and artificial intelligence (AI). These tools enable the identification of high-susceptibility areas, anticipation of fire spread, and optimization of resource allocation. Recent studies show that algorithms such as random forest or hybrid modeling approaches achieve accuracy levels exceeding 90%, making them essential tools for preventive management (Mishra et al., 2024; Bhadoria et al., 2021).

Postfire restoration is equally critical to ensure the functional recovery of ecosystems. Nature-based solutions (NBS), such as the application of biochar, mulching, microbial inoculation, and erosion control, enhance soil resilience, restore microbial communities, improve nutrient retention, and facilitate the regeneration of native species. In addition, they provide cobenefits, including biodiversity conservation and increased carbon sequestration (Wittenberg and Shtober-Zisu, 2023).

Finally, integrating these approaches into strategic decision-making frameworks that weigh environmental, social, and economic costs is essential for selecting context-appropriate combinations of measures (Tayari et al., 2023). Collectively, these strategies not only reduce the immediate risk of wildfires but also strengthen ecological and social resilience in the face of the growing threat posed by climate change.

1.6 Main challenges and future directions in wildfire research and management

Current and future research on wildfires focuses on several key challenges and emerging priorities. First, there is a need for accurate, spatially and temporally extensive wildfire databases, particularly in regions with difficult access due to orographic conditions. In this regard, satellite imagery, remote sensing, and advanced modeling techniques are considered essential tools for improving fire detection, monitoring, and management, as they provide critical spatio-temporal information to support decision-making.

Second, projections indicate that fire seasons will become longer and fire severity will increase, especially in regions undergoing significant climatic changes (Cunningham et al., 2024; Lindenmayer and Zylstra, 2024). This underscores the need to implement adaptive management strategies that account for future climate scenarios. These strategies must be dynamic and tailored to each specific region, considering shifts in not only temperature and precipitation but also vegetation types and their fire adaptive capacities (Nitschke and Innes, 2008).

Third, active participation of local communities in fire prevention efforts is considered essential for achieving effective mitigation. Traditional knowledge, involvement in fuel reduction activities, and awareness campaigns are fundamental in reducing human-caused ignitions and strengthening social resilience to wildfires (Chakraborty et al., 2024).

Fourth, technological innovation remains a top priority. Advances in early detection systems, including satellite imagery, geographic information systems, and computational predictive models, have significantly improved the capacity to detect smoke plumes, anticipate fire behavior, and coordinate emergency response. Among these innovations, early detection is the most critical variable, as it can help minimize ecological damage and reduce operational suppression costs. In recent years, the application of machine learning and AI to wildfire prediction and management has gained considerable importance (Reis and Turk, 2023; Fernández-Guisuraga et al., 2023). These technologies enable the processing of large volumes of spatio-temporal and environmental data, enhancing risk assessments, prediction accuracy, and operational response effectiveness. Current scientific evidence suggests that: (1) machine learning models can accurately predict wildfire risk based on climatic and environmental variables; (2) the integration of machine learning with remote sensing data significantly improves the accuracy of wildfire risk assessments; and (3) the combination of AI with Internet of Things (IoT) devices enhances real-time detection and response systems (Pandey et al., 2023).

In parallel, fuel treatment and landscape management strategies are recognized for their effectiveness in reducing fire risk (Lindenmayer and Zylstra, 2024). Various studies have demonstrated that prescribed burns, selective thinning, and other fuel load reduction techniques can significantly decrease fire severity, even under extreme weather conditions (Deak et al., 2024; Fernández-Guisuraga and Fernandes, 2024). However, their effectiveness varies across ecosystems, highlighting the need for management approaches that are adapted to local ecological contexts.

Beyond ecological and technological considerations, governance and public policy development also play a central role in risk mitigation. Evidence shows that fire management policies engaging forest-dependent populations can substantially reduce human-caused fires. Similarly, the establishment of strategic intervention zones and the implementation of proactive management plans are essential for reducing fire exposure in vulnerable regions.

AI disclosure

During the preparation of this work, the author(s) used ChatGPT in order to improve the scientific English. After using this tool/service, the author(s) reviewed and edited the content as needed and take(s) full responsibility for the content of the publication.

References

Abatzoglou, J.T., Kolden, C.A., Cullen, A.C., Sadegh, M., Williams, E.L., Turco, M., et al., 2025. Climate change has increased the odds of extreme regional forest fire years globally. Nature Communications 16 (1). Available at https://doi.org/10.1038/s41467-025-61608-1, https://www.nature.com/ncomms/.

Abt, K.L., Butry, D.T., Prestemon, J.P., Scranton, S., 2015. Effect of fire prevention programs on accidental and incendiary wildfires on tribal lands in the United States. International Journal of Wildland Fire 24 (6), 749–762. Available at https://doi.org/10.1071/WF14168, http://www.publish.csiro.au/nid/126.htm.

Balch, J.K., Iglesias, V., Braswell, A.E., Rossi, M.W., Joseph, M.B., Mahood, A.L., et al., 2020. Social-environmental extremes: rethinking extraordinary events as outcomes of interacting biophysical and social systems. Earth's Future 8 (7). Available at https://doi.org/10.1029/2019EF001319, http://onlinelibrary.wiley.com/journal/10.1002/(ISSN)2328-4277.

Barreiro, A., Díaz-Raviña, M., 2021. Fire impacts on soil microorganisms: mass, activity, and diversity. Current Opinion in Environmental Science & Health 22, 100264. Available at https://doi.org/10.1016/j.coesh.2021.100264.

de la Barrera, F., Reyes-Paecke, S., Banzhaf, E., 2018. Forest fires in chile: causes, impacts and resilience. Ecología Austral 28 (2), 243–256. Available at https://doi.org/10.25260/EA.18.28.2.0.719.

Barrett, K., 2018. The Full Community Costs Of Wildfire. Headwaters Economics. Available at https://headwaterseconomics.org/wildfire/homes-risk/full-community-costs-of-wildfire/.

Baïle, R., Muzy, J.-F., Silvani, X., 2021. Multifractal point processes and the spatial distribution of wildfires in French Mediterranean regions. Physica A: Statistical Mechanics and its Applications 568, 125697. Available at https://doi.org/10.1016/j.physa.2020.125697.

Beltrán-Marcos, D., Calvo, L., Fernández-Guisuraga, J.M., Fernández-García, V., Suárez-Seoane, S., 2023. Wildland-urban interface typologies prone to high severity fires in Spain. Science of The Total Environment 894, 165000. Available at https://doi.org/10.1016/j.scitotenv.2023.165000.

Bhadoria, R.S., Pandey, M.K., Kundu, P., 2021. RVFR: random vector forest regression model for integrated & enhanced approach in forest fires predictions. Ecological Informatics 66. Available at https://doi.org/10.1016/j.ecoinf.2021.101471, http://www.elsevier.com/wps/find/journaldescription.cws_home/705192/description#description.

Bowman, D.M.J.S., Balch, J.K., Artaxo, P., Bond, W.J., Carlson, J.M., Cochrane, M.A., et al., 2009. Fire in the earth system. Science (New York, N.Y.) 324 (5926), 481–484. Available at https://doi.org/10.1126/science.1163886.

Calvo, L., Tárrega, R., de Luis, E., 2002. Secondary succession after perturbations in a shrubland community. Acta Oecologica 23 (6), 393–404. Available at https://doi.org/10.1016/s1146-609x(02)01164-5.

Calvo, L., Baeza, L., Marcos, E., Santana, V., Papanastasis, V., 2012. Post-fire management of Mediterranean shrublands. In: Post-fire management and restoration of Southern European forests. Springer, Dordrecht, pp. 293–319. Available at https://doi.org/10.1007/978-94-007-2208-8_11

Chakraborty, D., Bhowmik, A., Mridha, N., Pyngrope, K., 2024. Forest Fire Detection and Monitoring for Its Management: Remote Sensing and GIS/Geospatial Technology Approach Sustainable Management and Conservation of Environmental Resources in

India. Apple Academic Press, India, pp. 79–104. Available at https://appleacademicpress.com/sustainable-and-conservation-management-of-environmental-resources-in-india-/9781774915929.

Cillis, G., Lanorte, A., Santarsiero, V., Nolè, G., 2023. Natural hazard impact in protected areas for resilience management: the case of wildfires in the basilicata region. Pollutants 3 (4), 437–450. Available at https://doi.org/10.3390/pollutants3040030.

Cunningham, C.X., Williamson, G.J., Bowman, D.M.J.S., 2024. Increasing frequency and intensity of the most extreme wildfires on Earth. Nature Ecology & Evolution 8 (8), 1420–1425. Available at https://doi.org/10.1038/s41559-024-02452-2.

Deak, A.L., Lucash, M.S., Coughlan, M.R., Weiss, S., Silva, L.C.R., 2024. Prescribed fire placement matters more than increasing frequency and extent in a simulated Pacific Northwest landscape. Ecosphere 15 (4). Available at https://doi.org/10.1002/ecs2.4827, http://esajournals.onlinelibrary.wiley.com/hub/journal/10.1002/(ISSN)2150-8925/.

Dmitrenko, A., Pavlova, V., Lesnykh, E., Buryanina, N., Lesnykh, A., 2022. Development of preventive measures to eliminate wildfires with the use of transport and other services. Transportation Research Procedia 63, 1758–1765. Available at https://doi.org/10.1016/j.trpro.2022.06.191.

Doerr, S.H., Santín, C., 2016. Global trends in wildfire and its impacts: perceptions versus realities in a changing world. Philosophical Transactions of the Royal Society B: Biological Sciences 371 (1696), 20150345. Available at https://doi.org/10.1098/rstb.2015.0345.

Ellis, T.M., Bowman, D.M.J.S., Jain, P., Flannigan, M.D., Williamson, G.J., 2022. Global increase in wildfire risk due to climate-driven declines in fuel moisture. Global Change Biology 28 (4), 1544–1559. Available at https://doi.org/10.1111/gcb.16006, http://onlinelibrary.wiley.com/journal/10.1111/(ISSN)1365-2486.

European Commission – Joint Research Centre, 2023. Forest fires in Europe, Middle East and North Africa 2022. EFFIS Report.

Feng, X., Merow, C., Liu, Z., Park, D.S., Roehrdanz, P.R., Maitner, B., et al., 2021. How deregulation, drought and increasing fire impact Amazonian biodiversity. Nature 597 (7877), 516–521. Available at https://doi.org/10.1038/s41586-021-03876-7, http://www.nature.com/nature/index.html.

Fernández-García, V., Alonso-González, E., 2023. Global patterns and dynamics of burned area and burn severity. Remote Sensing 15 (13), 3401. Available at https://doi.org/10.3390/rs15133401.

Fernández-García, V., Miesel, J., Baeza, M.J., Marcos, E., Calvo, L., 2019. Wildfire effects on soil properties in fire-prone pine ecosystems: indicators of burn severity legacy over the medium term after fire. Applied Soil Ecology 135, 147–156. Available at https://doi.org/10.1016/j.apsoil.2018.12.002, https://www.elsevier.com/inca/publications/store/5/2/4/5/1/8/index.htt.

Fernández-García, V., Marcos, E., Fulé, P.Z., Reyes, O., Santana, V.M., Calvo, L., 2020. Fire regimes shape diversity and traits of vegetation under different climatic conditions. Science of the Total Environment 716. Available at https://doi.org/10.1016/j.scitotenv.2020.137137, https://www.elsevier.com/locate/scitotenv.

Fernández-García, V., Beltrán-Marcos, D., Fernández-Guisuraga, J.M., Marcos, E., Calvo, L., 2022. Predicting potential wildfire severity across Southern Europe with global data sources. Science of the Total Environment 829, 154729. Available at https://doi.org/10.1016/j.scitotenv.2022.154729.

Fernández-Guisuraga, J.M., Fernandes, P.M., 2024. Enhanced post-wildfire vegetation recovery in prescribed-burnt Mediterranean shrubland: a regional assessment. Forest

Ecology and Management 561. Available at https://doi.org/10.1016/j.foreco.2024.121921, https://www.sciencedirect.com/science/journal/03781127.

Fernández-Guisuraga, J.M., Fernandes, P.M., Tárrega, R., Beltrán-Marcos, D., Calvo, L., 2023. Vegetation recovery drivers at short-term after fire are plant community-dependent in mediterranean burned landscapes. Forest Ecology and Management 539. Available at https://doi.org/10.1016/j.foreco.2023.121034, https://www.elsevier.com/inca/publications/store/5/0/3/3/1/0.

Gajendiran, K., Kandasamy, S., Narayanan, M., 2024. Influences of wildfire on the forest ecosystem and climate change: a comprehensive study. Environmental Research 240 (Pt 2), 117537. Available at https://doi.org/10.1016/j.envres.2023.117537.

García-Llamas, P., Suárez-Seoane, S., Taboada, A., Fernández-Manso, A., Quintano, C., Fernández-García, V., et al., 2019. Environmental drivers of fire severity in extreme fire events that affect Mediterranean pine forest ecosystems. Forest Ecology and Management 433, 24–32. Available at https://doi.org/10.1016/j.foreco.2018.10.051, https://www.elsevier.com/inca/publications/store/5/0/3/3/1/0.

García-Llamas, P., Suárez-Seoane, S., Fernández-Manso, A., Quintano, C., Calvo, L., 2020. Evaluation of fire severity in fire prone-ecosystems of Spain under two different environmental conditions. Journal of Environmental Management 271, 110706. Available at https://doi.org/10.1016/j.jenvman.2020.110706.

Hagmann, R.K., Hessburg, P.F., Prichard, S.J., Povak, N.A., Brown, P.M., Fulé, P.Z., et al., 2021. Evidence for widespread changes in the structure, composition, and fire regimes of western North American forests. Ecological Applications 31 (8). Available at https://doi.org/10.1002/eap.2431.

Hanson, D.A., Britney, E.M., Earle, C.J., Stewart, T.G., 2013. Adapting habitat equivalency analysis (HEA) to assess environmental loss and compensatory restoration following severe forest fires. Forest Ecology and Management 294, 166–177. Available at https://doi.org/10.1016/j.foreco.2012.12.032.

Hariri-Ardebili, M.A., Sattar, S., Johnson, K., Clavin, C., Fung, J., Ceferino, L., 2022. A perspective towards multi-hazard resilient systems: natural hazards and pandemics. Sustainability 14 (8), 4508. Available at https://doi.org/10.3390/su14084508.

Harris, R.M.B., Remenyi, T.A., Williamson, G.J., Bindoff, N.L., Bowman, D.M.J.S., 2016. Climate–vegetation–fire interactions and feedbacks: trivial detail or major barrier to projecting the future of the Earth system? Wiley Interdisciplinary Reviews: Climate Change 7 (6), 910–931. Available at https://doi.org/10.1002/wcc.428, http://onlinelibrary.wiley.com/journal/10.1002/(ISSN)1757-7799.

Houtman, R.M., Montgomery, C.A., Gagnon, A.R., Calkin, D.E., Dietterich, T.G., McGregor, S., et al., 2018. Allowing a wildfire to burn: estimating the effect on future fire suppression costs. International Journal of Wildland Fire 27 (10), 734–744. Available at https://doi.org/10.1071/WF17133.

Jurvélius, M., 2004. Community-based fire management in southern Africa. Unasylva 55 (217), 12–14.

Kalogiannidis, S., Chatzitheodoridis, F., Kalfas, D., Patitsa, C., Papagrigoriou, A., 2023. Socio-psychological, economic and environmental effects of forest fires. Fire 6 (7), 280. Available at https://doi.org/10.3390/fire6070280.

Keeley, J.E., 2009. Fire intensity, fire severity and burn severity: a brief review and suggested usage. International Journal of Wildland Fire 18 (1), 116. Available at https://doi.org/10.1071/wf07049.

Kreider, M.R., Higuera, P.E., Parks, S.A., Rice, W.L., White, N., Larson, A.J., 2024. Fire suppression makes wildfires more severe and accentuates impacts of climate change and fuel accumulation. Nature Communications 15 (1). Available at https://doi.org/10.1038/s41467-024-46702-0, https://www.nature.com/ncomms/.

Kumar, N., Singh, H., Kumar, A., Singh, A.K., Pandey, P.K., Kumar, A., 2024. Forest-fire-induced land degradation. Sustainable Land Management in India: Opportunities and Challenges. Springer Nature, India, pp. 51–68. Available at https://link.springer.com/book/9789819752232, http://10.1007/978-981-97-5223-2_4.

Lindenmayer, D., Zylstra, P., 2024. Identifying and managing disturbance-stimulated flammability in woody ecosystems. Biological Reviews 99 (3), 699–714. Available at https://doi.org/10.1111/brv.13041.

Mackes, K.H., Lynch, D.L., Kelly, S.K., Eckhoff, M., 2007. Missionary ridge fire cost assessment. Journal of Testing and Evaluation 35 (2), 167–171. Available at https://doi.org/10.1520/jte100044.

Mataix-Solera, J., Cerdà, A., Arcenegui, V., Jordán, A., Zavala, L.M., 2011. Fire effects on soil aggregation: a review. Earth-Science Reviews 109 (1-2), 44–60. Available at https://doi.org/10.1016/j.earscirev.2011.08.002.

McConnell, K., Fussell, E., DeWaard, J., Whitaker, S., Curtis, K.J., St. Denis, L., et al., 2024. Rare and highly destructive wildfires drive human migration in the U.S. Nature Communications 15 (1). Available at https://doi.org/10.1038/s41467-024-50630-4, https://www.nature.com/ncomms/.

Miller, J.D., Safford, H., 2012. Trends in wildfire severity: 1984 to 2010 in the Sierra Nevada, Modoc Plateau, and southern Cascades, California, USA. Fire Ecology 8 (3), 41–57. Available at https://doi.org/10.4996/fireecology.0803041, http://fireecology.org/docs/Journal/pdf/Volume08/Issue03/041.pdf.

Mishra, M., Guria, R., Baraj, B., Nanda, A.P., Santos, C.A.G., Silva, R.M.D., et al., 2024. Spatial analysis and machine learning prediction of forest fire susceptibility: a comprehensive approach for effective management and mitigation. Science of the Total Environment 926. Available at https://doi.org/10.1016/j.scitotenv.2024.171713, https://www.sciencedirect.com/science/journal/00489697.

Molinari, C., Lehsten, V., Blarquez, O., Carcaillet, C., Davis, B.A.S., Kaplan, J.O., et al., 2018. The climate, the fuel and the land use: long-term regional variability of biomass burning in boreal forests. Global Change Biology 24 (10), 4929–4945. Available at https://doi.org/10.1111/gcb.14380, http://onlinelibrary.wiley.com/journal/10.1111/(ISSN)1365-2486.

Moritz, M.A., Batllori, E., Bradstock, R.A., Gill, A.M., Handmer, J., Hessburg, P.F., et al., 2014. Learning to coexist with wildfire. Nature 515 (7525), 58–66. Available at https://doi.org/10.1038/nature13946, http://www.nature.com/nature/index.html.

Mueller, S.E., Thode, A.E., Margolis, E.Q., Yocom, L.L., Young, J.D., Iniguez, J.M., 2020. Climate relationships with increasing wildfire in the southwestern US from 1984 to 2015. Forest Ecology and Management 460. Available at https://doi.org/10.1016/j.foreco.2019.117861, https://www.elsevier.com/inca/publications/store/5/0/3/3/1/0.

Neary, D.G., Leonard, J., 2021. Fire management and soil protection. In: *Fire effects on soils and restoration strategies*. pp. 305–328. Available at https://doi.org/10.1016/B978-0-12-818528-8.00014-3

Nitschke, C.R., Innes, J.L., 2008. Climatic change and fire potential in South-Central British Columbia, Canada. Global Change Biology 14 (4), 841–855. Available at https://doi.org/10.1111/j.1365-2486.2007.01517.x.

Pandey, S., Singh, R., Kathuria, S., Negi, P., Chhabra, G., Joshi, K., 2023. Emerging technologies for prevention and monitoring of forest fire. In: *Proceedings of the International Conference on Innovative Data Communication Technologies and Application (ICIDCA 2023)*. Institute of Electrical and Electronics Engineers Inc.,

India, pp. 1115–1121. Available at http://ieeexplore.ieee.org/xpl/mostRecentIssue.jsp?punumber=10099475, https://doi.org/10.1109/ICIDCA56705.2023.10099572.

Parks, S.A., Miller, C., Abatzoglou, J.T., Holsinger, L.M., Parisien, M.A., Dobrowski, S.Z., 2016. How will climate change affect wildland fire severity in the western US? Environmental Research Letters 11 (3). Available at https://doi.org/10.1088/1748-9326/11/3/035002, http://iopscience.iop.org/article/10.1088/1748-9326/11/3/035002/pdf.

Pausas, J.G., Keeley, J.E., 2021. Wildfires and global change. Frontiers in Ecology and the Environment 19 (7), 387–395. Available at https://doi.org/10.1002/fee.2359, http://onlinelibrary.wiley.com/journal/10.1002/(ISSN)1540-9309.

Posavec, S., Barčić, D., Vuletić, D., Vučetić, V., Čavlina Tomašević, I., Pezdevšek Malovrh, Š., 2023. Forest fires, stakeholders' activities, and economic impact on state-level sustainable forest management. Sustainability 15 (22), 16080. Available at https://doi.org/10.3390/su152216080.

Price, O., Bradstock, R., 2013. Landscape scale influences of forest area and housing density on house loss in the 2009 victorian bushfires. PLoS One 8 (8). Available at https://doi.org/10.1371/journal.pone.0073421, http://www.plosone.org/article/fetchObjectAttachment.action;jsessionid=3713A55ADE848F6399B1E86AE8B0873C?uri=info%3Adoi%2F10.1371%2Fjournal.pone.0073421&representation=PDF.

Reis, H.C., Turk, V., 2023. Detection of forest fire using deep convolutional neural networks with transfer learning approach. Applied Soft Computing 143, 110362. Available at https://doi.org/10.1016/j.asoc.2023.110362.

Rielli, M., Pereira, M.G., Rodrigues, M., 2023. Assessing the influence of human and biophysical factors on wildfire occurrence in Portugal using machine learning techniques. Fire 6 (300). Available at https://doi.org/10.3390/fire6100300.

Robichaud, P.R., Lewis, S.A., Brown, R.E., Ashmun, L.E., 2009. Emergency post-fire rehabilitation treatment effects on burned area ecology and long-term restoration. Fire Ecology 5 (1), 115–128. Available at https://doi.org/10.4996/fireecology.0501115, http://www.fireecology.net/Journal/pdf/Volume05/Issue01/115.pdf.

Rodrigues, M., Jiménez, A., de la Riva, J., 2016. Analysis of recent spatial–temporal evolution of human driving factors of wildfires in Spain. Natural Hazards 84 (3), 2049–2070. Available at https://doi.org/10.1007/s11069-016-2533-4, https://www.wkap.nl/journalhome.htm/0921-030X.

Roshan, A., Biswas, A., 2023. Fire-induced geochemical changes in soil: Implication for the element cycling. Science of The Total Environment 868, 161714. Available at https://doi.org/10.1016/j.scitotenv.2023.161714.

Safford, H.D., Paulson, A.K., Steel, Z.L., Young, D.J.N., Wayman, R.B., 2022. The 2020 California fire season: a year like no other, a return to the past or a harbinger of the future? Global Ecology and Biogeography 31 (10), 2005–2025. Available at https://doi.org/10.1111/geb.13498.

Schüle, M., Domes, G., Schwanitz, C., Heinken, T., 2023. Early natural tree regeneration after wildfire in a Central European Scots pine forest: forest management, fire severity and distance matters. Forest Ecology and Management 539, 120999. Available at https://doi.org/10.1016/j.foreco.2023.120999.

Soontha, L., Bhat, M.Y., 2024. Preserving health, protecting economies: mitigating the impact of forest fires on healthcare expenditure and environmental sustainability. Sustainable Development 32 (3), 2066–2084. Available at https://doi.org/10.1002/sd.2764, http://onlinelibrary.wiley.com/journal/10.1002/(ISSN)1099-1719.

Tayari, S., Taghikhah, F., Bharathy, G., Voinov, A., 2023. Designing a conceptual framework for strategic selection of Bushfire mitigation approaches. Journal of Environmental Management 344, 118486. Available at https://doi.org/10.1016/j.jenvman.2023.118486.

Terzano, R., Rascio, I., Allegretta, I., Porfido, C., Spagnuolo, M., Khanghahi, M.Y., et al., 2021. Fire effects on the distribution and bioavailability of potentially toxic elements (PTEs) in agricultural soils. Chemosphere 281. Available at https://doi.org/10.1016/j.chemosphere.2021.130752, https://www.elsevier.com/locate/chemosphere.

Velichkova, R., Gieva, E., Simova, I., 2024. Assessment and analysis of forest fires in Bulgaria † Engineering Proceedings 60 (1). https://doi.org/10.3390/engproc2024060002, https://www.mdpi.com/journal/engproc.

Wasserman, T.N., Mueller, S.E., 2023. Climate influences on future fire severity: a synthesis of climate-fire interactions and impacts on fire regimes, high-severity fire, and forests in the western United States. Fire Ecology 19 (1). Available at https://doi.org/10.1186/s42408-023-00200-8, https://www.springer.com/journal/42408.

Wittenberg, L., Shtober-Zisu, N., 2023. Restoring fire-affected soils: the potential of nature-based solutions. Current Opinion in Environmental Science & Health 36, 100520. Available at https://doi.org/10.1016/j.coesh.2023.100520.

Zhang, Z., Tian, Y., Wang, G., Zheng, C., Zhao, F., 2024. A forest fire prediction method for lightning stroke based on remote sensing data. Forests 15 (4), 647. Available at https://doi.org/10.3390/f15040647.

Overview of the global wildfire scenario and its impacts on ecosystems

2

Jose Manuel Álvarez-Martínez[1,2] *and Susana Suárez-Seoane*[3,4]
[1]Area of Ecology, Department of Biodiversity and Environmental Management, Faculty of Biological and Environmental Sciences, University of León, León, Spain, [2]Institute for the Environment and Global Change (IMACG), University of León, León, Spain, [3]Biodiversity Research Institute (IMIB), University of Oviedo–CSIC–Principality of Asturias, Mieres, Asturias, Spain, [4]Department of Organisms and Systems Biology, University of Oviedo, Mieres, Asturias, Spain

2.1 General overview

Wildfires are ecological disturbances that have shaped terrestrial ecosystems for millennia, influencing their structure, composition and function across scales (Pechony and Shindell, 2010). In fire-prone biomes, such as the Mediterranean, fire plays a key role in maintaining biodiversity, regulating nutrient and carbon cycles, and driving succession (Bowman et al., 2011). Historically, indigenous populations and rural communities used fire locally to clear land, stimulate forage, or reduce fuel loads through practices adapted to specific ecological contexts Bilbao et al., 2025. Today, however, the emergence of a new fire regime, marked by larger, more severe, and often destructive events, contrasts sharply with those historical dynamics (Balch et al., 2024).

The intensification of fire activity in the 21st century results from the convergence of climate and land-use change (Jones et al., 2022). Rising temperatures, prolonged droughts, and more frequent heatwaves create conditions that favor greater fire severity and extent. At the same time, agricultural abandonment, rural depopulation and urban expansion alter fuel continuity and increase exposure in the wildland–urban interface (WUI) (Viedma et al., 2015) (Fig. 2.1). However, it is important to note that ignition remains the necessary trigger of any wildfire and, in most regions, it is predominantly of human origin through negligence, accidents or intentional burning. Thus, while climate and land-use change set the stage for extreme fire behavior, human activities largely determine when and where fires start (Bowman et al., 2011).

In many regions, long-term fire suppression policies have altered fuel dynamics and landscape structure, occasionally leading to denser and more continuous vegetation layers that can favor high-intensity fires under extreme weather conditions. However, these accumulations cannot always be regarded as "unnatural," since fire regimes have long been shaped by human activities, which determine both the frequency of burning and the spatial patterns of fuel continuity. Concurrently, a new scenario of large-scale passive rewilding is emerging,

Advances in Remote Sensing for Forest Fire Management. DOI: https://doi.org/10.1016/B978-0-443-41692-7.00007-X
© 2027 Elsevier Ltd. All rights are reserved, including those for text and data mining, AI training, and similar technologies.

Figure 2.1 Wildfire in Foyedo (Asturias, Spain), March 2023, which affected more than 10,000 ha. The image illustrates the spatial variations in fire impact across heterogeneous landscape mosaics in the Cantabrian Mountains (northern Spain).
Source: Photo: S. Suárez-Seoane.

particularly in developed countries where rural depopulation has left extensive areas without active land management (Lloret et al., 2024). This evolving context offers opportunities to restore ecological processes, but also poses new challenges for fire governance, calling for adaptive strategies that acknowledge shifting land-use dynamics and the intertwined role of humans and nature in shaping fire regimes. The ecological impacts of this shift are profound at all levels of organization. Large wildfires disrupt critical ecosystem services, including carbon sequestration, hydrological regulation and soil stabilization (Pausas and Keeley, 2019). They also fragment habitats, erode biodiversity and drive long-term changes in species composition and successional pathways (Holmquist et al., 2024). At the landscape scale, recurrent or extreme fires simplify mosaics, reduce ecological connectivity and promote alternative stable states dominated by flammable shrublands or grasslands (Sitters and Di Stefano, 2020).

These transformations weaken ecosystem resilience, diminishing the capacity of landscapes to buffer climate variability and to sustain the ecosystem services and resources upon which human livelihoods depend (Hamilton et al., 2019). Monitoring from patterns to processes requires tools that can track not only fire occurrence, but also its cascading effects on ecosystem structure and function (de la Riva et al., 2004). Remote sensing has become central in this regard, enabling consistent detection of fire activity over ecosystems, mapping of burned areas, and monitoring of postfire recovery across regions and biomes (Chuvieco and Congalton, 1989). When combined with ecological models and in-situ data, remote sensing supports the analysis of how wildfires reshape landscapes, modify connectivity and redefine the trajectories of ecosystem change. In this broad and multifaceted context, this chapter explores the global wildfire scenario, focusing on the main drivers behind its intensification and the resulting ecological consequences, contributing to understanding these dynamics and supporting integrated strategies for prevention, mitigation and adaptation in the face of accelerating global change.

2.2 Global patterns and trends in wildfire occurrence

The global wildfire landscape is shaped by complex regional patterns, multiscalar drivers and strong feedback with the climate system (Ruffault et al., 2018). Fire regimes are inherently dynamic: while some regions show declining burned areas, others face record-breaking extremes (Rodríguez Fernández-Blanco et al., 2025). These contrasting trajectories highlight the importance of understanding not only fire frequency, intensity and extent, but also their ecological consequences across landscapes and biomes as climate and land-use pressures intensify.

2.2.1 Historical context

Fire has shaped Earth's landscapes for over 400 million years, as shown by paleoecological evidence, such as charcoal deposits and fire scars in tree rings. Natural ignitions, mainly from lightning, have long interacted with climate and vegetation to sustain diverse fire regimes (Latham and Williams, 2001). With the emergence of human societies, fire became inseparable from landscape history. Early communities used it for hunting, clearing land and practicing agriculture, fundamentally altering ecosystem structure and composition (Bowman et al., 2009). These traditional regimes, often characterized by frequent, low-intensity burns, were generally well adapted to local conditions and played a key role in maintaining biodiversity, nutrient cycling and long-term resilience (Bignal and McCracken 1996). The industrial era marked a turning point. Rapid population growth, urban expansion and agricultural intensification transformed land use, while widespread fire suppression policies, aimed at protecting timber, infrastructure, and lives, became the norm (Pechony and Shindell, 2010). Although suppression reduced fire frequency in many regions, it disrupted long-established cycles that had maintained ecosystem balance for millennia. In fire-adapted landscapes, the absence of regular low-intensity burns has led to fuel accumulation, heightening the risk of large, severe wildfires when ignitions occur under extreme weather (Meyn et al., 2007).

This legacy of altered fire regimes poses complex challenges for contemporary fire management. In the current context, fire can no longer be regarded solely as a natural ecological process, since its frequency, timing and spatial patterns are now largely shaped by human activities. Understanding and integrating this dual ecological and anthropogenic dimension of fire into integrated landscape management is essential for sustaining ecosystem function and resilience under global change (Pausas and Keeley, 2021). At the same time, in many regions of the world, a new rural paradigm is unfolding, characterized by both widespread land abandonment and the intensification of forest use through large-scale, often monospecific plantations. While abandonment promotes fuel continuity and shrub encroachment, the expansion of homogeneous forest plantations increases flammability and alters fire behavior, frequently at a much larger spatial scale than the recovery of native forests. These processes are giving rise to "novel landscapes," that is, mosaics of recovering vegetation and unmanaged fuels that differ substantially from historical agricultural or forest systems. Managing such territories requires new approaches that combine ecological understanding with modern tools such as remote sensing, spatial modeling and adaptive governance frameworks (Seastedt et al., 2008). The goal is to design efficient and innovative methodologies that balance natural regeneration processes with the need to reduce wildfire risk and protect both biodiversity and human assets (Lindenmayer et al., 2008).

2.2.2 Contemporary trends

Because of the long history of fire and varying socio-ecological responses, wildfire-related landscapes are highly diverse, shaped by climate, vegetation, land management and socio-economic drivers. Unlike fire regimes of the past, modern fire patterns show stark regional contrasts and rapid shifts that challenge conventional management (Jones et al., 2022). Tropical savannas and grasslands account for nearly 70% of global burned area, with Sub-Saharan Africa and northern Australia representing the regions with the highest fire activity. Despite fires being ecologically necessary in these biomes, some regions may show declines linked to land conversion, intensification and grazing, which fragment fuels (Andela et al., 2017). By contrast, boreal and Arctic ecosystems show sharp increases, with longer seasons, more lightning and blazes reaching deep soils that release centuries-old carbon (Walker et al., 2020). Events in Siberia and Alaska during recent years illustrate climate-driven shifts. Temperate and Mediterranean forests of Southern Europe, California, Chile and Australia now face larger fires more frequently than in past decades. In the Mediterranean Basin, fire seasons are lengthening, and extreme weather events are also becoming more frequent (Dupuy et al., 2020). The "Black Summer" of Australia in 2019–20, which burned 18 million ha, epitomizes the ecological and social toll of extreme wildfires. Peatlands and tropical forests add complexity: Indonesian peat fires smolder for weeks, creating hazardous haze, while Amazon deforestation fires fragment habitats and push the biome toward a tipping point. Together, these patterns show how cumulative drivers heighten wildfire risk. Understanding them, from landscape mosaics to global climate feedback, is essential for strategies that recognize fire's ecological role while confronting its growing hazards in a warming world.

2.2.3 Climate change and fire weather intensification

There is a strong scientific consensus that climate change amplifies wildfire activity by intensifying fire-conducive weather (Jones et al., 2022). Rising global temperatures, more frequent heatwaves, prolonged droughts and shifting precipitation create conditions where wildfires ignite more easily, spread faster and burn more severely. Successive Intergovernmental Panel on Climate Change (IPCC) reports confirm the link between anthropogenic warming and extreme fire events, projecting increases in compound extremes such as droughts coinciding with heatwaves. In the Mediterranean, for example, days of "extreme fire weather" have risen sharply, extending fire seasons and heightening uncontrollable fire behavior. Boreal and Arctic regions also illustrate climate-fire feedbacks. Earlier snowmelt and drier fuels, combined with more lightning, increase ignition risk in remote forests. In Alaska, a 2004 lightning storm burned nearly three million hectares, releasing vast greenhouse gas stocks. Australia's 2019–20 "Black Summer" showed how prolonged drought and record heat can push ecosystems beyond resilience thresholds, generating unprecedented fire intensity and impacts.

These dynamics complicate climate policy frameworks such as land use, land-use change and forestry (LULUCF), which account for emissions and removals from land-based carbon pools (Schlamadinger et al., 2007). Wildfires can shift forests and peatlands from sinks to sources, challenging carbon neutrality targets and underscoring the need to integrate fire management into mitigation strategies. Effective responses must move beyond suppression, balancing ecological role of fire with the urgency of reducing emissions and adapting to new fire realities.

2.2.4 Growing fire risk at the wildland–urban interface

One of the most critical global wildfire trends is the expansion of the wildland-urban interface (WUI), where human settlements intermingle with flammable vegetation (Schug et al., 2023). Urban sprawl, rural depopulation and the spread of second homes in fire-prone landscapes expose more communities to wildfire hazards (Mell et al., 2010). Growth is pronounced in the Mediterranean Basin, western United States, and parts of Australia, where dense vegetation, long dry seasons, and extreme fire weather create volatile conditions. In California, the WUI expanded by over 40% between 1990 and 2010, placing hundreds of thousands of homes and infrastructure at risk. This has major implications for community safety, resilience and the escalating costs of suppression and recovery.

Recent disasters show how devastating WUI fires can be when coupled with severe weather (Guo et al., 2024). The 2018 Camp Fire in Paradise (California), which destroyed nearly 19,000 structures and killed at least 85 people, highlighted the speed and lethality of such events. Protecting dispersed housing requires immense resources, while expanding road networks and fragmented ownership limit the effectiveness of fuel reduction programs in decreasing fire spread and mitigating risk to people and property (Chen et al., 2024). Planning regulations often lag, allowing housing with insufficient defensible space or materials vulnerable to embers and radiant heat. In Spain, the summer of 2025 also witnessed massive wildfires crossing provincial borders in the Cantabrian Mountains, destroying dispersed housing and forcing large-scale evacuations (Fig. 2.2). The fire in Molezuelas de la Carballeda, between Zamora and León, burned more than 37,000 hectares and severely damaged rural settlements, illustrating how vulnerable scattered housing remains under extreme fire weather. As in other Mediterranean regions, planning regulations often lag, allowing construction with insufficient defensible space or materials highly vulnerable to embers and radiant heat. As WUI areas expand worldwide, bridging science, planning and community engagement is essential (Li et al., 2022). Reducing wildfire risk in these complex interfaces demands advanced monitoring, but also

Figure 2.2 Wildfire approaching the village of Fasgar (León, Spain) during the 2025 fire season. A large smoke plume rises above the wildland–urban interface, with flames advancing from the surrounding mountains toward rural settlements.
Source: Photo: Alejandro Fernández.

integrated policy approaches that balance housing, ecosystem management and nature conservation in an era where extreme fires are becoming the new norm. Human safety ranks at the top of priorities, but human activities remain, at the same time, the dominant source of ignitions, from agricultural burning and arson to land ownership and accidents involving power lines or vehicles.

2.3 Ecological impacts of wildfires

Wildfires emerge today as the outcome of intertwined climatic, ecological and social drivers, from rising temperatures and fuel accumulation to land-use transitions and growing human presence in fire-prone landscapes (Sayedi et al., 2024). Understanding how the new fire regimes impact biodiversity, ecosystem functioning and resilience is crucial for developing effective conservation strategies, guiding ecosystem management and ensuring long-term sustainability of landscapes in a changing world. The intensification of fire activity translates into profound ecological effects that cascade across ecosystem compartments, including vegetation and habitats, fauna, soils, hydrology and global carbon cycles. Wildfires firstly reshape vegetation structure and composition, often simplifying plant communities and reducing habitat heterogeneity. The loss of mature canopy and understory layers not only delays regeneration but also transforms habitat suitability for numerous species and faunal communities through direct mortality, habitat loss and altered food and shelter availability. While some mobile species can escape low-intensity fires, many less mobile organisms, such as soil invertebrates, amphibians, or small mammals, suffer high mortality rates. Soils are among the most vulnerable components, as intense burns remove organic layers that protect against erosion, diminish nutrient stocks and shift microbial communities (Agbeshie et al., 2022). The consequences extend to hydrology, with greater surface runoff, sediment flows and deteriorating water quality in streams and reservoirs (Mataix-Solera et al., 2011). At larger scales, wildfires influence the carbon cycle by releasing substantial amounts of CO_2, methane and aerosols (Lasslop et al., 2019). In carbon-rich systems, such as peatlands or boreal forests, repeated severe fires can shift ecosystems from long-term sinks to net sources of emissions, amplifying climate feedbacks. Taken together, these pathways illustrate how the new fire regimes are increasingly destructive, leaving persistent ecological legacies that undermine ecosystem stability in the context of climate and land-use change.

2.3.1 Impacts on vegetation composition, structure, and function

Wildfires directly reshape vegetation by consuming biomass, altering canopy architecture and modifying species composition (Lloret and Zedler, 2009). Their ecological effects depend on fire severity, frequency, seasonality and the adaptive traits of local flora. In many ecosystems, plant species have evolved functional strategies to persist under recurrent fire regimes, such as thick bark that protects cambial tissue, serotinous cones that release seeds after heat exposure, or strong resprouting ability from lignotubers and root systems (Midgley and Bond, 2013). However, the resilience conferred by these traits can be exceeded when fire activity surpasses historical norms or occurs at shorter intervals than recovery periods (McKenzie et al., 2011). Under such conditions, regeneration failure becomes increasingly common, leading to shifts from forest to shrubland or grassland states and, in extreme cases, to persistent nonvegetated areas. Repeated high-severity fires can simplify community structure, reduce vertical and horizontal heterogeneity and diminish functional diversity (Puig-Gironès et al., 2025). In turn, these changes alter microclimatic conditions, soil protection and nutrient cycling, generating feedbacks that further constrain vegetation recovery. Beyond structural loss, wildfires also modify successional trajectories. Early postfire stages are often dominated by opportunistic or invasive species, or fire-prone communities that take advantage of disturbed environments, sometimes delaying or preventing the return of late-successional communities (Fig. 2.3). These transitions, increasingly reported across temperate, boreal and Mediterranean biomes, exemplify how novel fire regimes are reshaping vegetation dynamics and eroding long-term ecosystem stability (Fernández-García and Alonso-González, 2023).

2.3.2 Impacts on biodiversity and ecological connectivity

Wildfires have a complex relationship with biodiversity. Under natural regimes, fire introduces heterogeneity across the landscape mosaic and promotes diversity by

Figure 2.3 Establishment of fire-prone heathland communities following recurrent wildfires in the Sierra de Ancares (León, NW Spain). The image illustrates a landscape-scale shift from prefire forested formations to shrub-dominated heathlands, largely composed of *Erica* and *Calluna* species, promoted by repeated fire events at short return intervals. Recurrent burning limits tree regeneration and favors resprouting and obligate seeding shrub species adapted to high fire frequency, leading to a simplified vegetation structure and altered successional trajectories. These fire-maintained heathlands represent an alternative stable state under persistent disturbance regimes, with implications for ecosystem function, fuel continuity and future fire feedbacks.
Source: Photo: Jose Manuel Álvarez-Martínez.

generating successional stages, open conditions and nutrient flushes. Many fire-adapted species exploit these opportunities, from plants that germinate after fire to animals that benefit from new foraging grounds. Yet when fire regimes intensify, the balance tilts. Species with narrow habitat requirements or limited dispersal may be pushed toward local extinction (Pastro et al., 2011). The 2019–20 Australian bushfires, which displaced or killed millions of animals, underscored how extreme events can surpass adaptive capacities and erode biodiversity. Severe or frequent fires also fragment habitats, reducing connectivity and gene flow while creating conditions for invasive species. Cheatgrass in North America exemplifies this: it rapidly colonizes after fire, increases fuel continuity, and drives feedback loops that suppress native regeneration. In Spain, recent megafires in Mediterranean forests have similarly revealed the vulnerability of biodiversity under extreme fire regimes. These events not only caused direct mortality of wildlife, including populations of Iberian capercaillies, bluethroats or partridges but also fragmented ecosystems already under pressure from rural abandonment and climate change. As in other Mediterranean landscapes, recurrent fires are favoring flammable shrublands over mature forests, eroding ecological connectivity and compromising the long-term conservation of endemic species.

2.3.3 Impacts on soil properties and hydrological processes

Wildfires profoundly affect soils and watersheds, with consequences that extend well beyond the fire perimeter (Fig. 2.4). High-severity burns often create hydrophobic layers, reducing infiltration and increasing surface runoff (Boyer et al., 2022). Combined with the loss of vegetation and litter cover, this makes soil highly vulnerable to erosion. Nutrient-rich topsoil can be stripped away, degrading site productivity and depositing sediments into rivers and reservoirs, where they reduce water quality and increase flooding and siltation risks. Nutrient cycling is also altered: combustion volatilizes nitrogen and other elements, while ash may enrich soils with phosphorus or potassium (Abbott et al., 2021). Whether fertility is enhanced or depleted depends on fire severity and the pace of vegetation recovery. Hydrological impacts include higher peak flows, flash floods and debris flows, especially in Mediterranean catchments, where severe postfire storms have triggered landslides and infrastructure damage (Belmar et al., 2018).

2.3.4 Impacts on atmospheric composition and climate

Wildfires are major sources of greenhouse gases and aerosols, with direct consequences for climate and air quality. As vegetation and soils burn, they release large amounts of CO_2, CH_4 and particulate matter, adding to the drivers of global warming (Urbanski, 2014). The magnitude of these impacts varies with ecosystem type. In grasslands and savannas, frequent but low-intensity fires are often balanced by rapid regrowth that reabsorbs carbon. In contrast, fires in carbon-rich peatlands

Figure 2.4 High-severity wildfire effects on *Erica*-dominated heathlands. The image shows extensive vegetation mortality and near-complete removal of surface organic layers, reflecting extreme soil heating, loss of soil organic carbon, reduced infiltration capacity and increased susceptibility to erosion, runoff generation and downstream sediment transport.
Source: Photo: Jose Manuel Álvarez Martínez.

and boreal forests mobilize ancient carbon stocks, creating long-term "carbon debts." The 2020 Siberian fires, for example, released nearly 600 million tons of CO_2, comparable to annual emissions of entire nations. These emissions also trigger feedback loops: warmer conditions promote fire-prone weather, which in turn drives more severe fires. Arctic and subarctic regions are particularly vulnerable, where thawing permafrost and peat combustion could release vast stores of carbon and methane. Beyond climate impacts, wildfire smoke degrades air quality over vast regions, with implications for human health, agriculture and even aviation safety (Filonchyk et al., 2022).

2.4 Wildfire-nature interactions under global change: toward resilient landscapes

The interaction between wildfires and landscape management has become one of the central themes in contemporary ecology. Wildfires today must be understood not only as ecological disturbances, but also as processes deeply intertwined with the spatial structure and dynamics of landscapes (Moreira et al., 2011). The patterns of ignition, spread and severity are strongly conditioned by landscape characteristics: composition, configuration, and land-use history.

2.4.1 Landscape determinants of emerging wildfire scenarios

In many regions, accumulation of fuels through secondary succession and expansion of flammable plantations have increased fuel continuity, producing landscapes that are more prone to large and severe fires (Barquín et al., 2022). At the same time, fires themselves feed back into landscape trajectories. Depending on their frequency and intensity, they may promote regeneration, open opportunities for colonization by new communities, or drive persistent type conversions that lock ecosystems into shrub- or grass-dominated states (Harrison et al., 2021). In this way, fire is regulated by landscape patterns, shaping mosaic diversity across space and time. Global change intensifies this coupling. Rising summer temperatures, prolonged droughts and more frequent heatwaves amplify landscape flammability, while vegetation shifts in altitude and latitude create novel fire-vegetation assemblages. The result is an escalating risk of extreme wildfires with cascading consequences not only for ecosystems but also for the flows of biodiversity, water and nutrients that operate at landscape to regional scales. Within this context, a multiscale ecological perspective becomes essential. Integrating land-use history, vegetation succession and fire dynamics into predictive models allows for the identification of areas most prone to change, the simulation of alternative scenarios and the design of adaptive strategies that enhance ecosystem resilience (Houet et al., 2022). Understanding wildfires as landscape processes highlights the need for spatially informed prevention and restoration policies: promoting heterogeneity, reducing fuel continuity and designing mosaics that reinforce the resilience of socio-ecological systems. Ultimately, the configuration and management of landscapes will shape the future of fire regimes

(Montiel Molina et al., 2019). By determining how, where and with what intensity wildfires occur, landscapes set the stage not only for ecological impacts but also for the capacity of ecosystems to sustain biodiversity, ecosystem functions and the services on which human societies depend, defining what can be called resilient landscapes.

Landscapes are therefore dynamic systems where fire interacts with land-use composition, configuration and history (Álvarez Martínez et al., 2011). These structural attributes largely determine ignition probability, fire spread and severity. Processes such as rural abandonment, secondary succession and the expansion of flammable plantations have increased fuel loads and continuity, connecting areas that were once part of heterogeneous, mosaic-like landscapes. However, not all fuels behave equally: while mature, humid forests tend to retain moisture and act as partial fire buffers, extensive plantations of fast-growing, resinous, or nonnative species, such as pines or eucalyptus, accumulate fine, dry biomass that markedly elevates flammability. The resulting homogenization of fuel types amplifies the likelihood of high-intensity, rapidly spreading fires, emphasizing the need to differentiate between fuel quantity and fuel quality when assessing landscape fire risk. Nevertheless, although these structural changes may not necessarily increase fire occurrence, they can actually facilitate faster and larger fire spread under extreme weather conditions (Mantero et al., 2020). In turn, fire shapes vegetation trajectories and mosaic dynamics, acting as both a disturbance and a regulator of landscape heterogeneity (López-Poma et al., 2014). Under global change, rising temperatures, droughts and vegetation shifts amplify these feedbacks, promoting novel fire-landscape assemblages with cascading effects on ecosystem resilience and biodiversity.

2.4.2 *Cumulative and cross-scale effects*

Building on this perspective, it is crucial to recognize that wildfires interact through time and space, generating cumulative effects that shape long-term landscape trajectories. Their cumulative impacts across time and space can drive large-scale transformations that fundamentally reshape entire biomes. When fires become more frequent, severe or extensive than ecosystems have evolved to withstand, they may push landscapes beyond critical tipping points, leading to type conversions, such as boreal forests turning into shrublands or tropical forests degrading into savanna-like states (Pezzatti et al., 2013). Once these thresholds are crossed, recovery to predisturbance conditions may become impossible under current or future climates, resulting in lasting losses of habitats, species and functions (Driscoll et al., 2021). These transformations inevitably erode ecosystem services that societies depend on, from water regulation and fertile soils to carbon storage, pollination, and cultural values, linking local impacts to global biodiversity and climate targets (McKenzie et al., 2004). Therefore, the intensifying impacts of wildfires highlight the need to bridge science, technology and practice. Only through integrated monitoring and proactive land management societies can anticipate ecosystem responses to extreme fire regimes across scales and implement evidence-based strategies that protect both nature and people in a warming world (Moritz et al., 2014).

2.4.3 Learning from the fire effects on ecosystems: the rise of nature-based solutions (NbS) and blue-green infrastructure networks (BGIN)

Understanding how ecosystems respond and recover after fire provides key lessons for building resilience. Wildfires reveal the thresholds and feedbacks that shape ecosystem functioning at the level of how structural diversity, connectivity and adaptive capacity determine postfire recovery or degradation. These ecological insights underpin the emerging role of nature-based solutions (NbS) in fire-prone landscapes (Lecina-Diaz et al., 2023). Defined as actions that protect, restore and sustainably manage ecosystems to address societal challenges while supporting biodiversity and human well-being (Sowińska-Świerkosz and García, 2022), NbS translate ecological understanding into practice by reinforcing the same mechanisms that sustain resilience after disturbance. Within this context, blue-green infrastructure networks (BGINs) offer a spatial framework to implement NbS at meaningful ecological scales (Debele et al., 2023; Langemeyer and Baró, 2021). By maintaining heterogeneous mosaics, ecological corridors and multifunctional land uses, these networks can reduce fuel continuity, buffer settlements and enhance ecosystem services (Houet et al., 2022). Integrating NbS into BGINs thus aligns biodiversity conservation and climate adaptation with fire risk reduction, transforming postfire landscapes into opportunities for ecological renewal and sustainable management.

Despite their potential, NbS face key challenges when applied to fire-prone ecosystems. Postfire studies show that recovery is shaped by trade-offs among ecosystem functions, carbon storage, grazing, biodiversity or soil protection, and by the social capacity to maintain them over time, that is, short political cycles and no funding continuity to reach their long-term ecological benefits (Deely et al., 2020). Learning from these dynamics is essential to avoid maladaptive actions, such as uniform forestation that increase flammability and reduce structural diversity. Ultimately, wildfire reminds us that resilience depends on diversity, connectivity and adaptive management strategies across scales. By incorporating these lessons, NbS and BGINs can move beyond restoration to become tools for coexisting with fire, designing landscapes that absorb disturbance while continuing to provide essential ecological and cultural services.

2.5 Final remark

Wildfires in the Anthropocene must be understood both as powerful drivers and consequences of socio-ecological transformation, reflecting the feedbacks between climate, land use and human activity. While fire has historically acted as a natural or human-managed disturbance that promoted ecosystem renewal and maintained resilience in many ecosystems, global fire regimes are now shifting beyond their historical variability, marked by longer fire seasons, larger burned areas and more frequent extreme events. These dynamics, shaped by climate change, land-use transitions and socio-economic pressures, are pushing ecosystems and societies toward critical thresholds where natural recovery becomes compromised. The ecological consequences of this intensification cascade across scales. At the ecosystem level, fires alter vegetation structure, erode biodiversity, degrade soil and water systems, and disrupt carbon balance. At the landscape level, recurrent disturbances simplify mosaics, reduce connectivity and trap systems in alternative stable states dominated by flammable shrubland or grassland communities. These transformations not only undermine ecological resilience but also diminish the capacity of ecosystems to provide essential services such as carbon sequestration, water regulation, soil fertility, pollination and cultural assets. Enhancing socio-ecological resilience requires analysing post-fire landscape reorganization, recovery trajectories and adaptive responses, translating scientific and cultural knowledge into management strategies that support biodiversity, ecosystem functioning and human well-being.

The following key messages synthesize the central insights of this chapter, highlighting how wildfires in the Anthropocene reflect intertwined socio-ecological dynamics:

1. Wildfires as socio-ecological processes. Historically, fire has often operated as a local and short-term disturbance within vegetation dynamics. In the Anthropocene, climate change, land-use change, fuel accumulation and socio-economic vulnerability are expanding the scale, intensity and consequences of wildfires across ecosystems.
2. From local disturbances to global transformations. Intensifying fire regimes, characterized by longer seasons and more severe and larger events, are driving cross-scale ecological shifts that erode biodiversity, disrupt carbon and water cycles and threaten the delivery of essential ecosystem services.
3. Designing fire-resilient landscapes through NbS. Restoration, rewilding and agroforestry practices can contribute to fire-resilient landscapes when embedded within connected blue–green infrastructure networks and systematic conservation planning frameworks. These approaches can reduce fuel continuity, increase landscape heterogeneity, sustain biodiversity and maintain ecosystem service flows. Their planning and evaluation require spatially continuous information on fire occurrence, severity, vegetation recovery and connectivity, which remote sensing and spatial modelling can provide to support adaptive management.
4. Toward integrated, multiscale fire governance. The future of wildfire management depends on bridging science, technology and policy to create spatially explicit, socially inclusive strategies that balance risk reduction with the conservation of biodiversity and ecosystem functions under global change.

References

Abbott, B.W., Rocha, A.V., Shogren, A., Zarnetske, J.P., Iannucci, F., Bowden, W.B., et al., 2021. Tundra wildfire triggers sustained lateral nutrient loss in Alaskan Arctic. Global Change Biology 27 (7), 1408–1430. Available at https://doi.org/10.1111/gcb.15507, http://onlinelibrary.wiley.com/journal/10.1111/(ISSN)1365-2486.

Agbeshie, A.A., Abugre, S., Atta-Darkwa, T., Awuah, R., 2022. A review of the effects of forest fire on soil properties. Journal of Forestry Research 33 (5), 1419–1441. Available at https://doi.org/10.1007/s11676-022-01475-4, https://www.springer.com/journal/11676.

Álvarez Martínez, J.-M., Suárez-Seoane, S., De Luis Calabuig, E., 2011. Modelling the risk of land cover change from environmental and socio-economic drivers in heterogeneous and changing landscapes: the role of uncertainty. Landscape and Urban Planning 101 (2), 108–119. Available at https://doi.org/10.1016/j.landurbplan.2011.01.009.

Andela, N., Morton, D.C., Giglio, L., Chen, Y., van der Werf, G.R., Kasibhatla, P.S., et al., 2017. A human-driven decline in global burned area. Science (New York, N.Y.) 356 (6345), 1356–1362. Available at https://doi.org/10.1126/science.aal4108.

Balch, J.K., Iglesias, V., Mahood, A.L., Cook, M.C., Amaral, C., DeCastro, A., et al., 2024. The fastest-growing and most destructive fires in the US (2001 to 2020). Science (New York, N.Y.) 386 (6720), 425–431. Available at https://doi.org/10.1126/science.adk5737, https://www.science.org/doi/10.1126/science.adk5737.

Belmar, O., Barquín, J., Álvarez-Martínez, J.M., Peñas, F.J., Del Jesus, M., 2018. The role of forest maturity in extreme hydrological events. Ecohydrology 11 (4). Available at https://doi.org/10.1002/eco.1947, http://onlinelibrary.wiley.com/journal/10.1002/(ISSN)1936-0592.

Barquín, J., Concostrina-Zubiri, L., Pérez-Silos, I., Hernández-Romero, G., Vélez-Martín, A., Álvarez-Martínez, J.M., 2022. Monoculture plantations fuel fires amid heat waves. Science 377, 1498. Available at https://doi.org/10.1126/science.ade5923.

Bignal, E.M., McCracken, D.I., 1996. Low-Intensity Farming Systems in the Conservation of the Countryside. The Journal of Applied Ecology 33, 413. Available at https://doi.org/10.2307/2404973.

Bilbao, B.A., Ferrero, B.G., Falleiro, R.M., Moura, L.C., Fagundes, G.M., McKenzie, D., et al., 2011. The Landscape Ecology of Fire. Springer Science & Business Media.

Bilbao, B.A., Ferrero, B.G., Falleiro, R.M., Moura, L.C., Fagundes, G.M., 2025. Traditional Fire Uses by Indigenous Peoples and Local Communities in South America. Fire in the South American Ecosystems. Springer, pp. 39–81.

Bowman, D.M.J.S., Balch, J.K., Artaxo, P., Bond, W.J., Carlson, J.M., Cochrane, M.A., et al., 2009. Fire in the earth system. Science (New York, N.Y.) 324 (5926), 481–484. Available at https://doi.org/10.1126/science.1163886.

Bowman, D.M.J.S., Balch, J., Artaxo, P., Bond, W.J., Cochrane, M.A., D'Antonio, C.M., et al., 2011. The human dimension of fire regimes on Earth. Journal of Biogeography 38 (12), 2223–2236. Available at https://doi.org/10.1111/j.1365-2699.2011.02595.x.

Boyer, E.W., Wagenbrenner, J.W., Zhang, L., 2022. Wildfire and hydrological processes. Hydrological Processes 36 (7). Available at https://doi.org/10.1002/hyp.14640, http://onlinelibrary.wiley.com/journal/10.1002/(ISSN)1099-1085.

Chen, B., Wu, S., Jin, Y., Song, Y., Wu, C., Venevsky, S., et al., 2024. Wildfire risk for global wildland–urban interface areas. Nature Sustainability 7 (4), 474–484. Available at https://doi.org/10.1038/s41893-024-01291-0.

Chuvieco, E., Congalton, R.G., 1989. Application of remote sensing and geographic information systems to forest fire hazard mapping. Remote Sensing of Environment 29 (2), 147–159. Available at https://doi.org/10.1016/0034-4257(89)90023-0.

Debele, S.E., Leo, L.S., Kumar, P., Sahani, J., Ommer, J., Bucchignani, E., et al., 2023. Nature-based solutions can help reduce the impact of natural hazards: a global analysis of NBS case studies. Science of The Total Environment 902, 165824. Available at https://doi.org/10.1016/j.scitotenv.2023.165824.

Deely, J., Hynes, S., Barquín, J., Burgess, D., Finney, G., Silió, A., et al., 2020. Barrier identification framework for the implementation of blue and green infrastructures. Land use Policy 99. Available at https://doi.org/10.1016/j.landusepol.2020.105108, www.elsevier.com/inca/publications/store/3/0/4/5/1/.

Driscoll, D.A., Armenteras, D., Bennett, A.F., Brotons, L., Clarke, M.F., Doherty, T.S., et al., 2021. How fire interacts with habitat loss and fragmentation. Biological Reviews 96 (3), 976–998. Available at https://doi.org/10.1111/brv.12687, http://onlinelibrary.wiley.com/journal/10.1111/(ISSN)1469-185X.

Dupuy, Jl, Fargeon, H., Martin-StPaul, N., Pimont, F., Ruffault, J., Guijarro, M., et al., 2020. Climate change impact on future wildfire danger and activity in southern Europe: a review. Annals of Forest Science 77 (2). Available at https://doi.org/10.1007/s13595-020-00933-5, http://www.springer.com.

Fernández-García, V., Alonso-González, E., 2023. Global patterns and dynamics of burned area and burn severity. Remote Sensing 15 (13). Available at https://doi.org/10.3390/rs15133401, http://www.mdpi.com/journal/remotesensing.

Filonchyk, M., Peterson, M.P., Sun, D., 2022. Deterioration of air quality associated with the 2020 US wildfires. Science of the Total Environment 826. Available at https://doi.org/10.1016/j.scitotenv.2022.154103, www.elsevier.com/locate/scitotenv.

Guo, Y., Wang, J., Ge, Y., Zhou, C., 2024. Global expansion of wildland-urban interface intensifies human exposure to wildfire risk in the 21st century. Science Advances 10 (45). Available at https://doi.org/10.1126/sciadv.ado9587.

Hamilton, M., Fischer, A.P., Ager, A., 2019. A social-ecological network approach for understanding wildfire risk governance. Global Environmental Change 54, 113–123. Available at https://doi.org/10.1016/j.gloenvcha.2018.11.007, www.elsevier.com/inca/publications/store/3/0/4/2/5.

Harrison, S.P., Prentice, I.C., Bloomfield, K.J., Dong, N., Forkel, M., Forrest, M., et al., 2021. Understanding and modelling wildfire regimes: an ecological perspective. Environmental Research Letters 16 (12). Available at https://doi.org/10.1088/1748-9326/ac39be, https://iopscience.iop.org/article/10.1088/1748-9326/ac39be.

Holmquist, A.J., Cody Markelz, R.J., Martinez, C.C., Gillespie, R.G., 2024. The importance of habitat type and historical fire regimes in arthropod community response following large-scale wildfires. Global Change Biology 30 (1). Available at https://doi.org/10.1111/gcb.17135, http://onlinelibrary.wiley.com/journal/10.1111/(ISSN)1365-2486.

Houet, T., Gaetan, P., Roberta, R., Hugues, B., Jacques, B., Xavier, P., et al., 2022. European blue and green infrastructure network strategy vs. the common agricultural policy. Insights from an integrated case study (Couesnon, Brittany). Land use Policy 120. Available at https://doi.org/10.1016/j.landusepol.2022.106277, http://www.elsevier.com/inca/publications/store/3/0/4/5/1/.

Jones, M.W., Abatzoglou, J.T., Veraverbeke, S., Andela, N., Lasslop, G., Forkel, M., et al., 2022. Global and regional trends and drivers of fire under climate change. Reviews of Geophysics 60 (3). Available at https://doi.org/10.1029/2020RG000726, http://agupubs.onlinelibrary.wiley.com/agu/journal/10.1002/(ISSN)1944-9208/.

Langemeyer, J., Baró, F., 2021. Nature-based solutions as nodes of green-blue infrastructure networks: a cross-scale, co-creation approach. Nature-Based Solutions 1, 100006. Available at https://doi.org/10.1016/j.nbsj.2021.100006.

Lasslop, G., Coppola, A.I., Voulgarakis, A., Yue, C., Veraverbeke, S., 2019. Influence of fire on the carbon cycle and climate. Current Climate Change Reports 5 (2), 112–123. Available at https://doi.org/10.1007/s40641-019-00128-9, https://rd.springer.com/journal/40641.

Latham, D., Williams, E., 2001. Lightning and Forest Fires. Elsevier BV, pp. 375–418. Available at 10.1016/b978-012386660-8/50013-1.

Lecina-Diaz, J., Campos, J.C., Pais, S., Carvalho-Santos, C., Azevedo, J.C., Fernandes, P., et al., 2023. Stakeholder perceptions of wildfire management strategies as nature-based solutions in two Iberian biosphere reserves. Ecology and Society 28 (1). Available at https://doi.org/10.5751/ES-13907-280139, https://ecologyandsociety.org/vol28/iss1/art39/ES-2022-13907.pdf.

Li, S., Dao, V., Kumar, M., Nguyen, P., Banerjee, T., 2022. Mapping the wildland-urban interface in California using remote sensing data. Scientific Reports 12 (1). Available at https://doi.org/10.1038/s41598-022-09707-7.

Lindenmayer, D.B., Fischer, J., Felton, A., Crane, M., Michael, D., Macgregor, C., et al., 2008. Novel ecosystems resulting from landscape transformation create dilemmas for modern conservation practice. Conservation Letters 1 (3), 129–135. Available at https://doi.org/10.1111/j.1755-263x.2008.00021.x.

Lloret, F., Zedler, P.H., 2009. The Effect of Forest Fire on Vegetation Fire Effects on Soils and Restoration Strategies. CRC Press, Spain, pp. 257–295. Available at https://www.taylorfrancis.com/books/e/9781439843338.

Lloret, F., Escudero, A., Lloret, J., Valladares, F., 2024. An ecological perspective for analysing rural depopulation and abandonment. People and Nature 6 (2), 490–506. Available at https://doi.org/10.1002/pan3.10606.

López-Poma, R., Orr, B.J., Bautista, S., 2014. Successional stage after land abandonment modulates fire severity and post-fire recovery in a Mediterranean mountain landscape. International Journal of Wildland Fire 23 (7), 1005–1015. Available at https://doi.org/10.1071/WF13150, http://www.publish.csiro.au/nid/126.htm.

Mantero, G., Morresi, D., Marzano, R., Motta, R., Mladenoff, D.J., Garbarino, M., 2020. The influence of land abandonment on forest disturbance regimes: a global review. Landscape Ecology 35 (12), 2723–2744. Available at https://doi.org/10.1007/s10980-020-01147-w, http://www.springerlink.com/content/103025/.

Mataix-Solera, J., Cerdà, A., Arcenegui, V., Jordán, A., Zavala, L.M., 2011. Fire effects on soil aggregation: a review. Earth-Science Reviews 109 (1-2), 44–60. Available at https://doi.org/10.1016/j.earscirev.2011.08.002.

McKenzie, D., Gedalof, Z., Peterson, D.L., Mote, P., 2004. Climatic change, wildfire, and conservation. Conservation Biology 18 (4), 890–902. Available at https://doi.org/10.1111/j.1523-1739.2004.00492.x.

McKenzie, D., Miller, C., & Falk, D.A. (2011). The landscape ecology of fire. Springer Science & Business Media.

Mell, W.E., Manzello, S.L., Maranghides, A., Butry, D., Rehm, R.G., 2010. The wildland-urban interface fire problem - current approaches and research needs. International Journal of Wildland Fire 19 (2), 238–251. Available at https://doi.org/10.1071/WF07131.

Meyn, A., White, P.S., Buhk, C., Jentsch, A., 2007. Environmental drivers of large, infrequent wildfires: the emerging conceptual model. Progress in Physical Geography 31 (3), 287–312. Available at https://doi.org/10.1177/0309133307079365, https://journals.sagepub.com/home/PPG.

Midgley, J.J., Bond, W.J., 2013. Plant Adaptations to Fire: An Evolutionary Perspective Fire Phenomena and the Earth System: An Interdisciplinary Guide to Fire Science. John Wiley and Sons, South Africa, pp. 125–134. Available at http://onlinelibrary.wiley.com/book/10.1002/9781118529539, 10.1002/9781118529539.ch7.

Montiel Molina, C., Karlsson Martín, O., Galiana Martín, L., 2019. Regional fire scenarios in spain: linking landscape dynamics and fire regime for wildfire risk management. Journal of Environmental Management 233, 427–439. Available at https://doi.org/10.1016/j.jenvman.2018.12.066.

Moreira, F., Viedma, O., Arianoutsou, M., Curt, T., Koutsias, N., Rigolot, E., et al., 2011. Landscape – wildfire interactions in southern Europe: implications for landscape management. Journal of Environmental Management 92 (10), 2389–2402. Available at https://doi.org/10.1016/j.jenvman.2011.06.028.

Moritz, M.A., Batllori, E., Bradstock, R.A., Gill, A.M., Handmer, J., Hessburg, P.F., et al., 2014. Learning to coexist with wildfire. Nature 515 (7525), 58–66. Available at https://doi.org/10.1038/nature13946, http://www.nature.com/nature/index.html.

Pastro, L.A., Dickman, C.R., Letnic, M., 2011. Burning for biodiversity or burning biodiversity? Prescribed burn vs. wildfire impacts on plants, lizards, and mammals. Ecological Applications 21 (8), 3238–3253. Available at https://doi.org/10.1890/10-2351.1Australia, http://www.esajournals.org/doi/pdf/10.1890/10-2351.1.

Pausas, J.G., Keeley, J.E., 2019. Wildfires as an ecosystem service. Frontiers in Ecology and the Environment 17 (5), 289–295. Available at https://doi.org/10.1002/fee.2044, http://onlinelibrary.wiley.com/journal/10.1002/(ISSN)1540-9309.

Pausas, J.G., Keeley, J.E., 2021. Wildfires and global change. Frontiers in Ecology and the Environment 19 (7), 387–395. Available at https://doi.org/10.1002/fee.2359, http://onlinelibrary.wiley.com/journal/10.1002/(ISSN)1540-9309.

Pechony, O., Shindell, D.T., 2010. Driving forces of global wildfires over the past millennium and the forthcoming century. Proceedings of the National Academy of Sciences 107 (45), 19167–19170. Available at https://doi.org/10.1073/pnas.1003669107.

Pezzatti, G.B., Zumbrunnen, T., Bürgi, M., Ambrosetti, P., Conedera, M., 2013. Fire regime shifts as a consequence of fire policy and socio-economic development: an analysis based on the change point approach. Forest Policy and Economics 29, 7–18. Available at https://doi.org/10.1016/j.forpol.2011.07.002.

Puig-Gironès, R., Palmero-Iniesta, M., Fernandes, P.M., Oliveras Menor, I., Ascoli, D., Kelly, L.T., et al., 2025. The use of fire to preserve biodiversity under novel fire regimes. Philosophical Transactions of the Royal Society B: Biological Sciences 380 (1924). Available at https://doi.org/10.1098/rstb.2023.0449, http://rstb.royalsocietypublishing.org/.

de la Riva, J., Pérez-Cabello, F., Lana-Renault, N., Koutsias, N., 2004. Mapping wildfire occurrence at regional scale. Remote Sensing of Environment 92 (3), 363–369. Available at https://doi.org/10.1016/j.rse.2004.06.022.

Rodríguez Fernández-Blanco, C., Muys, B., Winkel, G., Parra, C., 2025. Revisiting wildfire resilience from a territorial perspective: insights from Mediterranean Spain. Journal of Environmental Planning and Management 68 (13), 3091–3120. Available at https://doi.org/10.1080/09640568.2024.2342333.

Ruffault, J., Curt, T., Martin-Stpaul, N.K., Moron, V., Trigo, R.M., 2018. Extreme wildfire events are linked to global-change-type droughts in the northern Mediterranean. Natural Hazards and Earth System Sciences 18 (3), 847–856. Available at https://doi.org/10.5194/nhess-18-847-2018, http://www.nat-hazards-earth-syst-sci.net/volumes_and_issues.html.

Sayedi, S.S., Abbott, B.W., Vannière, B., Leys, B., Colombaroli, D., Romera, G.G., et al., 2024. Assessing changes in global fire regimes. Fire Ecology 20 (1). Available at https://doi.org/10.1186/s42408-023-00237-9, https://www.springer.com/journal/42408.

Schlamadinger, B., Bird, N., Johns, T., Brown, S., Canadell, J., Ciccarese, L., et al., 2007. A synopsis of land use, land-use change and forestry (LULUCF) under the Kyoto Protocol and Marrakech Accords. Environmental Science & Policy 10 (4), 271–282. Available at https://doi.org/10.1016/j.envsci.2006.11.002.

Schug, F., Bar-Massada, A., Carlson, A.R., Cox, H., Hawbaker, T.J., Helmers, D., et al., 2023. The global wildland–urban interface. Nature 621 (7977), 94–99. Available at https://doi.org/10.1038/s41586-023-06320-0, https://www.nature.com/nature/.

Seastedt, T.R., Hobbs, R.J., Suding, K.N., 2008. Management of novel ecosystems: are novel approaches required? Frontiers in Ecology and the Environment 6 (10), 547–553. Available at https://doi.org/10.1890/070046.

Sitters, H., Di Stefano, J., 2020. Integrating functional connectivity and fire management for better conservation outcomes. Conservation Biology 34 (3), 550–560. Available at https://doi.org/10.1111/cobi.13446.

Sowińska-Świerkosz, B., García, J., 2022. What are nature-based solutions (NBS)? Setting core ideas for concept clarification. Nature-Based Solutions 2, 100009. Available at https://doi.org/10.1016/j.nbsj.2022.100009.

Urbanski, S., 2014. Wildland fire emissions, carbon, and climate: emission factors. Forest Ecology and Management 317, 51–60. Available at https://doi.org/10.1016/j.foreco.2013.05.045.

Viedma, O., Moity, N., Moreno, J.M., 2015. Changes in landscape fire-hazard during the second half of the 20th century: agriculture abandonment and the changing role of driving factors. Agriculture, Ecosystems and Environment 207, 126–140. Available at https://doi.org/10.1016/j.agee.2015.04.011, www.elsevier.com/inca/publications/store/5/0/3/2/9/8.

Walker, X.J., Baltzer, J.L., Bourgeau-Chavez, L., Day, N.J., Dieleman, C.M., Johnstone, J.F., et al., 2020. Patterns of ecosystem structure and wildfire carbon combustion across six ecoregions of the North American boreal forest. Frontiers in Forests and Global Change 3. Available at https://doi.org/10.3389/ffgc.2020.00087, www.frontiersin.org/journals/forests-and-global-change#.

Fundamentals of remote sensing applied to wildfire studies

3

Mariano Garcia, Patricia Oliva, and Amin Khairoun
Universidad de Alcalá, Departamento de Geología, Geografía y Medio Ambiente, Environmental Remote Sensing Research Group, Alcalá de Henares, Madrid, Spain

3.1 Introduction

Forest fires are a natural disturbance that have been ubiquitous in all vegetated terrestrial ecosystems for millions of years, shaping their distribution and dynamics (Bond et al., 2005; Bowman et al., 2009). They also play a paramount role in the functioning of terrestrial ecosystems, affecting vegetation structure and composition, and therefore biodiversity; the global carbon and water cycles; the radiative balance; air quality; or human health (Cascio, 2018; Kelly et al., 2020; van der Werf et al., 2025). Every year more than 4,000,000 km^2 are burned (Giglio et al., 2018; Lizundia-Loiola et al., 2020), although these estimates are conservative, as global products underestimate burned area (BA) due to the inability to capture small fires (Chuvieco et al., 2002; Ramo et al., 2021).

The spatio-temporal characteristics of forest fires can be aggregated to define the so-called fire regimes or pyromes (Archibald et al., 2013; García et al., 2022). The variability of fire regimes is crucial for maintaining ecosystems' functioning. However, fire regimes are changing globally due to the combined effects of climate change and human activities (Kelly et al., 2020; Krawchuk et al., 2009). It is expected that the shift in fire regimes will be diverse, with some regions experiencing an increase in fire frequency and intensity, and others a decrease due to changes in precipitation and agricultural activities (Artés et al., 2019; Chuvieco et al., 2008; Pechony and Shindell, 2010).

Understanding forest fires involves considering different stages, which together form the so-called fire disturbance continuum (FDC) (Jain et al., 2004), a conceptual framework that describes fire events as a process occurring across three distinct phases: prefire, fire (or active fire), and postfire environments. The prefire phase involves the climate, short-term weather, and fuel characteristics. The fire stage refers to the environmental characteristics that determine fire intensity and spread, and first-order fire effects. Finally, the postfire phase refers to the impacts of fire on vegetation and soil, that is, burn severity, including the ecosystem response.

Remote sensing (RS) has become a pivotal technique to understand each stage of the FDC, due to the ability of remote sensors to provide observations of the Earth in a comprehensive, systematic way, in different regions of the electromagnetic spectrum and at multiple scales (Chuvieco, 2020), enabling a thorough analysis of different components of the FDC. In the prefire phase, RS has proved particularly useful for characterizing fuels. In this sense, while life fuel moisture content (FMC) has been successfully estimated using multispectral and Radio Detection and Ranging (RaDAR) sensors (Forkel et al., 2023;

Advances in Remote Sensing for Forest Fire Management. DOI: https://doi.org/10.1016/B978-0-443-41692-7.00001-9
© 2027 Elsevier Ltd. All rights are reserved, including those for text and data mining, AI training, and similar technologies.

García et al., 2020a; Rao et al., 2020; Yebra et al., 2013), light detection and ranging (LiDAR) data has proved its potential to estimate fuel load and structure (Aragoneses et al., 2024; Aragoneses et al., 2025; García et al., 2012). Moreover, the integration of multispectral and LiDAR data offers unprecedented accuracy for fuel type mapping (García et al., 2011a; Marino et al., 2016; Stefanidou et al., 2020).

RS also contributes to the analysis of the fire stage, particularly in relation to the detection of active fires and emitted energy. There are currently various thermal anomaly (hotspot) products used to detect active fires, taking advantage of the distinct temperature of a fire compared to its surroundings (Wooster et al., 2005). Although these products represent coarse-resolution data, they are able to detect thermal anomalies even when the fire is much smaller than the pixel size, which allows for the development of early fire-detection systems (Koltunov et al., 2016). In addition, the energy emitted by the fire, Fire Radiative Power (FRP), can be estimated from these products, which is essential for the estimation of biomass burning emissions (Di Giuseppe et al., 2018; Wooster et al., 2005).

In the final stage of the FDC, effective monitoring of fire impacts can be achieved using RS data. The primary application of RS in this phase is the quantification of BA, with a multitude of products developed at different scales (Giglio et al., 2025; Lizundia-Loiola et al., 2020; Otón et al., 2021; Wei et al., 2021). Burn severity, defined as the ecological damage caused by fire (Lentile et al., 2006), has been widely quantified using multispectral data across diverse biomes (De Santis and Chuvieco, 2007; Gibson et al., 2020; Quintano et al., 2018; Soverel et al., 2010). Although less extensively, synthetic aperture radar (SAR) and LiDAR data have proven strong potential to estimate burn severity by capturing the structural changes in the forest (García et al., 2020b; Tanase et al., 2015). These structural changes are particularly important for assessing biomass consumed by fire and, therefore, the CO_2 emitted by fires (Garcia et al., 2017; McCarley et al., 2020). Moreover, vegetation recovery, an essential indicator of ecosystem resilience to wildfires, can be effectively monitored using RS time-series data (Viana-Soto et al., 2020; Viana-Soto et al., 2022a, b).

This chapter examines the fundamental contributions of RS across the various stages of the FDC. It begins with an overview of RS applications for fuel characterization, with particular emphasis on fuel moisture and structural attributes. Afterwards, it focuses on the use of RS during the active fire stage, particularly on fire detection and the quantification of radiative energy. Subsequently, the chapter considers postfire applications, including burn severity assessment and the evaluation of ecosystem responses to wildfire disturbance. The chapter concludes with a synthesis of key findings and their broader implications.

3.2 Forest fuel characterization using remote sensing

Forest fuels are considered any material that can burn in a forest fire and are defined by the characteristics of live and dead biomass loads that contribute to the spread, intensity, and severity of a forest fire (Burgan and Rothermel, 1984). Forest fuels are a critical component of both the fire fundamentals triangle, which represents the elements required for fire occurrence (Pyne et al., 1996), and the fire environment triangle, which describes the elements whose interactions govern fire behavior (Countryman, 1972). Therefore, accurate and spatially explicit information on these components is essential for fire management strategies aimed at reducing forest fire risk. Forest fuels can be characterized by their status (dead or live), moisture content, and structure, which describes their spatial arrangement and load.

3.2.1 Fuel moisture content

FMC of both live and dead fuels is an important variable for fire danger estimation, as it affects fire ignition and fire propagation, as well as determining fuel consumption and fire effects. FMC is defined as:

$$\mathrm{FMC} = \frac{W_f - W_d}{W_d}$$

where W_f represents the wet weight and W_d the dry weight of the sample after oven drying. While some authors recommend drying the fuel at 60°C–100°C for 24–48 hours (Viegas et al., 1992), other authors suggest drying samples at 105°C to ensure more complete drying of the samples (Matthews, 2010).

Dead fuel moisture content (DFMC) has been successfully estimated from meteorological danger indices relying on current and past weather conditions (Aguado et al., 2007; Dragozi et al., 2021); however, estimation of live fuel moisture content (LFMC) is further complicated due to plant adaptation strategies to drought, including physiological and structural traits that influence fuel flammability (Jolly and Johnson, 2018). This section will focus on the estimation of LFMC using RS data.

3.2.2 Physical basis for the estimation of life fuel moisture content from remote sensing

LFMC from RS data has been mainly estimated from passive RS data based on the spectral changes produced by variations in vegetation water content (Fig. 3.1). In the solar domain, these changes in water content are reflected in the near infrared (NIR) and shortwave infrared (SWIR) regions of the electromagnetic spectrum, where water presents important absorption features, allowing direct estimation of leaf moisture. Additionally, depletion of chlorophyll and wilting of leaves due to dehydration produce changes in the visible spectrum, particularly in the red and NIR regions (Yebra et al., 2013), that can be correlated with LFMC. At the canopy level, leaf area index (LAI) also plays a significant role in the spectral signal that needs to be accounted for.

At longer wavelengths, variation in moisture content can also be estimated in the thermal-infrared region because of the effect of water content on plant evapotranspiration rates, which is a function of stomatal conductance (Gerhards et al., 2019; Jones, 1999). Water stress reduces evapotranspiration due to stomatal closure, resulting in an increase in canopy surface temperature (Chaerle and Van Der Straeten, 2000; Jones, 2004). Passive microwave sensors are also able to measure surface brightness temperature, providing information on soil and vegetation emissions as a function of their temperature.

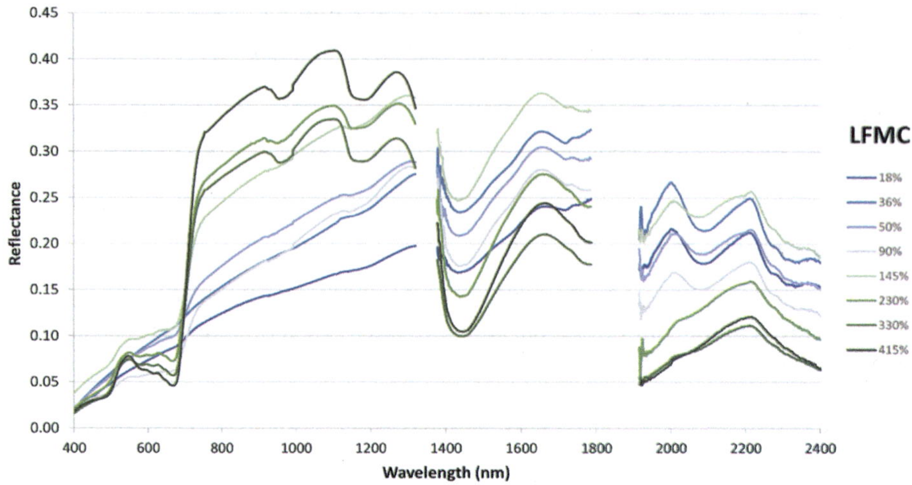

Figure 3.1 Spectral signature of grasses with different live fuel moisture contents (%). *Source*: Data provided by M. Pilar Martín, Environmental Remote Sensing and Spectroscopy Laboratory (SpecLab), Spanish National Research Council (CSIC). https://speclab.csic.es.

Estimation of LFMC from passive microwave systems relies on the attenuation of the signal by the canopy and the dielectric characteristics of water (Jackson and Schmugge, 1991; Wigneron et al., 2003). The microwave signal is attenuated as it passes through the vegetation layer. This attenuation is described by the vegetation optical depth (VOD) (Jackson and Schmugge, 1991; Wigneron et al., 2003), which is related to vegetation water content and, therefore, to dry biomass and LFMC (Konings et al., 2019). Liquid water presents a high dielectric constant; therefore, moist soil and vegetation are poor emitters of microwave radiation, resulting in lower brightness temperature. The microwave polarization difference index (MPDI) was proposed to derive information on soil and vegetation water content independently of surface temperature (Owe et al., 2001; Paloscia and Pampaloni, 1988). Finally, changes in canopy water content affect the dielectric constant of vegetation, which in turn affects the radar backscatter signal recorded by active radar systems (Ulaby and Jedlicka, 1984). These changes in backscatter are the basis for estimating LFMC from SAR measurements.

An important consideration in the estimation of LFMC through RS is the trade-off between spatial and temporal resolutions. On the one hand, LFMC monitoring requires high temporal resolution, ideally daily, which can be achieved by low-resolution sensors such as the Advanced Very High Resolution Radiometer (AVHRR), Sentinel-3, or Moderate Resolution Imaging Spectroradiometer (MODIS), although cloud coverage may reduce their nominal spatial resolution. On the other hand, the spatial resolution of these sensors may be insufficient to capture the inherent variability of forest fuels, which can be better achieved using medium- and high-spatial resolution sensors such as Landsat or Sentinel-2, at the expense of a lower temporal resolution, although the integration of the Sentinel-2 and Landsat sensors could provide nearly daily information at mid-latitudes (Whitcraft et al., 2015).

3.2.3 Modeling approaches

The most accurate method to estimate LFMC is based on field measurements, which are costly and provide limited spatial and temporal information. Nevertheless, field measurements are paramount to develop RS methods both to calibrate models and to validate the LFMC estimates. Despite the importance of these datasets, there are few datasets collected at large scales. The National Fuel Moisture Database (NFMD) provides LFMC measurements collected for more than 40 years across more than 900 sites, following standard protocols (Pollet & Brown, 2007). A global database was developed by Yebra et al. (2019) with 1383 sampling sites in 11 countries, designed for the calibration and validation of RS algorithms. This database was subsequently updated, incorporating 800 new sampling sites, covering a total of 15 countries and spanning almost 50 years (Yebra et al., 2024).

Two different approaches have been used to estimate LFMC from passive RS data: empirical modeling and radiative transfer model (RTM) inversion (see (Yebra et al., 2013) for a detailed review). The former establishes statistical models relating field-gathered samples of LFMC to the spectral information recorded by satellite sensors, commonly synthesized using spectral indices; or integrating remotely sensed data with meteorological and other ancillary data, using parametric or machine learning frameworks (Costa-Saura et al., 2021; García et al., 2020a; Marino et al., 2016; Qi et al., 2012; Rodriguez-Jimenez et al., 2023; Yebra et al., 2008). The second approach is based on physical models to simulate the interaction of the electromagnetic radiation with the vegetation using radiative transfer equations. These RTMs can be run forward to simulate the spectral response of vegetation, based on the parameterization of their optical and structural properties. This allows for generating look-up tables (LUTs) corresponding to different scenarios. Alternatively, the RTM can be inverted to estimate vegetation parameters at each pixel by finding the spectrum within the LUT most similar to the observed reflectance of the pixel using a cost function. This RTM inversion approach has been successfully applied for the estimation of LFMC (Pampanoni et al., 2023; Quan et al., 2017; Yang et al., 2025; Quan et al., 2021; Yebra and Chuvieco, 2009).

Physical-based approaches are considered more robust and generalizable than empirical models, which makes them more suitable for regional or global estimates (Quan et al., 2021); however, their performance is subject to appropriate parameterization, which can be more complex than fitting the parameters of empirical models (Yebra et al., 2008). Moreover, at local scales, empirical models have outperformed RTM approaches, particularly when remote-sensed data were integrated with meteorological and auxiliary data, with R^2 values above 0.5, reaching up to 0.9, and root mean square error (RMSE) between 25% and 49% for microwave sensors and between 8% and 18% for passive optical data. RTM performance, on the other hand, was slightly lower with R^2 values above 0.45, up to 0.71, and RMSE values between 6% and 40%. Regarding the performance of methods using different spectral regions, the most accurate estimates of LFMC have been found using information in the visible, near infrared (VNIR) and SWIR regions with R^2 up to 0.91 and RMSE = 8%. Results using microwave sensors showed lower performance with R^2 up to 0.70 (0.64 for passive microwave) and RMSE = 10% (30% for passive microwave)

(Costa-Saura et al., 2021; Forkel et al., 2023; García et al., 2008; García et al., 2020a; Pampanoni et al., 2023; Quan et al., 2021; Quan et al., 2024; Rao et al., 2020; Rodriguez-Jimenez et al., 2023; Santos et al., 2025; Tanase et al., 2015; Yang et al., 2025; Yebra et al., 2018). Regardless of the spectral region and the modeling approach used, results showed high variability associated with the vegetation type considered, with better results generally obtained for shrubs and grassland, whereas the LFMC of forests is estimated less accurately.

3.2.4 Fuel structure

Fuel structure refers to the spatial and temporal distribution of fuels, which is commonly described by variables such as fuel load, height, cover, or bulk density. These variables are important for fire behavior and fire effects modeling (Gajardo and García, 2014). Depending on their vertical position in the strata, fuels can be classified as surface, ladder, and canopy fuels.

Initial attempts to characterize fuels' structure focused on mapping fuel models using passive data (Riaño et al., 2002); however, passive optical data are limited to capture surface fuel properties due to canopy occlusion (Keane et al., 2001). They are also limited to discriminate fuel types that only differ in their height, as well as to estimate canopy fuel properties since there is no direct relation between reflectance and vegetation structure. Active sensors such as LiDAR and SAR are more sensitive to vegetation structure and therefore can provide more accurate information on fuel structure. SAR data have been used to estimate fuel parameters including fuel load and canopy bulk density (CBD) (Saatchi et al., 2007), as well as canopy height (CH) using polarimetric interferometric SAR data and polarimetric decomposition (Pourshamsi et al., 2018; Pourshamsi et al., 2021). Nevertheless, LiDAR has been more widely used to characterize fuel structure given its ability to provide a detailed 3D description of the forest and to overcome the limitations of passive optical data (Gajardo et al., 2014). Thus, we will focus our discussion on fuel structure characterization from LiDAR data.

LiDAR systems can be classified according to the platform into terrestrial, airborne (plane, helicopter, and drone), or satellite. Additionally, considering the recording capabilities of the sensor, they can be classified into discrete return or full waveform systems. All these different LiDAR sensors have been successfully applied to estimate various fuel properties (Andersen et al., 2005; Aragoneses et al., 2024; Engelstad et al., 2019; García et al., 2011b; García et al., 2012; Hermosilla et al., 2014; Narine et al., 2022).

3.2.5 Surface fuel properties

Accurately characterizing surface fuels is critical for fire management since most fires originate from this fuel layer (Chávez-Durán et al., 2024). Despite the importance of this fuel layer, few studies have aimed at estimating surface fuel properties. The spatially heterogeneous nature of this layer, with some components varying at very fine scales, makes it difficult to capture in the field and therefore hampers the establishment of LiDAR-based models to estimate surface fuel parameters. The most promising approach is based on terrestrial and handheld LiDAR systems due to their scan location under the canopy, thus reducing the occlusion caused by overstory fuel layers. Surface fuels have generally been characterized by the height of the different layers considered and their load, including litter, grass, forbs, shrubs, and dead woody fuels (1, 10, 100 hours) (Batista et al., 2025; Loudermilk et al., 2023; Rowell et al., 2020; Rowell et al., 2016; Tenny et al., 2025).

Although common metrics derived from height distribution of the returns, such as height percentiles, kurtosis, and density of returns in height strata, have been applied, the extremely high point density of terrestrial laser scanning (TLS) systems allows for characterizing fuel structure at the voxel level and estimating more detailed metrics related to occupied volume by surface fuel components, the total surface area, or the porosity of the layer, which represent the packing ratio of the fuel bed (Post et al., 2025; Rowell et al., 2020; Tenny et al., 2025).

The heterogeneity in surface fuels is also reflected in the wide variability of model performance achieved depending on the variables estimated. Thus, the load of surface fuels in the 0–20 or 0–30 cm layers has been successfully estimated with R^2 values larger than 0.7 (Loudermilk et al., 2023; Rowell et al., 2020). Fine woody debris and downed surface fuel load showed moderate results with R^2 = 0.61 and R^2 = 0.55, respectively, although the latter varied significantly with vegetation type with R^2 values ranging between 0.02 and 0.49 (Tenny et al., 2025). Similarly, standing surface fuel load showed R^2 values of 0.53 when different vegetation types were pooled in the model, ranging between 0.07 and 0.2 for individual vegetation types (Tenny et al., 2025). An important factor in the results achieved is the scanning configuration. Multiple scans reduced the occlusion effect, enabling better sampling of the fuel layers and improving results, whereas single scans resulted in poorer performance. For instance, Batista et al. (2025) obtained poor results in the estimation of the biomass of surface fuels (R^2 = 0.11) and ground fuels (R^2 = 0.25), the latter showing R^2 values between 0.14 and 0.3 for the different components using a single scan.

In the case of airborne LiDAR, the occlusion caused by the upper canopy reduces the number of returns reaching the lower vegetation strata, and therefore, the information on surface fuels. Airborne LiDAR-based models showed moderate performance in the estimation of surface fuel parameters with R^2 between 0.27 and 0.71 even with point densities above 70 points m^{-2}. Since parameters like litter or dead 1 hour fuels depend on the canopy present in the area, they have been found to be better explained by canopy metrics than by lower return LiDAR metrics (Labenski et al., 2023; Lin et al., 2024), although they were still poorly estimated due to their dependence on other environmental variables that cannot be inferred from LiDAR data and affect the decomposition rates.

3.2.6 Canopy fuel properties

Information on canopy fuel parameters is critical to reducing fire risk, particularly those spreading through the canopy, which are almost impossible to control, leading to catastrophic effects (Scott and Reinhardt, 2001). Canopy parameters required to model fire behavior and effects include CH, canopy base height (CBH), canopy cover (CC), canopy fuel load (CFL), and CBD. All these parameters have been successfully estimated from LiDAR data onboard terrestrial, airborne, and satellite platforms.

TLS systems have shown potential to robustly describe canopy fuels (Alonso-Rego et al., 2021; Batista et al., 2025; García et al., 2011b; Tenny et al., 2025) across different environments. Despite observing the canopy from below, CH has been estimated accurately (R^2 >0.8 and RMSE <10%) at the plot level. The location of the system allows for estimating CBH, which is an important parameter to determine crown fire danger, with around 80% of the variance explained and errors RMSE = 3 m or 12% in relative terms. Importantly, TLS systems enabled the accurate determination of the distance between the surface and canopy fuels through the fuel strata gap, which is critical to evaluate the danger of transition from surface to crown fire, with R^2 values greater than 0.75 and RMSE = 1.5 or below 10% when expressed in relative terms (Alonso-Rego et al., 2021; García et al., 2011b). Nevertheless, moderate results have been achieved to estimate CFL ($0.4 < R^2 < 0.6$, relative 20% < RMSE < 60%) and density, ($0.46 < R^2 < 0.51$, relative 25% < RMSE < 60%) (Alonso-Rego et al., 2021; Batista et al., 2025).

Airborne LiDAR systems are better suited for estimating canopy fuel parameters, particularly those concerning the upper canopy, such as CH ($R^2 > 0.7$ and RMSE < 2 m) or CFL ($0.49 < R^2 < 0.92$, RMSE = 0.3) (Andersen et al., 2005; Bright et al., 2017; Skowronski et al., 2011). Nevertheless, the estimation of parameters related to the lower parts of the canopy such as CBH is cumbersome, with models showing much more variable performance, with R^2 ranging between 0.3 and 0.78, and RMSE values between 1.4 and 3.5 m (Arkin et al., 2021; Bright et al., 2017; Engelstad et al., 2019; González-Ferreiro et al., 2017). The estimation of CBD has proved to be also difficult, with the amount of variance explained ranging between 0.48 and 0.94 and RMSE between 0.01 and 0.27 kg m^{-3} (Andersen et al., 2005; Bright et al., 2017; Engelstad et al., 2019; Skowronski et al., 2011), although the lack of accurate field methods makes it difficult to precisely evaluate this parameter. Using high point density data, fuel properties at the individual tree level have also been successfully estimated, including CBH, CBD, or CH (Hillman et al., 2021; Riaño et al., 2004).

Satellite LiDAR has also been used to estimate CH, CC, CBH, CFL, and CBD. First attempts were based on ICESat/GLAS (García et al., 2012; Peterson et al., 2013), although more recently ICESat-2/ATLAS (Narine et al., 2022; Neuenschwander et al., 2022) and particularly the Global Ecosystem Dynamics Investigation Lidar (GEDI) (Aragoneses et al., 2024, 2025) have proven their ability to estimate canopy fuel parameters. A detailed description of the parameters estimated, methods, and results achieved based on satellite LiDAR can be found in (Garcia and Aragoneses, 2026).

3.3 Fundamentals of remote sensing for active fire detection

RS of active fires relies on Planck's law of blackbody radiation, which states that the spectral radiance emitted by a blackbody is a function of its temperature and wavelength. Actively burning fires can exhibit a wide range of temperatures spanning several hundred Kelvin, associated with the flaming and smoldering phases of combustion. Typically, smoldering fires show cooler temperatures between 450K and

850K, whereas flaming fires have higher temperatures ranging from 800 to upwards of 1200K (Wooster et al., 2003). Simulating the spectral radiance emitted by black bodies at different temperatures, we can observe the wavelengths at which the curves peak (Fig. 3.2). The Earth's surface temperature, represented at 300K, shows the peak radiance at 9.6 μm, while a burning fire at 1000K has its peak at 2.9 μm. Long-wave infrared bands (LWIR), typically located at 10–12 μm, are close to the land surface's peak radiance. Thermal-infrared bands help discriminate between actively burning pixels and the unburned pixels around them, which are considered background pixels. Smoldering fires at 600K have their peak radiance at 4.8 μm, and flaming fires at 800K have their peak at 3.6 μm. The emitted spectral radiance from a 1000K flaming fire in the LWIR (8–14 μm) is an order of magnitude higher than from the background land surface, but in the Mid-infrared (MIR, 3–5 μm) it can be three orders of magnitude higher (Wooster et al., 2021). These bands help identify burning pixels because they are sensitive to the radiance emitted from the burning fires (Dozier, 1981; Giglio et al., 2003; Kant et al., 2000). In addition, the spectral radiance captured by the sensors can be used to estimate the radiant energy of a fire, which can be a measure of fire severity and intensity (Giglio et al., 2009).

Active fire (AF) detection algorithms use contextual tests to detect thermal anomalies by estimating the background temperature and analyzing the strong contrast in brightness temperature between the potential fire pixel and the low background temperature. MIR bands allow detection of pixels with higher brightness temperatures, as they are extremely sensitive to even subpixel-level high-temperature objects. By this means, intense active fires (>1000K) can be detected even when the

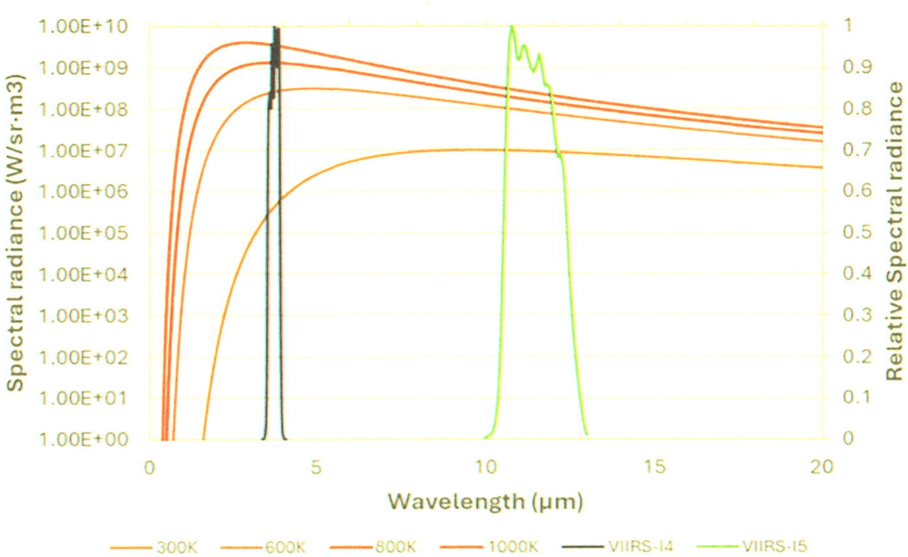

Figure 3.2 Spectral radiance of black bodies at different temperatures, and the spectral response functions of I4 (blue) and I5 (green) bands of the Visible Infrared Imaging Radiometer Suite sensor.

fire occupies less than 1% of the pixel area (Wooster et al., 2021; Schroeder et al., 2014b). The thermal bands provide a more stable comparison between the pixel and the background signal, which represents the basal brightness temperature. Hence, solar-reflective and long-wave infrared channels are used to reduce the probability of false alarms, for example, caused by sunglint from water bodies or by solar-heated warm ground (Schroeder et al., 2014a; Wooster et al., 2021).

Active fire products also provide estimates of FRP, which is the rate at which a fire emits thermal energy within a pixel, integrated over all angles and wavelengths. Estimates of the FRP are obtained by assessing the increase in the middle-infrared channel signal relative to the surrounding ambient background pixels. The algorithm used is the MIR radiance FRP approach, developed by Wooster et al. (2003, 2005) and first applied operationally in the Spinning Enhanced Visible and Infrared Imager (SEVIRI) FRP product (Roberts and Wooster, 2014). The same MIR radiance FRP approach is now used in many operational fire products, including the Collection 6 MODIS active fire products (Giglio et al., 2016) and the Visible Infrared Imaging Radiometer Suite (VIIRS) active fire products (Schroeder et al., 2014b). The FRP measures derived using the MIR radiance method (Wooster et al., (2003) showed a decrease of 16% in comparison with the FRP estimates computed using the Dozier method.

Several polar-orbiting sensors, such as MODIS, Along Track Scanning Radiometer (ASTR), AVHRR, and VIIRS, have well-defined Infrared (IR) fire-detection algorithms for the mid- and thermal-IR ranges (Giglio et al., 2009; Kant et al., 2000; Lentile et al., 2006; San-Miguel-Ayanz and Ravail, 2005; Schroeder et al., 2014b), based on contextual algorithms. In all these sensors, the MIR band's sensitivity is an important factor to consider, as saturation can occur when an active fire releases more energy than a pixel can register, which can be problematic for detection (Kant et al., 2000). For example, in the VIIRS 375 m active fire product, a saturation problem was identified in gas flares and large wildfires. In that case, saturated pixels were usually located in the center of a large flaming area, showing low radiance values, leading to omission errors (Schroeder et al., 2014b). The bandwidth of the MIR band can also affect FRP retrieval, as it is influenced by atmospheric CO_2 absorption from 3.9 μm (Oliva and Schroeder, 2015a). Nevertheless, the VIIRS active fire product offers the finest delineation of the fire front due to the higher spatial resolution and the increased number of daily acquisitions (Oliva and Schroeder, 2015b; Schroeder et al., 2014b).

In addition, algorithms developed on a global scale are designed conservatively to avoid false alarms, which may result in a high AF omission rate in areas with lower brightness temperatures. To improve the AF detection, several authors developed regional algorithms focused on detecting the low-intensity thermal anomalies occurring in those regions, such as Alaska (Waigl et al., 2019) or China (Xia et al., 2018). On the other hand, novel AF algorithms incorporate multitemporal rules into their classification process (Lin et al., 2018; Schroeder et al., 2016). The temporal dimension allows separation between permanent heat sources and new fire events, facilitating earlier fire detection and reducing false alarms caused by bright surfaces.

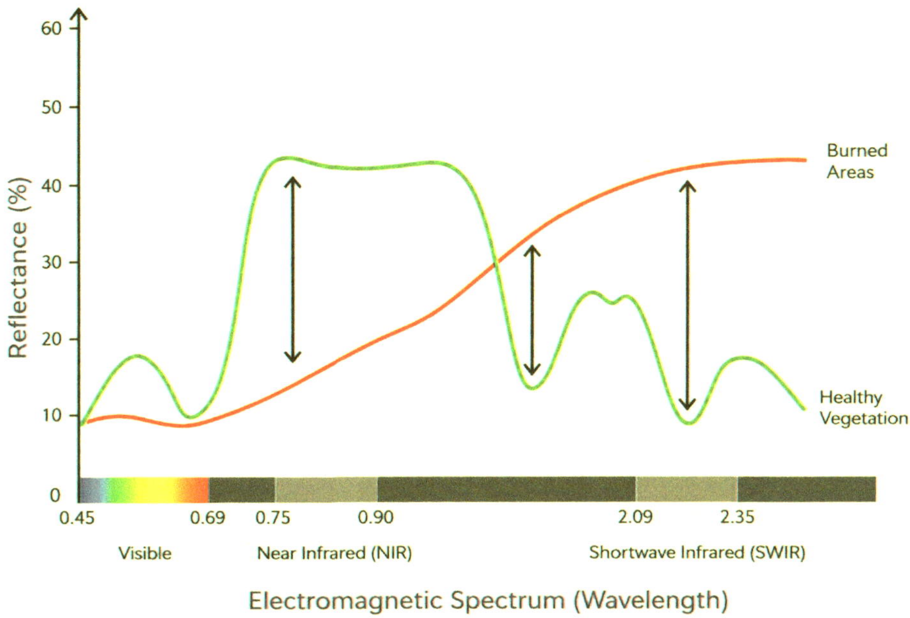

Figure 3.3 Simulated reflectance spectra of healthy vegetation and burned areas.
Source: The figure is adapted from an illustration of the United States Department of Agriculture (USDA) Forest Service.

3.4 Fundamentals of remote sensing for postfire analysis

3.4.1 Fundamentals of burned area detection

Burned vegetation exhibits unique spectral characteristics that contrast sharply with prefire conditions. The combined effects of structural damage, moisture loss, and ash/char deposition create distinctive reflectance patterns, particularly in the NIR and SWIR portions of the electromagnetic spectrum. Based on the spectral signatures shown in Fig. 3.3, three distinct local extrema can be identified in vegetation reflectance. There are two peaks—one in the NIR and another at the beginning of the SWIR region (around 1.6 μm)—followed by a minimum in the larger SWIR region (around 2.2 μm). In contrast, burned biomass lacks these sharp spectral peaks and instead shows a relatively uniform and strong reflectance across the SWIR region.

Early efforts in BA mapping heavily relied on these bands to develop spectral indices that highlight the differences between the reflectance signals of healthy vegetation and those from fire residues (i.e., ash, charcoal, etc.). The normalized ratio of the NIR and SWIR bands—proposed by López García and Caselles (López-García and Caselles, 1991), called the normalized burned ratio

(NBR)—was the first spectral index specifically designed for BA detection, following the formula:

$$NBR = \frac{\rho_{NIR} - \rho_{SWIRL}}{\rho_{NIR} + \rho_{SWIRL}} \tag{3.1}$$

Subsequent studies introduced additional indices to improve BA detection under varying conditions. For instance, the NBR2 (Giglio et al., 2009) replaces the NIR band with a SWIR band and is computed as the normalized difference between the short (S) and long (L) SWIR bands:

$$NBR2 = \frac{\rho_{NIRS} - \rho_{SWIRL}}{\rho_{NIRS} + \rho_{SWIRL}}$$

Other notable indices include the Mid-Infrared Burn Index (MIRBI) (Trigg and Flasse, 2001), which employs a linear combination of short- and long-wave SWIR reflectance, and the Burned Area Index (BAI) (Chuvieco et al., 2002), which is based on the distance between observed reflectance values and empirically derived burned-surface reflectance in the red–NIR spectral space. These indices are defined as:

$$MIRBI = 10 \times \rho_{SWIRS} - 9.8 \times \rho_{SWIRL} + 2\ BAI$$
$$= \frac{1}{(\rho_{NIR} - 0.06)^2 + (\rho_{Red} - 0.1)^2}$$

3.4.2 Burned area detection through multitemporal change detection

Since the emergence of operational, consistent, and systematic daily (e.g., MODIS) or biweekly (e.g., Landsat) satellite observations, the detection of burned land surfaces has largely relied on temporal changes in surface reflectance or spectral indices (Giglio et al., 2009; Giglio et al., 2018; Lizundia-Loiola et al., 2022; Lizundia-Loiola et al., 2020). These changes can be directly linked to the transition from intact vegetation to postfire conditions, such as burned scars, ash deposits, vegetation loss, and fire-induced drought stress (Chuvieco et al., 2019; Lentile et al., 2006). Consequently, most BA mapping algorithms are based on time-series analyses of spectral change, which have become the standard approach for accurate BA detection (Chuvieco et al., 2019; Mouillot et al., 2014). An additional advantage of this methodology is its ability to identify the timing of the first detectable spectral change associated with fire occurrence.

Some BA mapping algorithms compare each new satellite observation with a set of preceding observations to detect fire-induced changes (Bastarrika et al., 2024; Giglio et al., 2018; Giglio et al., 2009; Lizundia-Loiola et al., 2022; Lizundia-Loiola et al., 2020; Roteta et al., 2019). Regardless of the specific spectral indices employed, this approach is generally expected to perform well for BA detection because it explicitly targets abrupt spectral changes and seeks to confirm their persistence over time, thereby reducing false detections caused by noisy or unreliable observations. However, this strategy is computationally demanding, as it requires repeated pairwise comparisons across long-image time series. Such requirements can be excessive and difficult to manage, both in terms of storage and processing, particularly for medium- and high-resolution BA mapping over large spatial extents. To overcome these limitations, several studies at medium spatial resolution have leveraged dense satellite time series to produce large-scale regional BA maps using systematic temporal compositing. Instead of iteratively comparing each individual acquisition, this approach performs the change detection only once by contrasting two temporal composite mosaics, typically generated around the date of the strongest fire signal (Alencar et al., 2022; Fernández-García et al., 2024; Huang et al., 2025; Khairoun and Solano, 2023; Kouachi et al., 2024; Neves et al., 2023; Roteta et al., 2021). This strategy enables more efficient use of the dense Landsat and Sentinel-2 data archives and has been greatly facilitated by cloud-based platforms such as Google Earth Engine (GEE), which support scalable long-term time-series processing at large regional scales.

3.4.3 Review of remote sensing–based burned area datasets

3.4.3.1 Coarse-resolution global burned area products

Over the last two decades, several methodologies have been developed to assess global BA using coarse-resolution Earth Observation (EO) sensors (≥250 m). In 2004, two benchmark global BA datasets were released. The first, GBA2000, was a pioneering prototype generated to identify BA for the year 2000 using a synthesis of regional algorithms based on the Satellite Pour l'Observation de la Terre (SPOT) VEGETATION data (Grégoire et al., 2003). This product was later extended by the European Commission's Joint Research Centre to create the seven-year L3JRC dataset (Tansey et al., 2008). In parallel, the Globscar dataset (Simon et al., 2004) was developed by the European Space Agency (ESA) using daytime data from the Along-Track Scanning Radiometer (ATSR-2) instrument onboard European Remote Sensing Satellite (ERS-2). This dataset played a key role in the generation of the Globcarbon product (Plummer et al., 2006) that was mainly designed to be integrated into Earth system models, as it covers a longer period from 1998 to 2007. ESA further developed another global BA dataset within the framework of the FireCCI project of the Climate Change Initiative, known as FireCCI41 (Alonso-Canas and Chuvieco, 2015). FireCCI41 was derived from the 300-m resolution Medium Resolution Imaging Spectrometer (MERIS) images onboard the Environmental Satellite (EnviSat) satellite, covering the period from 2005 to 2011. Its BA detection algorithm followed a two-phase hybrid approach. Initially, thermal anomalies from the MODIS were used to identify the most distinctly burned pixels, referred to as "seeds." Then, multitemporal MERIS reflectance data were incorporated into a contextual criteria approach to reduce omission errors.

Among coarse-resolution EO sensors, MODIS has been the most widely used for fire applications. The first global product of BA based on MODIS was the MCD45A1 (Roy et al., 2008), which was generated using a multitemporal change approach that related the estimated bi-directional reflectance of each pixel (based on previous observations) to the actual observed reflectance. This was the standard BA product of the National Aeronautics and Space Administration (NASA) for the Collection 5 products of MODIS. After the release of Collection 6, a new BA product, now called MCD64A1, replaced the previous algorithm. The new one relies on a hybrid approach combining fire anomalies and multitemporal change detection (Giglio et al., 2018). This product is currently the operational BA product of NASA, covering the period from 2000 onwards. Global independent stratified validation indicates that MCD64A1 has an omission error of 49.3% and a commission error of 18.4%, resulting in a dice coefficient (DC) of 0.62 (Franquesa et al., 2022).

Based on the same sensor, the FireCCI project of ESA developed the successor of FireCCI41, named FireCCI51 (Lizundia-Loiola et al., 2020), along with its preliminary publication, FireCCI50 (Chuvieco et al., 2018). Although it follows the same basic philosophy as its predecessor, the spectral information used for the change detection process was derived from the MODIS 250-mm NIR band (Lizundia-Loiola et al., 2020), and the contextual growing criteria are enhanced to improve accuracy. This allowed for extending the original time series to the 2001–22 period. In terms of accuracy, FireCCI51 achieves a DC of 0.67, with omission and commission errors of 43.1% and 19.4%, respectively (Franquesa et al., 2022).

The FireCCI51 algorithm was further adapted to the 300-m NIR band of the Ocean and Land Colour Instrument (OLCI) sensor onboard the Sentinel-3 satellite to develop the BA component of the Copernicus Climate Change Service (C3S), officially named C3SBA11 (Lizundia-Loiola et al., 2021), expanding for the period 2017–23. The longest coarse-resolution BA dataset of the ESA Fire project was named FireCCILT11 (Otón et al., 2021), preceded by its preliminary version, FireCCILT10 (Otón et al., 2019). This product was based on spectral information derived from the Land Long Term Data Record (LTDR) of a degraded resolution dataset of the images acquired by the AVHRR sensor, covering the period 1982–2018 at a 5-km pixel size. Its algorithm combines two random forest (RF) models: one trained on a global sample and another fitted specifically to boreal regions. More recently, Lizundia-Loiola et al. (2022) used a methodology combining the NBR2 index derived from the SWIR bands of the Sea and Land Surface Temperature Radiometer (SLSTR) sensor of Sentinel-3 with the 375-m active fires derived from the VIIRS to develop the FireCCIS310 product (and its second release, FireCCIS311). This is ESA's most recent global BA product, covering the period from 2019 onward, and it offered the most accurate coarse-resolution BA product among the existing ones, with a DC of 0.68, and omission and commission errors of 41.2% and 19.2%, respectively (Lizundia-Loiola et al., 2022).

3.4.3.2 Medium-resolution burned area datasets

With advances in computational resources and change detection techniques, multispectral medium-resolution satellite imagery (≤30 m) has become widely used in BA mapping. The NASA Landsat mission has launched nine satellites since 1972, incorporating the Thematic Mapper (TM) sensor in 1982 (Landsat-4), which has been the most useful of the satellite series, extended by the Enhanced Thematic Mapper Plus (ETM+) on board the Landsat-6 (which failed) and 7, and the Operational Land

Imager (OLI) instrument on board the Landsat-8 and 9. Alongside the most recent ESA's Sentinel-2 missions (launched in 2015), these medium-resolution satellites are commonly utilized for detailed and more accurate BA mapping.

Early efforts to detect BA using Landsat data date back to the 1980s following the launch of Landsat-4 and Landsat-5 TM instruments (Chuvieco and Congalton, 1988; Furyaev, 1985; Milne, 1986; Smith and Woodgate, 1985). Most algorithms leverage spectral indices derived from the NIR and SWIR bands, particularly the NBR and NBR2, to implement change detection methods that have evolved significantly over recent years, making long-term Landsat archives especially effective for BA mapping. The Monitoring Trends in Burn Severity (MTBS) project was the first large-scale operational initiative to systematically process and generate a continuous BA dataset, in this case covering the whole US territory, using dense, multiyear Landsat data (Eidenshink et al., 2007). Although based on 30-m medium-resolution imagery, MTBS primarily focused on large wildfires due to inherent data gaps. Two subsequent works were considered benchmarks for later designs to develop regional long-term BA records from Landsat data. In Queensland, Australia, Goodwin and Collett (2014) demonstrated that algorithms that include decision rules in combination with contextual region-growing can be highly effective for accurate delineation of BA patches using the full available archive of Landsat TM/ETM+ imagery. Then, Boschetti et al. (2015) fused a multitemporal change detection approach based on spectral decision rules applied to Landsat ETM+ data with MODIS active fire detections to map BA in western the United States. Based on the experience learned from these benchmark studies, several national BA databases have been developed. For instance, Hawbaker et al. (2017) implemented the US Geological Survey's (USGS) Landsat Burned Area Essential Climate Variable (BAECV) to monitor all fires larger than 4 ha across the Conterminous United States (CONUS), which was later updated with further enhancements (Hawbaker et al., 2020). Several other Landsat-derived national fire datasets have since been developed using various techniques in different countries, including Canada (Hall et al., 2020), Brazil (Alencar et al., 2022), Portugal (Neves et al., 2023), and Australia (Canadell et al., 2021). Up until now, only a single global-scale Landsat-based BA product has been developed, known as global annual burned area mapping (GABAM) (Long et al., 2019).

The growing availability of public Sentinel-2 Multi-Spectral Instrument (MSI) data, owing to a notably shorter revisit time than Landsat satellites (10 days at the equator, or 5 days when both satellites are used, versus 16 days for Landsat, reduced to 8 days with two satellites), has driven the development of large-scale fire inventories that rely on Sentinel-2 imagery to delineate fire perimeters. For instance, the European Forest Fire Information System (EFFIS) uses a combination of semi-supervized classification and expert visual interpretation of satellite imagery, initially using MODIS data (San-Miguel-Ayanz et al., 2012) and later integrating Sentinel-2 imagery, once available, to delineate pan-European fires in near real time. Under the framework of the FireCCI project, a major fire dataset covering the whole Sub-Saharan Africa was generated using Sentinel-2A (the first satellite of the constellation) data for the year 2016, known as FireCCISFD11 (Roteta et al., 2019). This dataset was later updated for 2019 using imagery from both Sentinel-2A and 2B, which is referred to as FireCCISFD20 (Chuvieco et al., 2022). In Arctic Siberia, Descals et al. (2022) used Sentinel-2 data to map BA for 2019 and 2020, revealing unprecedented fire activity in carbon-rich peatlands during these years. Additionally, Landsat and Sentinel-2 data have been extensively utilized to generate reference BA perimeters serving for validation purposes (Boschetti et al., 2019; Franquesa et al., 2022; Padilla et al., 2015; Vanderhoof et al., 2017).

In general, the accuracies of medium-resolution BA datasets show higher accuracy than coarse-resolution products, particularly in their ability to capture small fires (Ramo et al., 2021). For instance, FireCCISFD20 achieves a DC of 0.88, with omission and commission no higher than 8.5% and 15%, respectively (Chuvieco et al., 2022). Similarly, validation of the MapBiomas product over Brazil for 2019 reported omission and commission errors below 11% (Alencar et al., 2022).

3.4.4 Fundamentals of burn severity mapping

Fires burn unevenly across landscapes; therefore, analyzing fire impacts requires the discrimination of fire damage levels. RS methods have proven useful, providing comprehensive estimates of postfire effects. Two types of fire effects can be distinguished: first- and second-order effects, which occur in the temporal continuum in which fire can be analyzed as a disturbance (Epting et al., 2005). First-order effects are related to the immediate impact on preexisting components or conditions and are a direct result of combustion. First-order effects are referred to as fire severity. The interval for their measurement is limited to a few months after the fire, as many effects are worn away and modified by subsequent biophysical processes. Second-order effects refer to the ecosystem's response after the fire, primarily through observations of plant regeneration and hydrogeomorphological dynamics (Jain, T.B., Graham, R.T. and Pilliod, D.S., 2004). Second-order effects are referred to as burn severity.

The differenced normalized burn ratio (dNBR) (Key and Benson, 2006) is a standard for quantifying and mapping fire severity using RS data such as Landsat or Sentinel-2 (Soverel et al., 2010). The dNBR exploits fire-induced changes in vegetation and soils by using reflectance changes in the (NIR, 0.7–1.2 um) and (SWIR, 1.2–2.5 um) spectral regions within the same sector across different time periods (Yin et al., 2020). Although dNBR has proven successful in many studies, it has been criticized for not necessarily reflecting the effects of fire on plants and soils in areas with scattered vegetation or in some types of forests (Soverel et al., 2010) and thus being dependent on the prefire vegetation type. Several indices have been developed to improve the discrimination of fire severity, such as the relative dNBR (RdNBR) (Miller and Thode, 2007), the SWIR-MIR index (SMI) (Veraverbeke et al., 2011), or the Relativized Burn Ratio (RBR) (Parks et al., 2014). The calibration and validation of these indices' performance is done by comparing them to field measurements. Field estimation of fire and burn severity is performed by applying the Composite Burn Index (CBI) (Key and Benson, 2006). CBI is a semiquantitative field-sampling approach based on an expert-judgment procedure, developed as an operational methodology for validating remotely sensed assessments of fire/burn severity in the United States, and has since been extended as a global protocol. Later, De Santis and Chuvieco (2009) developed the Geospatial CBI

(GeoCBI), a modified version of the CBI that accounts for the fraction of coverage (FCOV) of the different vegetation strata. RBR performance was evaluated with 3000 plots located in Western US covering different vegetation types. RBR showed the highest overall accuracy (70.5%), followed by the RdNBR (69.2%) and the dNBR (68.4%) (Parks et al., 2014).

Some efforts have been made to provide a more ecologically sound interpretation of satellite-retrieved burn severity, using RTMs. The use of biophysical RTMs (De Santis and Chuvieco, 2009; Fernández-Guisuraga et al., 2023; Yin et al., 2020), coupled with geometric canopy models, allows for the production of a set of spectra that can be directly related to postfire damage levels as (Vogeler et al., 2016) measured by the GeoCBI (Fig. 3.4) (De Santis and Chuvieco, 2009; Yin et al., 2020). Those studies obtained high accuracy with R^2 higher than 0.7, with inconsistencies in the estimation of intermediate fire severity. Another approach used the ProSAIL-D RTM to retrieve the fraction cover from Sentinel-2 images. They compute the change in fraction cover as an estimation of fire severity. They found that the change in fraction cover showed an inverse linear relation to the CBI ($R^2 > 0.85$, RMSE = 10%) in a more consistent manner than the commonly used spectral indices (Fernández-Guisuraga et al., 2023). Other works applied RF models to estimate CBI from satellite data (Parks et al., 2014). On the other hand, some studies have combined passive optical data with LiDAR data to assess the relevance of postfire effects on vegetation height and biomass (Garcia et al., 2017; Hoffman et al., 2018; Sato et al., 2016) or to the distribution of snags of different sizes (Casas et al., 2016; Vogeler et al., 2016).

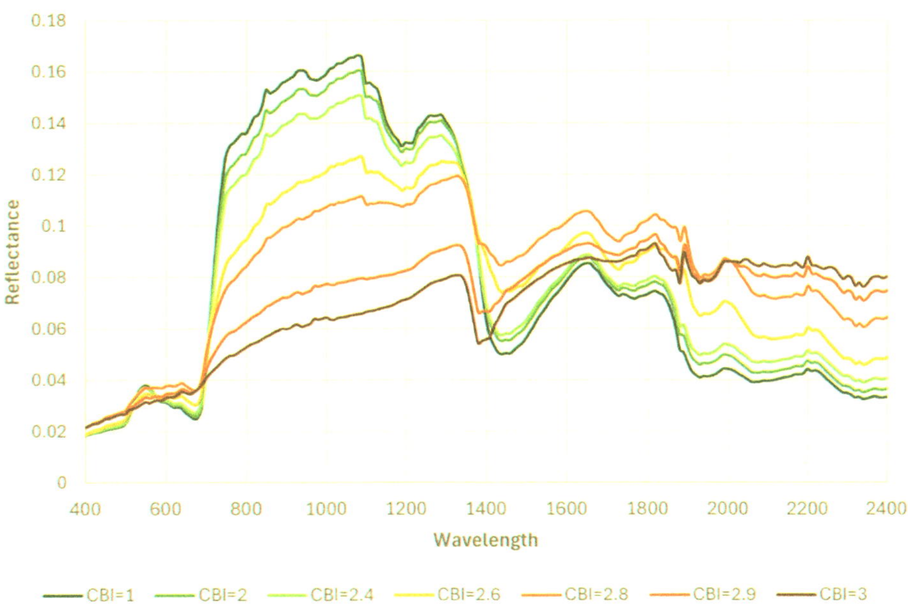

Figure 3.4 Spectra representing different levels of Composite Burn Index (CBI), low fire severity (CBI = 1 and CBI = 2), intermediate fire severity levels (CBI = 2.4, 2.6, 2.8), and high severity levels (CBI = 2.9 and 3). The spectrum of CBI = 3 represents the spectral signature of burned surf.

Fire/burn severity estimation is also linked to postfire regeneration monitoring, as both are relevant for ecological and management purposes. Several papers have provided a review of concepts, methods, and sensors related to mapping forest disturbance and recovery (Bartels et al., 2016; Chu and Guo, 2014; Hirschmugl et al., 2017; Kennedy et al., 2018; Bartels et al., 2016; Chu and Guo, 2014; Hirschmugl et al., 2017). Since 2008, there has been a trend toward using Landsat data for monitoring fire recovery, driven by the long time series of RS data. Several algorithms have been developed and applied to Landsat Time Series (LTS). The most extensively used algorithm for time-series analysis of Landsat data is the Landsat-based Detection of Trends in Disturbance and Recovery (LandTrendr) (Bright et al., 2019; Frazier et al., 2018). LandTrendr was specifically designed for Landsat data, but it can be adapted for Sentinel 2 to extend the baseline data. In addition, this algorithm has been adapted to GEE, thus facilitating the application over large areas and long periods of time (Kennedy et al., 2018). Recent studies have used the LandTrendr to analyze postfire forest recovery (Viana-Soto et al., 2022a).

Those time-series algorithms use spectral vegetation indexes (SVI) such as the Normalised Difference Vegetation Index (NDVI), NBR (or its derivatives: dNBR, RdNBR), or the components of the Tasseled Cap Transformation as input data (Bright et al., 2019; Frazier et al., 2015; Morresi et al., 2019; Storey et al., 2016). New indices have also been developed, such as the Vegetation Structure Perpendicular Index (VSPI) (Massetti et al., 2019), based on the divergence from a linear regression between two SWIR bands. The VSPI outperformed the NDVI and NBR, showing lower interannual variability and stronger postfire monitoring capability in Australian eucalypt forests.

LiDAR data have also been used to estimate fire impacts, particularly postfire changes in forest structure. Few attempts have been made to estimate burn severity from LiDAR data alone. Hu et al. (2019) proposed a single metric to estimate fire impacts on vegetation using multitemporal LiDAR, based on the area of the height percentile profile computed as Profile Area Change (PAC), showing good results to estimate changes in basal area and LAI (R^2>0.65). García et al. (2020a) also proposed a single metric based on the relative change in the area enclosed by the LiDAR waveform (WARC). The metric was derived using LiDAR simulated data with a 3D RTM and subsequently applied to airborne LiDAR data, showing strong agreement with CBI estimates (R^2=0.8 and RMSE=0.22 for the simulated data; R^2=0.78 and RMSE=0.37 for real data collected over the King Fire). Despite these promising results based on LiDAR data, most approaches have explored the potential of integrating LiDAR and multispectral data. Matasci et al. (2018) combined LiDAR and Landsat data using a nearest-neighbor imputation approach with a RF-based distance metric to estimate height, basal area, volume, and biomass changes in the Canadian forest for the period 1984–2016. These estimations were used to derive annual forest structure dynamics, detecting both undisturbed and regenerating stands after disturbances. Viana-Soto et al. (2022a) also monitored forest structure recovery by integrating airborne LiDAR and LTS, considering structural metrics related to vegetation height, cover, and heterogeneity using a machine learning framework.

Models were proven to robustly estimate these variables (R^2=0.63–0.78; rRMSE=24%–36%), being also temporally consistent (R^2=0.51–0.74; rRMSE=31%–44%). Radar can also be useful for assessing postfire regeneration, as backscattered microwave energy has proven sensitive to changes in vegetation structure. Radar sensors have the advantage over LiDAR of having different satellite systems in place, particularly Sentinel-1, which provides systematic global coverage. However, only a few studies have been conducted using radar data to assess postfire forest dynamics. Chen et al. (2017) compared different restoration treatments, including natural regeneration, using ALOS/PALSAR backscattering coefficients and two radar indices, with good results. On the other hand, Polychronaki et al. (2013) combined the NDVI index and ERS backscatter coefficients to study the effect of forest fires in a Mediterranean area. They found that radar data could provide information related to changes in dense coniferous forests, whereas the NDVI was sensitive to the understory vegetation recovery.

3.5 Concluding remarks

RS represents an unparalleled technology for providing critical information across all phases of the FDC, and at multiple spatial and temporal scales. In the prefire phase, active and passive sensors enable comprehensive characterization of fuels, providing detailed information on fuel structure and moisture content that is essential for assessing fire ignition potential and spread. During the active fire phase, passive sensors operating in the mid- and thermal-infrared regions of the electromagnetic spectrum allow for the detection and monitoring of active fires and the measurement of the energy released, which underpins estimates of biomass burning emissions. In the postfire phase, both active and passive sensors offer valuable insights into the spatial patterns of fire severity and vegetation recovery, supporting the design and evaluation of postfire management and restoration strategies.

Despite substantial advances, several key challenges remain. For fuel characterization, the temporal resolution of moderate- and high-spatial-resolution sensors is still insufficient to capture rapid changes in FMC, whereas low-spatial-resolution sensors, although temporally rich, lack the ability to resolve fine-scale variations. In terms of fuel structure, research efforts have largely focused on canopy fuels, whereas surface fuels, through which most fires ignite and spread, remain comparatively understudied. During the active fire phase, polar-orbiting sensors can reliably detect fires, but their limited temporal sampling (typically one to two observations per day) constrains accurate quantification of total fire energy release and, consequently, biomass burning emissions. In this context, geostationary sensors represent a promising opportunity to explore. Finally, the integration of active and passive RS observations holds considerable potential for improving assessments of burn severity and postfire vegetation dynamics, including species composition and structural recovery. Advancing our capacity to monitor the full FDC will be critical for deepening our understanding of fire's role in the Earth system and for informing more effective fire management under changing fire regimes.

References

Aguado, I., Chuvieco, E., Borén, R., Nieto, H., 2007. Estimation of dead fuel moisture content from meteorological data in Mediterranean areas. Applications in fire danger assessment. International Journal of Wildland Fire 16 (4), 390–397. Available at https://doi.org/10.1071/wf06136.

Alencar, A.A.C., Arruda, V.L.S., Silva, W.V.da, Conciani, D.E., Costa, D.P., Crusco, N., et al., 2022. Long-term landsat-based monthly burned area dataset for the Brazilian biomes using deep learning. Remote Sensing 14 (11). Available at https://doi.org/10.3390/rs14112510, https://www.mdpi.com/2072-4292/14/11/2510/pdf?version=1653388197.

Alonso-Canas, I., Chuvieco, E., 2015. Global burned area mapping from ENVISAT-MERIS and MODIS active fire data. Remote Sensing of Environment 163, 140–152. Available at https://doi.org/10.1016/j.rse.2015.03.011.

Alonso-Rego, C., Arellano-Pérez, S., Guerra-Hernández, J., Molina-Valero, J.A., Martínez-Calvo, A., Pérez-Cruzado, C., et al., 2021. Estimating stand and fire-related surface and canopy fuel variables in pine stands using low-density airborne and single-scan terrestrial laser scanning data. Remote Sensing 13 (24). Available at https://doi.org/10.3390/rs13245170, https://www.mdpi.com/2072-4292/13/24/5170/pdf.

Andersen, H.E., McGaughey, R.J., Reutebuch, S.E., 2005. Estimating forest canopy fuel parameters using LIDAR data. Remote Sensing of Environment 94 (4), 441–449. Available at https://doi.org/10.1016/j.rse.2004.10.013.

Aragoneses, E., García, M., Ruiz-Benito, P., Chuvieco, E., 2024. Mapping forest canopy fuel parameters at European scale using spaceborne LiDAR and satellite data. Remote Sensing of Environment 303, 114005. Available at https://doi.org/10.1016/j.rse.2024.114005.

Aragoneses, E., García, M., Tang, H., Chuvieco, E., 2025. A multi-sensor approach allows confident mapping of forest canopy fuel load and canopy bulk density to assess wildfire risk at the European scale. Remote Sensing of Environment 318, 114578. Available at https://doi.org/10.1016/j.rse.2024.114578.

Archibald, S., Lehmann, C.E.R., Gómez-Dans, J.L., Bradstock, R.A., 2013. Defining pyromes and global syndromes of fire regimes. Proceedings of the National Academy of Sciences 110 (16), 6442–6447. Available at https://doi.org/10.1073/pnas.1211466110.

Arkin, J., Coops, N.C., Daniels, L.D., Plowright, A., 2021. Estimation of vertical fuel layers in tree crowns using high density lidar data. Remote Sensing 13 (22). Available at https://doi.org/10.3390/rs13224598, https://www.mdpi.com/2072-4292/13/22/4598/pdf.

Artés, T., Oom, D., de Rigo, D., Durrant, T.H., Maianti, P., Libertà, G., et al., 2019. A global wildfire dataset for the analysis of fire regimes and fire behaviour. Scientific Data 6 (1). Available at https://doi.org/10.1038/s41597-019-0312-2, www.nature.com/sdata/.

Bartels, S.F., Chen, H.Y.H., Wulder, M.A., White, J.C., 2016. Trends in post-disturbance recovery rates of Canada's forests following wildfire and harvest. Forest Ecology and Management 361, 194–207. Available at https://doi.org/10.1016/j.foreco.2015.11.015, www.elsevier.com/inca/publications/store/5/0/3/3/1/0.

Bastarrika, A., Rodriguez-Montellano, A., Roteta, E., Hantson, S., Franquesa, M., Torre, L., et al., 2024. An automatic procedure for mapping burned areas globally using Sentinel-2 and VIIRS/MODIS active fires in Google Earth Engine. ISPRS Journal of Photogrammetry and Remote Sensing 218, 232–245. Available at https://doi.org/10.1016/j.isprsjprs.2024.08.019, https://www.sciencedirect.com/science/journal/09242716.

Batista, E.K.L., Hudak, A.T., Atkins, J.W., Broadbent, E.N., Brock, K.M., Campbell, M.J., et al., 2025. Comparing terrestrial and mobile laser scanning approaches for multi-layer fuel load prediction in the western United States. Remote Sensing 17 (16). Available at https://doi.org/10.3390/rs17162757, http://www.mdpi.com/journal/remotesensing.

Bond, W.J., Woodward, F.I., Midgley, G.F., 2005. The global distribution of ecosystems in a world without fire. New Phytologist 165 (2), 525–538. Available at https://doi.org/10.1111/j.1469-8137.2004.01252.x.

Boschetti, L., Roy, D.P., Justice, C.O., Humber, M.L., 2015. MODIS-Landsat fusion for large area 30m burned area mapping. Remote Sensing of Environment 161, 27–42. Available at https://doi.org/10.1016/j.rse.2015.01.022, www.elsevier.com/inca/publications/store/5/0/5/7/3/3.

Boschetti, L., Roy, D.P., Giglio, L., Huang, H., Zubkova, M., Humber, M.L., 2019. Global validation of the collection 6 MODIS burned area product. Remote Sensing of Environment 235. Available at https://doi.org/10.1016/j.rse.2019.111490, www.elsevier.com/inca/publications/store/5/0/5/7/3/3.

Bowman, D.M.J.S., Balch, J.K., Artaxo, P., Bond, W.J., Carlson, J.M., Cochrane, M.A., et al., 2009. Fire in the earth system. Science (New York, N.Y.) 324 (5926), 481–484. Available at https://doi.org/10.1126/science.1163886.

Bright, B.C., Hudak, A.T., Meddens, A.J.H., Hawbaker, T.J., Briggs, J.S., Kennedy, R.E., 2017. Prediction of forest canopy and surface fuels from lidar and satellite time series data in a bark beetle-affected forest. Forests 8 (9). Available at https://doi.org/10.3390/f8090322, http://www.mdpi.com/1999-4907/8/9/322/pdf.

Bright, B.C., Hudak, A.T., Kennedy, R.E., Braaten, J.D., Henareh Khalyani, A., 2019. Examining post-fire vegetation recovery with Landsat time series analysis in three western North American forest types. Fire Ecology 15 (1). Available at https://doi.org/10.1186/s42408-018-0021-9, http://fireecology.org/journal/journal-archive/.

Burgan, R.E., Rothermel, R.C., 1984. BEHAVE: Fire Behavior Prediction and Fuel Modeling System–FUEL Subsystem. In. U.S. Department of Agriculture, Forest Service, Intermountain Forest and Range Experiment Station.

Canadell, J.G., Meyer, C.P., Cook, G.D., Dowdy, A., Briggs, P.R., Knauer, J., et al., 2021. Multi-decadal increase of forest burned area in Australia is linked to climate change. Nature Communications 12 (1). Available at https://doi.org/10.1038/s41467-021-27225-4, http://www.nature.com/ncomms/index.html.

Casas, Á., García, M., Siegel, R.B., Koltunov, A., Ramírez, C., Ustin, S., 2016. Burned forest characterization at single-tree level with airborne laser scanning for assessing wildlife habitat. Remote Sensing of Environment 175, 231–241. Available at https://doi.org/10.1016/j.rse.2015.12.044, www.elsevier.com/inca/publications/store/5/0/5/7/3/3.

Cascio, W.E., 2018. Wildland fire smoke and human health. Science of the Total Environment 624, 586–595. Available at https://doi.org/10.1016/j.scitotenv.2017.12.086, www.elsevier.com/locate/scitotenv.

Chaerle, L., Van Der Straeten, D., 2000. Imaging techniques and the early detection of plant stress. Trends in Plant Science 5 (11), 495–501. Available at https://doi.org/10.1016/s1360-1385(00)01781-7.

Chávez-Durán, Á.A., Olvera-Vargas, M., Aguado, I., Figueroa-Rangel, B.L., Trucíos-Caciano, R., Rubio-Camacho, E.A., et al., 2024. High-resolution mapping of litter and duff fuel loads using multispectral data and random forest modeling. Fire 7 (11). Available at https://doi.org/10.3390/fire7110408, www.mdpi.com/journal/fire.

Chen, W., Jiang, H., Moriya, K., Sakai, T., Cao, C., 2017. Monitoring of post-fire forest regeneration under different restoration treatments based on ALOS/PALSAR data. New Forests 49, 105–121. Available at https://doi.org/10.1007/s11056-017-9608-2.

Chu, T., Guo, X., 2014. Remote sensing techniques in monitoring post-fire effects and patterns of forest recovery in boreal forest regions: a review. Remote Sensing 6 (1), 470–520. Available at https://doi.org/10.3390/rs6010470.

Chuvieco, E., 2020. Fundamentals of Satellite Remote Sensing. An Environmental Approach.

Chuvieco, E., Congalton, R.G., 1988. Mapping and inventory of forest fires from digital processing of tm data. Geocarto International 3 (4), 41–53. Available at https://doi.org/10.1080/10106048809354180.

Chuvieco, E., Martín, M.P., Palacios, A., 2002. Assessment of different spectral indices in the red-near-infrared spectral domain for burned land discrimination. International Journal of Remote Sensing 23 (23), 5103–5110. Available at https://doi.org/10.1080/01431160210153129, https://www.tandfonline.com/loi/tres20.

Chuvieco, E., Giglio, L., Justice, C., 2008. Global characterization of fire activity: toward defining fire regimes from Earth observation data. Global Change Biology 14 (7), 1488–1502. Available at https://doi.org/10.1111/j.1365-2486.2008.01585.x.

Chuvieco, E., Lizundia-Loiola, J., Pettinari, M.L., Ramo, R., Padilla, M., Tansey, K., et al., 2018. Generation and analysis of a new global burned area product based on MODIS 250 m reflectance bands and thermal anomalies. Earth System Science Data 10 (4), 2015–2031. Available at https://doi.org/10.5194/essd-10-2015-2018.

Chuvieco, E., Mouillot, F., van der Werf, G.R., San Miguel, J., Tanasse, M., Koutsias, N., et al., 2019. Historical background and current developments for mapping burned area from satellite Earth observation. Remote Sensing of Environment 225, 45–64. Available at https://doi.org/10.1016/j.rse.2019.02.013, www.elsevier.com/inca/publications/store/5/0/5/7/3/3.

Chuvieco, E., Roteta, E., Sali, M., Stroppiana, D., Boettcher, M., Kirches, G., et al., 2022. Building a small fire database for Sub-Saharan Africa from Sentinel-2 high-resolution images. Science of The Total Environment 845, 157139. Available at https://doi.org/10.1016/j.scitotenv.2022.157139.

Costa-Saura, J.M., Balaguer-Beser, Á., Ruiz, L.A., Pardo-Pascual, J.E., Soriano-Sancho, J.L., 2021. Empirical models for spatio-temporal live fuel moisture content estimation in mixed mediterranean vegetation areas using sentinel-2 indices and meteorological data. Remote Sensing 13 (18). Available at https://doi.org/10.3390/rs13183726, https://www.mdpi.com/2072-4292/13/18/3726/pdf.

Countryman, C.M., 1972. The Fire Environment Concept. USDA Forest Service, Pacific Southwest Range and Experiment Station, Berkeley, California, USA.

De Santis, A., Chuvieco, E., 2007. Burn severity estimation from remotely sensed data: performance of simulation versus empirical models. Remote Sensing of Environment 108 (4), 422–435. Available at https://doi.org/10.1016/j.rse.2006.11.022.

De Santis, A., Chuvieco, E., 2009. GeoCBI: a modified version of the Composite Burn Index for the initial assessment of the short-term burn severity from remotely sensed data. Remote Sensing of Environment 113 (3), 554–562. Available at https://doi.org/10.1016/j.rse.2008.10.011.

Descals, A., Gaveau, D.L.A., Verger, A., Sheil, D., Naito, D., Peñuelas, J., 2022. Unprecedented fire activity above the Arctic Circle linked to rising temperatures. Science (New York, N.Y.) 378 (6619), 532–537. Available at https://doi.org/10.1126/science.abn9768, https://www.science.org/doi/epdf/10.1126/science.abn9768.

Di Giuseppe, F., Rémy, S., Pappenberger, F., Wetterhall, F., 2018. Using the fire weather index (FWI) to improve the estimation of fire emissions from fire radiative power (FRP) observations. Atmospheric Chemistry and Physics 18 (8), 5359–5370. Available at https://doi.org/10.5194/acp-18-5359-2018.

Dozier, J., 1981. A method for satellite identification of surface temperature fields of subpixel resolution. Remote Sensing of Environment 11 (C), 221–229. Available at https://doi.org/10.1016/0034-4257(81)90021-3.

Dragozi, E., Giannaros, T.M., Kotroni, V., Lagouvardos, K., Koletsis, I., 2021. Dead fuel moisture content (Dfmc) estimation using modis and meteorological data: The case of Greece. Remote Sensing 13 (21). Available at https://doi.org/10.3390/rs13214224, https://www.mdpi.com/2072-4292/13/21/4224/pdf.

Eidenshink, J., Schwind, B., Brewer, K., Zhu, Z.-L., Quayle, B., Howard, S., 2007. A project for monitoring trends in burn severity. Fire Ecology 3 (1), 3–21. Available at https://doi.org/10.4996/fireecology.0301003.

Engelstad, P.S., Falkowski, M., Wolter, P., Poznanovic, A., Johnson, P., 2019. Estimating canopy fuel attributes from low-density lidar. Fire 2 (3), 1–19. Available at https://doi.org/10.3390/fire2030038, https://www.mdpi.com/2571-6255/2/3/38/pdf.

Epting, J., Verbyla, D., Sorbel, B., 2005. Evaluation of remotely sensed indices for assessing burn severity in interior Alaska using Landsat TM and ETM+. Remote Sensing of Environment 96 (3-4), 328–339. Available at https://doi.org/10.1016/j.rse.2005.03.002.

Fernández-García, V., Franquesa, M., Kull, C.A., 2024. Madagascar's burned area from Sentinel-2 imagery (2016–2022): four times higher than from lower resolution sensors. Science of The Total Environment 914, 169929. Available at https://doi.org/10.1016/j.scitotenv.2024.169929.

Fernández-Guisuraga, J.M., Calvo, L., Quintano, C., Fernández-Manso, A., Fernandes, P.M., 2023. Fractional vegetation cover ratio estimated from radiative transfer modeling outperforms spectral indices to assess fire severity in several Mediterranean plant communities. Remote Sensing of Environment 290. Available at https://doi.org/10.1016/j.rse.2023.113542, www.elsevier.com/inca/publications/store/5/0/5/7/3/3.

Forkel, M., Schmidt, L., Zotta, R.M., Dorigo, W., Yebra, M., 2023. Estimating leaf moisture content at global scale from passive microwave satellite observations of vegetation optical depth. Hydrology and Earth System Sciences 27 (1), 39–68. Available at https://doi.org/10.5194/hess-27-39-2023, http://www.hydrol-earth-syst-sci.net/volumes_and_issues.html.

Franquesa, M., Lizundia-Loiola, J., Stehman, S.V., Chuvieco, E., 2022. Using long temporal reference units to assess the spatial accuracy of global satellite-derived burned area products. Remote Sensing of Environment 269. Available at https://doi.org/10.1016/j.rse.2021.112823, www.elsevier.com/inca/publications/store/5/0/5/7/3/3.

Frazier, R.J., Coops, N.C., Wulder, M.A., 2015. Boreal Shield forest disturbance and recovery trends using Landsat time series. Remote Sensing of Environmet 170, 317–327.

Frazier, R.J., Coops, N.C., Wulder, M.A., Hermosilla, T., White, J.C., 2018. Analyzing spatial and temporal variability in short-term rates of post-fire vegetation return from Landsat time series. Remote Sensing of Environment 205, 32–45. Available at https://doi.org/10.1016/j.rse.2017.11.007, www.elsevier.com/inca/publications/store/5/0/5/7/3/3.

Furyaev, V.V., 1985. The use of aerospace imagery to examine and assess the consequences of forest fires. Soviet Journal of Remote Sensing 269, 773–782.

Gajardo, J., García, M., Riaño, D., 2014. Applications of airborne laser scanning in forest fuel assessment and fire prevention. In: Maltamo, M., Næsset, E., Vauhkonen, J. (Eds.), Forestry Applications of Airborne Laser Scanning. Managing Forest Ecosystems. Springer, Dordrecht, pp. 439–462.

Garcia, M., Aragoneses, E., 2026. Chapter 6 - Mapping forest fuels: how satellite LiDAR transforms our understanding. In: Rodríguez Gonzálvez, P., Guerra-Hernández, J., González Ferreiro, E.M. (Eds.), Satellite Remote Sensing for Forest and Environmental Monitoring. Elsevier, pp. 155–184.

García, M., Chuvieco, E., Nieto, H., Aguado, I., 2008. Combining AVHRR and meteorological data for estimating live fuel moisture content. Remote Sensing of Environment 112 (9), 3618–3627. Available at https://doi.org/10.1016/j.rse.2008.05.002.

García, M., Riaño, D., Chuvieco, E., Salas, J., Danson, F.M., 2011a. Multispectral and LiDAR data fusion for fuel type mapping using support vector machine and decision rules. Remote Sensing of Environment 115 (6), 1369–1379. Available at https://doi.org/10.1016/j.rse.2011.01.017.

García, M., Danson, F.M., Riaño, D., Chuvieco, E., Ramirez, F.A., Bandugula, V., 2011b. Terrestrial laser scanning to estimate plot-level forest canopy fuel properties. International Journal of Applied Earth Observation and Geoinformation 13 (4), 636–645. Available at https://doi.org/10.1016/j.jag.2011.03.006, http://www.sciencedirect.com/.

García, M., Popescu, S., Riaño, D., Zhao, K., Neuenschwander, A., Agca, M., et al., 2012. Characterization of canopy fuels using ICESat/GLAS data. Remote Sensing of Environment 123, 81–89. Available at https://doi.org/10.1016/j.rse.2012.03.018.

Garcia, M., Saatchi, S., Casas, A., Koltunov, A., Ustin, S., Ramirez, C., et al., 2017. Quantifying biomass consumption and carbon release from the California Rim fire by integrating airborne LiDAR and Landsat OLI data. Journal of Geophysical Research: Biogeosciences 122 (2), 340–353. Available at https://doi.org/10.1002/2015jg003315.

García, M., Riaño, D., Yebra, M., Salas, J., Cardil, A., Monedero, S., et al., 2020a. A live fuel moisture content product from landsat TM satellite time series for implementation in fire behavior models. Remote Sensing 12 (11). Available at https://doi.org/10.3390/rs12111714, https://res.mdpi.com/d_attachment/remotesensing/remotesensing-12-01714/article_deploy/remotesensing-12-01714-v2.pdf.

García, M., North, P., Viana-Soto, A., Stavros, N.E., Rosette, J., Martín, M.P., et al., 2020b. Evaluating the potential of LiDAR data for fire damage assessment: A radiative transfer model approach. Remote Sensing of Environment 247. Available at https://doi.org/10.1016/j.rse.2020.111893, www.elsevier.com/inca/publications/store/5/0/5/7/3/3.

García, M., Pettinari, M.L., Chuvieco, E., Salas, J., Mouillot, F., Chen, W., et al., 2022. Characterizing global fire regimes from satellite-derived products. Forests 13 (5). Available at https://doi.org/10.3390/f13050699, https://www.mdpi.com/1999-4907/13/5/699/pdf.

Gerhards, M., Schlerf, M., Mallick, K., Udelhoven, T., 2019. Challenges and future perspectives of multi-/hyperspectral thermal infrared remote sensing for crop water-stress detection: a review. Remote Sensing 11 (10), 1240. Available at https://doi.org/10.3390/rs11101240.

Gibson, R., Danaher, T., Hehir, W., Collins, L., 2020. A remote sensing approach to mapping fire severity in south-eastern Australia using sentinel 2 and random forest. Remote Sensing of Environment 240, 111702. Available at https://doi.org/10.1016/j.rse.2020.111702.

Giglio, L., Descloitres, J., Justice, C.O., Kaufman, Y.J., 2003. An enhanced contextual fire detection algorithm for MODIS. Remote Sensing of Environment 87 (2-3), 273–282. Available at https://doi.org/10.1016/S0034-4257(03)00184-6, www.elsevier.com/inca/publications/store/5/0/5/7/3/3.

Giglio, L., Loboda, T., Roy, D.P., Quayle, B., Justice, C.O., 2009. An active-fire based burned area mapping algorithm for the MODIS sensor. Remote Sensing of Environment 113 (2), 408–420. Available at https://doi.org/10.1016/j.rse.2008.10.006.

Giglio, L., Schroeder, W., Justice, C.O., 2016. The collection 6 MODIS active fire detection algorithm and fire products. Remote Sensing of Environment 178, 31–41. Available at https://doi.org/10.1016/j.rse.2016.02.054, www.elsevier.com/inca/publications/store/5/0/5/7/3/3.

Giglio, L., Boschetti, L., Roy, D.P., Humber, M.L., Justice, C.O., 2018. The collection 6 MODIS burned area mapping algorithm and product. Remote Sensing of Environment

217, 72–85. Available at https://doi.org/10.1016/j.rse.2018.08.005, www.elsevier.com/inca/publications/store/5/0/5/7/3/3.

Giglio, L., Boschetti, L., Roy, D.P., Hall, J.V., Zubkova, M., Humber, M., et al., 2025. The NASA VIIRS burned area product, global validation, and intercomparison with the NASA MODIS burned area product. Remote Sensing of Environment 331. Available at https://doi.org/10.1016/j.rse.2025.115006, https://www.sciencedirect.com/science/journal/00344257.

González-Ferreiro, E., Arellano-Pérez, S., Castedo-Dorado, F., Hevia, A., Vega, J.A., Vega-Nieva, D., et al., 2017. Modelling the vertical distribution of canopy fuel load using national forest inventory and low-density airbone laser scanning data. PLoS One 12 (4), e0176114. Available at https://doi.org/10.1371/journal.pone.0176114.

Goodwin, N.R., Collett, L.J., 2014. Development of an automated method for mapping fire history captured in Landsat TM and ETM+ time series across Queensland, Australia. Remote Sensing of Environment 148, 206–221. Available at https://doi.org/10.1016/j.rse.2014.03.021, www.elsevier.com/inca/publications/store/5/0/5/7/3/3.

Grégoire, J.M., Tansey, K., Silva, J.M.N., 2003. The GBA2000 initiative: developing a global burnt area database from SPOT-VEGETATION imagery. International Journal of Remote Sensing 24 (6), 1369–1376. Available at https://doi.org/10.1080/0143116021000044850, https://www.tandfonline.com/loi/tres20.

Hall, R.J., Skakun, R.S., Metsaranta, J.M., Landry, R., Fraser, R.H., Raymond, D., et al., 2020. Generating annual estimates of forest fire disturbance in Canada: the National Burned Area Composite. International Journal of Wildland Fire 29 (10), 878–891. Available at https://doi.org/10.1071/wf19201.

Hawbaker, T.J., Vanderhoof, M.K., Beal, Y.J., Takacs, J.D., Schmidt, G.L., Falgout, J.T., et al., 2017. Mapping burned areas using dense time-series of Landsat data. Remote Sensing of Environment 198, 504–522. Available at https://doi.org/10.1016/j.rse.2017.06.027, www.elsevier.com/inca/publications/store/5/0/5/7/3/3.

Hawbaker, T.J., Vanderhoof, M.K., Schmidt, G.L., Beal, Y.J., Picotte, J.J., Takacs, J.D., et al., 2020. The Landsat Burned Area algorithm and products for the conterminous United States. Remote Sensing of Environment 244. Available at https://doi.org/10.1016/j.rse.2020.111801, www.elsevier.com/inca/publications/store/5/0/5/7/3/3.

Hermosilla, T., Ruiz, L.A., Kazakova, A.N., Coops, N.C., Moskal, L.M., 2014. Estimation of forest structure and canopy fuel parameters from small-footprint full-waveform LiDAR data. International Journal of Wildland Fire 23 (2), 224–233. Available at https://doi.org/10.1071/WF13086.

Hillman, S., Wallace, L., Reinke, K., Jones, S., 2021. A comparison between TLS and UAS LiDAR to represent eucalypt crown fuel characteristics. ISPRS Journal of Photogrammetry and Remote Sensing 181, 295–307. Available at https://doi.org/10.1016/j.isprsjprs.2021.09.008.

Hirschmugl, M., Gallaun, H., Dees, M., Datta, P., Deutscher, J., Koutsias, N., et al., 2017. Methods for mapping forest disturbance and degradation from optical earth observation data: a review. Current Forestry Reports 3 (1), 32–45. Available at https://doi.org/10.1007/s40725-017-0047-2.

Hoffman, K.M., Trant, A.J., Nijland, W., Starzomski, B.M., 2018. Ecological legacies of fire detected using plot-level measurements and LiDAR in an old growth coastal temperate rainforest. Forest Ecology and Management 424, 11–20. Available at https://doi.org/10.1016/j.foreco.2018.04.020, www.elsevier.com/inca/publications/store/5/0/3/3/1/0.

Hu, T., Ma, Q., Su, Y., Battles, J.J., Collins, B.M., Stephens, S.L., Kelly, M., Guo, Q., 2019. A simple and integrated approach for fire severity assessment using bi-temporal airborne

LiDAR data. International Journal of Applied Earth Observation and Geoinformation 78, 25–38. Available at https://doi.org/10.1016/j.jag.2019.01.007.

Huang, S., Long, T., Zhang, Z., He, G., Wang, G., 2025. Generation of 30 m resolution monthly burned area product in Africa based on Landsat 8/9 and Sentinel-2 data. ISPRS Journal of Photogrammetry and Remote Sensing 230, 170–191. Available at https://doi.org/10.1016/j.isprsjprs.2025.09.012.

Jackson, T.J., Schmugge, T.J., 1991. Vegetation effects on the microwave emission of soils. Remote Sensing of Environment 36 (3), 203–212. Available at https://doi.org/10.1016/0034-4257(91)90057-d.

Jain, T.B., Graham, R.T., Pilliod, D.S., 2004. Tongue-tied: confused meanings for common fire terminology can lead to fuels mismanagement. Wildfire 22–26 July/August.

Jolly, W.M., Johnson, D.M., 2018. Pyro-Ecophysiology: Shifting the Paradigm of Live Wildland Fuel Research. Fire 1 (1), 8. Available at https://doi.org/10.3390/fire1010008.

Jones, H.G., 1999. Use of infrared thermometry for estimation of stomatal conductance as a possible aid to irrigation scheduling. Agricultural and Forest Meteorology 95 (3), 139–149. Available at https://doi.org/10.1016/S0168-1923(99)00030-1.

Jones, H.G., 2004. Application of Thermal Imaging and Infrared Sensing in Plant Physiology and Ecophysiology. Advances in Botanical Research. Academic Press, pp. 107–163.

Kant, Y., Prasad, V.K., Badarinath, K.V.S., 2000. Algorithm for detection of active fire zones using NOAA AVHRR data. Infrared Physics and Technology 41 (1), 29–34. Available at https://doi.org/10.1016/S1350-4495(99)00053-5.

Keane, R.E., Burgan, R., Van Wagtendonk, J., 2001. Mapping wildland fuels for fire management across multiple scales: Integrating remote sensing, GIS, and biophysical modeling. International Journal of Wildland Fire 10 (3-4), 301–319. Available at https://doi.org/10.1071/wf01028, http://www.publish.csiro.au/nid/126.htm.

Kelly, L.T., Giljohann, K.M., Duane, A., Aquilué, N., Archibald, S., Batllori, E., et al., 2020. Fire and biodiversity in the Anthropocene. Science (New York, N.Y.) 370 (6519). Available at https://doi.org/10.1126/science.abb0355, https://science.sciencemag.org/content/sci/370/6519/eabb0355.full.pdf.

Kennedy, R.E., Yang, Z., Gorelick, N., Braaten, J., Cavalcante, L., Cohen, W.B., et al., 2018. Implementation of the LandTrendr algorithm on Google Earth Engine. Remote Sensing 10 (5). Available at https://doi.org/10.3390/rs10050691, http://www.mdpi.com/2072-4292/10/5/691/pdf.

Key, C.H., Benson, N.C., 2006. Landscape Assessment: Ground measure of severity, the Composite Burn Index; and Remote sensing of severity, the Normalized Burn Ratio. In. Ogden, UT.

Khairoun, A., Solano, E., 2023. In: Agency, E.S. (Ed.), ESA CCI ECV Fire Disturbance: D2.2.2 Algorithm Theoretical Basis Document-SFDL. European Space Agency.

Koltunov, A., Ustin, S.L., Quayle, B., Schwind, B., Ambrosia, V.G., Li, W., 2016. The development and first validation of the GOES Early Fire Detection (GOES-EFD) algorithm. Remote Sensing of Environment 184, 436–453. Available at https://doi.org/10.1016/j.rse.2016.07.021, www.elsevier.com/inca/publications/store/5/0/5/7/3/3.

Konings, A.G., Rao, K., Steele-Dunne, S.C., 2019. Macro to micro: microwave remote sensing of plant water content for physiology and ecology. New Phytologist 223 (3), 1166–1172. Available at https://doi.org/10.1111/nph.15808, http://onlinelibrary.wiley.com/journal/10.1111/(ISSN)1469-8137.

Krawchuk, M.A., Moritz, M.A., Parisien, M.A., Van Dorn, J., Hayhoe, K., 2009. Global pyrogeography: the current and future distribution of wildfire. PLoS One 4 (4). Available at https://doi.org/10.1371/journal.pone.0005102, http://www.plosone.org/article/fetchObjectAttachment.action?uri=info%3Adoi%2F10.1371%2Fjournal.pone.0005102&representation=PDF.

Kouachi, M.E., Khairoun, A., Moghli, A., Rahmani, S., Mouillot, F., Baeza, M.J., et al., 2024. Forty-year fire history reconstruction from landsat data in mediterranean ecosystems of algeria following international standards. Remote Sensing 16 (13). Available at https://doi.org/10.3390/rs16132500, http://www.mdpi.com/journal/remotesensing.

Labenski, P., Ewald, M., Schmidtlein, S., Heinsch, F.A., Fassnacht, F.E., 2023. Quantifying surface fuels for fire modelling in temperate forests using airborne lidar and Sentinel-2: potential and limitations. Remote Sensing of Environment 295. Available at https://doi.org/10.1016/j.rse.2023.113711, www.elsevier.com/inca/publications/store/5/0/5/7/3/3.

Lentile, L.B., Holden, Z.A., Smith, A.M.S., Falkowski, M.J., Hudak, A.T., Morgan, P., et al., 2006. Remote sensing techniques to assess active fire characteristics and post-fire effects. International Journal of Wildland Fire 15 (3), 319–345. Available at https://doi.org/10.1071/WF05097.

Lin, Z., Chen, F., Niu, Z., Li, B., Yu, B., Jia, H., et al., 2018. An active fire detection algorithm based on multi-temporal FengYun-3C VIRR data. Remote Sensing of Environment 211, 376–387. Available at https://doi.org/10.1016/j.rse.2018.04.027, www.elsevier.com/inca/publications/store/5/0/5/7/3/3.

Lin, D., Giannico, V., Lafortezza, R., Sanesi, G., Elia, M., 2024. Use of airborne LiDAR to predict fine dead fuel load in Mediterranean forest stands of Southern Europe. Fire Ecology 20 (1). Available at https://doi.org/10.1186/s42408-024-00287-7.

Lizundia-Loiola, J., Franquesa, M., Boettcher, M., Kirches, G., Pettinari, M.L., Chuvieco, E., 2021. Implementation of the burned area component of the copernicus climate change service: from modis to olci data. Remote Sensing 13 (21). Available at https://doi.org/10.3390/rs13214295, https://www.mdpi.com/2072-4292/13/21/4295/pdf.

Lizundia-Loiola, J., Otón, G., Ramo, R., Chuvieco, E., 2020. A spatio-temporal active-fire clustering approach for global burned area mapping at 250 m from MODIS data. Remote Sensing of Environment 236, 111493. Available at https://doi.org/10.1016/j.rse.2019.111493.

Lizundia-Loiola, J., Franquesa, M., Khairoun, A., Chuvieco, E., 2022. Global burned area mapping from Sentinel-3 Synergy and VIIRS active fires. Remote Sensing of Environment 282, 113298. Available at https://doi.org/10.1016/j.rse.2022.113298.

Long, T., Zhang, Z., He, G., Jiao, W., Tang, C., Wu, B., et al., 2019. 30 m resolution global annual burned area mapping based on landsat images and Google earth engine. Remote Sensing 11 (5), 489. Available at https://doi.org/10.3390/rs11050489.

López-García, M.J., Caselles, 1991. Mapping burns and natural reforestation using thematic mapper data. Geocarto International 6 (1), 31–37. Available at https://doi.org/10.1080/10106049109354290.

Loudermilk, E.L., Pokswinski, S., Hawley, C.M., Maxwell, A., Gallagher, M., Skowronski, N., et al., 2023. Terrestrial Laser Scan Metrics Predict Surface Vegetation Biomass and Consumption in a Frequently Burned Southeastern U.S. Ecosystem. United States bioRxiv. Available at https://www.biorxiv.org, https://doi.org/10.1101/2023.01.15.524107.

Marino, E., Ranz, P., Tomé, J.L., Noriega, M.Á., Esteban, J., Madrigal, J., 2016. Generation of high-resolution fuel model maps from discrete airborne laser scanner and Landsat-8 OLI: a low-cost and highly updated methodology for large areas. Remote Sensing of Environment 187, 267–280. Available at https://doi.org/10.1016/j.rse.2016.10.020, www.elsevier.com/inca/publications/store/5/0/5/7/3/3.

Massetti, A., Rüdiger, C., Yebra, M., Hilton, J., 2019. The Vegetation Structure Perpendicular Index (VSPI): A forest condition index for wildfire predictions. Remote Sensing of Environment 224, 167–181. Available at https://doi.org/10.1016/j.rse.2019.02.004.

Matasci, G., Hermosilla, T., Wulder, M.A., White, J.C., Coops, N.C., Hobart, G.W., Bolton, D.K., Tompalski, P., Bater, C.W., 2018. Three decades of forest structural dynamics over Canada's forested ecosystems using Landsat time-series and lidar plots. Remote Sensing of Environment 216, 697–714. Available at https://doi.org/10.1016/j.rse.2018.07.024.

Matthews, S., 2010. Effect of drying temperature on fuel moisture content measurements. International Journal of Wildland Fire 19 (6), 800–802. Available at https://doi.org/10.1071/wf08188.

McCarley, T.R., Hudak, A.T., Sparks, A.M., Vaillant, N.M., Meddens, A.J.H., Trader, L., et al., 2020. Estimating wildfire fuel consumption with multitemporal airborne laser scanning data and demonstrating linkage with MODIS-derived fire radiative energy. Remote Sensing of Environment 251. Available at https://doi.org/10.1016/j.rse.2020.112114, www.elsevier.com/inca/publications/store/5/0/5/7/3/3.

Miller, J.D., Thode, A.E., 2007. Quantifying burn severity in a heterogeneous landscape with a relative version of the delta Normalized Burn Ratio (dNBR). Remote Sensing of Environment 109 (1), 66–80. Available at https://doi.org/10.1016/j.rse.2006.12.006.

Milne, A.K., 1986. The use of remote sensing in mapping and monitoring vegetational change associated with bushfire events in Eastern Australia. Geocarto International 1 (1), 25–32. Available at https://doi.org/10.1080/10106048609354022.

Morresi, D., Vitali, A., Urbinati, C., Garbarino, M., 2019. Forest Spectral Recovery and Regeneration Dynamics in Stand-Replacing Wildfires of Central Apennines Derived from Landsat Time Series. Remote Sensing 11, 308. Available at https://doi.org/10.3390/rs11030308.

Mouillot, F., Schultz, M.G., Yue, C., Cadule, P., Tansey, K., Ciais, P., et al., 2014. Ten years of global burned area products from spaceborne remote sensing-A review: analysis of user needs and recommendations for future developments. International Journal of Applied Earth Observation and Geoinformation 26 (1), 64–79. Available at https://doi.org/10.1016/j.jag.2013.05.014, http://www.elsevier.com/locate/jag.

Narine, L., Malambo, L., Popescu, S., 2022. Characterizing canopy cover with ICESat-2: a case study of southern forests in Texas and Alabama, USA. Remote Sensing of Environment 281, 113242. Available at https://doi.org/10.1016/j.rse.2022.113242.

Neuenschwander, A., Pitts, K., Jelley, B., Robbins, J., Markel, J., Popescu, S., et al., 2022. Ice, Cloud, and Land Elevation Satellite (ICESat-2) Project Algorithm Theoretical Basis Document (ATBD) for Land - Vegetation Along-Track Products (ATL08), Version 6. In.

Neves, A.K., Campagnolo, M.L., Silva, J.M.N., Pereira, J.M.C., 2023. A Landsat-based atlas of monthly burned area for Portugal, 1984–2021. International Journal of Applied Earth Observation and Geoinformation 119. Available at https://doi.org/10.1016/j.jag.2023.103321, http://www.elsevier.com/locate/jag.

Oliva, P., Schroeder, W., 2015a. Atmospheric correction of VIIRS and MODIS fire radiative power retrievals for multi-sensor comparison 2015-International Geoscience and Remote Sensing Symposium (IGARSS), pp. 2038-2040. https://doi.org/10.1109/IGARSS.2015.7326200.

Oliva, P., Schroeder, W., 2015b. Assessment of VIIRS 375m active fire detection product for direct burned area mapping. Remote Sensing of Environment 160, 144–155. Available at https://doi.org/10.1016/j.rse.2015.01.010.

Otón, G., Ramo, R., Lizundia-Loiola, J., Chuvieco, E., 2019. Global detection of long-term (1982–2017) burned area with AVHRR-LTDR data. Remote Sensing 11 (18), 2079. Available at https://doi.org/10.3390/rs11182079.

Otón, G., Lizundia-Loiola, J., Pettinari, M.L., Chuvieco, E., 2021. Development of a consistent global long-term burned area product (1982–2018) based on AVHRR-LTDR data. International Journal of Applied Earth Observation and Geoinformation 103. Available at https://doi.org/10.1016/j.jag.2021.102473, http://www.elsevier.com/locate/jag.

Owe, M., de Jeu, R., Walker, J., 2001. A methodology for surface soil moisture and vegetation optical depth retrieval using the microwave polarization difference index. IEEE Transactions on Geoscience and Remote Sensing 39 (8), 1643–1654. Available at https://doi.org/10.1109/36.942542.

Padilla, M., Stehman, S.V., Ramo, R., Corti, D., Hantson, S., Oliva, P., et al., 2015. Comparing the accuracies of remote sensing global burned area products using stratified random sampling and estimation. Remote Sensing of Environment 160, 114–121. Available at https://doi.org/10.1016/j.rse.2015.01.005, www.elsevier.com/inca/publications/store/5/0/5/7/3/3.

Paloscia, S., Pampaloni, P., 1988. Microwave polarization index for monitoring vegetation growth. IEEE Transactions on Geoscience and Remote Sensing 26 (5), 617–621. Available at https://doi.org/10.1109/36.7687.

Pampanoni, V., Laneve, G., Viegas, D.X., Alves, D., Ribeiro, L.M., Pellizzaro, G., et al., 2023. Early validation of a live fuel moisture content product based on sentinel-2 and sentinel-3 images 2023-international geoscience and remote sensing symposium (IGARSS), 3042-3045. https://doi.org/10.1109/IGARSS52108.2023.10281970 .

Parks, S.A., Dillon, G.K., Miller, C., 2014. A new metric for quantifying burn severity: the relativized burn ratio. Remote Sensing 6 (3), 1827–1844. Available at https://doi.org/10.3390/rs6031827.

Pechony, O., Shindell, D.T., 2010. Driving forces of global wildfires over the past millennium and the forthcoming century. Proceedings of the National Academy of Sciences 107 (45), 19167–19170. Available at https://doi.org/10.1073/pnas.1003669107.

Peterson, B., Nelson, K., Wylie, B., 2013. Towards integration of GLAS into a national fuel mapping program. Photogrammetric Engineering & Remote Sensing 79 (2), 175–183. Available at https://doi.org/10.14358/pers.79.2.175.

Plummer, S., Arino, O., Simon, M., Steffen, W., 2006. Establishing a earth observation product service for the terrestrial carbon community: the globcarbon initiative. Mitigation and Adaptation Strategies for Global Change 11 (1), 97–111. Available at https://doi.org/10.1007/s11027-006-1012-8.

Pollet, J., Brown, A., 2007. Fuel moisture sampling guide. (Ed.)S.L.C. Bureau of Land Management. Utah State Office, UT, USA.

Polychronaki, A., Gitas, I.Z., Minchella, A., 2013. Monitoring post-fire vegetation recovery in the Mediterranean using SPOT and ERS imagery. International Journal of Wildland Fire 23, 631–642. Available at https://doi.org/10.1071/wf12058.

Post, A.J., Forbes, B., Cooper, Z., Faro, K., Seel, C., Clark, M., et al., 2025. Using handheld mobile laser scanning to quantify fine-scale surface fuels and detect changes post-disturbance in northern California forests. Ecological Indicators 172. Available at https://doi.org/10.1016/j.ecolind.2025.113276, https://www.sciencedirect.com/science/journal/1470160X.

Pourshamsi, M., Garcia, M., Lavalle, M., Balzter, H., 2018. A machine-learning approach to PolInSAR and LiDAR data fusion for improved tropical forest canopy height estimation using NASA AfriSAR campaign data. IEEE Journal of Selected Topics in Applied Earth Observations and Remote Sensing 11 (10), 3453–3463. Available at https://doi.org/10.1109/jstars.2018.2868119.

Pourshamsi, M., Xia, J., Yokoya, N., Garcia, M., Lavalle, M., Pottier, E., et al., 2021. Tropical forest canopy height estimation from combined polarimetric SAR and LiDAR using machine-learning. ISPRS Journal of Photogrammetry and Remote Sensing 172, 79–94. Available at https://doi.org/10.1016/j.isprsjprs.2020.11.008.

Pyne, S.J., Andrews, P.L., Laven, R.D., 1996. Introduction to Wildland Fire. John Wiley & Sons, New York, USA.

Qi, Y., Dennison, P.E., Spencer, J., Riaño, D., 2012. Monitoring live fuel moisture using soil moisture and remote sensing proxies. Fire Ecology 8 (3), 71–87. Available at https://doi.org/10.4996/fireecology.0803071, http://fireecology.org/docs/Journal/pdf/Volume08/Issue03/071.pdf.

Quan, X., He, B., Yebra, M., Yin, C., Liao, Z., Li, X., 2017. Retrieval of forest fuel moisture content using a coupled radiative transfer model. Environmental Modelling & Software 95, 290–302. Available at https://doi.org/10.1016/j.envsoft.2017.06.006.

Quan, X., Yebra, M., Riaño, D., He, B., Lai, G., Liu, X., 2021. Global fuel moisture content mapping from MODIS. International Journal of Applied Earth Observation and Geoinformation 101, 102354. Available at https://doi.org/10.1016/j.jag.2021.102354.

Quan, X., Chen, R., Yebra, M., Riaño, D., Resco de Dios, V., Li, X., et al., 2024. Sub-daily live fuel moisture content estimation from Himawari-8 data. Remote Sensing of Environment 308. Available at https://doi.org/10.1016/j.rse.2024.114170, https://www.sciencedirect.com/science/journal/00344257.

Quintano, C., Fernández-Manso, A., Fernández-Manso, O., 2018. Combination of Landsat and Sentinel-2 MSI data for initial assessing of burn severity. International Journal of Applied Earth Observation and Geoinformation 64, 221–225. Available at https://doi.org/10.1016/j.jag.2017.09.014.

Ramo, R., Roteta, E., Bistinas, I., van Wees, D., Bastarrika, A., Chuvieco, E., et al., 2021. African burned area and fire carbon emissions are strongly impacted by small fires undetected by coarse resolution satellite data. Proceedings of the National Academy of Sciences 118 (9). Available at https://doi.org/10.1073/pnas.2011160118.

Rao, K., Williams, A.P., Flefil, J.F., Konings, A.G., 2020. SAR-enhanced mapping of live fuel moisture content. Remote Sensing of Environment 245. Available at https://doi.org/10.1016/j.rse.2020.111797, www.elsevier.com/inca/publications/store/5/0/5/7/3/3.

Riaño, D., Chuvieco, E., Salas, J., Palacios-Orueta, A., Bastarrika, A., 2002. Generation of fuel type maps from Landsat TM images and ancillary data in Mediterranean ecosystems. Canadian Journal of Forest Research 32 (8), 1301–1315. Available at https://doi.org/10.1139/x02-052.

Riaño, D., Chuvieco, E., Condés, S., González-Matesanz, J., Ustin, S.L., 2004. Generation of crown bulk density for *Pinus sylvestris* L. from lidar. Remote Sensing of Environment 92 (3), 345–352. Available at https://doi.org/10.1016/j.rse.2003.12.014, www.elsevier.com/inca/publications/store/5/0/5/7/3/3.

Roberts, G., Wooster, M.J., 2014. Development of a multi-temporal Kalman filter approach to geostationary active fire detection & fire radiative power (FRP) estimation. Remote Sensing of Environment 152, 392–412. Available at https://doi.org/10.1016/j.rse.2014.06.020.

Rodriguez-Jimenez, F., Lorenzo, H., Novo, A., Acuña-Alonso, C., Alvarez, X., 2023. Modelling of live fuel moisture content in different vegetation scenarios during dry periods using meteorological data and spectral indices. Forest Ecology and Management 546, 121378. Available at https://doi.org/10.1016/j.foreco.2023.121378.

Roteta, E., Bastarrika, A., Padilla, M., Storm, T., Chuvieco, E., 2019. Development of a Sentinel-2 burned area algorithm: generation of a small fire database for sub-Saharan Africa. Remote Sensing of Environment 222, 1–17. Available at https://doi.org/10.1016/j.rse.2018.12.011.

Roteta, E., Bastarrika, A., Franquesa, M., Chuvieco, E., 2021. Landsat and sentinel-2 based burned area mapping tools in Google earth engine. Remote Sensing 13 (4), 816. Available at https://doi.org/10.3390/rs13040816.

Rowell, E., Loudermilk, E.L., Seielstad, C., O'Brien, J.J., 2016. Using simulated 3D surface fuelbeds and terrestrial laser scan data to develop inputs to fire behavior models. Canadian

Journal of Remote Sensing 42 (5), 443–459. Available at https://doi.org/10.1080/07038992.2016.1220827, http://www.tandfonline.com/loi/ujrs20.

Rowell, E., Loudermilk, E.L., Hawley, C., Pokswinski, S., Seielstad, C., Queen, L.L., et al., 2020. Coupling terrestrial laser scanning with 3D fuel biomass sampling for advancing wildland fuels characterization. Forest Ecology and Management 462. Available at https://doi.org/10.1016/j.foreco.2020.117945, www.elsevier.com/inca/publications/store/5/0/3/3/1/0.

Roy, D.P., Boschetti, L., Justice, C.O., Ju, J., 2008. The collection 5 MODIS burned area product—global evaluation by comparison with the MODIS active fire product. Remote Sensing of Environment 112 (9), 3690–3707. Available at https://doi.org/10.1016/j.rse.2008.05.013.

Saatchi, S., Halligan, K., Despain, D.G., Crabtree, R.L., 2007. Estimation of forest fuel load from radar remote sensing. IEEE Transactions on Geoscience and Remote Sensing 45 (6), 1726–1740. Available at https://doi.org/10.1109/TGRS.2006.887002.

San-Miguel-Ayanz, J., Ravail, N., 2005. Active fire detection for fire emergency management: potential and limitations for the operational use of remote sensing. Natural Hazards 35 (3), 361–376. Available at https://doi.org/10.1007/s11069-004-1797-2.

San-Miguel-Ayanz, J., Schulte, E., Schmuck, G., Camia, A., Strobl, P., Liberta, G., et al., 2012. Comprehensive Monitoring of Wildfires in Europe: The European Forest Fire Information System (EFFIS). InTech. Available at https://doi.org/10.5772/28441.

Santos, F.L.M., Couto, F.T., Monteiro, M., Ribeiro, N.A., Le Moigne, P., Salgado, R., 2025. A hybrid framework to estimate live fuel moisture content through land surface modelling. Modeling Earth Systems and Environment 11 (6). Available at https://doi.org/10.1007/s40808-025-02561-2, https://www.springer.com/journal/40808.

Sato, L.Y., Gomes, V.C.F., Shimabukuro, Y.E., Keller, M., Arai, E., dos-Santos, M.N., et al., 2016. Post-fire changes in forest biomass retrieved by airborne LiDAR in Amazonia. Remote Sensing 8 (10). Available at https://doi.org/10.3390/rs8100839, http://www.mdpi.com/journal/remotesensing.

Schroeder, W., Ellicott, E., Ichoku, C., Ellison, L., Dickinson, M.B., Ottmar, R.D., et al., 2014a. Integrated active fire retrievals and biomass burning emissions using complementary near-coincident ground, airborne and spaceborne sensor data. Remote Sensing of Environment 140, 719–730. Available at https://doi.org/10.1016/j.rse.2013.10.010.

Schroeder, W., Oliva, P., Giglio, L., Csiszar, I.A., 2014b. The New VIIRS 375m active fire detection data product: algorithm description and initial assessment. Remote Sensing of Environment 143, 85–96. Available at https://doi.org/10.1016/j.rse.2013.12.008.

Schroeder, W., Oliva, P., Giglio, L., Quayle, B., Lorenz, E., Morelli, F., 2016. Active fire detection using Landsat-8/OLI data. Remote Sensing of Environment 185, 210–220. Available at https://doi.org/10.1016/j.rse.2015.08.032.

Scott, J.H., Reinhardt, E.D., 2001. Assessing Crown Fire Potential By Linking Models of Surface and Crown Fire Behavior. US Department of Agriculture, Forest Service, Rocky Mountain Research Station, Fort Collins, CO, pp. 59 -59.

Simon, M., Plummer, S., Fierens, F., Hoelzemann, J.J., Arino, O., 2004. Burnt area detection at global scale using ATSR-2: The GLOBSCAR products and their qualification. Journal of Geophysical Research: Atmospheres 109 (D14). Available at https://doi.org/10.1029/2003jd003622.

Skowronski, N.S., Clark, K.L., Duveneck, M., Hom, J., 2011. Three-dimensional canopy fuel loading predicted using upward and downward sensing LiDAR systems. Remote Sensing of Environment 115 (2), 703–714. Available at https://doi.org/10.1016/j.rse.2010.10.012.

Smith, R.B., Woodgate, P.W., 1985. Appraisal of fire damage and inventory for timber salvage by remote sensing in mountain ash forests in Victoria. Australian Forestry 48 (4), 252–263. Available at https://doi.org/10.1080/00049158.1985.10674453.

Soverel, N.O., Perrakis, D.D.B., Coops, N.C., 2010. Estimating burn severity from Landsat dNBR and RdNBR indices across western Canada. Remote Sensing of Environment 114 (9), 1896–1909. Available at https://doi.org/10.1016/j.rse.2010.03.013.

Stefanidou, A., Z. Gitas, I., Korhonen, L., Georgopoulos, N., Stavrakoudis, D., 2020. Multispectral LiDAR-based estimation of surface fuel load in a dense coniferous forest. Remote Sensing 12 (20), 3333. Available at https://doi.org/10.3390/rs12203333.

Storey, E.A., Stow, D.A., O'Leary, J.F., 2016. Assessing postfire recovery of chamise chaparral using multi-temporal spectral vegetation index trajectories derived from Landsat imagery. Remote Sensing of Environment 183, 53–64. Available at https://doi.org/10.1016/j.rse.2016.05.018.

Tanase, M.A., Panciera, R., Lowell, K., Aponte, C., 2015. Monitoring live fuel moisture in semiarid environments using L-band radar data. International Journal of Wildland Fire 24 (4), 560–572. Available at https://doi.org/10.1071/wf14149.

Tansey, K., Grégoire, J.M., Defourny, P., Leigh, R., Pekel, J.F., van Bogaert, E., et al., 2008. A new, global, multi-annual (2000-2007) burnt area product at 1 km resolution. Geophysical Research Letters 35 (1). Available at https://doi.org/10.1029/2007GL031567.

Tenny, J.T., Sankey, T.T., Munson, S.M., Sánchez Meador, A.J., Goetz, S.J., 2025. Canopy and surface fuels measurement using terrestrial lidar single-scan approach in the Mogollon Highlands of Arizona. International Journal of Wildland Fire 34 (7). Available at https://doi.org/10.1071/WF24221, http://www.publish.csiro.au/nid/126.htm.

Trigg, Flasse, 2001. An evaluation of different bi-spectral spaces for discriminating burned shrub-savannah. International Journal of Remote Sensing 22 (13), 2641–2647. Available at https://doi.org/10.1080/01431160110053185.

Ulaby, F.T., Jedlicka, R.P., 1984. Microwave dielectric properties of plant materials. IEEE Transactions on Geoscience and Remote Sensing **GE** 22 (4), 406–415. Available at https://doi.org/10.1109/TGRS.1984.350644.

Vanderhoof, M.K., Fairaux, N., Beal, Y.J.G., Hawbaker, T.J., 2017. Validation of the USGS landsat burned area essential climate variable (BAECV) across the conterminous United States. Remote Sensing of Environment 198, 393–406. Available at https://doi.org/10.1016/j.rse.2017.06.025, www.elsevier.com/inca/publications/store/5/0/5/7/3/3.

Veraverbeke, S., Harris, S., Hook, S., 2011. Evaluating spectral indices for burned area discrimination using MODIS/ASTER (MASTER) airborne simulator data. Remote Sensing of Environment 115 (10), 2702–2709. Available at https://doi.org/10.1016/j.rse.2011.06.010.

Viana-Soto, A., Aguado, I., Salas, J., García, M., 2020. Identifying post-fire recovery trajectories and driving factors using landsat time series in fire-prone Mediterranean pine forests. Remote Sensing 12 (9), 1499. Available at https://doi.org/10.3390/rs12091499.

Viana-Soto, A., García, M., Aguado, I., Salas, J., 2022a. Assessing post-fire forest structure recovery by combining LiDAR data and Landsat time series in Mediterranean pine forests. International Journal of Applied Earth Observation and Geoinformation 108, 102754. Available at https://doi.org/10.1016/j.jag.2022.102754.

Viana-Soto, A., Okujeni, A., Pflugmacher, D., García, M., Aguado, I., Hostert, P., 2022b. Quantifying post-fire shifts in woody-vegetation cover composition in Mediterranean pine forests using Landsat time series and regression-based unmixing. Remote Sensing of Environment 281, 113239. Available at https://doi.org/10.1016/j.rse.2022.113239.

Viegas, D.X., Viegas, M.T.S.P., Ferreira, A.D., 1992. Moisture content of fine forest fuels and fire occurrence in central Portugal. International Journal of Wildland Fire 2 (2), 69–86. Available at https://doi.org/10.1071/wf9920069.

Vogeler, J.C., Yang, Z., Cohen, W.B., 2016. Mapping post-fire habitat characteristics through the fusion of remote sensing tools. Remote Sensing of Environment 173, 294–303. Available at https://doi.org/10.1016/j.rse.2015.08.011, www.elsevier.com/inca/publications/store/5/0/5/7/3/3.

Waigl, C.F., Prakash, A., Stuefer, M., Verbyla, D., Dennison, P., 2019. Fire detection and temperature retrieval using EO-1 Hyperion data over selected Alaskan boreal forest fires. International Journal of Applied Earth Observation and Geoinformation 81, 72–84. Available at https://doi.org/10.1016/j.jag.2019.03.004, http://www.elsevier.com/locate/jag.

Wei, M., Zhang, Z., Long, T., He, G., Wang, G., 2021. Monitoring landsat based burned area as an indicator of sustainable development goals. Earth's Future 9 (6). Available at https://doi.org/10.1029/2020ef001960.

van der Werf, G.R., Randerson, J.T., van Wees, D., Chen, Y., Giglio, L., Hall, J., et al., 2025. Landscape fire emissions from the 5th version of the Global Fire Emissions Database (GFED5). Scientific Data 12 (1). Available at https://doi.org/10.1038/s41597-025-06127-w, https://www.nature.com/sdata/.

Whitcraft, A.K., Becker-Reshef, I., Killough, B.D., Justice, C.O., 2015. Meeting earth observation requirements for global agricultural monitoring: an evaluation of the revisit capabilities of current and planned moderate resolution optical earth observing missions. Remote Sensing 7 (2), 1482–1503. Available at https://doi.org/10.3390/rs70201482, http://www.mdpi.com/2072-4292/7/2/1482/pdf.

Wigneron, J.-P., Calvet, J.-C., Pellarin, T., Van de Griend, A.A., Berger, M., Ferrazzoli, P., 2003. Retrieving near-surface soil moisture from microwave radiometric observations: current status and future plans. Remote Sensing of Environment 85 (4), 489–506. Available at https://doi.org/10.1016/s0034-4257(03)00051-8.

Wooster, M.J., Zhukov, B., Oertel, D., 2003. Fire radiative energy for quantitative study of biomass burning: derivation from the BIRD experimental satellite and comparison to MODIS fire products. Remote Sensing of Environment 86 (1), 83–107. Available at https://doi.org/10.1016/S0034-4257(03)00070-1, www.elsevier.com/inca/publications/store/5/0/5/7/3/3.

Wooster, M.J., Roberts, Kaufman, P., 2005. Retrieval of biomass combustion rates and totals from fire radiative power observations: FRP derivation and calibration relationships between biomass consumption and fire radiative energy release. Journal of Geophysical Research Atmospheres 110 (24), 1–24. Available at https://doi.org/10.1029/2005JD006318.

Wooster, M.J., Roberts, G.J., Giglio, L., Roy, D., Freeborn, P., Boschetti, L., et al., 2021. Satellite remote sensing of active fires: history and current status, applications and future requirements. Remote Sensing of Environment 267. Available at https://doi.org/10.1016/j.rse.2021.112694, www.elsevier.com/inca/publications/store/5/0/5/7/3/3.

Xia, H., Chen, Y., Quan, J., 2018. A simple method based on the thermal anomaly index to detect industrial heat sources. International Journal of Applied Earth Observation and Geoinformation 73, 627–637. Available at https://doi.org/10.1016/j.jag.2018.08.003, http://www.elsevier.com/locate/jag.

Yang, S., Chen, R., He, B., Zhang, Y., 2025. Low-variance estimation of live fuel moisture content using VIIRS data through radiative transfer model. International Journal of Applied Earth Observation and Geoinformation 136, 104311. Available at https://doi.org/10.1016/j.jag.2024.104311.

Yebra, M., Chuvieco, E., 2009. Linking ecological information and radiative transfer models to estimate fuel moisture content in the Mediterranean region of Spain: solving the ill-posed inverse problem. Remote Sensing of Environment 113 (11), 2403–2411. Available at https://doi.org/10.1016/j.rse.2009.07.001.

Yebra, M., Chuvieco, E., Riaño, D., 2008. Estimation of live fuel moisture content from MODIS images for fire risk assessment. Agricultural and Forest Meteorology 148 (4), 523–536. Available at https://doi.org/10.1016/j.agrformet.2007.12.005.

Yebra, M., Dennison, P.E., Chuvieco, E., Riaño, D., Zylstra, P., Hunt, E.R., et al., 2013. A global review of remote sensing of live fuel moisture content for fire danger assessment: moving towards operational products. Remote Sensing of Environment 136, 455–468. Available at https://doi.org/10.1016/j.rse.2013.05.029.

Yebra, M., Quan, X., Riaño, D., Rozas Larraondo, P., van Dijk, A.I.J.M., Cary, G.J., 2018. A fuel moisture content and flammability monitoring methodology for continental Australia based on optical remote sensing. Remote Sensing of Environment 212, 260–272. Available at https://doi.org/10.1016/j.rse.2018.04.053, www.elsevier.com/inca/publications/store/5/0/5/7/3/3.

Yebra, M., Scortechini, G., Badi, A., Beget, M.E., Boer, M.M., Bradstock, R., et al., 2019. Globe-LFMC, a global plant water status database for vegetation ecophysiology and wildfire applications. Scientific Data 6 (1). Available at https://doi.org/10.1038/s41597-019-0164-9, www.nature.com/sdata/.

Yebra, M., Scortechini, G., Adeline, K., Aktepe, N., Almoustafa, T., Bar-Massada, A., et al., 2024. Globe-LFMC 2.0, an enhanced and updated dataset for live fuel moisture content research. Scientific Data 11 (1). Available at https://doi.org/10.1038/s41597-024-03159-6, https://www.nature.com/sdata/.

Yin, C., He, B., Yebra, M., Quan, X., Edwards, A.C., Liu, X., et al., 2020. Improving burn severity retrieval by integrating tree canopy cover into radiative transfer model simulation. Remote Sensing of Environment 236, 111454. Available at https://doi.org/10.1016/j.rse.2019.111454.

Validation of remote sensing products for wildfire monitoring and modeling

4

Víctor Fernández-García
Department of Engineering and Agricultural Sciences, School of Agricultural and Forestry Engineering, Universidad de León, Ponferrada, León, Spain

4.1 Introduction to wildfire monitoring and modeling: sources, products, and errors

4.1.1 Sources of remote sensing data and major applications in wildfire monitoring and modeling

Remote sensing has become an indispensable tool for wildfire monitoring, providing systematic, repeatable, and large-scale observations that are often unattainable through ground-based methods alone. A wide variety of remote sensing data sources are now available, each with distinct and potentially complementary capabilities. The suitability of a particular dataset for a given monitoring objective ultimately depends on balancing costs (including acquisition, personnel, processing, and computational resources) against the expected benefits (notably in terms of product performance, which is central to this chapter). These cost–benefit considerations often translate into choices about the technical characteristics of the data. Specifically, three parameters usually govern the efficiency of a dataset in wildfire science: the modality of the sensor (passive vs active), the part of the spectrum it measures (optical, thermal infrared [TIR], or microwave), and its resolution regimes (spatial, temporal, spectral, and radiometric) (Chuvieco, 2008; Chen et al., 2024).

Passive sensors detect natural radiation, either reflected sunlight or emitted energy. They include optical systems, which operate in the visible, near-infrared (VNIR), shortwave infrared (SWIR), and TIR domains, as well as microwave radiometers that sense natural microwave emissions, although the latter are not common in wildfire science (Chen et al., 2024; Lee et al., 2024). Active sensors such as radar and light detection and ranging (LiDAR), by contrast, emit their own signal and record its interaction with the surface (Andersen et al., 2005; Campbell, 2002; Meyer, 2019). They can retrieve information on the three-dimensional structure of fuels (Fernández-Guisuraga et al., 2025; Monzón-González et al., 2025), and radar has the additional advantage of being able to partially penetrate smoke and clouds, thus increasing the continuity of observations during fire events and in cloudy regions (Tanase et al., 2020).

The different resolution regimes are equally important. Spatial and temporal resolution are typically coupled, since coarse-resolution sensors (from 100 m to 1 km

Advances in Remote Sensing for Forest Fire Management. DOI: https://doi.org/10.1016/B978-0-443-41692-7.00004-4
© 2027 Elsevier Ltd. All rights are reserved, including those for text and data mining, AI training, and similar technologies.

according to recent references such as Chen et al. (2024) or Pleniou and Koutsias (2025)) generally provide frequent revisits, moderate-resolution sensors (about 30 m) offer a balance between detail and revisit suitable for event- and regional-scale analyses, and high or very-high-resolution systems (<5 m) enable site-specific mapping, often with restrictions in spatial or temporal coverage. Spectral resolution, defined by the number, width, and placement of spectral bands (Chuvieco et al., 2007; Campbell et al., 2022), determines the degree to which different materials can be spectrally distinguished. Panchromatic sensors acquire a single broad band with high spatial detail but limited spectral information. Multispectral sensors provide a handful of relatively broad bands, whereas hyperspectral sensors acquire hundreds of narrow, contiguous bands that improve material discrimination, although at greater cost and complexity (Lewis et al., 2007; Veraverbeke et al., 2018; Quintano et al., 2023). Radiometric resolution, which reflects the quantization level and signal-to-noise ratio, further influences the ability to detect subtle changes. Narrower bands and a higher number of them can only provide superior information if radiometric precision is sufficient to distinguish fine contrasts such as moisture or ash-related changes (Chuvieco, 2008; Campbell et al., 2022).

Among all these data sources, the most used are optical imagery from passive sensors, so we will focus on it in this chapter. Structured in terms of spatial resolution, we can highlight several passive sensors central to applications in wildfire science, spanning the fire event timeline: monitoring and modeling related to prefire conditions, active-fire detection, burned area, burn severity, and postfire situations and trajectories (Chen et al., 2024). In the passive optical domain, Moderate Resolution Imaging Spectroradiometer (MODIS) (aboard Terra and Aqua), Visible Infrared Imaging Radiometer Suite (VIIRS) (aboard Suomi National Polar-orbiting Partnership and NOAA-20), or Ocean and Land Colour Instrument (OLCI), and Sea and Land Surface Temperature Radiometer (SLSTR) (aboard Sentinel-3) are examples of sensors providing coarse-resolution imagery in the VNIR, SWIR, and TIR domains, with daily or subdaily coverage. Thus, the use of these sensors has allowed the assessment of prefire fuel conditions (Yebra et al., 2008), the detection of active fires (Schroeder et al., 2014; Giglio et al., 2016), or the analyses of the global patterns and trends in burned area (Andela et al., 2017), intensity (Cunningham et al., 2024), and severity (Fernández-García and Alonso-González, 2023). Moreover, they have improved our understanding of fire regimes across the globe (Archibald et al., 2013) and facilitated the estimation of carbon emissions due to fire (van der Werf et al., 2017) and postfire recovery (Maillard et al., 2025) among many other aspects.

At finer spatial resolution, imagery from the Thematic Mapper (TM), Enhanced Thematic Mapper (ETM+), and OLI/TIRS sensors from Landsat missions (30 m for most bands), and from MultiSpectral Instrument (MSI) on board Sentinel-2 (20 m for most bands) has become the reference for regional burned-area mapping, including many small fires and fires with complex spatial patterns usually missed in coarse-resolution data (Roteta et al., 2019; Hawbaker et al., 2020; Gaveau et al., 2021; Fernández-García et al., 2024). They are also the most common imagery used to map burn severity through a variety of methods (Miller et al., 2009; Quintano et al., 2013; Fernández-García et al., 2018; García-Llamas et al., 2019a; Howe et al., 2022), as well as for monitoring postfire situations and trajectories (Huerta et al., 2022; Viana-Soto et al., 2022). Both aspects, burn-severity assessments and postfire trajectories, are critical for identifying priority areas for postfire restoration.

Although less commonly used due to acquisition and computational costs, commercial missions such as PlanetScope (~3–5 m) or WorldView (≤2 m) offer extended capabilities related to their higher spatial resolution. This enables mapping of burned area (Stroppiana et al., 2022) and severity at finer spatial detail (Warner et al., 2016; Hamilton et al., 2023), as well as the development of ecological models requiring greater spatial detail, such as those in spatially heterogeneous areas (e.g., Fernández-Guisuraga et al., 2021; Simioni et al., 2022) or working with small elements (e.g., trees or shrubs) (Fernández-Guisuraga et al., 2019). Sensors on aircraft and unmanned aerial vehicles (UAVs) offer very high spatial resolution but are especially suited for site-specific analyses due to limited spatial coverage (Chen et al., 2024). UAV panchromatic and multispectral imagery is increasingly used for assessing burn-severity indicators (Beltrán-Marcos et al., 2021), postfire soil properties (Beltrán-Marcos et al., 2023), or early vegetation regeneration in specific fire events (Pádua et al., 2019) in areas of special research or management interest.

4.1.2 Maps, models, and the need for validation

The common goal of using optical imagery from passive sensors in wildfire science is to derive two-dimensional geospatial products (maps) that represent observed surface properties. Such maps may be obtained directly from spectral data via simple calculations or through the application of models.

Models using optical imagery from passive sensors generally fall into three main categories (Verrelst et al., 2015): (1) Physical models, such as radiative transfer models (RTMs), describe energy interactions with the atmosphere and surface using physical laws (Chuvieco et al., 2007; Fernández-Guisuraga et al., 2025). (2) Empirical or statistical models instead rely on calibration with reference data to learn relationships between predictor variables (e.g., spectral bands, indices, ancillary data, etc.) and target variables (e.g., burned area, burn severity, etc.). These can be built using learning approaches such as regression or classification models (Miller et al., 2009; Fernández-García et al., 2018; Beltrán-Marcos et al., 2023). Spectral unmixing techniques are often regarded as semiempirical approaches with a physical basis, since they rely on linear mixing assumptions grounded in physics but are implemented through statistical decomposition (Quintano et al., 2013; Fernández-Guisuraga et al., 2025). (3) Finally, hybrid models combine both perspectives, for instance, applying a physically based index but calibrating it statistically against reference measurements (e.g., hybrid RTM; Fernández-Guisuraga et al., 2021). In wildfire science, models have also been classified according to the fire event timeline: susceptibility or risk models, fire ignition and behavior models, fire severity models, and ecosystem response models. In addition, the fire regime has also been subject to modeling (Seidl et al., 2011).

Maps and model outputs can be either categorical or continuous. Categorical products divide the landscape into discrete classes (e.g., burned versus unburned areas, categories of burn severity, or classes of fuel type), whereas continuous products assign each pixel a numerical value (e.g., continuous values of fire radiative power [FRP], burn severity, percentage of ash, or postfire tree density). In any case, these spatial products are rarely perfect due to errors from multiple sources (Olofsson et al., 2014; Duncanson et al., 2021). In the case of those based on the information from passive sensors, errors include (1) those related to the imagery and auxiliary data itself, which are limited in terms of spatial, temporal, spectral, and radiometric resolution. In addition, imagery may have uncertainties in positioning, as well as sensor and atmospheric noise; (2) errors arising during the mapping process, including preprocessing errors, the capacity of the mapping algorithm, the training quality of the algorithm, and adequacy to the mapped region. In the case of categorical maps, we can add the potential errors coming from the number and separability of classes; (3) and errors linked to the intrinsic characteristics of the target region, such as persistency of cloudiness, complexity of background spectral properties, and spatio-temporal variability in landscape and fire patterns, including the persistence of burn signals over time, which can further complicate product performance (Lunetta et al., 1991; Powell et al., 2004; Franquesa et al., 2022a; Fernández-García and Kull, 2023).

The magnitude and type of errors vary across products, affecting their reliability and interpretation. Thus, the assessment of the performance of these remote sensing products, called validation, is essential to understand the confidence in their use (Congalton, 2001; Olofsson et al., 2014). This validation is generally conducted by comparing the results on the map with reference data through sampling a portion of the population (a limited number of locations and times).

4.2 Sampling design for validation purposes

The sampling design defines the procedure for choosing a statistically representative subset of the population from both the product and the reference data, enabling the calculation of validation metrics. Good practice recommendations, generally included in National Aeronautics and Space Administration (NASA) and European Space Agency (ESA) validation handbooks (Boschetti et al., 2009; Olofsson et al., 2014; Guillevic et al., 2018; Duncanson et al., 2021; ESA, 2025), encourage probability-based sampling approaches. These include simple random, stratified random, or systematic sampling, which ensure unbiased validation estimates while enabling the calculation of uncertainty estimates. Stratification is particularly important when classes of interest occupy only a small fraction of the study area or, at regional scales, when they are spatially heterogeneous. This is often the case for burned areas, where stratified sampling has demonstrated better validation estimates (Boschetti et al., 2016; Padilla et al., 2017). For burned-area validation, strata are commonly defined by combining ecological factors (e.g., biomes) with fire incidence patterns derived from coarse-resolution datasets, and stratification can also be applied to the temporal dimension (Giglio et al., 2018). For continuous products, it is also recommended to sample a representative range of conditions, which can be facilitated with stratification (e.g., Guillevic et al., 2018).

Sampling units in remote sensing product validation can be defined at the point level, the object level (e.g., individual trees), or as area-based units (contiguous spatial areas or regions). Point-level validation compares individual map pixels with reference data, often collected in situ or from very high-resolution imagery, using plots that typically match or approximate the size of the pixels in the map. Nonetheless, it is recommended not to use single-pixel units when they cannot be properly located on the ground and in the imagery (Congalton, 2001). This approach is common for continuous variables such as burn-severity indices or fractional vegetation cover, but it can also be applied to categorical data. In contrast, area-based validation focuses on the agreement between polygons or regions and is generally applied to validate categorical maps, such as burned-area products (Giglio et al., 2018; Boschetti et al., 2019; Franquesa et al., 2022a).

The number of sampling units to be established should balance what is statistically sound and what is practically attainable. For categorical maps, a generally accepted rule of thumb is to use a minimum of 50–100 sampling units for each category (Congalton, 1991; Congalton and Green, 2019). However, to estimate an appropriate sample size (n) that meets desired confidence and error margins, several formulas are used (Chuvieco, 2008; Olofsson et al., 2014). The most common are those based on Cochran's formulas (Cochran, 1977), expressed in their large-population form (infinite population), since the population in remote sensing products (number of pixels or polygons) is often very large (Table 4.1).

Once the overall sample size is determined, samples must be allocated. In the case of categorical maps with stratified designs, several strategies have been proposed, including (1) proportional allocation that assigns samples according to the area of each class, which can leave rare classes underrepresented; (2) equal allocation, which assigns the same number of samples to each class, supporting class-level accuracy but at the expense of overall estimates; (3) optimal (Neyman) allocation, which minimizes variance but is difficult to apply when multiple accuracy objectives are of interest. Olofsson et al. (2014) suggest a practical compromise in which a sufficient number of sample units (e.g., 50–100) is first guaranteed for rare classes, while the remaining samples are distributed proportionally to class area. This procedure can be iterated until the anticipated standard errors of the class-level and overall accuracy (OA) estimates are satisfactory. The allocation scheme influences the precision of these estimates but not their unbiasedness.

A critical aspect of sampling design for map and model validation is spatial autocorrelation, the tendency for values in geographically close locations to be more similar than those farther apart (Congalton, 2001). Ignoring spatial autocorrelation can lead to over or underestimation of product performance, particularly when reference data are clustered or when models exploit spatially structured variables (Ploton et al., 2020), which is usually the case for burned areas. Metrics such as Moran's I or semivariograms can quantify spatial dependency in residuals between predicted (or mapped) and reference values, guiding sampling design and validation strategies (e.g., García-Llamas et al., 2019b). Possible remedies include using

Table 4.1 Formulas based on commonly used methods to estimate the sample size for validating categorical and continuous products using simple or stratified random designs.

Variable type	Sampling design	Formula	Terms
Categorical	Simple random	$n = \frac{Z^2 O(1-O)}{d^2}$	*Z*: *Z*-score for the desired confidence level *O*: expected overall accuracy *d*: tolerated error (half-width of CI) in proportion units
Categorical	Stratified random	$n = \frac{(\Sigma WiSi)^2}{S(\hat{O})^2}$	*Wi*: mapped proportion of stratum *i*; *Si*: std. dev. in stratum *i* ($Si=\sqrt{(Ui\,(1-Ui))}$); *Ui*: user's accuracy expected for stratum *i* $S(\hat{O})$: target standard error of overall accuracy
Continuous	Simple random	$n = \frac{Z^2 s^2}{L^2}$	*Z*: *Z*-score for the desired confidence level; s^2: estimated variance of variable; *L*: tolerated error (half-width of CI) in the units of the continuous variable
Continuous	Stratified random	$n = \frac{(\Sigma WiSi)^2}{S(\bar{y})^2}$	*Wi*: proportion of population in stratum *h*; *Si*: std. dev. in stratum i; $S(\bar{y})$: target standard error of overall mean

* Expected overall accuracy (*O*), standard deviation in strata (*Si*), and estimated variance (s^2) can be obtained from prior studies or pilot data. The calculation of *O* and *Ui* is explained in section 4.4.4.
Cochran, W.G., 1977. Sampling techniques, 3rd Ed., John Wiley & Sons, p. 428.

spatially balanced sampling designs (e.g., Generalized Random Tessellation Stratified [GRTS]; Burkle et al., 2015) or applying block bootstrapping and spatial cross-validation to account for spatial dependence when estimating validation metrics (e.g., Benito, 2021; Fernández-García et al., 2025).

4.3 Reference data for validation in wildfire monitoring and modeling

4.3.1 Defining reference data

Reference data (also referred to as validation data or ground truth) are observations or measurements that approximate reality and serve as a benchmark for assessing the performance of maps and models. The requirements for reference data differ depending on the sampling strategy. Thus, point-level validation demands very high positional and temporal accuracy, while area-based validation also requires spatial completeness and representative coverage. According to best practices (Olofsson et al., 2014; NASA-CEOS-LPV, 2025), valid reference data must satisfy several conditions: independence (collected separately from the product being validated), relevance (measuring the same variable as the product), representativeness (capturing the variability of the phenomenon across space, time, and classes or ranges of the mapped variable), and quality (being demonstrably more reliable than the product under evaluation).

In wildfire monitoring and modeling, spatial representativeness and quality are especially sensitive to temporal alignment between mapped and reference data (Franquesa et al., 2022b). This is critical for active fire, burned area, and severity validation, since the spectral signal of fire, as well as the changes caused by fire, can be short-lived. Finer spatial resolution does not necessarily ensure a better approximation of spatial reality, as other factors should also be considered (temporal, spectral, or radiometric resolution, geolocation, or collection and interpretation protocols). For example, Landsat imagery (30 m, 16-day revisit) has been shown to underestimate burned area in much the same way as MODIS (500 m, near-daily revisit) in persistently cloudy tropical regions, where cloudiness limits the expected advantage of Landsat's finer resolution (Fernández-García et al., 2024).

In parallel, reference data used for temporal validations (i.e., date validation) may be affected by spatial accuracy. In this sense, although geostationary satellites provide more continuous monitoring (a desirable feature for such applications), their spatial resolution makes them less suitable than MODIS or VIIRS active fires, which offer subdaily revisits and higher spatial resolution for validating the timing of burned area from MODIS, Landsat, or Sentinel-2 (Giglio et al., 2018; Roteta et al., 2019; Chuvieco et al., 2022). Thus, different reference data can be used for different validation purposes (e.g., spatial versus temporal validations).

4.3.2 Common methods and relevant datasets

Although field observations can be considered a form of proximal sensing (our eyes and certain measuring tools being, in essence, short-range remote sensors), it is useful to distinguish between two main sources of reference data: field-collected measurements and remotely sensed observations from airborne or satellite platforms. In remote sensing applications, reference data for model development and validation are often derived from the same collection effort but should be partitioned into independent subsets to avoid bias.

4.3.2.1 Field-collected reference data

Field-collected data are considered paradigmatic reference data and typically use field plots and standardized protocols, which allow reproducibility and cross-study comparability. The primary strengths of field-collected data lie in their high reliability, thanks to the direct measurement of ground conditions. However, the acquisition is generally costly and time-consuming. This leads to limitations in spatial coverage, with allocation of sampling units usually restricted to accessible areas, as well as limitations in temporal coverage, potentially leading to scarce representativeness and temporal mismatch (e.g., Key and Benson, 2006; Lewis et al., 2007; Miller et al., 2009). Thus, field reference data are more often used in validating prefire conditions (Fernández-Guisuraga et al., 2025), burn severity (Fernández-

García et al., 2018; Howe et al., 2022; Monzón-González et al., 2025) and postfire variables (Pádua et al., 2019; Beltrán-Marcos et al., 2023) than in validating active fires, burned-area data, or models requiring wide spatio-temporal training.

Notably, in the prefire context, forest inventory data provide baselines against which to build or validate fuel maps and models, risk models, and fire-behavior predictions (Reeves et al., 2009; Puliti et al., 2020; Fernández-Guisuraga et al., 2025; Frank et al., 2025). Among them, the long-term national forest inventories are particularly noteworthy. These are generally plot-based, statistically designed, and remeasured on multiyear cycles, providing rich detail on ecosystem structure and composition. This makes them especially powerful for being used as a reference for live and dead biomass, stand structure, canopy attributes, and, in many programs, surface and ladder fuels. Examples include the US Forest Service's Forest Inventory and Analysis (FIA) (USFS, 2025), Canada's National Forest Inventory (NFIS, 2025), the national inventories in most European countries, such as Spain (IFN, MITECO, 2025) or Sweden (NFI, SLU, 2025), as well as forest inventories in other regions, including Mexico (INFyS; CONAFOR, 2025) or Australia (NFI ABARES, 2025).

Burn-severity reference data are most often collected with standardized field protocols that integrate multiple indicators across strata. The Composite Burn Index (CBI) is the de facto standard (Key and Benson, 2006; Miller et al., 2009; MTBS, 2025), along with its adaptations (De Santis and Chuvieco, 2009; Fernández-García et al., 2018). In essence, CBI approaches score the changes caused by fire in soil/substrate and vegetation strata on a 0–3 scale in 30 m diameter plots, yielding site, vegetation, and soil severity metrics. These metrics are then used for both validation and calibration (understood here as the generation of categories through regression-informed thresholding) of remotely sensed severity (usually spectral indices such as differenced Normalized Burn Ratio [dNBR], Relative dNBR [RdNBR], or dNBR-EVI). As originally conceived, "initial assessments" target the first months postfire, whereas in "extended assessments" CBI is estimated in the first postfire growing season and includes indicators of delayed mortality and ecosystem responses (Key and Benson, 2006). Importantly, field reference for burn severity is neither limited to CBI nor to 30 m plots. Tree mortality fractions (e.g., percentage change in tree basal area or percent change in tree canopy cover) have been used explicitly as validation data for severity (Miller et al., 2009). Likewise, soil-focused indicators (such as ash depth/cover, fine and coarse debris cover, or unstructured soil depth), either combined or used individually, provide robust ground references for the substrate component of severity (Beltrán-Marcos et al., 2021). Among the most relevant field burn-severity reference datasets are the USGS/USDA CBI data for the Conterminous United States (Picotte et al., 2019), which are point-level CBI plots collected from 1994 to 2018 across the United States, compiled from multiple agencies for calibrating and validating spectral indices. Reference burn-severity data is not limited to field data. Thus, Landsat burn-severity spectral indices have been used as reference to check the performance of these indices in the global burn severity product MOSEV, obtained from MODIS (Alonso-González and Fernández-García, 2021), expert-interpreted severity maps from Sentinel-2 imagery are sometimes considered as a reference (Copernicus CEMS/European Forest Fire Information System [EFFIS] burn-severity maps; EFFIS, 2025), and orthophotography-based assessments have been considered as reference data for crown damage (Thompson and Spies, 2009).

Third, numerous field studies target specific postfire conditions and trajectories, providing independent references for validating maps and models of recovery, erosion risk, or fuel re-accumulation. In this case, project-related inventories are the most common source for reference field data. For example, four years after three megafires along an Atlantic–Transition–Mediterranean gradient, Fernández-Guisuraga et al. (2019) used 120 2×2 m field plots in each fire to measure pine seedling density, woody cover, and dead plant material. This dataset served as the reference to validate the corresponding models derived from WorldView-2 at 2 m. Pádua et al. (2019) used field measurements of tree height in 35 individuals to perform an object-based validation of segmented trees obtained from UAV images. Beyond vegetation, postfire soil properties such as topsoil organic carbon from smaller plots (0.5 × 0.5 m) have also been used to train and validate UAV-derived models at 0.5 m (Beltrán-Marcos et al., 2023).

4.3.2.2 Remotely sensed reference data

Remote sensing data can also be used as a reference because, in certain cases, it can be of higher quality than field data due to spatial or temporal completeness and coincidence, positioning accuracy, or systematic acquisition, among other factors. This is generally the case for active fire and burned area, as well as for some modeling approaches.

Active-fire products from MODIS (MOD14 and MYD14, 1 km) and VIIRS (VNP14 at 750 m and VNP14IMG at 375 m) are among the most widely used to flag hotspots associated with ongoing combustion as well as to retrieve FRP, which is proportional to the rate of energy release. However, the acquisition of reference data for these products is challenging due to the dynamic, ephemeral, and hazardous nature of fires (Ramsey et al., 2024). On the one hand, for hotspot detection, the standard validation approach compares MODIS/VIIRS fire pixels with near-simultaneous, higher-resolution imagery. For example, validation exercises of Terra MODIS (MOD14) used coorbital imagery at 30m from the advanced spaceborne thermal emission and reflection radiometer (ASTER) as reference data (Giglio et al., 2016), whereas earlier regional studies combined reference data from ASTER, Landsat, and fieldwork (Schroeder et al., 2008). In addition, comparisons between both MODIS and VIIRS hotspots have also provided information on the detection capacities and limitations of both products (Csiszar et al., 2014; Schroeder et al., 2014). On the other hand, reference data for FRP validation is also very limited, as apart from coincident, measurements should be thermally unsaturated, ideally ground-based or airborne. Consequently, MODIS FRP has been validated through opportunistic checks using airborne thermal imagers and Bispectral Infrared Detector (BIRD) satellite data (Ruecker et al., 2013). Intercomparisons between VIIRS and MODIS FRP have also been conducted to assess product consistencies (Li et al., 2018).

Focusing on reference burned area, it has been reported that remotely sensed data are preferred to field data, not only due to costs but also because of accuracy and spatial completeness issues. In general, field-delineated perimeters may miss fires or isolated unburned islands that are better detected by moderate-resolution images of about 30 m (Flasse et al., 2004). For the spatial validation of burned-area products derived from coarse-resolution sensors like MODIS, good practices recommend building reference burned-area data based on higher-resolution satellites (e.g., Landsat or Sentinel-2). These practices recommend the use of pairs of images (before and after the fire) rather than single scenes or mosaics to improve temporal alignment and reliability (Boschetti et al., 2016; Padilla et al., 2017; Giglio et al., 2018; Boschetti et al., 2019; Franquesa et al., 2022a). In practice, the reference perimeters are produced in area-based units where the complete sampling region is mapped, not just the scars. The mapping for reference data often uses semiautomatic classifiers trained on the pre/postfire pair (often Random Forests over postfire reflectance plus pre-to-post differences), followed by rigorous visual editing and documentation of unburned and unobserved categories (Roteta et al., 2021). Moreover, Franquesa et al. (2022b) emphasized long temporal reference units rather than the interval between only consecutive scenes to reduce confounding between spatial and dating errors in global burned-area validation. When it comes to the spatial validation of burned-area products from Sentinel-2 and Landsat, a recurrent challenge is the limited availability of accessible imagery with enough spatial detail and temporal frequency to be used as reference. Thus, validations often rely on reference data derived from the other sensor (Roteta et al., 2019), or from the same one but generated through more robust approaches, following the good practices mentioned above (Chuvieco et al., 2022; Fernández-García et al., 2024). Nevertheless, there are some validation exercises using reference burned area from higher-resolution imagery (Vanderhoof et al., 2017; Stroppiana et al., 2022). The largest publicly available dataset of reference burned-area data is the Burned Area Reference Database (BARD) (Franquesa et al., 2020), which is updated and available at Franquesa et al. (2025).

Temporal assessments of burned area are often conducted separately from spatial validations, but they take advantage of the reference perimeters used in the spatial assessment. MODIS (MOD14/MYD14) and VIIRS (VNP14/VNP14IMG) active-fire products are the primary references for temporal validations. Active-fire pixels falling within (or immediately adjacent to) the reference scar are collated, and their detection times in MODIS or VIIRS are used to define the bounds for the burn date. These timestamps are then compared with the burned-area product's reported burn date (Giglio et al., 2018; Roteta et al., 2019; Chuvieco et al., 2022).

Other models relying on burned-area data often require a wide spatial and/or temporal training, thus using reference data from remote sensing sources (e.g., Mermoz et al., 2005; Fernández-García et al., 2022; Fernández-García and Kull, 2023). A case where the need for spatial and temporal breadth and completeness is particularly acute is in models predicting fire regimes or their attributes. An example is found in Fernández-García et al. (2025), who used fire frequency, seasonality, and burned-area size data calculated from seven years of MODIS and Sentinel-2 imagery as reference to validate predictive models of these fire regime attributes.

4.4 Validation analysis of maps and models

4.4.1 Validation analysis for categorical and continuous products

Validation in wildfire monitoring is conducted through the quantification of metrics that can be of different types. These include accuracy or error metrics (measuring the closeness of the product estimates to the reference), association/fit metrics (strength of the relationship between the product estimates and the reference), uncertainty (how much the estimates could reasonably vary), and less frequently, precision (understood as repeatability; this is the closeness among repeated estimates under the same conditions) (Guillevic et al., 2018; Duncanson et al., 2021). The procedures and metrics for validation differ depending on whether the map represents discrete categories or continuous variables, requiring distinct metrics and interpretation approaches.

The basis for evaluating categorical maps is the error (or confusion) matrix that compares the mapped classes against reference data (Congalton, 2001; Congalton and Green, 2019). From this matrix, several metrics can be calculated (Table 4.2). OA expresses the proportion of all correctly classified samples, providing a simple and intuitive measure of performance. However, OA can be misleading if classes are unbalanced, so class-specific metrics are also important. In this sense, producer's accuracy (PA) measures the probability that a reference pixel of a given class has been correctly mapped, and is the complement of the errors of omission, whereas user's accuracy (UA) indicates the probability that a pixel mapped as belonging to a given class truly corresponds to that class in the reference data, and is the complement of the errors of commission. The Dice coefficient (or Sørensen–Dice index) provides a class-specific measure of spatial overlap between map and reference data, and is increasingly used in assessing the accuracy of binary maps, such as those of burned versus unburned (Franquesa et al., 2022a; Fernández-García et al., 2024). Some studies also report the kappa coefficient, which estimates agreement beyond chance. However, kappa has been increasingly criticized, and current recommendations favor reporting overall and class-specific accuracies (PA, UA, Dice) accompanied by confidence intervals to characterize statistical uncertainty in both accuracy and area estimates (Olofsson et al., 2014).

For continuous maps, the mean error, or bias, captures systematic deviation by preserving the sign of differences between mapped and reference values, indicating whether the map tends to over or underestimate reality on average. It can be divided by the mean observed value in the reference data to calculate the relative bias (rBias), which is more comparable across different scales and units. In contrast, the mean absolute error (MAE) removes the sign, thus reflecting the average magnitude of deviations regardless of direction. The root mean square error (RMSE) also removes the sign but squares the deviations before averaging, which gives more weight to larger errors, making it more sensitive to outliers than MAE. To enable comparison across datasets or variables expressed in different units or scales, errors are often relativized. A common approach is the relative RMSE (rRMSE) that expresses error as a percentage of the mean observed value in the reference data. The coefficient of

Table 4.2 Common validation metrics for categorical and continuous maps.

Variable type	Metric	Formula	Interpretation
Categorical	Overall accuracy	$\frac{\sum_{i=1}^{K} TPi}{N}$	Proportion of reference samples correctly classified in the map.
	Producer's accuracy	$\frac{TPi}{TPi + FNi}$	Proportion of reference samples of class i that are correctly mapped.
	Omission error	$\frac{FNi}{TPi + FNi}$	Proportion of reference samples of class i that are mapped as another class.
	User's accuracy	$\frac{TPi}{TPi + FPi}$	Proportion of samples mapped as class i that are class i in the reference data.
	Commission error	$\frac{FPi}{TPi + FPi}$	Proportion of samples mapped as class i that belong to another class in the reference data.
	Dice coefficient	$\frac{2TPi}{2TPi + FPi + FNi}$	Spatial overlap between the mapped and reference samples of class i.
	Kappa coefficient	$\kappa = \frac{Po - Pe}{1 - Pe}$	Agreement between mapped and reference classes after correcting for agreement expected from the marginal class distributions.
Continuous	Bias (mean error)	$\frac{1}{N}\sum(y\_map - y\_ref)$	Average signed difference between mapped and reference values (systematic error).
	Relative Bias	$\frac{\text{Bias}}{\bar{y}\_ref} \times 100$	Average signed error expressed as a percentage of the mean reference value.
	Mean absolute error (MAE)	$\frac{1}{N}\sum \lvert y\_map - y\_ref\rvert$	Average absolute difference between mapped and reference values.
	Root mean square error (RMSE)	$\sqrt{\frac{\sum(y\_map - y\_ref)^2}{N}}$	Square root of the average squared difference between mapped and reference values, penalizes large errors more strongly than MAE.
	Relative RMSE	$\frac{\text{RMSE}}{\bar{y}\_ref} \times 100$	RMSE expressed as a percentage of the mean reference value.
	R^2 (coefficient of determination)	$1 - \frac{\sum(y\_map - y\_ref)^2}{\sum(y\_ref - \bar{y}\_ref)^2}$	Proportion of reference variance accounted for by the map.

Notation: K, indexes classes; *TPi*, True positives for class i; *FPi*, false positives for class i; *FNi*, false negatives for class i; *TNi*, true negatives for class i; *y_map*, mapped value; *y_ref*, reference value; $\bar{y}$*_ref*, mean of reference values; N, number of observations; *Po*, observed agreement; *Pe*, expected agreement by chance.

determination (R^2) is a fit metric that indicates the proportion of variance in the reference data that is explained by the mapped estimates, thus reflecting the strength of association between them (Table 4.2). In practice, using a combination of these metrics allows for a more nuanced understanding of map or model performance, highlighting not only the average size of errors but also whether the map systematically over or underpredicts (e.g., Beltrán-Marcos et al., 2023; Fernández-García and Kull, 2023; Fernández-García et al., 2024).

Spatial representations of uncertainty in the map or model estimates help highlight areas where mapped estimates are more or less reliable. These uncertainties are expressed in multiple ways. For example, some burned area and severity products, such as the MCD64A1 or MOSEV, include a band indicating the uncertainty in burn date, expressed in days (Giglio et al., 2020; Alonso-González and Fernández-García, 2021), whereas fraction images from spectral unmixing, indicative of burn severity, are often accompanied by an error band reflecting the model performance on a per-pixel basis through RMSE (Quintano et al., 2013).

In addition, an increasingly recommended fact is to provide uncertainty estimates for the accuracy metrics (e.g., RMSE or R^2) (Olofsson et al., 2014). This is commonly addressed by providing confidence intervals or other variability metrics around the accuracy metrics, which indicate the statistical reliability of the estimates. Resampling and repetition-based approaches, such as bootstrapping or Monte Carlo techniques, are widely used to quantify variability in these metrics, especially when reference data are limited (Duncanson et al., 2021; Fernández-García et al., 2025).

When using stratified sampling to distribute the reference data, reference points may be forced to fall into rare classes (like burned area, which may cover only a small fraction of the total landscape). In this case, the sample may not be proportional to the true map distribution, and appropriate inference methods are needed to extrapolate from the sample to the population. Specifically, weighting is used to correct for the difference between the sample distribution and the true map distribution. Each reference unit (pixel or polygon) is assigned a weight proportional to the proportion of the population it represents, and when computing accuracy metrics, each sampled unit's contribution is scaled by its weight, ensuring the results are unbiased and representative of the full mapped area (Olofsson et al., 2014; Boschetti et al., 2016; Padilla et al., 2017).

4.4.2 Validation of model performance

Validation of models in remote sensing focuses on assessing how well the model's predictions match reality, and in practice, it almost always involves comparing model outputs (predictions that can generally be represented in the form of maps) to reference data using the metrics mentioned in the previous heading, depending on the type of model output (categorical or continuous). However, some particularities for model validation should be highlighted, depending on the model type.

The validation of physical models typically involves comparing model outputs, such as simulated surface reflectance, with field or airborne spectroradiometer measurements collected under comparable conditions. Sensitivity analyses are often performed to evaluate how variations in input parameters (e.g., leaf area index, soil properties, atmospheric conditions) affect model outputs, ensuring that the model behaves consistently with physical expectations. Additionally, RTMs can be validated by applying them across multiple sites or sensors to confirm the robustness of their predictions, or by comparing derived biophysical variables (e.g., burn severity, vegetation fraction) against independent measurements (Verrelst et al., 2015). For empirical and hybrid models, which are trained on reference data, performance at a given site and time depends heavily on the data used for training, and validation can be framed in terms of internal, interpolation, and extrapolation assessments (Fernández-Guisuraga et al., 2019; Fernández-García et al., 2022).

Internal validation examines the model's fit to the training data itself. This provides information on internal consistency or the degree of association between variables (e.g., RMSE or R^2 from a linear model between dNBR and CBI), but fit does not guarantee predictive accuracy on new data due to risks of overfitting (see example in García-Llamas et al., 2019a; Howe et al., 2022).

Interpolation validation evaluates the model's predictive ability within the range of conditions represented during training but applied to independent observations, either spatially (e.g., applying the model to neighboring areas with similar vegetation) or temporally (e.g., testing performance in subsequent years). When independent reference data are limited, interpolation performance is often estimated through resampling strategies such as cross-validation approaches (Olofsson et al., 2014). Interpolation analyses can be conducted with data relatively close to the training data (e.g., using independent data from the same wildfire or using out-of-bag error measurements; García-Llamas et al., 2019c), or with data relatively more distant from the training dataset (e.g., using validation data from independent wildfires within the same region; Miller et al., 2009).

Extrapolation validation, in contrast, tests the model's robustness when applied beyond the training domain, such as transferring a model calibrated in one ecosystem to a different type of ecosystem, region, biome, or a distant epoch (Yu et al., 2024). This last step is critical for assessing model transferability and determining the limits of predictive reliability across diverse landscapes and environmental conditions (Ploton et al., 2020). Assessments of extrapolation performance can be found in Fernández-García et al. (2018, 2022, 2025).

4.4.3 Maturity levels in validation

Leading space agencies, such as NASA and ESA, have defined protocols to validate their products. In general, both agencies structure validation stages into formal maturity levels, reflecting the rigor and extent of validation applied to a product (ESA, 2025; NASA, 2025) that also applies to fire data (e.g., Stroppiana et al., 2021). Both agencies differentiate Beta products (not validated) from validated products with four validation stages (ESA, 2025; NASA, 2025):

Stage 1: Accuracy estimated using a small number of independent reference measurements (typically <30).

Stage 2: Accuracy assessed over a broader set of locations and time periods with in situ or other reference data; spatial and temporal consistency evaluated.

Stage 3: Uncertainties quantified robustly across multiple locations and periods representing global conditions; spatial and temporal consistency confirmed.

Stage 4: Stage 3 validation results are systematically updated as new product versions and extended time series become available.

AI disclosure

During the preparation of this work, the author(s) used OpenAI in order to review writing errors and writing quality in general. After using this tool/service, the author (s) reviewed and edited the content as needed and take(s) full responsibility for the content of the publication.

References

ABARES, 2025. Australia's National Forest Inventory. Department of Agriculture, Fisheries and Forestry ABARES. Australian Government Available at https://www.agriculture.gov.au/abares/forestsaustralia/australias-national-forest-inventory.

Alonso-González, E., Fernández-García, V., 2021. MOSEV: a global burn severity database from MODIS (2000–2020). Earth System Science Data 13 (5), 1925–1938. Available at https://doi.org/10.5194/essd-13-1925-2021.

Andela, N., Morton, D.C., Giglio, L., Chen, Y., van der Werf, G.R., Kasibhatla, P.S., et al., 2017. A human-driven decline in global burned area. Science (New York, N.Y.) 356 (6345), 1356–1362. Available at https://doi.org/10.1126/science.aal4108.

Andersen, H.-E., McGaughey, R.J., Reutebuch, S.E., 2005. Estimating forest canopy fuel parameters using LIDAR data. Remote Sensing of Environment 94 (4), 441–449. Available at https://doi.org/10.1016/j.rse.2004.10.013.

Archibald, S., Lehmann, C.E.R., Gómez-Dans, J.L., Bradstock, R.A., 2013. Defining pyromes and global syndromes of fire regimes. Proceedings of the National Academy of Sciences 110 (16), 6442–6447. Available at https://doi.org/10.1073/pnas.1211466110.

Beltrán-Marcos, D., Suárez-Seoane, S., Fernández-Guisuraga, J.M., Fernández-García, V., Pinto, R., García-Llamas, P., et al., 2021. Mapping soil burn severity at very high spatial resolution from unmanned aerial vehicles. Forests 12 (2), 179. Available at https://doi.org/10.3390/f12020179.

Beltrán-Marcos, D., Suárez-Seoane, S., Fernández-Guisuraga, J.M., Fernández-García, V., Marcos, E., Calvo, L., 2023. Relevance of UAV and sentinel-2 data fusion for estimating topsoil organic carbon after forest fire. Geoderma 430, 116290. Available at https://doi.org/10.1016/j.geoderma.2022.116290.

Benito, B.M., 2021. R package spatialRF: Easy Spatial Regression with Random Forest. https://doi.org/10.5281/zenodo.4745208.

Boschetti, L., Roy, D.P., Justice, C.O., 2009. International Global Burned Area Satellite Product Validation Protocol. Part I – production and standardization of validation reference data. LPVS/GSFC/NSAA. Available at https://lpvs.gsfc.nasa.gov/PDF/BurnedAreaValidationProtocol.pdf

Boschetti, L., Stehman, S.V., Roy, D.P., 2016. A stratified random sampling design in space and time for regional to global scale burned area product validation. Remote Sensing of Environment 186, 465–478. Available at https://doi.org/10.1016/j.rse.2016.09.016.

Boschetti, L., Roy, D.P., Giglio, L., Huang, H., Zubkova, M., Humber, M.L., 2019. Global validation of the collection 6 MODIS burned area product. Remote Sensing of Environment 235, 111490. Available at https://doi.org/10.1016/j.rse.2019.111490.

Burkle, L.A., Myers, J.A., Belote, R.T., 2015. Wildfire disturbance and productivity as drivers of plant species diversity across spatial scales. Ecosphere 6 (10), 1–14. Available at https://doi.org/10.1890/es15-00438.1.

Campbell, B.A., 2002. Radar Remote Sensing of Planetary Surfaces. Cambridge University Press, pp. 350.

Campbell, J.B., Wynne, R.H., Thomas, V.A., 2022. Introduction to Remote Sensing, 6th Ed. Guilford Press, pp. 634.

Chen, Y., Morton, D.C., Randerson, J.T., 2024. Remote sensing for wildfire monitoring: insights into burned area, emissions, and fire dynamics. One Earth 7 (6), 1022–1028. Available at https://doi.org/10.1016/j.oneear.2024.05.014.

Chuvieco, E., 2008. Teledetección ambiental. La observación de la Tierra desde el Espacio, 3rd Ed. Ariel Ciencia, pp. 594.

Chuvieco, E., De Santis, A., Riaño, D., Halligan, K., 2007. Simulation approaches for burn severity estimation using remotely sensed images. Fire Ecology 3 (1), 129–150. Available at https://doi.org/10.4996/fireecology.0301129.

Chuvieco, E., Roteta, E., Sali, M., Stroppiana, D., Boettcher, M., Kirches, G., et al., 2022. Building a small fire database for Sub-Saharan Africa from Sentinel-2 high-resolution images. Science of the Total Environment 845, 157139. Available at https://doi.org/10.1016/j.scitotenv.2022.157139.

Cochran, W.G., 1977. Sampling Techniques, 3rd Ed. John Wiley & Sons, pp. 428.

CONAFOR, 2025. Inventario Nacional Forestal y de Suelos. CONAFOR. Gobierno de México Available at https://snmf.cnf.gob.mx/infys/.

Congalton, R.G., 1991. A review of assessing the accuracy of classifications of remotely sensed data. Remote Sensing of Environment 37 (1), 35–46. Available at https://doi.org/10.1016/0034-4257(91)90048-b.

Congalton, R.G., 2001. Accuracy assessment and validation of remotely sensed and other spatial information. International Journal of Wildland Fire 10 (4), 321. Available at https://doi.org/10.1071/wf01031.

Congalton, R.G., Green, K., 2019. Assessing the Accuracy of Remotely Sensed Data. CRC Press. Available at https://doi.org/10.1201/9780429052729.

Csiszar, I., Schroeder, W., Giglio, L., Ellicott, E., Vadrevu, K.P., Justice, C.O., et al., 2014. Active fires from the Suomi NPP visible infrared imaging radiometer suite: product status and first evaluation results. Journal of Geophysical Research: Atmospheres 119 (2), 803–816. Available at https://doi.org/10.1002/2013jd020453.

Cunningham, C.X., Williamson, G.J., Bowman, D.M.J.S., 2024. Increasing frequency and intensity of the most extreme wildfires on Earth. Nature Ecology & Evolution 8 (8), 1420–1425. Available at https://doi.org/10.1038/s41559-024-02452-2.

De Santis, A., Chuvieco, E., 2009. GeoCBI: a modified version of the composite burn index for the initial assessment of the short-term burn severity from remotely sensed data. Remote Sensing of Environment 113 (3), 554–562. Available at https://doi.org/10.1016/j.rse.2008.10.011.

Duncanson, L., Armston, J., Disney, M., Avitabile, V., Barbier, N., Calders, K., et al., 2021. Aboveground woody biomass product validation good practices protocol. version 1.0. In: Duncanson, L., Disney, M., Armston, J., Nickeson, J., Minor, D., Camacho, F. (Eds.), Good Practices for Satellite Derived Land Product Validation. Land Product Validation Subgroup (WGCV/CEOS), pp. 236. Available at https://doi.org/10.5067/doc/ceoswgcv/lpv/agb.001.

EFFIS, 2025. European Forest Fire Information System. EFFIS. Available at https://forest-fire.emergency.copernicus.eu/

ESA, 2025. Biomass Products Verification and Validation Plan. Biomass Science Team. The European Space Agency Available at https://earth.esa.int/eogateway/documents/d/earth-online/biomass-products-verification-and-validation-plan.

Fernández-García, V., Alonso-González, E., 2023. Global patterns and dynamics of burned area and burn severity. Remote Sensing 15 (13), 3401. Available at https://doi.org/10.3390/rs15133401.

Fernández-García, V., Kull, C.A., 2023. Refining historical burned area data from satellite observations. International Journal of Applied Earth Observation and Geoinformation 120, 103350. Available at https://doi.org/10.1016/j.jag.2023.103350.

Fernández-García, V., Santamarta, M., Fernández-Manso, A., Quintano, C., Marcos, E., Calvo, L., 2018. Burn severity metrics in fire-prone pine ecosystems along a climatic gradient using Landsat imagery. Remote Sensing of Environment 206, 205–217. Available at https://doi.org/10.1016/j.rse.2017.12.029.

Fernández-García, V., Beltrán-Marcos, D., Fernández-Guisuraga, J.M., Marcos, E., Calvo, L., 2022. Predicting potential wildfire severity across Southern Europe with global data sources. Science of the Total Environment 829, 154729. Available at https://doi.org/10.1016/j.scitotenv.2022.154729.

Fernández-García, V., Franquesa, M., Kull, C.A., 2024. Madagascar's burned area from Sentinel-2 imagery (2016–2022): four times higher than from lower resolution sensors. Science of the Total Environment 914, 169929. Available at https://doi.org/10.1016/j.scitotenv.2024.169929.

Fernández-García, V., Phelps, L.N., Strydom, T., Muando, P.J., Ranaivonasy, J., Lehmann, C.E.R., et al., 2025. High-resolution satellite data improve insights into landscape fires and their drivers in Southeastern Africa. Journal of Geophysical Research: Biogeosciences 130 (9), e2024JG008635. Available at https://doi.org/10.1029/2024JG008635.

Fernández-Guisuraga, J.M., Calvo, L., Fernández-García, V., Marcos-Porras, E., Taboada, Á., Suárez-Seoane, S., 2019. Efficiency of remote sensing tools for post-fire management along a climatic gradient. Forest Ecology and Management 433, 553–562. Available at https://doi.org/10.1016/j.foreco.2018.11.045.

Fernández-Guisuraga, J.M., Verrelst, J., Calvo, L., Suárez-Seoane, S., 2021. Hybrid inversion of radiative transfer models based on high spatial resolution satellite reflectance data improves fractional vegetation cover retrieval in heterogeneous ecological systems after fire. Remote Sensing of Environment 255, 112304. Available at https://doi.org/10.1016/j.rse.2021.112304.

Fernández-Guisuraga, J.M., Monzón-González, A., Fernández-García, V., Peña-Pérez, S.A., Calvo, L., 2025. Estimation of Prometheus fuel types using physically based remote sensing techniques. Fire Ecology 21 (1). Available at https://doi.org/10.1186/s42408-025-00373-4.

Flasse, S.P., Trigg, S.N., Ceccato, P.N., Perryman, A.H., Hudak, A.T., Thompson, M.W., et al., 2004. Remote Sensing of Vegetation Fires and its Contribution to a fire Management

Information System. Columbia University. Available at https://doi.org/10.7916/D8959Q87.

Frank, B., Strunk, J.L., Fried, J.S., Wolken, K., McKenzie, S.C., 2025. Comparison of digital aerial photogrammetry, lidar, and Sentinel-2 for evaluating forest fire effects. Forest Ecology and Management 595, 123002. Available at https://doi.org/10.1016/j.foreco.2025.123002.

Franquesa, M., Vanderhoof, M.K., Stavrakoudis, D., Gitas, I.Z., Roteta, E., Padilla, M., et al., 2020. Development of a standard database of reference sites for validating global burned area products. Earth System Science Data 12 (4), 3229–3246. Available at https://doi.org/10.5194/essd-12-3229-2020.

Franquesa, M., Stehman, S.V., Chuvieco, E., 2022a. Assessment and characterization of sources of error impacting the accuracy of global burned area products. Remote Sensing of Environment 280, 113214. Available at https://doi.org/10.1016/j.rse.2022.113214.

Franquesa, M., Lizundia-Loiola, J., Stehman, S.V., Chuvieco, E., 2022b. Using long temporal reference units to assess the spatial accuracy of global satellite-derived burned area products. Remote Sensing of Environment 269, 112823. Available at https://doi.org/10.1016/j.rse.2021.112823.

Franquesa, M., Bastarrika, A., Boschetti, M., Buschetto, L., Chuvieco, E., Fernández-García, V., et al. (2025). Burned Area Reference Database (BARD). eCiencia Datos. Available at https://edatos.consorciomadrono.es/dataverse/BARD

García-Llamas, P., Suárez-Seoane, S., Fernández-Guisuraga, J.M., Fernández-García, V., Fernández-Manso, A., Quintano, C., et al., 2019a. Evaluation and comparison of Landsat 8, Sentinel-2 and Deimos-1 remote sensing indices for assessing burn severity in Mediterranean fire-prone ecosystems. International Journal of Applied Earth Observation and Geoinformation 80, 137–144. Available at https://doi.org/10.1016/j.jag.2019.04.006.

García-Llamas, P., Suárez-Seoane, S., Taboada, A., Fernández-Manso, A., Quintano, C., Fernández-García, V., et al., 2019b. Environmental drivers of fire severity in extreme fire events that affect Mediterranean pine forest ecosystems. Forest Ecology and Management 433, 24–32. Available at https://doi.org/10.1016/j.foreco.2018.10.051.

García-Llamas, P., Suárez-Seoane, S., Taboada, A., Fernández-García, V., Fernández-Guisuraga, J.M., Fernández-Manso, A., et al., 2019c. Assessment of the influence of biophysical properties related to fuel conditions on fire severity using remote sensing techniques: a case study on a large fire in NW Spain. International Journal of Wildland Fire 28 (7), 512. Available at https://doi.org/10.1071/wf18156.

Gaveau, D.L.A., Descals, A., Salim, M.A., Sheil, D., Sloan, S., 2021. Refined burned-area mapping protocol using Sentinel-2 data increases estimate of 2019 Indonesian burning. Earth System Science Data 13 (11), 5353–5368. Available at https://doi.org/10.5194/essd-13-5353-2021.

Giglio, L., Schroeder, W., Justice, C.O., 2016. The collection 6 MODIS active fire detection algorithm and fire products. Remote Sensing of Environment 178, 31–41. Available at https://doi.org/10.1016/j.rse.2016.02.054.

Giglio, L., Boschetti, L., Roy, D.P., Humber, M.L., Justice, C.O., 2018. The collection 6 MODIS burned area mapping algorithm and product. Remote Sensing of Environment 217, 72–85. Available at https://doi.org/10.1016/j.rse.2018.08.005.

Giglio, L., Boschetti, L., Roy, D.P., Hoffmann, A.A., Humber, M.L., Hall, J.V., 2020. Collection 6 MODIS Burned Area Product User's Guide. Version 1.3. NASA. Available at https://lpdaac.usgs.gov/documents/875/MCD64_User_Guide_V6.pdf.

Guillevic, P., Göttsche, F., Nickeson, J., Hulley, G., Ghent, D., Yu, Y., et al., 2018. Land surface temperature product validation best practice protocol. version 1.1. In: Guillevic, P., Göttsche, F., Nickeson, J., Román, M. (Eds.), Good Practices for Satellite-Derived Land Product Validation. Land Product Validation Subgroup (WGCV/CEOS), pp. 58. Available at https://doi.org/10.5067/doc/ceoswgcv/lpv/lst.001.

Hamilton, D., Gibson, W., Harris, D., McGath, C., 2023. Evaluation of multi-spectral band efficacy for mapping wildland fire burn severity from planetscope imagery. Remote Sensing 15 (21), 5196. Available at https://doi.org/10.3390/rs15215196.

Hawbaker, T.J., Vanderhoof, M.K., Schmidt, G.L., Beal, Y.-J., Picotte, J.J., Takacs, J.D., et al., 2020. The landsat burned area algorithm and products for the conterminous United States. Remote Sensing of Environment 244, 111801. Available at https://doi.org/10.1016/j.rse.2020.111801.

Howe, A.A., Parks, S.A., Harvey, B.J., Saberi, S.J., Lutz, J.A., Yocom, L.L., 2022. Comparing sentinel-2 and landsat 8 for burn severity mapping in western North America. Remote Sensing 14 (20), 5249. Available at https://doi.org/10.3390/rs1420524.

Huerta, S., Marcos, E., Fernández-García, V., Calvo, L., 2022. Short-term effects of burn severity on ecosystem multifunctionality in the northwest Iberian Peninsula. Science of The Total Environment 844, 157193. Available at https://doi.org/10.1016/j.scitotenv.2022.157193.

Key, C.H., Benson, N.C., 2006. Landscape Assessment (LA). Sampling and Analysis Methods. USDA Forest Service Gen. Tech. Rep. RMRS-GTR-164-CD. Available at https://www.fs.usda.gov/rm/pubs_series/rmrs/gtr/rmrs_gtr164/rmrs_gtr164_13_land_assess.pdf.

Lee, J.H., Veraverbeke, S., Rogers, B., Kerr, Y.H., 2024. L-band microwave-retrieved fuel temperature predicts million-hectare-scale destructive wildfires. International Journal of Applied Earth Observation and Geoinformation 129, 103776. Available at https://doi.org/10.1016/j.jag.2024.103776.

Lewis, S.A., Lentile, L.B., Hudak, A.T., Robichaud, P.R., Morgan, P., Bobbitt, M.J., 2007. Mapping ground cover using hyperspectral remote sensing after the 2003 simi and old wildfires in southern California. Fire Ecology 3 (1), 109–128. Available at https://doi.org/10.4996/fireecology.0301109.

Li, F., Zhang, X., Kondragunta, S., Csiszar, I., 2018. Comparison of fire radiative power estimates from VIIRS and MODIS observations. Journal of Geophysical Research: Atmospheres 123 (9), 4545–4563. Available at https://doi.org/10.1029/2017jd027823.

Lunetta, R.S., Congalton, R.G., Fenstermaker, L.K., Jensen, J.R., McGwire, K., Tinney, L.R., 1991. Remote sensing and geographic information system data integration: error sources and research issues. Photogrammetric Engineering & Remote Sensing 57, 677–687. Available at https://www.asprs.org/wp-content/uploads/pers/1991journal/jun/1991_jun_677-687.pdf.

Maillard, O., Ribeiro, N., Armstrong, A., Ribeiro-Barros, A.I., Andrew, S.M., Amissah, L., et al., 2025. Seasonal spatial-temporal trends of vegetation recovery in burned areas across Africa. PLoS One 20 (2), e0316472. Available at https://doi.org/10.1371/journal.pone.0316472.

Mermoz, M., Kitzberger, T., Veblen, T.T., 2005. Landscape influences on occurrence and spread of wildfires in Patagonian forests and shrublands. Ecology 86 (10), 2705–2715. Available at https://doi.org/10.1890/04-1850.

Meyer, F., 2019. Spaceborne synthetic aperture radar: principles, data access, and basic processing techniques. In: Herndon, F.A., Thapa, K., Cherrington, R. (Eds.), The SAR Handbook. NASA. Available at https://doi.org/10.25966/ez4f-mg98.

Miller, J.D., Knapp, E.E., Key, C.H., Skinner, C.N., Isbell, C.J., Creasy, R.M., et al., 2009. Calibration and validation of the relative differenced Normalized Burn Ratio (RdNBR) to three measures of fire severity in the Sierra Nevada and Klamath Mountains, California, USA. Remote Sensing of Environment 113 (3), 645–656. Available at https://doi.org/10.1016/j.rse.2008.11.009.

MITECO, 2025. Inventario Forestal Nacional. Ministerio para la Transición Ecológica y el Reto Demográgico. Gobierno de España. Available at https://www.miteco.gob.es/es/biodiversidad/temas/inventarios-nacionales/inventario-forestal-nacional.html.

Monzón-González, A., Calvo, L., Fernández-García, V., Fernández-Guisuraga, J.M., 2025. Upscaling wildfire consumption using UAV-LiDAR and Sentinel-2 data: a Mediterranean case study. GIScience & Remote Sensing 62 (1). Available at https://doi.org/10.1080/15481603.2025.2555626.

MTBS, 2025. Mapping Methods. MTBS. USDA-USFS-DOI-USGS Available from. Available at https://mtbs.gov/mapping-methods.

NASA, 2025. Data Maturity Levels. EarthData. NASA Available from. Available at https://www.earthdata.nasa.gov/learn/earth-observation-data-basics/data-maturity-levels.

NASA-CEOS-LPV, 2025. Documents Related to LPV and CEOS. NASA-CEOS Working Group on Calibration and Validation-Land Product Validation Subgroup. Available at https://lpvs.gsfc.nasa.gov/documents.html.

NFIS, 2025. Canada's National Forest Inventory. NFIS Available at https://nfi.nfis.org/.

Olofsson, P., Foody, G.M., Herold, M., Stehman, S.V., Woodcock, C.E., Wulder, M.A., 2014. Good practices for estimating area and assessing accuracy of land change. Remote Sensing of Environment 148, 42–57. Available at https://doi.org/10.1016/j.rse.2014.02.015.

Padilla, M., Olofsson, P., Stehman, S.V., Tansey, K., Chuvieco, E., 2017. Stratification and sample allocation for reference burned area data. Remote Sensing of Environment 203, 240–255. Available at https://doi.org/10.1016/j.rse.2017.06.041.

Pádua, L., Adão, T., Guimarães, N., Sousa, A., Peres, E., Sousa, J.J., 2019. Post-fire forestry recovery monitoring using high-resolution multispectral imagery from unmanned aerial vehicles. The International Archives of the Photogrammetry, Remote Sensing and Spatial Information Sciences XLII-3/W8, 301–305. Available at https://doi.org/10.5194/isprs-archives-xlii-3-w8-301-2019.

Picotte, J.J., Arkle, R., Bastian, H., Benson, N., Cansler, A., Caprio, T., et al., 2019. Composite Burn Index (CBI) Data for the Conterminous US, Collected Between 1996 and 2018 [Dataset]. U.S. Geological Survey. Available at https://doi.org/10.5066/P91BH1BZ.

Pleniou, M., Koutsias, N., 2025. The role of spectral vs spatial resolution of satellite data on the accuracy of mapping unburned vegetation within fire scar perimeters. Science of Remote Sensing 11, 100241. Available at https://doi.org/10.1016/j.srs.2025.100241.

Ploton, P., Mortier, F., Réjou-Méchain, M., Barbier, N., Picard, N., Rossi, V., et al., 2020. Spatial validation reveals poor predictive performance of large-scale ecological mapping models. Nature Communications 11 (1). Available at https://doi.org/10.1038/s41467-020-18321-y.

Powell, R.L., Matzke, N., de Souza Jr., C., Clark, M., Numata, I., Hess, L.L., et al., 2004. Sources of error in accuracy assessment of thematic land-cover maps in the Brazilian Amazon. Remote Sensing of Environment 90 (2), 221–234. Available at https://doi.org/10.1016/j.rse.2003.12.007.

Puliti, S., Hauglin, M., Breidenbach, J., Montesano, P., Neigh, C.S.R., Rahlf, J., et al., 2020. Modelling above-ground biomass stock over Norway using national forest inventory data with ArcticDEM and Sentinel-2 data. Remote Sensing of Environment 236, 111501. Available at https://doi.org/10.1016/j.rse.2019.111501.

Quintano, C., Fernández-Manso, A., Roberts, D.A., 2013. Multiple endmember spectral mixture analysis (MESMA) to map burn severity levels from Landsat images in Mediterranean countries. Remote Sensing of Environment 136, 76–88. Available at https://doi.org/10.1016/j.rse.2013.04.017.

Quintano, C., Calvo, L., Fernández-Manso, A., Suárez-Seoane, S., Fernandes, P.M., Fernández-Guisuraga, J.M., 2023. First evaluation of fire severity retrieval from PRISMA hyperspectral data. Remote Sensing of Environment 295, 113670. Available at https://doi.org/10.1016/j.rse.2023.113670.

Ramsey, S., Jones, S., Reinke, K., 2024. Review of approaches and challenges for the validation of satellite-based active fire products in savannah ecosystems. International Journal of Wildland Fire 33 (10). Available at https://doi.org/10.1071/wf23202.

Reeves, M.C., Ryan, K.C., Rollins, M.G., Thompson, T.G., 2009. Spatial fuel data products of the LANDFIRE Project. International Journal of Wildland Fire 18 (3), 250. Available at https://doi.org/10.1071/wf08086.

Roteta, E., Bastarrika, A., Padilla, M., Storm, T., Chuvieco, E., 2019. Development of a Sentinel-2 burned area algorithm: generation of a small fire database for sub-Saharan Africa. Remote Sensing of Environment 222, 1–17. Available at https://doi.org/10.1016/j.rse.2018.12.011.

Roteta, E., Bastarrika, A., Ibisate, A., Chuvieco, E., 2021. A preliminary global automatic burned-area algorithm at medium resolution in google earth engine. Remote Sensing 13 (21), 4298. Available at https://doi.org/10.3390/rs13214298.

Ruecker, G., Tiemann, J., Leimbach, D., Lorenz, E., 2013. Assessment of the fire radiative power retrieved by BIRD and the MODIS active fire products from near coincident data acquisitions. Geophysical Research Abstracts 15 EGU General Assembly EGU2013-11142-1. Available at https://meetingorganizer.copernicus.org/EGU2013/EGU2013-11142-1.pdf.

Schroeder, W., Oliva, P., Giglio, L., Csiszar, I.A., 2014. The New VIIRS 375 m active fire detection data product: algorithm description and initial assessment. Remote Sensing of Environment 143, 85–96. Available at https://doi.org/10.1016/j.rse.2013.12.008.

Schroeder, W., Prins, E., Giglio, L., Csiszar, I., Schmidt, C., Morisette, J., et al., 2008. Validation of GOES and MODIS active fire detection products using ASTER and ETM+ data. Remote Sensing of Environment 112 (5), 2711–2726. Available at https://doi.org/10.1016/j.rse.2008.01.005.

Seidl, R., Fernandes, P.M., Fonseca, T.F., Gillet, F., Jönsson, A.M., Merganičová, K., et al., 2011. Modelling natural disturbances in forest ecosystems: a review. Ecological Modelling 222 (4), 903–924. Available at https://doi.org/10.1016/j.ecolmodel.2010.09.040.

Simioni, J.P.D., Guasselli, L.A., Belloli, T.F., Ramos, R.A., 2022. Assessment of fire resilience in subtropical wetlands using high spatial resolution images. Environmental Monitoring and Assessment 194 (6). Available at https://doi.org/10.1007/s10661-022-09985-8.

SLU, 2025. The Swedish National Forest Inventory. Sweedish University of Agricultural Sciences. SLU Available at https://www.slu.se/en/about-slu/organisation/departments/forest-resource-management/miljoanalys/nfi/.

Stroppiana, D., Busetto, L., Boschetti, M., Ranghetti, L., Sali, M., 2021. ESA CCI ECV Fire Disturbance: D1.3 Product Validation Plan, version 2.0. Available at https://climate.esa.int/en/projects/fire/key-documents/

Stroppiana, D., Sali, M., Busetto, L., Boschetti, M., Ranghetti, L., Franquesa, M., et al., 2022. Sentinel-2 sampling design and reference fire perimeters to assess accuracy of Burned Area products over Sub-Saharan Africa for the year 2019. ISPRS Journal of Photogrammetry and Remote Sensing 191, 223–234. Available at https://doi.org/10.1016/j.isprsjprs.2022.07.015.

Tanase, M.A., Belenguer-Plomer, M.A., Roteta, E., Bastarrika, A., Wheeler, J., Fernández-Carrillo, Á., et al., 2020. Burned area detection and mapping: intercomparison of sentinel-1 and Sentinel-2 based algorithms over tropical Africa. Remote Sensing 12 (2), 334. Available at https://doi.org/10.3390/rs12020334.

Thompson, J.R., Spies, T.A., 2009. Vegetation and weather explain variation in crown damage within a large mixed-severity wildfire. Forest Ecology and Management 258 (7), 1684–1694. Available at https://doi.org/10.1016/j.foreco.2009.07.031.

USFS, 2025. Nationwide Forest Inventory (NFI). U.S. Forest Service. USDA Available at https://research.fs.usda.gov/programs/nfi.

van der Werf, G.R., Randerson, J.T., Giglio, L., van Leeuwen, T.T., Chen, Y., Rogers, B.M., et al., 2017. Global fire emissions estimates during 1997–2016. Earth System Science Data 9 (2), 697–720. Available at https://doi.org/10.5194/essd-9-697-2017.

Vanderhoof, M.K., Fairaux, N., Beal, Y.-J.G., Hawbaker, T.J., 2017. Validation of the USGS landsat burned area essential climate variable (BAECV) across the conterminous United States. Remote Sensing of Environment 198, 393–406. Available at https://doi.org/10.1016/j.rse.2017.06.025.

Veraverbeke, S., Dennison, P., Gitas, I., Hulley, G., Kalashnikova, O., Katagis, T., et al., 2018. Hyperspectral remote sensing of fire: state-of-the-art and future perspectives. Remote Sensing of Environment 216, 105–121. Available at https://doi.org/10.1016/j.rse.2018.06.020.

Verrelst, J., Camps-Valls, G., Muñoz-Marí, J., Rivera, J.P., Veroustraete, F., Clevers, J.G.P.W., et al., 2015. Optical remote sensing and the retrieval of terrestrial vegetation bio-geophysical properties – a review. ISPRS Journal of Photogrammetry and Remote Sensing 108, 273–290. Available at https://doi.org/10.1016/j.isprsjprs.2015.05.005.

Viana-Soto, A., Okujeni, A., Pflugmacher, D., García, M., Aguado, I., Hostert, P., 2022. Quantifying post-fire shifts in woody-vegetation cover composition in Mediterranean pine forests using Landsat time series and regression-based unmixing. Remote Sensing of Environment 281, 113239. Available at https://doi.org/10.1016/j.rse.2022.113239.

Warner, T.A., Skowronski, N.S., Gallagher, M.R., 2016. High spatial resolution burn severity mapping of the New Jersey Pine Barrens with WorldView-3 near-infrared and shortwave infrared imagery. International Journal of Remote Sensing 38 (2), 598–616. Available at https://doi.org/10.1080/01431161.2016.1268739.

Yebra, M., Chuvieco, E., Riaño, D., 2008. Estimation of live fuel moisture content from MODIS images for fire risk assessment. Agricultural and Forest Meteorology 148 (4), 523–536. Available at https://doi.org/10.1016/j.agrformet.2007.12.005.

Yu, M., Zhou, Y.-N., Wang, Q., Yan, F., 2024. Extrapolation validation (EV): a universal validation method for mitigating machine learning extrapolation risk. Digital Discovery 3 (5), 1058–1067. Available at https://doi.org/10.1039/d3dd00256j.

The role of fuel in driving extreme wildfire events

5

Paulo M. Fernandes[1] *and Víctor Resco de Dios*[2]
[1]CITAB - Centre for the Research and Technology of Agro-Environmental and Biological Sciences, University of Trás-os-Montes e Alto Douro, Vila Real, Portugal, [2]Department of Forest and Agricultural Science and Engineering, University of Lleida, Lleida, Spain

5.1 Introducing extreme wildfire events

Extreme wildfire events are the major concern of the current fire regimes in temperate and Mediterranean biomes, which are those where human assets, fire-prone vegetation, and conducive weather are more likely to intersect (Bowman et al., 2017). The definition and concept of what an extreme fire is vary and are open to interpretation (Tedim et al., 2018), but they are related directly or indirectly to extreme (or the upper-end) fire behavior. In this regard, the focus can be on control difficulty, whereby fireline intensity precludes effective suppression by direct attack (Hirsch and Martell, 1996); nonsteady fire behavior (Werth et al., 2016), the variant of high-intensity fire when rapid and often unexpected increases are involved: "blow-up fire," "fire storm," "conflagration," "erratic", and, largely as a consequence of the former, wildfire size, for example, "extremely large fire" (Burton et al., 2008) or "megafire," increasingly used in science (Linley et al., 2022) and in the media (Stoof et al., 2024). Thus, generically, an extreme wildfire is especially large or displays unusual behavior or impacts, and both features tend to align (Keane et al., 2008; Buonanduci et al., 2023; Fernández-Guisuraga et al., 2023). Whether a fire is extreme can be established through relative or absolute numerical thresholds. For example, extreme wildfire events in (Bowman et al., 2017) are defined as the daily clusters of fire radiative power (FRP) exceeding the 99.997th percentile of the ΣFRP 100 km^{-2} in the Moderate Resolution Imaging Spectroradiometer (MODIS) record, which may be more indicative of fire size or clustered fires than of fire intensity per se (Schütze and Resco de Dios, 2025). After Stephens et al. (2014) and Linley et al. (2022) proposed 10,000 ha as the size threshold for a megafire. Nonetheless, it should be acknowledged that the thresholds for extreme fires are context dependent and often physically arbitrary (Stoof et al., 2024).

The extreme nature of a particular wildfire can and should also be gaged through its impacts, usually on people, buildings, and infrastructure, arguably the most relevant criterion from the perspective of risk management (Berchtold et al., 2025). As such, and especially when considerable loss of human life occurs (e.g., Lagouvardos et al., 2019), these are referred to as "catastrophic" or "disaster" fires, even "unprecedented" if warranted by the degree of perceived novelty in impacts (Duane et al., 2021).

Advances in Remote Sensing for Forest Fire Management. DOI: https://doi.org/10.1016/B978-0-443-41692-7.00013-5
© 2027 Elsevier Ltd. All rights are reserved, including those for text and data mining, AI training, and similar technologies.

Extreme wildfire events threaten people's livelihoods and safety, ecosystem services, and integrity (Berchtold et al., 2025), and ultimately affect the Earth system by impacting on biogeochemical cycles and the atmosphere through mechanisms that are not completely understood (Bowman et al., 2009). While rising globally (Coop et al., 2022; Abatzoglou et al., 2025; Jones et al., 2024), extreme wildfires are far from new. For example, seven wildfire conflagrations in the United States burned more than 2.5 million hectares and killed 2667 people between 1871 and 1918, of which ~1500 in the Pestigo fire alone (Haines and Sando, 1969). This extent of human life loss is now unthinkable, but as early as 1949, a firestorm in Landes de Gascoigne, France, was reported to burn 6000 ha in just 20 minutes and cause 82 fatalities (Chevrou, 2005). This resembles contemporary disasters commonly perceived as novel in Europe, like the Pedrogão Grande and Mati wildfires in 2017 and 2018 in Portugal and Greece, respectively (Boer et al., 2017; Lagouvardos et al., 2019).

Remote sensing techniques have greatly expanded our ability to describe, monitor, and understand vegetation fires, and their drivers, and impacts, particularly at spatial and temporal scales relevant to extreme wildfires. As most remote-sensing products directly or indirectly reflect fuel-related information (before, during, and after fire), this chapter addresses fuels and their contribution to the characteristics of extreme wildfire events.

5.2 Drivers and dynamics of fire activity

Fire activity, from individual events to their long-term expression—the fire regime—has been conceptualized as driven by four switches that need to be turned on concurrently (Bradstock, 2010). First, sufficient biomass is needed, a climate-determined and management-influenced variable that decreases from mesic to arid conditions. Whether such biomass will become fuel, that is, will be available to burn, depends on its moisture content, which in turn is a function of plant phenology, prior rainfall, and atmospheric conditions. Then, an ignition source should be present. Finally, favorable weather is needed in the form of adequate combinations of wind speed, air temperature, and relative humidity for a fire to spread significantly. It follows that the most extreme fire behavior occurs in biomass-rich environments during a drought period and under the influence of elevated fire weather, that is, strong wind, high ambient temperature, and low relative humidity. However, the nature and prevalence of extreme wildfires are not explicit attributes of fire regimes, which reflect the typical spatiotemporal features (Krebs et al., 2010). Advances in remote sensing and the resulting databases (e.g., Andela et al., 2019) are now allowing analyses of the characteristics of individual fire events, which are increasingly seen as crucial for understanding fire regimes (Mahood et al., 2025).

Fire regimes are changing due to the mixed effect of top-down (climate change) and bottom-up (land use and land cover, fire management) influences, as described in Chapters 1 and 2. Isolating and quantifying the relative effects of these drivers can be challenging, as illustrated by the case of Portugal (Fig. 5.1). Burned area composition

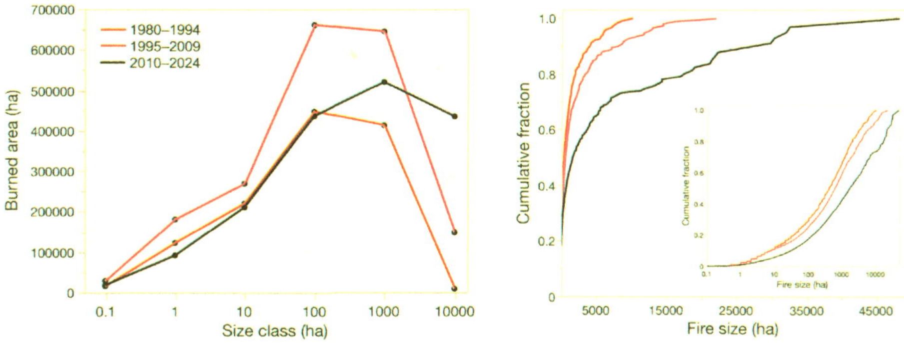

Figure 5.1 Burned area in Portugal for three consecutive 15-year periods: distribution by fire size class and cumulative fraction as a function of fire size. The depicted fire size intervals are [0.1–1[, [1–10[, [10–100[, [100–1000[, [1000–10,000[, and ≥10,000 ha.

in Portugal in terms of fire size distribution has substantially changed from 1980–94 to 2010–24, with fires ≥1000 ha and especially ≥10,000 ha increasingly more represented, and an increase in maximum wildfire size (see also Stoof et al., 2024). Implicit in Fig. 5.1 is the transition from a fuel-limited to a weather-limited fire regime, as fuel connectivity and amount have increased due to rural abandonment, forest, and shrubland expansion (Pausas and Fernández-Muñoz, 2012; Fernandes et al., 2014) and the establishment of protected areas (Resco de Dios et al., 2025). This effect is compounded by policies centered on fire suppression that further promote landscape-level flammability (Fernandes et al., 2020a), and a synergy is expected with climate change (Turco et al., 2019), leading to abnormally large and fast-spreading fires (Benali et al., 2023), identified as those having the highest societal impacts (Balch et al., 2024).

5.3 Extreme fire behavior phenomena

Wildfires exhibiting extreme fire behavior do not necessarily grow to become particularly large or pose a societal or environmental threat. However, at least one facet of extreme fire behavior is implied in extreme fire impacts: fast spread, high intensity, active crowning, generalized spotting, plume dominance, and fire-atmosphere coupling (Werth et al., 2016).

Byram's fireline intensity (Byram, 1959) and FRP—a partial proxy acquired by remote sensing (Johnston et al., 2017)—are customarily used to describe how extreme a fire is. Fireline intensity is the product (in units of kW m^{-1}) of fire-spread rate, fuel consumption by continuous flaming, and heat of combustion, and it combines the movement and energy attributes of a flame front into a single number. Fireline intensity reflects the combined influences of atmospheric conditions, recent rainfall, and fuel structure and accumulation on fire behavior. Extreme fire behavior sensu fire control difficulty is assumed from fireline intensities higher than

4000–10,000 kW m^{-1} (Hirsch and Martell, 1996), equivalent to flame lengths of 5–9 and 2–3 m, respectively, in forests and shrublands and in grasslands (Rossa et al., 2024).

Fireline intensity has limitations for comparison purposes, namely because contrasting combinations of rate of spread and energy release arising from distinct fuel structures result in the same intensity (Cheney, 1990). Nonetheless, it remains a useful metric, especially in the context of fire danger rating and to assess the effectiveness of fire control operations, thus classifying a wildfire event as extreme from these perspectives. However, because variability in rate of spread can be much larger than variability in fuel consumption, rate of spread and fireline intensity are highly correlated for a given fuel type or landscape context (Alexander, 1982). Wildfire case studies are selective for the most impactful events and typically report sustained fire runs of 3–12 km h^{-1} in forests and shrublands, usually corresponding to fireline intensities >30,000 kW m^{-1}, up to extremes of >15 km h^{-1} in grasslands (Rego et al., 2021).

Several fire behavior phenomena can contribute to extreme wildfire events as defined by their movement and energy release. In forests, and depending on the fire environment, surface fires can become crown fires, whether the fire spreads through the tree canopy or the trees burn individually, respectively termed active and passive crowning (Van Wagner, 1977). Transition from a surface to a crown fire typically increases rate of spread by a factor of two to four (Alexander and Cruz, 2006), mostly because of increased exposure to ambient wind and enhanced energy release and transfer (Taylor et al., 2004).

Spotting, whereby burning material travels and ignites areas beyond the perimeter of the main fire, is a prominent feature of extreme wildfire behavior, and one that substantially challenges wildfire containment, firefighter safety, and the protection of wildland-urban interfaces (Werth et al., 2016). Spotting enables fire spread through discontinuous fuels and creates interactions between flame fronts (that eventually merge), increasing burning and heat release rates, flame size, and causing strong indrafts and tall convection columns (Finney and McAllister, 2011).

The longest spotting distances are an outcome of crown fires (Albini et al., 2012), illustrating that the manifestations of extreme fire behavior seldom occur in isolation. Spot fires initiating at distances of ~5 km from the fire front are common in conifer forests (Albini et al., 2012), but forests of eucalypt species with aerodynamic bark streams feature the highest long-range spotting potential, up to ~30 km (Cruz et al., 2012). However, short-distance spotting caused by ember showers can substantially increase the rate of fire spread and cause mass (or area) ignition, resulting in sudden or periodic bursts in fire growth, for example, Wade and Ward (1973).

The most extreme wildfire events result from fire-atmosphere interactions, which, in terms of fire size, are signaled by cooccurring elevated fire weather and atmospheric instability (Fernandes et al., 2016b; Resco de Dios et al., 2022). In plume-driven fire behavior, the upward force related to buoyancy (power of the fire) exceeds the strength of horizontal wind (power of the wind) (Nelson, 1993). In these three-dimensional fires, a significant fraction of the heat energy is converted "into kinetic energy which appears in the form of strong turbulent updrafts, indrafts, convection column eddies, and whirlwinds which can carry burning material aloft" (Byram, 1957). Rapid fire growth and deep flaming are required to develop vigorous and tall pyroconvection (Badlan et al., 2021). However, a firestorm and its potential catastrophic consequences, namely as a result of strong downdrafts, will ensue only if plume dynamics interact with a favorable atmospheric profile in terms of instability, moisture, and shear (Fromm et al., 2010; Potter, 2012).

5.4 Fuels as drivers of extreme wildfire events

Fire weather enables extreme fire events, which are then shaped by topography and landscape-level flammability as determined by vegetation fuels. Increasingly larger fires are possible as the number of environmental drivers reaching critical levels increases (Resco de Dios et al., 2022) and fire-spread state transitions involving weather-fuel interactions are established (Fernandes et al., 2016b).

The effect of fuels on fire behavior is a matter of fuel load (mass per unit area), but also of fuel composition, structure, and spatial organization. Fuel elements vary in size, shape, and condition (dead or alive, i.e., physiologically active), and form distinctive layers—litter, downed wood (slash), grass, shrubs, tree canopy—variable in cover, depth or height, bulk density (fuel mass per unit volume), and vertical continuity (Rego et al., 2021). Fire moves across the top of the fuel bed and then down through it (Cheney, 1990), meaning that the combustion interface is not vertical and that fire characteristics other than rate of spread, flame length, and fireline intensity—that is, flame depth, flaming duration, and postfrontal combustion—are determined by the coarser and deeper (more compact) fuel components.

The nature of the fuel broadly defines the nature of the fire, despite the substantial variability in fire behavior characteristics among fuel types and under various fire weather circumstances (Cruz et al., 2025). Forests are multilayered in their fuel structure and more variable in fire behavior across the fire weather spectrum (Cruz et al., 2025), mostly reflecting transitions from surface to crown fire spread determined by fireline intensity, vertical discontinuity, and canopy density (Van Wagner, 1977). Shrublands are essentially single-layered and display high fire intensity under low fire weather conditions (Anderson et al., 2015). In grasslands, flammability is contingent on degree of curing, that is, the fraction of dead fuel (Cruz et al., 2015a). Higher fuel loads comprised of loosely arranged finer and drier elements generically promote more extreme fire behavior, especially if shrub or crown fuel layers are present and burn. However, the interacting effects of individual fuel properties on fire spread and their collective influence are complex, as shown by laboratory experiments under controlled conditions (Catchpole et al., 1998).

Worldwide patterns in extreme fire behavior characteristics (Fig. 5.2) indicate that grasslands and arid open shrublands combine maximum fire-spread rate with minimum heat release, owing to highly aerated and dry fuel layer, but comparatively low and/or discontinuous fuel load. Deciduous broadleaved forests, characterized by surface fires in compact litter, are comparable in heat output to grasslands but show the slowest-spreading fires. Most other forest types and shrublands occupy an

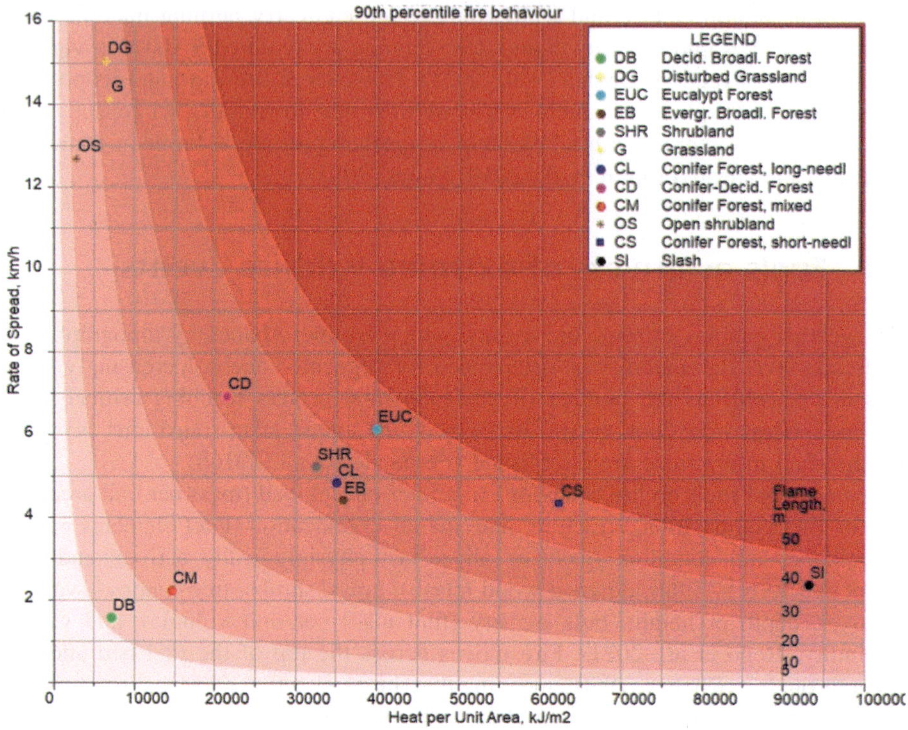

Figure 5.2 Extreme fire behavior characteristics (90th percentiles) for different fuel types in a worldwide fire-behavior database comprised of observations from experimental fires and wildfires (Fernandes et al., 2020b) and graphed by the fire characteristics chart software (Andrews et al., 2011).

intermediate position, that is, they are characterized by high-intensity fast fires in comparatively fuel-rich environments, with short-needled conifer forests releasing a substantially greater heat output, mainly as an outcome of the fuel structure and type of fire depicted by boreal crown fires (Cruz et al., 2025). Finally, slash fuels exhibit the highest heat per unit area, as expected from a layer of downed dead woody fuel, with rates of spread in-between deciduous broadleaved forests and most of the other forest types. Notwithstanding, Fig. 5.2 is far from representing the isolated effect of fuel characteristics, because the weather-related component of the fire environment differs between fuel types.

The amount of energy released by a wildfire is largely a combined function of the amount and moisture content of the existing fuels, defining the available fuel load (Ottmar, 2014), modulated by fire weather (Cruz et al., 2022a). Fuel moisture content is decisive to fire behavior because it increases the heat demand for ignition and combustion, and slows down combustion velocity, thus determining the likelihood and sustainability of fire spread, as well as the movement and energy attributes of the flame front (Nelson, 2001). The damping effect of fuel moisture content on fire-spread rate is strongly nonlinear and is usually modeled using a decaying exponential function (Cruz et al., 2015b).

Large fires in Mediterranean Europe arise once dead fuel moisture content (DFMC) falls to 10%–12% (Resco de Dios et al., 2022), which is consistent with rate-of-spread patterns in various fuel types worldwide (Cruz et al., 2025). Identifying live fuel moisture content (LFMC) thresholds associated with critical transitions in fire activity has often been performed by its relation to patterns of cumulative burned area (Dennison and Moritz, 2009), which in southwestern Europe shows large increases after crossing the thresholds of 100% for trees and 70% for shrubs (Rodrigues et al., 2023; see Fig. 5.3. In Portugal, 98% of the wildfires larger than 500 ha and 92% of those faster than 1 km h^{-1} occurred with estimated shrub LFMC below 100% (Benali et al., 2025).

Quantifying the fire-spread influence of fuel characteristics other than moisture content under experimental field conditions is a challenging endeavor. The usual situation is to find several fuel descriptors having a significant statistical effect on rate of spread after accounting for the effects of wind speed and fuel moisture content. However, correlation between those fuel properties makes the source of the effect uncertain e.g. Anderson et al. (2015) present alternative equations for rate of fire spread in shrubland based on the inversely correlated shrub height and bulk density, which in turn are also expected to reflect the influences of fuel load and dead fuel content. As for wildfires, while linking observed or reconstructed fire propagation to prior knowledge and mapping of quantitative fuel metrics is possible (Burrows et al., 2000), it is seldom attained. However, studies of wildfire spatial patterns resort to time since fire and metrics of fuel connectivity and diversity as proxies (Lloret et al., 2002; Duane et al., 2019; Davim et al., 2023; Curt et al., 2013; Fernandes et al., 2016a,b).

Increasingly extreme fire weather conditions progressively diminish the role of fuel load and structure on fire-spread rate (Cruz et al., 2022a) or on fire severity (Fernández-Guisuraga and Calvo, 2025), up to a threshold where such role ceases or at least is no longer detectable and quantifiable. However, fire spread models assume a constant fuel effect on rate of fire spread, which leads to unrealistic expectations regarding the contribution of fuel-reduction to decrease burned area. A notable exception is the recent model of (Cruz et al., 2022b) for eucalypt dry forest, distinguishing three fire-spread phases and making rates of spread in excess of 1.5 km h^{-1} a sole function of wind speed, after earlier work that established rules of thumb based on wind speed for fire-spread rate under critically low fuel moisture (Cruz and Alexander, 2019; Cruz et al., 2020).

While fast fires are possible in grassland carrying very low fuel loads (e.g., Noble, 1991), extremely large fires are typically selective for heavy fuel accumulation, for example, Fernandes et al. (2016b). There should be minimum fuel requirements for particularly extreme fire behavior (plume-driven blow-up fires), expected to depend on surface fire weather, that is, wind speed and the combined effects of atmospheric conditions and drought on fuel availability. Wendel et al. (1962) state that the probability of a blow-up fire is low when fuel load is lesser than 15 t ha^{-1} and very

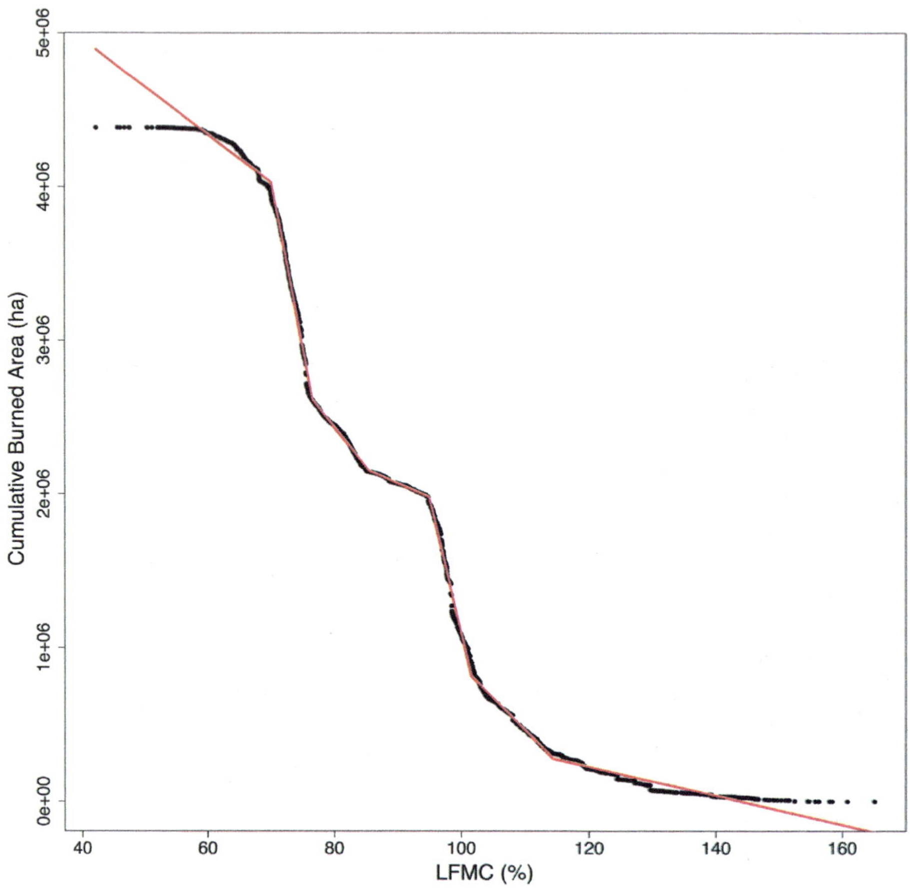

Figure 5.3 Relationship between cumulative burned area and live fuel moisture content (LFMC) for southwestern Europe. The red line indicates the results from segmented regression. The growth in cumulative burned area around 100% and 70% LFMC corresponds with critical moisture transitions in forests and shrublands, respectively.
Source: Based on data from Resco de Dios, Víctor, Cunill Camprubí, Àngel, Pérez-Zanón, Núria, Peña, Juan Carlos, Martínez del Castillo, Edurne, Rodrigues, Marcos, Yao, Yinan, Yebra, Marta, Vega-García, Cristina & Boer, Matthias M. (2022). Convergence in critical fuel moisture and fire weather thresholds associated with fire activity in the pyroregions of Mediterranean Europe. Science of The Total Environment, 806(Pt 4). https://doi.org/10.1016/j.scitotenv.2021.151462.

low below 10 t ha^{-1}. As energy release is directly proportional to available fuel, lower fuel loads across the landscape should diminish the impact of fire on soils and vegetation, as well as the likelihood of plume-driven fire behavior (Cruz et al., 2022a). As a vivid example of the consequences of heavy fuel loads, Stephens et al. (2022) describe how a mass fire with persistent heat release away from the flaming front was originated by large dead fuels (created by drought and insect attacks) combined with high canopy density.

5.5 Assessing fuel characteristics through remote sensing

Remote sensing has emerged as the main tool to assess key fuel characteristics—fuel moisture and fuel structure—over broad geographical scales. A plethora of products is now available, but their precision and accuracy vary widely, and thus remote sensing has become both a curse and a blessing. End-users often trust a product once it has been published, but validation of the product before use is recommended (Resco de Dios et al., 2022) because the goals of the database developer may not match those of the end-user.

As previously shown, fuel moisture content is especially relevant for the onset and behavior of extreme wildfires. Traditionally, seasonal LFMC patterns were tracked by remote sensing through mechanistic radiative transfer models (RTMs) or through empirical spectral indices (Yebra et al., 2013). RTMs hold promise over global scales, but their application has been limited by the large number of parameters that need to be included. Consequently, their application at regional scales has sometimes led to undesirable outcomes (Resco De Dios and Boer, 2025). Empirical models, on the other hand, using spectral indices and land surface temperature (LST), are better suited for regional or subcontinental scales and have shown a superior performance over RTMs at those scales (Cunill Camprubí et al., 2022). This is an example of how a validation of available products is recommended before usage, particularly now that global LFMC datasets have become available.

More recently, some studies have assessed the potential of synthetic aperture radar (SAR) backscatter or soil moisture products, yielding similar results to those previously mentioned for regional empirical models (Rao et al., 2020). The application of machine learning techniques, such as random forests, can enhance predictions toward the middle range, but are weaker at predicting the extreme values associated with extreme wildfires (Cunill Camprubí et al., 2022).

Estimation of fuel moisture through hybrid modes, meaning those combining remotely sensed products with other types of data, have been gaining traction and shows enhanced potential for operational purposes. For instance, the combination of empirical models and the Drought Code of the Canadian fire danger rating system Fire Weather Index (FWI) has led to significantly improved predictions in Portugal (Benali et al., 2025). Data assimilation techniques have also been developed to couple remotely sensed products with process-based models, similarly enhancing LFMC predictions (Jia et al., 2026).

Spatial predictions of DFMC have received much less attention than LFMC, and remotely sensed products became available only in 2016 (Nolan et al., 2016). DFMC is often estimated from weather data, but it can also be remotely sensed from LST. However, given the coarse resolution of LST sensors, typically between 25 and

100 ha, substantial spatial variability nuance is lost. Likewise, quantifying the degree of grass curing remains uncertain (Collins et al., 2018).

Remote sensing provides crucial information on biomass, but not necessarily on fuel load. Optical indices of greenness (e.g., Normalized Difference Vegetation Index [NDVI]; Enhanced Vegetation Index [EVI]) correlate moderately with aboveground biomass, especially in shrublands and forests, but mix the signal from different layers, resolve dead fuels poorly, can saturate in dense forests, and are limited in their sensitivity to vertical structure (Huete et al., 2002). LiDAR (airborne and spaceborne, e.g., Global Ecosystem Dynamics Investigation [GEDI]) and radar, as well as fused multisensor approaches, are able to describe fuel structure and identify fuel models, and allow for retrieving the metrics used to quantify crown fire likelihood and behavior (Abdollahi and Yebra, 2025). However, remote-sensing methods fall short at assessing surface fuels and the specific fuel properties that are known to drive fire behavior (Gale et al., 2021).

Extreme wildfire events are enabled by fire weather, shaped by the landscape, and amplified by fire management policies focused on controlling fires rather than fuels. While it is impossible to know where and when a catastrophic fire will occur, it is decisive and possible to know that it can happen (e.g., Johnson et al., 2014). Remote sensing is useful to face and avoid their impacts and can reveal when fuels are approaching functional thresholds beyond which extreme fire behavior becomes likely. However, the current limitations imply that the use of remote sensing is more in the domain of risk analysis than in that of deterministic prediction.

References

Abdollahi, A., Yebra, M., 2025. Challenges and opportunities in remote sensing-based fuel load estimation for wildfire behavior and management: a comprehensive review. Remote Sensing 17 (3), 415. Available at https://doi.org/10.3390/rs17030415.

Abatzoglou, J.T., Kolden, C.A., Cullen, A.C., Sadegh, M., Williams, E.L., Turco, M., Jones, M.W., 2025. Climate change has increased the odds of extreme regional forest fire years globally. Nature Communications 16. Available at https://doi.org/10.1038/s41467-025-61608-1.

Albini, F.A., Alexander, M.E., Cruz, M.G., 2012. A mathematical model for predicting the maximum potential spotting distance from a crown fire. International Journal of Wildland Fire 21 (5), 609–627. Available at https://doi.org/10.1071/WF11020.

Alexander, M.E., 1982. Calculating and interpreting forest fire intensities. Canadian Journal of Botany 60 (4), 349–357. Available at https://doi.org/10.1139/b82-048.

Alexander, M.E., Cruz, M.G., 2006. Evaluating a model for predicting active crown fire rate of spread using wildfire observations. Canadian Journal of Forest Research 36 (11), 3015–3028. Available at https://doi.org/10.1139/X06-174.

Andela, N., Morton, D.C., Giglio, L., Paugam, R., Chen, Y., Hantson, S., et al., 2019. The global fire Atlas of individual fire size, duration, speed and direction. Earth System Science Data 11 (2), 529–552. Available at https://doi.org/10.5194/essd-11-529-2019, www.earth-system-science-data.net.

Anderson, W.R., Cruz, M.G., Fernandes, P.M., McCaw, L., Vega, J.A., Bradstock, R.A., et al., 2015. A generic, empirical-based model for predicting rate of fire spread in shrublands.

International Journal of Wildland Fire 24 (4), 443–460. Available at https://doi.org/10.1071/WF14130, http://www.publish.csiro.au/nid/126.htm.

Andrews P.L., Heinsch F.A., Schelvan, L., 2011. How to generate and interpret fire characteristics charts for surface and crown fire behavior. General Technical Report RMRS-GTR-253. Fort Collins (CO): US Department of Agriculture, Forest Service, Rocky Mountain Research Station.

Badlan, R.L., Sharples, J.J., Evans, J.P., McRae, R.H.D., 2021. Factors influencing the development of violent pyroconvection. Part I: fire size and stability. International Journal of Wildland Fire 30 (7), 484–497. Available at https://doi.org/10.1071/WF20040, http://www.publish.csiro.au/nid/126.htm.

Balch, J.K., Iglesias, V., Mahood, A.L., Cook, M.C., Amaral, C., DeCastro, A., et al., 2024. The fastest-growing and most destructive fires in the US (2001 to 2020). Science 386 (6720), 425–431. Available at https://doi.org/10.1126/science.adk5737, https://www.science.org/doi/10.1126/science.adk5737.

Benali, A., Guiomar, N., Gonçalves, H., Mota, B., Silva, F., Fernandes, P.M., et al., 2023. The Portuguese large wildfire spread database (PT-FireSprd). Earth System Science Data 15 (8), 3791–3818. Available at https://doi.org/10.5194/essd-15-3791-2023, www.earth-system-science-data.net.

Benali, A., Baldassarre, G., Loureiro, C., Briquemont, F., Fernandes, P.M., Rossa, C., et al., 2025. A near-real-time operational live fuel moisture content (LFMC) product to support decision-making at the national level. Fire 8 (5), 178. Available at https://doi.org/10.3390/fire8050178.

Berchtold, C., Plana, E., Costa Saura, J.M., Kalapodis, N., Sakkas, G., Handmer, J., et al., 2025. Disaster risk management tipping points: impacts of extreme wildfire events and the resulting need for layered disaster risk management solutions. International Journal of Disaster Risk Reduction 131, 105894. Available at https://doi.org/10.1016/j.ijdrr.2025.105894.

Boer, M.M., Nolan, R.H., Resco De Dios, V., Clarke, H., Price, O.F., Bradstock, R.A., 2017. Changing weather extremes call for early warning of potential for catastrophic fire. Earth's Future 5 (12), 1196–1202. Available at https://doi.org/10.1002/2017EF000657, https://onlinelibrary.wiley.com/journal/10.1002/(ISSN)2328-4277.

Bowman, D.M.J.S., Balch, J.K., Artaxo, P., Bond, W.J., Carlson, J.M., Cochrane, M.A., et al., 2009. Fire in the earth system. Science 324 (5926), 481–484. Available at https://doi.org/10.1126/science.1163886.

Bowman, D.M.J.S., Williamson, G.J., Abatzoglou, J.T., Kolden, C.A., Cochrane, M.A., Smith, A.M.S., 2017. Human exposure and sensitivity to globally extreme wildfire events. Nature Ecology & Evolution 1 (3). Available at https://doi.org/10.1038/s41559-016-0058.

Bradstock, R.A., 2010. A biogeographic model of fire regimes in Australia: current and future implications. Global Ecology and Biogeography 19 (2), 145–158. Available at https://doi.org/10.1111/j.1466-8238.2009.00512.x.

Buonanduci, M.S., Donato, D.C., Halofsky, J.S., Kennedy, M.C., Harvey, B.J., 2023. Consistent spatial scaling of high-severity wildfire can inform expected future patterns of burn severity. Ecology Letters 26 (10), 1687–1699. Available at https://doi.org/10.1111/ele.14282, http://onlinelibrary.wiley.com/journal/10.1111/(ISSN)1461-0248.

Burrows, N., Ward, B., Robinson, A., 2000. Behaviour and some impacts of a large wildfire in the Gnangara maritime pine (*Pinus pinaster*) plantation, Western Australia. CALMScience 3 (2), 251–260.

Burton, P.J., Parisien, M.A., Hicke, J.A., Hall, R.J., Freeburn, J.T., 2008. Large fires as agents of ecological diversity in the North American boreal forest. International Journal of Wildland Fire 17 (6), 754–767. Available at https://doi.org/10.1071/WF07149.

Byram, G.M., 1959 Forest fire: control and useCombustion of forest fuels. In: K.P. Davis. McGraw-Hill, New York.

Byram, G.M., 1957. Some principles of combustion and their significance in forest fire behavior. Fire Management Today 18, 47–57.

Catchpole, W.R., Catchpole, E.A., Butler, B.W., Rothermel, R.C., Morris, G.A., Latham, D.J., 1998. Rate of spread of free-burning fires in woody fuels in a wind tunnel. Combustion Science and Technology 131 (1-6), 1–37. Available at https://doi.org/10.1080/00102209808935753, http://www.tandfonline.com/toc/gcst20/current.

Cheney, N.P., 1990. Quantifying bushfires. Mathematical and Computer Modelling 13 (12), 9–15. Available at https://doi.org/10.1016/0895-7177(90)90094-4.

Chevrou, R.B., 2005. Pourquoi les incendies de forêts sont-ils si meurtriers? EDP Sciences, Paris.

Collins, L., Bradstock, R.A., Resco de Dios, V., Duursma, R.A., Velasco, S., Boer, M.M., 2018. Understorey productivity in temperate grassy woodland responds to soil water availability but not to elevated [CO_2]. Global Change Biology 24 (6), 2366–2376. Available at https://doi.org/10.1111/gcb.14038, http://onlinelibrary.wiley.com/journal/10.1111/(ISSN)1365-2486.

Coop, J.D., Parks, S.A., Stevens-Rumann, C.S., Ritter, S.M., Hoffman, C.M., 2022. Extreme fire spread events and area burned under recent and future climate in the western USA. Global Ecology and Biogeography 31 (10), 1949–1959. Available at https://doi.org/10.1111/geb.13496, http://onlinelibrary.wiley.com/journal/10.1111/(ISSN)1466-8238.

Cruz, M.G., Alexander, M.E., 2019. The 10% wind speed rule of thumb for estimating a wildfire's forward rate of spread in forests and shrublands. Annals of Forest Science 76 (2). Available at https://doi.org/10.1007/s13595-019-0829-8, http://www.springer.com.

Cruz, M.G., Sullivan, A.L., Gould, J.S., Sims, N.C., Bannister, A.J., Hollis, J.J., et al., 2012. Anatomy of a catastrophic wildfire: the Black Saturday Kilmore East fire in Victoria, Australia. Forest Ecology and Management 284, 269–285. Available at https://doi.org/10.1016/j.foreco.2012.02.035.

Cruz, M.G., Gould, J.S., Alexander, M.E., Sullivan, A.L., McCaw, W.L., Matthews, S., 2015a. Empirical-based models for predicting head-fire rate of spread in Australian fuel types. Australian Forestry 78, 118–158.

Cruz, M.G., Gould, J.S., Kidnie, S., Bessell, R., Nichols, D., Slijepcevic, A., 2015b. Effects of curing on grassfires: II. Effect of grass senescence on the rate of fire spread. International Journal of Wildland Fire 24, 838–848.

Cruz, M.G., Alexander, M.E., Fernandes, P.M., Kilinc, M., Sil, Â., 2020. Evaluating the 10% wind speed rule of thumb for estimating a wildfire's forward rate of spread against an extensive independent set of observations. Environmental Modelling and Software 133. Available at https://doi.org/10.1016/j.envsoft.2020.104818, www.elsevier.com/inca/publications/store/4/2/2/9/2/1.

Cruz, M.G., Alexander, M.E., Fernandes, P.M., 2022a. Evidence for lack of a fuel effect on forest and shrubland fire rates of spread under elevated fire danger conditions: implications for modelling and management. International Journal of Wildland Fire 31, 471–479.

Cruz, M.G., Cheney, N.P., Gould, J.S., Mccaw, W.L., Kilinc, M., Sullivan, A.L., 2022b. An empirical-based model for predicting the forward spread rate of wildfires in eucalypt forests. CSIRO, Australia International Journal of Wildland Fire 31 (1), 81–95. Available at https://doi.org/10.1071/WF21068, http://www.publish.csiro.au/nid/126.htm.

Cruz, M.G., Hoffman, C.M., Fernandes, P.M., 2025. Global synthesis of quantification of fire behaviour characteristics in forests and shrublands: recent progress. Current Forestry Reports 11 (1). Available at https://doi.org/10.1007/s40725-024-00241-5, https://www.springer.com/journal/40725.

Cunill Camprubí, À., González-Moreno, P., Resco de Dios, V., 2022. Live fuel moisture content mapping in the mediterranean basin using random forests and combining MODIS spectral and thermal data. Remote Sensing 14 (13), 3162. Available at https://doi.org/10.3390/rs14133162.

Curt, T., Borgniet, L., Bouillon, C., 2013. Wildfire frequency varies with the size and shape of fuel types in southeastern France: implications for environmental management. Journal of Environmental Management 117, 150–161. Available at https://doi.org/10.1016/j.jenvman.2012.12.006.

Davim, D.A., Rossa, C.G., Pereira, J.M.C., Guiomar, N., Fernandes, P.M., 2023. The effectiveness of past wildfire at limiting reburning is short-lived in a Mediterranean humid climate. Fire Ecology 19 (1). Available at https://doi.org/10.1186/s42408-023-00227-x.

Dennison, P.E., Moritz, M.A., 2009. Critical live fuel moisture in chaparral ecosystems: a threshold for fire activity and its relationship to antecedent precipitation. International Journal of Wildland Fire 18 (8), 1021–1027. Available at https://doi.org/10.1071/WF08055.

Duane, A., Kelly, L., Giljohann, K., Batllori, E., McCarthy, M., Brotons, L., 2019. Disentangling the influence of past fires on subsequent fires in Mediterranean landscapes. Ecosystems 22 (6), 1338–1351. Available at https://doi.org/10.1007/s10021-019-00340-6.

Duane, A., Castellnou, M., Brotons, L., 2021. Towards a comprehensive look at global drivers of novel extreme wildfire events. Climatic Change 165 (3-4). Available at https://doi.org/10.1007/s10584-021-03066-4.

Fernandes, P.M., Loureiro, C., Guiomar, N., Pezzatti, G.B., Manso, F.T., Lopes, L., 2014. The dynamics and drivers of fuel and fire in the Portuguese public forest. Journal of Environmental Management 146, 373–382. Available at https://doi.org/10.1016/j.jenvman.2014.07.049, https://www.sciencedirect.com/journal/journal-of-environmental-management.

Fernandes, P.M., Barros, A.G., Pinto, A., Santos, J.A., 2016a. Characteristics and controls of extremely large wildfires in the western Mediterranean Basin. Journal of Geophysical Research: Biogeosciences 121, 2141–2157.

Fernandes, P.M., Monteiro-Henriques, T., Guiomar, N., Loureiro, C., Barros, A., 2016b. Bottom-up variables govern large-fire size in Portugal. Ecosystems 19, 1362–1375.

Fernandes, P.M., Delogu, G.M., Leone, V., Ascoli, D., 2020a. Wildfire policies contribution to foster extreme wildfires. In: Tedim, F., Leone, V., McGee, T. (Eds.), Extreme Wildfire Events and Disasters. Elsevier, pp. 187–200.

Fernandes P.M., Sil A., Rossa C.G., Ascoli D., Cruz M.G., Alexander M.E., 2020b. Characterizing fire behavior across the globe. In: Hood, S., Drury, S., Steelman, T., Steffens, R., technical editors. The Fire Continuum—Preparing for the Future of Wildland Fire: Proceedings of the Fire Continuum Conference. Proceedings RMRS-P-78. Fort Collins (CO): US Department of Agriculture, Forest Service, Rocky Mountain Research Station, pp. 258–263.

Fernández-Guisuraga, J.M., Calvo, L., 2025. Fire severity shows limited dependence on fuel structure under adverse fire weather conditions: a case study of two extreme wildfire events. Fire Ecology 21 (1). Available at https://doi.org/10.1186/s42408-025-00371-6, https://www.springer.com/journal/42408.

Fernández-Guisuraga, J.M., Martins, S., Fernandes, P.M., 2023. Characterization of biophysical contexts leading to severe wildfires in Portugal and their environmental controls. Science of the Total Environment 875, 162575.

Finney, M.A., McAllister, S.S., 2011. A review of fire interactions and mass fires. Journal of Combustion 2011. Available at https://doi.org/10.1155/2011/548328.

Fromm, M., Lindsey, D.T., Servranckx, R., Yue, G., Trickl, T., Sica, R., et al., 2010. The untold story of pyrocumulonimbus. Bulletin of the American Meteorological Society 91 (9), 1193–1209. Available at https://doi.org/10.1175/2010BAMS3004.1, http://journals.ametsoc.org/doi/pdf/10.1175/2010BAMS3004.1.

Gale, M.G., Cary, G.J., Van Dijk, A.I.J.M., Yebra, M., 2021. Forest fire fuel through the lens of remote sensing: review of approaches, challenges and future directions in the remote sensing of biotic determinants of fire behaviour. Remote Sensing of Environment 255, 112282. Available at https://doi.org/10.1016/j.rse.2020.112282.

Haines, D.A., Sando, R.W. 1969. Climatic conditions preceding historically great fires in the North Central Region. Research Paper NC-34. St. Paul (MN): US Department of Agriculture, Forest Service, North Central Forest Experiment Station.

Hirsch, K.G., Martell, D.L., 1996. A review of initial attack fire crew productivity and effectiveness. International Journal of Wildland Fire 6 (4), 199–215. Available at https://doi.org/10.1071/wf9960199.

Huete, A., Didan, K., Miura, T., Rodriguez, E.P., Gao, X., Ferreira, L.G., 2002. Overview of the radiometric and biophysical performance of the MODIS vegetation indices. Remote Sensing of Environment 83 (1-2), 195–213. Available at https://doi.org/10.1016/s0034-4257(02)00096-2.

Jia, Q., Quan, X., Resco de Dios, V., Yebra, M., He, B., Li, X., et al., 2026. Enhancing two-week live fuel moisture content forecasts through biophysical modelling and remote sensing data assimilation. Remote Sensing of Environment 335, 115267. Available at https://doi.org/10.1016/j.rse.2026.115267.

Johnson, R.H., Schumacher, R.S., Ruppert, J.H., Lindsey, D.T., Ruthford, J.E., Kriederman, L., 2014. The role of convective outflow in the waldo canyon fire. Monthly Weather Review 142 (9), 3061–3080. Available at https://doi.org/10.1175/mwr-d-13-00361.1.

Johnston, J.M., Wooster, M.J., Paugam, R., Wang, X., Lynham, T.J., Johnston, L.M., 2017. Direct estimation of Byram's fire intensity from infrared remote sensing imagery. International Journal of Wildland Fire 26 (8), 668–684. Available at https://doi.org/10.1071/WF16178, http://www.publish.csiro.au/nid/126.htm.

Jones, M.W., Kelley, D.I., Burton, C.A., Di Giuseppe, F., Barbosa, M.L.F., Brambleby, E., et al., 2024. State of wildfires 2023–24. Earth System Science Data 16, 3601–3685.

Keane, R.E., Agee, J.K., Fuĺ, P., Keeley, J.E., Key, C., Kitchen, S.G., et al., 2008. Ecological effects of large fires on US landscapes: benefit or catastrophe? International Journal of Wildland Fire 17 (6), 696–712. Available at https://doi.org/10.1071/WF07148.

Krebs, P., Pezzatti, G.B., Mazzoleni, S., Talbot, L.M., Conedera, M., 2010. Fire regime: history and definition of a key concept in disturbance ecology. Theory in Biosciences 129 (1), 53–69. Available at https://doi.org/10.1007/s12064-010-0082-z.

Lagouvardos, K., Kotroni, V., Giannaros, T.M., Dafis, S., 2019. Meteorological conditions conducive to the rapid spread of the deadly wildfire in eastern attica, Greece. Bulletin of the American Meteorological Society 100 (11), 2137–2145. Available at https://doi.org/10.1175/BAMS-D-18-0231.1, https://journals.ametsoc.org/doi/pdf/10.1175/BAMS-D-18-0231.1.

Linley, G.D., Jolly, C.J., Doherty, T.S., Geary, W.L., Armenteras, D., Belcher, C.M., et al., 2022. What do you mean, 'megafire'? Global Ecology and Biogeography 31 (10),

1906–1922. Available at https://doi.org/10.1111/geb.13499, http://onlinelibrary.wiley.com/journal/10.1111/(ISSN)1466-8238.

Lloret, F., Calvo, E., Pons, X., Díaz-Delgado, R., 2002. Wildfires and landscape patterns in the Eastern Iberian Peninsula. Landscape Ecology 17 (8), 745–759. Available at https://doi.org/10.1023/A:1022966930861.

Mahood, A.L., Tuff, T., Cattau, M.E., Iglesias, V., Hanson, T.E., Stephens, J., et al., 2025. The properties of individual fire events are essential for understanding global fire regimes. Global Ecology and Biogeography 34 (11). Available at https://doi.org/10.1111/geb.70145, http://onlinelibrary.wiley.com/journal/10.1111/(ISSN)1466-8238.

Nelson, R.M., 1993. Byram derivation of the energy criterion for forest and wildland fires. International Journal of Wildland Fire 3 (3), 131–138. Available at https://doi.org/10.1071/wf9930131.

Nelson, R.M., 2001. Water relations of forest fuels. In: Johnson, E.A., Miyanishi, K. (Eds.), Forest fires: Behavior and ecological effects. Academic Press, San Diego (CA), pp. 79–149.

Noble, J.C., 1991. Behaviour of a very fast grassland wildfire on the riverine plain of southeastern Australia. International Journal of Wildland Fire 1 (3), 189–196. Available at https://doi.org/10.1071/wf9910189.

Nolan, R.H., Resco de Dios, V., Boer, M.M., Caccamo, G., Goulden, M.L., Bradstock, R.A., 2016. Predicting dead fine fuel moisture at regional scales using vapour pressure deficit from MODIS and gridded weather data. Remote Sensing of Environment 174, 100–108. Available at https://doi.org/10.1016/j.rse.2015.12.010, www.elsevier.com/inca/publications/store/5/0/5/7/3/3.

Ottmar, R.D., 2014. Wildland fire emissions, carbon, and climate: Modeling fuel consumption. Forest Ecology and Management 317, 41–50. Available at https://doi.org/10.1016/j.foreco.2013.06.010, www.elsevier.com/inca/publications/store/5/0/3/3/1/0.

Pausas, J.G., Fernández-Muñoz, S., 2012. Fire regime changes in the Western Mediterranean basin: From fuel-limited to drought-driven fire regime. Climatic Change 110 (1-2), 215–226. Available at https://doi.org/10.1007/s10584-011-0060-6.

Potter, B.E., 2012. Atmospheric interactions with wildland fire behaviour - II. Plume and vortex dynamics. International Journal of Wildland Fire 21 (7), 802–817. Available at https://doi.org/10.1071/WF11129.

Rao, K., Williams, A.P., Flefil, J.F., Konings, A.G., 2020. SAR-enhanced mapping of live fuel moisture content. Remote Sensing of Environment 245, 111797. Available at https://doi.org/10.1016/j.rse.2020.111797.

Rego, F.C., Morgan, P., Fernandes, P., Hoffman, C., 2021. Fire Science from Chemistry to Landscape Management. Springer International Publishing, Cham.

Resco De Dios, V., Boer, M.M., 2025. EU forest monitoring should combine up-to-date science with best practice. Nature Ecology & Evolution 9 (5), 743–744. Available at https://doi.org/10.1038/s41559-025-02672-0.

Resco de Dios, V., Cunill Camprubí, À., Pérez-Zanón, N., Peña, J.C., Martínez del Castillo, E., Rodrigues, M., et al., 2022. Convergence in critical fuel moisture and fire weather thresholds associated with fire activity in the pyroregions of Mediterranean Europe. Science of the Total Environment 806 (Pt 4), 151462. Available at https://doi.org/10.1016/j.scitotenv.2021.151462.

Resco de Dios, V., Schütze, S.J., Cunill Camprubí, À., Balaguer-Romano, R., Boer, M.M., Fernandes, P.M., 2025. Protected areas as hotspots of wildfire activity in fire-prone Temperate and Mediterranean biomes. Journal of Environmental Management 385, 125669. Available at https://doi.org/10.1016/j.jenvman.2025.125669.

Rodrigues, M., Cunill Camprubí, À., Balaguer-Romano, R., Coco Megía, C.J., Castañares, F., Ruffault, J., et al., 2023. Drivers and implications of the extreme 2022 wildfire season in

Southwest Europe. Science of The Total Environment 859 (Pt 2), 160320. Available at https://doi.org/10.1016/j.scitotenv.2022.160320.

Rossa, C.G., Davim, D.A., Sil, Â., Fernandes, P.M., 2024. Field-based generic empirical flame length-fireline intensity relationships for wildland surface fires. International Journal of Wildland Fire 33 (1). Available at https://doi.org/10.1071/WF23127, http://www.publish.csiro.au/nid/126.htm.

Schütze, S.J., Resco de Dios, V., 2025. Increases in the world's most extreme wildfire events probably driven by fire size and simultaneity. Nature Ecology & Evolution 9, 1345–1348.

Stephens, S.L., Burrows, N., Buyantuyev, A., Gray, R.W., Keane, R.E., Kubian, R., et al., 2014. Temperate and boreal forest mega-fires: characteristics and challenges. Frontiers in Ecology and the Environment 12 (2), 115–122. Available at https://doi.org/10.1890/120332, http://www.esajournals.org/doi/pdf/10.1890/120332.

Stephens, S.L., Bernal, A.A., Collins, B.M., Finney, M.A., Lautenberger, C., Saah, D., 2022. Mass fire behavior created by extensive tree mortality and high tree density not predicted by operational fire behavior models in the southern Sierra Nevada. Forest Ecology and Management 518. Available at https://doi.org/10.1016/j.foreco.2022.120258, www.elsevier.com/inca/publications/store/5/0/3/3/1/0.

Stoof, C.R., de Vries, J.R., Castellnou Ribau, M., F. Fernández, M., Flores, D., Galarza Villamar, J., et al., 2024. Megafire: an ambiguous and emotive term best avoided by science. Global Ecology and Biogeography 33 (2), 341–351. Available at https://doi.org/10.1111/geb.13791.

Taylor, S.W., Wotton, B.M., Alexander, M.E., Dalrymple, G.N., 2004. Variation in wind and crown fire behaviour in a northern jack pine – black spruce forest. Canadian Journal of Forest Research 34 (8), 1561–1576. Available at https://doi.org/10.1139/x04-116.

Tedim, F., Leone, V., Amraoui, M., Bouillon, C., Coughlan, M., Delogu, G., et al., 2018. Defining extreme wildfire events: difficulties, challenges, and impacts. Fire 1 (1), 9. Available at https://doi.org/10.3390/fire1010009.

Turco, M., Jerez, S., Augusto, S., Tarín-Carrasco, P., Ratola, N., Jiménez-Guerrero, P., et al., 2019. Climate drivers of the 2017 devastating fires in Portugal. Scientific Reports 9 (1). Available at https://doi.org/10.1038/s41598-019-50281-2.

Van Wagner, C.E., 1977. Conditions for the start and spread of crown fire. Canadian Journal of Forest Research 7 (1), 23–34. Available at https://doi.org/10.1139/x77-004.

Wade, D.D., Ward, D.E., 1973. An analysis of the Airforce Bomb Range Fire. Research Paper SE-105. Asheville (NC): US Department of Agriculture, Forest Service, Southeastern Forest Experiment Station.

Wendel, G.W., Storey, T.G., Byram, G.M., 1962. Forest fuels on organic and associated soils in the coastal plain of North Carolina. Station Paper SFES-SP-144. Asheville (NC): US Department of Agriculture, Forest Service, Southeastern Forest Experiment Station.

Werth, P.A., Potter, B.E., Alexander, M.E., Clements, C.B., Cruz, M.G., Finney, M.A., et al., 2016. Synthesis of knowledge of extreme fire behavior: volume 2 for fire behavior specialists, researchers, and meteorologists. Gen. Tech. Rep. PNW-GTR-891. Portland, OR: US Department of Agriculture, Forest Service, Pacific Northwest Research Station. 258 p.

Yebra, M., Dennison, P.E., Chuvieco, E., Riaño, D., Zylstra, P., Hunt, E.R., et al., 2013. A global review of remote sensing of live fuel moisture content for fire danger assessment: moving towards operational products. Remote Sensing of Environment 136, 455–468. Available at https://doi.org/10.1016/j.rse.2013.05.029.

Part II

Technological innovations across the forest fire management cycle

Forest fire ecology and ecosystem dynamics: foundations for integrated management strategies

6

Alfonso Fernández-Manso[1] and Carmen Quintano[2]
[1]Agrarian Science and Engineering Department, University of León, Ponferrada, Spain,
[2]Electronic Technology Department, IuFOR, University of Valladolid, Valladolid, Spain

6.1 Introduction

Forest fires are one of the main agents of disturbance in terrestrial ecosystems, with profound effects on biodiversity, biogeochemical cycles, carbon dynamics, and human safety. They have also been the most recurrent disturbance in many of the planet's ecosystems for millennia (Pausas and Keeley, 2021). In recent decades, their frequency, extent, and intensity have increased significantly in large regions of the world as a result of the interaction between climate change, changes in land use, and the abandonment of traditional land management practices (Chuvieco et al., 2020). In this context, remote sensing has established itself as a key tool for the comprehensive analysis of the forest fire cycle, providing spatially and temporally explicit information on all its phases: prefire, active fire, and postfire. From a contemporary perspective, forest fires are no longer understood as isolated events, but as dynamic and recurring processes integrated into complex socio-ecological systems (McLauchlan et al., 2020).

This view has driven the development of fire cycle management approaches that require consistent, repeatable, and comparable data at multiple spatial and temporal scales. Remote sensing, using optical, thermal, radar, and light detection and ranging (LiDAR) sensors, offers a unique ability to meet these needs, acting as a backbone between fire science, preventive planning, operational response, and postfire restoration (Guiop-Servan et al., 2025).

In the prefire phase, remote sensing research has focused on characterizing forest fuel and assessing the structural and functional hazards of the landscape. Variables such as fuel load and continuity, species composition, canopy vertical structure, and, particularly relevant, fuel moisture content (FMC) directly influence the probability of ignition and the potential behavior of the fire (Chowdhury and Hassan, 2015). Medium spatial resolution multispectral sensors, such as Moderate Resolution Imaging Spectroradiometer (MODIS) or Sentinel-2, have made it possible to exploit the spectral response of vegetation in the near-infrared and shortwave to infer the water and physiological states of live fuel. However, several studies have pointed out the limitations of empirical approaches, promoting the use of models based on radiative transfer due to their greater capacity for generalization across ecosystems (Guiop-Servan et al., 2025).

Advances in Remote Sensing for Forest Fire Management. DOI: https://doi.org/10.1016/B978-0-443-41692-7.00008-1
© 2027 Elsevier Ltd. All rights are reserved, including those for text and data mining, AI training, and similar technologies.

At the same time, microwave and synthetic aperture radar (SAR) sensors have emerged as a robust alternative for estimating the water status of vegetation, as they are sensitive to the dielectric constant of fuel and less dependent on atmospheric or lighting conditions. This capability is particularly relevant in regions with high cloud cover or during prolonged episodes of smoke, where optical sensors have significant limitations (Cao and Lei, 2024). In terms of fuel structure, the integration of LiDAR data has been a substantial advance, allowing direct quantification of canopy height, crown density, and vertical continuity, which are determining factors in the spread of crown fires. Although the operational use of airborne LiDAR remains constrained by its cost, recent space missions have facilitated the extrapolation of structural metrics to regional scales (Chen et al., 2024).

The active fire phase represents one of the greatest technological challenges in fire cycle management, due to the pressing need for early detection and near-real-time monitoring. Satellite systems offer a synoptic view of vast forest regions, including remote or inaccessible areas, overcoming the limitations inherent in ground-based surveillance systems (Muthulakshmi et al., 2022). The complementarity between low Earth orbit (LEO) satellites, such as MODIS or Visible Infrared Imaging Radiometer Suite (VIIRS), and geostationary satellites (GEO), such as Himawari-8 or GOES-R, has proven to be key to balancing spatial resolution and temporal frequency (Xie et al., 2018; Tian et al., 2022).

A central concept in active fire monitoring is Fire Radiative Power (FRP), which allows the intensity of the fire and biomass emissions to be estimated almost instantaneously. Advances in high-resolution thermal sensors have significantly reduced the underdetection of small fires, which account for a substantial fraction of the cumulative burned area (BA) in many ecosystems (Chen et al., 2024). Similarly, the integration of thermal, infrared, and LiDAR sensors has improved the tracking of fire fronts, propagation speed, and spatial heterogeneity of fires (Pang et al., 2020).

In recent years, the use of artificial intelligence and deep learning techniques has expanded real-time detection and monitoring capabilities. Algorithms based on convolutional neural networks (CNNs) and object detection models, such as YOLO, have proven highly effective in automatically identifying fire and smoke from video streams and multisensory data, with direct applications in early warning systems (Madkar et al., 2022). These advances reinforce the transition from reactive systems to more proactive approaches in emergency management.

The postfire phase has concentrated a substantial part of scientific production in remote sensing due to the need to quantify impacts, assess burn severity, and analyze the ecological recovery processes of affected ecosystems. In this context, the Normalized Burn Ratio (NBR) and its differential Normalized Burn Ratio (dNBR) variant have become established as reference tools for assessing severity, allowing fires to be compared across different regions and time periods (Makineci, 2024). However, several authors have emphasized the importance of conceptually differentiating between burn severity, fire intensity, and ecological response in order to avoid simplistic interpretations that could lead to errors in postfire management (Pérez-Cabello et al., 2021). In this sense, traditional spectral indices, such as dNBR or Normalized Difference Vegetation Index (NDVI), are considered proxy or latent variables, lacking standardized units and intrinsic ecological meaning. In contrast, variables derived from the energy balance, such as evapotranspiration (ET), land surface temperature (LST), and land surface albedo (LSA), allow a more direct interpretation of fire-induced changes in ecosystem energy and water flows, providing a more robust biophysical basis for assessing severity and postfire response (Quintano et al., 2020; Fernández-Manso et al., 2016; Quintano et al., 2015).

Monitoring vegetation regeneration using indices such as NDVI or the Forest Recovery Index has made it possible to identify patterns of resilience and recovery trajectories across different severity gradients (Chu et al., 2016; Meng et al., 2018b). Studies in boreal and temperate forests have shown that areas affected by moderate fires tend to recover more quickly than those subjected to extreme fires, a finding with direct implications for the planning of restoration actions (Meng et al., 2018b; Nedkov et al., 2020). The analysis of long time series, supported by cloud computing platforms, has also made it possible to evaluate changes in species composition and possible transitions to alternative ecosystem states (Chu and Guo, 2013).

Another critical aspect of postfire is the increased risk of soil erosion, especially in mountainous and Mediterranean regions. The loss of vegetation cover and the alteration of soil physical properties increase susceptibility to erosion processes, with direct impacts on hydrology and water quality. The integration of remote sensing data with erosion models, such as RUSLE, has proven effective in identifying priority areas for intervention and evaluating the effectiveness of mitigation measures (Dursun et al., 2025).

Despite the advances described, significant challenges remain related to the integration of multisource data, the harmonization of time series, and the validation of derived products. The coexistence of sensors with different spatial, temporal, and spectral resolutions requires increasingly sophisticated data fusion strategies (Cao and Lei, 2024). Furthermore, the transition between satellite missions raises long-term consistency issues that must be addressed to ensure robust analyses of global change (Guiop-Servan et al., 2025).

Overall, the scientific literature agrees that remote sensing has evolved from an observation tool to a central component of integrated forest fire cycle management systems. Its ability to connect biophysical processes, spatial dynamics, and decision-making positions it as a strategic field of research and development in a scenario of accelerated global change (Chen et al., 2024; Chuvieco et al., 2020).

In this context, the objective of this chapter is to provide an integrated and updated review of advances in remote sensing applied to the entire forest fire cycle, coherently addressing the prefire, active fire, and postfire phases from a multiscale and multisource perspective, as conceptually summarized in Fig. 6.1, which illustrates the fire cycle from a 360-degree management approach. The chapter analyzes the main sensors, biophysical variables, indices, and methodological approaches developed in recent decades, with particular emphasis on their conceptual foundations, operational capabilities, and limitations. It also discusses the emerging role of artificial intelligence, data fusion, and cloud computing platforms as key elements for integrated fire management. The scope of the chapter is not limited to a technical

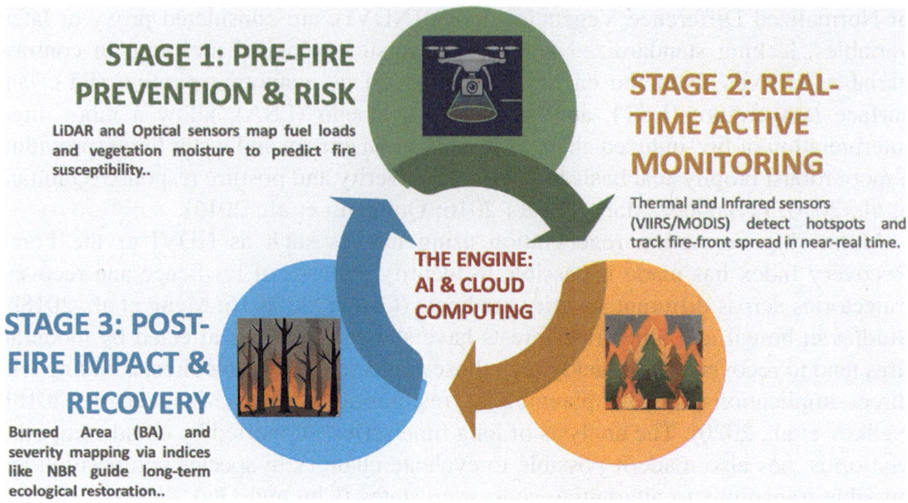

Figure 6.1 The remote sensing fire cycle: a 360-degree management framework: advances in remote sensing applied to the forest fire cycle.

review but aims to connect scientific advances with their application in preventive planning, operational response, and postfire restoration, thus contributing to a more systemic understanding of fire as a socio-ecological process and to the formulation of adaptive land management strategies in a scenario of global change.

6.2 Fuel types and remote sensing

The risk of forest fire and its emergent behavior depend on the interaction between meteorological conditions, topography, and fuel properties (Countryman, 1972; Chuvieco et al., 2023). Among these factors, plant fuel is the only element that can be directly managed at the landscape scale, making its characterization a strategic component of risk prevention and planning. Over large areas, prevention policies are based primarily on fuel management strategies aimed at reducing fuel loads, modifying their horizontal and vertical continuity, and creating structural discontinuities such as auxiliary strips and firebreaks.

From a conceptual and operational point of view, forest fuels are described using the complementary concepts of fuel types and fuel models (Arroyo et al., 2008). Fuel types group relatively homogeneous categories of vegetation according to their fire spread properties, while fuel models provide a quantitative parameterization of these categories, differentiating between surface and canopy fuels. These parameters constitute the fundamental input for fire behavior models, which estimate variables such as propagation speed, flame length, intensity, and the ecological and atmospheric effects of the fire.

The coherent combination of fuel type maps and quantitative models has been identified as a key requirement for improving fire monitoring and risk assessment at multiple scales (Aragoneses and Chuvieco, 2021; Aragoneses et al., 2024; Chuvieco et al., 2020; Gale et al., 2022). In particular, the move toward operational fire management systems requires consistent spatial representations of forest canopy structure, given its decisive role in the development of crown fires, one of the most extreme and difficult scenarios to control.

Recent literature agrees on identifying a small set of canopy parameters as particularly relevant for describing the behavior of crown fires and their associated impacts (Gale, 2021). These include mean canopy height (Hm), crown cover fraction (CC), crown base height (CBH), crown fuel load, and canopy bulk density (CBD) (Keane et al., 2005).

Of these five parameters, Hm, CC, and CBH provide a particularly informative spectrum for modeling fuel properties. Mean canopy height influences crown fire spread, biomass consumption, and greenhouse gas emissions, as it is related to the internal microclimate of the forest and total fuel availability (Albini, 1976). Canopy cover controls the horizontal continuity of fuel, conditioning the probability of sustained crown fire spread. For its part, CBH determines the critical threshold for the transition between surface fires and crown fires, which is one of the key processes in forest fire dynamics (Ferraz et al., 2009).

At the continental scale, the growing availability of forest information on Hm and CC has opened up new opportunities for their systematic characterization, as demonstrated by recent pan-European initiatives based on forest inventories and remote sensing (Ruiz-Benito et al., 2020). However, CBH remains one of the most difficult parameters to estimate consistently due to the scarcity of direct data and its high spatial variability, which limits its operational incorporation into large-scale fire behavior models.

Traditional methods for estimating structural canopy parameters include field inventories, aerial photogrammetry, and ecological models. Although the first two offer high local accuracy, their applicability is restricted to small scales, while ecological models present high levels of uncertainty when extrapolated to large regions (Arroyo et al., 2008; Gale, 2021; Chuvieco et al., 2020). These limitations have driven, over the last two decades, the systematic use of remote sensing data calibrated with field information for the spatially explicit estimation of fuel parameters.

A wide range of sensors has been used, including medium- and high-resolution multispectral imagery, as well as active sensors such as LiDAR and radar, either independently or in combination (Gale, 2021). Among these technologies, LiDAR has established itself as the most powerful tool for structural fuel characterization, thanks to its ability to directly capture the vertical structure of the forest canopy. In particular, airborne LiDAR Airborne Laser Scanning (ALS) has demonstrated high accuracy in estimating key parameters such as Hm (Riaño et al., 2003), CC (Arumäe and Lang, 2018), CBH (Stefanidou et al., 2020), and CBD (Riaño et al., 2003).

Most previous studies have been based on high-density ALS observations, generally exceeding four pulses per square meter, which allow for a detailed reconstruction of the forest canopy. However, this approach has operational limitations associated with acquisition costs and spatial data availability. In contrast, studies based on terrestrial or low-density LiDAR have been comparatively scarce, although they have shown promising results in experimental and validation contexts (Alonso-Rego et al., 2021).

A recent and particularly relevant advance is the integration of spatial LiDAR data with multispectral satellite information for large-scale fuel parameter mapping. The use of LiDAR carried on space platforms, combined with optical images and advanced statistical models, has made it possible to partially overcome the coverage limitations of ALS and move toward operational products on a continental scale. A notable example is the work of (Aragoneses et al., 2024), which demonstrates the feasibility of mapping forest canopy fuel parameters at the European scale by integrating spatial LiDAR and satellite data.

Such approaches represent a paradigm shift in fire management, enabling a consistent structural characterization of the fuel that directly feeds into fire behavior models and risk assessment systems. Overall, remote sensing has evolved from a descriptive support tool to a central component of integrated forest fire cycle management systems, providing the quantitative basis necessary for more effective and adaptive prevention in a context of accelerated global change.

6.3 Wildfire risk assessment, forecasting, and remote sensing

Wildfires pose a growing threat to ecosystems, infrastructure, and human safety, especially in the context of climate change, urban-forest interface expansion, and rapid changes in land use. Fire risk assessment and forecasting have therefore become strategic components of modern fire management, as they enable the anticipation of conditions favorable to ignition and spread before a fire occurs. Remote sensing plays a central role in this area by providing spatially explicit, consistent, and up-to-date information on the state of vegetation, climate, and land surface, which, combined with advanced modeling and machine learning techniques, underpins operational prevention and early warning systems.

From a conceptual point of view, fire risk and behavior depend on the interaction between weather conditions, topography, and fuel properties (Countryman, 1972). Of these three components, fuel is the only factor that can be directly managed at the landscape scale, which underlines the importance of its correct characterization in the prefire phase. However, even with a detailed description of the fuel, risk assessment requires the integration of dynamic information on the water status of vegetation, surface temperature, and atmospheric conditions, variables that remote sensing allows to be monitored systematically at multiple scales.

Many fire risk assessment systems are based on indices derived from remote sensors, especially those related to vegetation vigor and dryness. Satellite platforms such as MODIS have been widely used to calculate indicators such as the NDVI, LST, and Temperature Vegetation Dryness Index (TVDI). These indices provide an indirect but effective approximation of the water status of fuel and, therefore, its potential flammability (Chen et al., 2023; Kim et al., 2019). While NDVI provides information on greenness and photosynthetic activity, LST is closely associated with thermal stress and fuel desiccation; TVDI combines both dimensions to improve the spatial discrimination of fire-prone areas.

At more local scales, the use of high spatial resolution imagery, such as that provided by the National Agriculture Imagery Program (NAIP), has proven particularly useful for characterizing urban-forest interface areas, where anthropogenic pressure significantly increases the risk of ignition (Shen et al., 2023). The combination of medium- and high-resolution data allows for the simultaneous capture of regional hazard patterns and local risk hotspots, an essential requirement for preventive planning and efficient resource allocation.

The volume, diversity, and frequency of remote sensing data used in fire risk assessment require advanced processing frameworks capable of handling temporal, spatial, and spectral variability. In this context, machine learning and deep learning algorithms have taken on an increasingly prominent role. Deep learning techniques are used, for example, to improve the detection and correction of cloudiness in optical images, increasing the reliability of derived indices (Li et al., 2024).

In the prefire phase, machine learning algorithms, such as random forest and logistic regression, are systematically applied to classify fire-prone areas based on historical patterns, integrating remote sensing, topography, and climatology variables (Brys et al., 2025). These approaches enable the generation of probabilistic risk maps that help managers prioritize preventive actions, such as prescribed burns, silvicultural treatments, or the reinforcement of defense lines before ignition occurs.

A recent advance in this field is the use of ensemble learning methods, which combine multiple classifiers to improve the robustness and accuracy of predictions. Krishna Reddy et al. (2025) show that these approaches are particularly effective in integrating heterogeneous data from remote sensors, topographic information, and meteorological records, reducing the sensitivity of the model to individual biases in each data source.

Among the most widely used fire risk prediction systems at the operational level is the Fire Weather Index (FWI), a central component of the Canadian Forest Fire Danger Rating System (CFFDRS). This index integrates meteorological variables such as temperature, relative humidity, wind, and precipitation to estimate daily fire danger levels (Wang et al., 2017). Although it has traditionally been based on meteorological observations, recent research has shown that integration with numerical weather prediction models substantially improves short-term forecasting capabilities, including hourly scales relevant to operational planning (Rodell et al., 2024).

More recently, hybrid models have emerged that combine advanced deep learning architectures, such as graph neural networks (GNN), temporal convolutional networks (TCN), and long short-term memory (LSTM) models, for the spatiotemporal forecasting of fire danger indices. These models allow for the capture of complex spatial dependencies and nonlinear temporal dynamics, offering high-resolution predictions in regions with heterogeneous climates and topographies (Shahriar et al., 2025). Their application opens up new possibilities for anticipating extreme risk events under climate change scenarios.

Fire risk assessment is operationalized in many contexts through integrated decision support systems. The Wildland Fire Risk Assessment System (WFRAS) is an early example of a platform that combines spectral data, topography, fuel load, and meteorological variables to map fire hazard and potential severity (Smith et al., 2007). Such systems make it possible to link remote sensing information with preventive planning and postfire mitigation, closing the fire management cycle.

The incorporation of Internet of Things (IoT) technologies and artificial intelligence systems has further expanded the capabilities of these systems. On-site sensor networks, integrated with satellite data and machine learning models, enable near-real-time monitoring of critical conditions and the issuance of early warnings to emergency management centers, especially in remote regions where direct observation is limited (Janardhan et al., 2025).

A key aspect highlighted in the literature is that poor risk assessment in the prefire phase compromises the entire fire cycle. During active fire, inaccurate mapping of fuel or hazards leads to errors in real-time propagation models, increasing the risk to firefighting crews and reducing the effectiveness of suppression operations. In the postfire phase, prior knowledge of the structure and condition of vegetation is an indispensable reference point for assessing the severity of damage and the resilience of the ecosystem during recovery. In this sense, remote sensing-based prevention is not an isolated phase, but the foundation on which comprehensive and adaptive fire cycle management is built.

Despite the progress made, significant challenges remain for the full operation of fire risk assessment and forecasting systems. The integration of diverse data sets—satellite, ground sensors, and historical records—remains complex due to differences in formats, resolutions, and temporal coverage (Ejaz and Choudhury, 2025). Furthermore, the growing complexity of deep learning-based models raises issues of interpretability, which has driven the development of explainable artificial intelligence frameworks that allow predictions to be understood and justified to managers and policymakers (Xu et al., 2025).

Finally, system scalability and real-time processing are emerging as critical requirements in a scenario marked by exponential growth in the volume and heterogeneity of available data. In this context, architectures based on cloud computing and edge computing are considered essential to ensure resilient remote sensing systems that are adaptable to different geographical contexts and capable of providing timely and operational information for decision-making in forest fire risk management (Janardhan et al., 2025). This technological evolution is giving rise to new-generation methodologies for regional risk mapping, which integrate advanced remote sensing data and field information sources.

6.4 Fire detection, monitoring, and remote sensing

Fire detection and monitoring using remote sensing have become a structural component of modern fire management, enabling early identification of hotspots, near real-time monitoring of their evolution, and operational integration of multiscale information to support suppression and postevent assessment (E. Chuvieco et al., 2020; Wooster et al., 2021). In a context of intensifying fire regimes under climate change and land-use transformations, the added value of remote sensing lies in its ability to provide repeatable and comparable observations over large areas, overcoming the limitations of access, coverage, and continuity inherent in ground-based surveillance (Mohapatra and Trinh, 2022; Szpakowski and Jensen, 2019). Recent advances in sensors, satellite constellations, airborne platforms, and algorithms, including the rise of deep learning, have reconfigured the state of the art, shifting the field from predominantly threshold-based approaches toward multimodal architectures geared toward automation and false alarm reduction (Vasconcelos et al., 2024; Yang et al., 2024).

From an operational standpoint, remote fire detection relies on three families of complementary platforms: ground-based, airborne, and satellite systems. Ground-based systems include fixed optical cameras (sometimes with 360-degree coverage) and, in some deployments, networks of environmental sensors that monitor conditions associated with fire initiation. Their main strength is very early detection in environments close to critical infrastructure or high-risk areas, but their spatial coverage is limited, and their performance is sensitive to lighting conditions, weather, and occlusions by terrain or vegetation (Barmpoutis et al., 2020a,b). The technological trend in this area focuses on computer vision applied to video sequences to discriminate smoke and flames by color, texture, and dynamics, with the aim of reducing false positives and increasing robustness in complex scenarios (Barmpoutis et al., 2020b).

Airborne systems, manned aircraft, and above all, unmanned aerial vehicles (UAVs) play a key role when high spatial resolution and tactical flexibility are required. The deployment of RGB, multispectral, and thermal (mid-infrared [MIR]/thermal infrared [TIR]) cameras makes it possible to locate small hotspots, identify hotspots, delimit active perimeters, and characterize propagation patterns with a level of detail that is difficult to achieve by satellite in critical time windows (Allison et al., 2016). The availability of UAV RGB/IR datasets specifically designed for detection and monitoring has driven algorithmic developments in classification and segmentation, opening the door to near real-time operational analysis chains to support firefighting (Chen et al., 2022). However, UAVs have logistical and regulatory limitations (autonomy, flight regulations, wind, and smoke conditions), so their maximum value emerges as an element of "operational zoom" within a multiscale system, rather than as a substitute for the satellite component (Allison et al., 2016; Mohapatra and Trinh, 2022).

Satellites are at the heart of regional and global fire monitoring due to their systematic coverage and historical continuity. LEO sensors have supported operational products on active fires and, complementarily, on BA for decades; their advantage is moderate spatial resolution with the ability to characterize relatively fine patterns, while their main limitation is revisit frequency (Wooster et al., 2021). In parallel, geostationary (GEO) sensors provide temporal cadences of minutes, essential for capturing intraday dynamics, albeit at the cost of coarser spatial resolution. This dichotomy has consolidated the GEO–LEO integration strategy as an emerging standard for balancing early detection and spatial accuracy (Wooster et al., 2021; Tian et al., 2022). The trend toward integrated multisource architectures is consistent with recent reviews documenting improvements in sensitivity to small fires, reduced uncertainty, and greater robustness against clouds and smoke by combining constellations and sensors with different spectral windows (Tian et al., 2022; Yang et al., 2024).

In terms of physical fundamentals, remote fire detection relies primarily on two signatures: (1) thermal radiation emitted by combustion, and (2) optical signals associated with flames and smoke. In the thermal domain, algorithms exploit brightness temperature anomalies in the MIR and TIR ranges, where the contrast between burning pixels and the background is usually at its maximum. Historically, thermal detection has evolved from fixed threshold methods and bispectral approaches to spatial contextual approaches, which are more robust to background variability and surface heterogeneity, and to multitemporal strategies based on sudden changes in observation series (Wooster et al., 2021; Hua and Shao, 2017). The operational maturity of these approaches is reflected in the consolidation of global active fire products and their widespread use for fire activity analysis and response support (Chuvieco et al., 2020; Wooster et al., 2021).

In the optical domain, algorithms focus on smoke and flame detection using spectral characteristics and spatial patterns, especially in ground-based and airborne systems, where resolution facilitates the discrimination of textures and dynamics. Specific reviews on early detection with optical remote sensing highlight the role of smoke movement modeling, flame flickering, and segmentation in variable lighting conditions, although they emphasize the persistence of false positives in the presence of low clouds, dust, or unrelated thermal sources (Barmpoutis et al., 2020a; Sulthana et al., 2023). Along these lines, the development of wide-angle aerial sensors and their use with deep networks oriented toward dynamic textures has expanded early detection capabilities in complex scenarios (Barmpoutis et al., 2020a).

Beyond detecting the "presence of fire," modern remote sensing integrates derived variables that connect detection with quantification. The estimation of FRP has been established as a proxy for combustion and emission rates, providing an operational link between observation and atmospheric/ecological modeling (Wooster et al., 2021). Likewise, the combination of pre and postfire observations allows for the mapping of BA and severity, completing the monitoring cycle and providing essential context for interpreting the activity detected during the active fire (Chuvieco et al., 2019; Tian et al., 2022). In particular, historical developments and methodological advances in BA mapping provide an essential reference framework for assessing consistency between active hotspots and final footprints, as well as for validating detection products (Chuvieco et al., 2019).

In recent years, deep learning has emerged as one of the most dynamic and disruptive lines of research. Recent reviews document rapid growth in work applying CNNs, object detection approaches, semantic segmentation, and more complex architectures to satellite, UAV, and terrestrial video data, with notable improvements in sensitivity and false alarm reduction compared to classical methods (Ghali and Akhloufi, 2023; Vasconcelos et al., 2024). The literature clearly distinguishes three main tasks: classification (fire/no fire), detection (location of hotspots or smoke columns), and segmentation (pixel-by-pixel delimitation of active regions). In the satellite field, deep learning methods tend to improve the detection of small fires and capture complex spatial patterns that challenge traditional contextual thresholds (Ghali and Akhloufi, 2023; Yang et al., 2024). In UAVs and aerial platforms, RGB/IR combinations and segmentation models have proven highly effective in simultaneously discriminating fire and smoke with high spatial accuracy, supported by specific datasets and reproducible evaluation protocols (Allison et al., 2016; Chen et al., 2024).

Another high-impact trend is the fusion of multisource and multiscale data. The integration of sensors with different spatial/temporal resolutions and different spectral domains (optical, thermal, and even hyperspectral) seeks to overcome the individual limitations of each platform and move toward a continuous surveillance system. Studies aimed at monitoring fire spread and dynamics highlight the value of integrating multiple sources to reconstruct spatiotemporal trajectories, extract active fronts, and link fire evolution to fuel and weather conditions (Tian et al., 2022). On a more exploratory level, hyperspectral remote sensing is emerging as a relevant scientific frontier: its ability to discriminate fine spectral signatures opens up possibilities for characterizing emissions, estimating temperatures, and improving the separation between fire, smoke, and background signals; however, its operational adoption depends on availability, acquisition frequency, and costs (Veraverbeke et al., 2018).

Despite progress, technical and methodological challenges remain. The detection of small or low-intensity fires, especially in heterogeneous mosaics or under dense cloud cover/smoke, remains an open problem that affects the completeness of global products (Hua and Shao, 2017; Wooster et al., 2021). The trade-off between spatial and temporal resolution also remains: GEO platforms offer high cadence but coarse resolution, while LEO platforms provide more detail with less frequent revisits, reinforcing the need for integration strategies (Tian et al., 2022; Wooster et al., 2021). From a data perspective, the scarcity of high-quality references and geographical/biome bias in training sets limits the generalization of deep models, emphasizing the need for curated datasets and standardized benchmarks, as well as reproducible and comparable assessments across regions (Graham et al., 2023; Yang et al., 2024). In particular, reviews focusing on African contexts and veld vegetation fires highlight both methodological advances and persistent gaps in data coverage, validation, and operational transfer (Graham et al., 2023).

Finally, bibliometric analyses and thematic reviews suggest that scientific output on fire detection and monitoring using remote sensing, especially at the intersection with deep learning, will continue to grow rapidly, driven both by the urgency of the problem and by the expansion of Earth observation missions and computational capabilities (Feitosa et al., 2024; Zhang et al., 2025). This expansion, however, increases the demand for rigor in validation, uncertainty quantification, and operational transfer: the challenge is no longer just to detect fire, but to do so in a traceable, comparable, and useful way for real-time decisions and robust scientific inferences about fire dynamics, impacts, and climate feedbacks (Chuvieco et al., 2019; Szpakowski and Jensen, 2019).

6.5 Remote sensing and fire effects: burned area and severity

Remote sensing is currently the main source of spatially and temporally consistent information for quantifying the effects of fire using two core variables: BA and burn severity. Both are essential for estimating ecological impacts, assessing emissions, and guiding postfire restoration, as they provide metrics that can be compared between fires, regions, and time periods (Chuvieco et al., 2020; Chuvieco et al., 2019; Kurbanov et al., 2022). The development of new sensors (e.g., Sentinel-2), the refinement of near-infrared (NIR)–shortwave infrared (SWIR)-based spectral indices, and the consolidation of machine learning and deep learning algorithms have significantly improved the ability to map burn patterns and their "effective intensity" from local to global scales, in time windows increasingly closer to operational ones (Hu et al., 2021; Han et al., 2024; Wu et al., 2024).

The term BA refers to the surface area actually affected by fire. In practice, it does not correspond to a continuous perimeter; it incorporates the internal heterogeneity inherent in fires, with mosaics of burned and unburned patches within the overall outline, a feature that is particularly relevant in Mediterranean landscapes, savannahs, and agroforestry mosaics (Chuvieco et al., 2019; Kurbanov et al., 2022). Burn severity, on the other hand, describes the degree of change induced by the fire in the vegetation and soil, and is usually assessed using integrated indicators of damage to above-ground biomass and the edaphic stratum, with the Composite Burn Index (CBI) being one of the reference metrics for calibration and validation of remote sensing products (Miller and Thode, 2007; Key and Benson, 2006).

Remote sensing connects these biophysical changes with variations in reflectance, thermal emission, and 3D structure of vegetation, allowing both the extent of the fire footprint and internal severity gradients to be mapped. In spectral terms, the most exploited changes are concentrated in NIR and SWIR: the loss of chlorophyll and leaf structure reduces NIR reflectance, while surface char, ash, and desiccation markedly alter SWIR, providing a coherent physical mechanism for indices such as NBR/dNBR and variants (Epting et al., 2005; Roy et al., 2006; Kurbanov et al., 2022).

At the regional and global scales, BA products are mostly based on moderate- and medium-resolution multispectral sensors (e.g., advanced very high resolution radiometer [AVHRR], MODIS, VIIRS, Sentinel-2, Landsat), which balance spatial coverage, repeatability, and radiometric stability (Chuvieco et al., 2019; Kurbanov et al., 2022). Early global algorithms combined MODIS time series with spectral thresholds, change detection, and contextual rules to identify fire scars and generate operational maps (Loboda et al., 2007). Subsequent developments have incorporated adaptive approaches and machine learning, with notable results in the global classification of BA using random forest on MODIS series and derived variables (Ramo and Chuvieco, 2017), as well as data mining strategies to improve consistency and reduce regional biases (Ramo et al., 2018). Currently, BA mapping based on Landsat and Sentinel-2 has promoted multidecadal series with greater spatial detail, supported by annual compositions, NBR minimization, or bitemporal change detection, together with classifiers (e.g., random forest) and regional growth techniques (Pulvirenti et al., 2020; Chen et al., 2025). The switch to Sentinel-2 has been particularly relevant for the problem of small fires (Roteta et al., 2019). Complementarily, synergistic strategies that integrate high-dimensional spectral information with active thermal hotspots (e.g., VIIRS) are expanding the methodological repertoire for BA by improving detection consistency in complex scenes and supporting training or scar verification (Fernández-Manso and Quintano, 2020).

The assessment of postfire severity is central to the application of remote sensing to fire management, as it allows the degree of ecological alteration to be quantified, restoration actions to be guided, and decision-making in operational contexts to be supported. Postfire damage mapping methodologies have undergone a progressive evolution, from the use of simple spectral indices, which have served as the operational standard for decades, to increasingly complex and integrative approaches. This transition includes the shift from empirical models to process-based physical models, the incorporation of machine learning techniques and species distribution information, the use of active sensors to characterize the three-dimensional structure of fuel, and, more recently, the integration of energy balance variables as a new frontier for severity estimation. Taken together, this methodological framework reflects a paradigm shift from indirect, one-dimensional approaches to biophysically grounded, multidimensional assessments that are capable of more realistically capturing the complexity of fire impacts on ecosystems.

Severity assessment is almost universally based on indices derived from NIR and SWIR. The NBR and its pre/postfire difference (dNBR) are the de facto standard for estimating changes in vegetation and surface carbon/ash, and have been widely validated against field measurements such as the CBI (Key and Benson, 2006; Roy et al., 2006; Kurbanov et al., 2022). Beyond multispectral optics, the rise of hyperspectral data is reopening the discussion on "optimal bands" for severity, allowing the exploration of narrow spectral combinations that maximize separability (Quintano et al., 2023).

Forest fires not only alter the radiation reflected and emitted by the Earth's surface, but also cause substantial changes in the energy balance of the ecosystem, directly affecting key biophysical variables. Among these, evapotranspiration (ET), LST, and LSA undergo significant changes after a fire, and their analysis provides valuable information for assessing both the severity of the damage and postfire recovery processes. Fire induces a generalized reduction in ET (Quintano et al., 2024), which translates into an increase in surface runoff and a decrease in soil moisture, while LST tends to increase markedly in areas affected by high-severity fires (Quintano et al., 2015).

Although most BA and severity mapping comes from satellites, UAV and LiDAR platforms are gaining prominence in fine analysis and validation. Multispectral sensors carried on UAVs, with green, red, NIR, and red-edge bands, allow for pre/postfire differential indices (dNDVI, dNDRE, etc.) at centimetric-metric resolutions, with the ability to distinguish patches of high and low severity and to capture patterns that are diluted in 10–30 m pixels, especially if the effect of shadows is controlled (Carvajal-Ramírez et al., 2019; McKenna et al., 2017). These approaches are particularly useful in high-priority restoration contexts or when detailed impact mapping is required. Aerial and spaceborne LiDAR add a distinct dimension: direct quantification of structural changes in 3D (height, canopy density, vertical profiles), allowing severity to be disaggregated by strata (canopy vs understory) and capturing damage that may not be fully reflected in spectral indices. Along these lines, the integration of high-resolution spectral attributes with spatial LiDAR (e.g., GEDI) has made it possible to propose hybrid HCBI-type metrics, with the ability to map stratified severity even in scenarios of predominantly understory burning (Lin et al., 2024).

Advances in machine learning have improved both severity classification and the explicit incorporation of environmental covariates (topography, climate, previous structure). Random forest models trained with Sentinel-2 bands, indices, and topographic variables can achieve high accuracies (Wu et al., 2024). In parallel, deep learning models have accelerated their adoption for both BA and multiclass severity (Han et al., 2024; Wu et al., 2024; Fernández-Guisuraga and Fernandes, 2024).

BA and severity maps derived from remote sensing are critical inputs for quantifying impacts on biodiversity, soils, and biogeochemical cycles, as well as for estimating emissions and prioritizing restoration (Szpakowski and Jensen, 2019; Chuvieco et al., 2020; Kurbanov et al., 2022). At long scales, BA series have made it possible to study regional trends and changes in fire regimes, while severity provides an essential interpretative layer for differentiating between recurrent low-intensity fires and extreme high-severity events, with direct implications for resilience and risk of state changes (Chuvieco et al., 2019; Chen et al., 2025). At the landscape scale, initial severity strongly influences recovery trajectories: areas of high-severity tend to show slower recovery and more persistent changes in structure and functionality, especially in closed forests, while low-moderate severities are usually associated with faster recoveries, although modulated by climate, altitude, soils, and composition (Liu et al., 2025; Fernández-Manso et al., 2016; Fernández-Guisuraga et al., 2023).

Finally, the effective transfer of these advances to society is materialized in initiatives such as the FIREMAP project (Fernández-Guisuraga et al., 2024), which integrates state-of-the-art remote sensing, cloud computing, and artificial intelligence to optimize decision-making processes in postfire recovery, representing a decisive step toward democratizing access to advanced tools for monitoring and planning the restoration of ecosystems affected by forest fires.

Despite the advances, methodological and operational challenges remain. The detection of small BAs and the mapping of low severity remain problematic in agricultural mosaics, dense forests, or under cloud cover, with implications for global BA accounting and the assessment of cumulative impacts (Chuvieco et al., 2019; Sismanis et al., 2023; Roteta et al., 2019). The transfer of dNBR/dNDVI thresholds between regions introduces uncertainty, reinforcing the need for regional calibrations and adaptive approaches based on machine learning and systematic validation (Tran et al., 2018; Wu et al., 2024). Furthermore, the scarcity of CBI field data and stratified severity limits the validation of deep models and hybrid metrics, especially in understory scenarios (Key and Benson, 2006; Lin et al., 2024).

In summary, remote sensing has transformed the quantification of BA and severity: from general perimeter maps to detailed descriptions of spatial patterns and vertical effects of fire, with increasingly automated and scalable products. The immediate direction of the field points to multisensor solutions (optical, thermal, SAR, LiDAR), cloud-based flows, and more explainable AI models that integrate physical-ecological knowledge, so that BA and severity feed not only fire science but also adaptive management and restoration in a changing climate (Kurbanov et al., 2022; Fernández-Guisuraga and Fernandes, 2024; Han et al., 2024).

6.6 Ecosystem recovery: monitoring ecosystem recovery and regrowth through remote sensing

The recovery of ecosystems after a forest fire is a critical phase in the fire cycle, as it determines ecological resilience, carbon sequestration capacity, soil stability, and the restoration of essential ecosystem services such as hydrological regulation, biodiversity, and habitat provision (Chuvieco et al., 2020; Pérez-Cabello et al., 2021). From a contemporary ecological perspective, recovery is not conceived as a simple return to a previous state, but as a dynamic trajectory conditioned by the severity of the fire, the recurrence of fire, the postfire climate, soil properties, and land-use history (Meng et al., 2018a; Jain et al., 2020). In this context, remote sensing has established itself as an essential tool for observing, quantifying, and comparing these recovery trajectories consistently across space and time, overcoming the limitations inherent in traditional field inventories (Stankova, 2023; Kurbanov et al., 2022). In this sense, postfire remote sensing approaches are increasingly integrated into holistic fire cycle frameworks, connecting the assessment of BAs and emissions with the monitoring of ecological recovery and long-term resilience (Chen et al., 2024).

Current conceptual frameworks for fire ecology and ecological restoration place recovery within the paradigm of resilience, understood as the speed and trajectory by which an ecosystem reorganizes itself after a disturbance and maintains its essential functions (McKenna et al., 2023). At the global level, and in line with the objectives of the United Nations Decade on Ecosystem Restoration, there is a growing need for comparable metrics to assess the structural, compositional, and functional recovery of ecosystems (Wang et al., 2024; de Almeida et al., 2025). In this regard, the ecological recovery wheel proposed by the Society for Ecological Restoration has been a significant conceptual advance, synthesizing desirable attributes of restored ecosystems. Remote sensing has been shown to directly or indirectly measure up to 11 of the 18 subattributes defined in this framework, particularly those related to canopy structure, vegetation cover, functional composition, and primary productivity (McKenna et al., 2023; Kurbanov et al., 2022).

The scientific literature of the last decade shows exponential growth in studies using satellite sensor time series to analyze postfire recovery in a wide range of biomes, from boreal and temperate forests to savannahs, Mediterranean scrublands, and arid grasslands (Pérez-Cabello et al., 2021; Sun et al., 2025). These studies agree that the integration of field data, airborne sensors, and satellite platforms, combined with temporal analysis approaches, is key to capturing the spatial heterogeneity of recovery and its interannual variability (White et al., 2022). In particular, the availability of historical Landsat archives, now harmonized with Sentinel-2, has made it possible to reconstruct recovery trajectories over several decades, providing an unprecedented perspective on return times and possible state changes induced by recurrent fires (Huang et al., 2010; Kurbanov et al., 2022).

Vegetation indices derived from optical sensors continue to be the most widely used tool for monitoring the recovery of plant regrowth. Indices such as normalized difference vegetation index (NDVI), enhanced vegetation index (EVI), normalized burn ratio (NBR), normalized difference infrared index (NDII), and their differential variants differenced normalized difference vegetation index (dNDVI), differenced normalized burn ratio (dNBR) are routinely used to describe the dynamics of greenness, vegetation moisture, and postfire regeneration (Pérez-Cabello et al., 2021; Zahabnazouri et al., 2025). Their widespread use is due to their well-established relationship with biophysical variables such as leaf area index, green biomass, and primary productivity (Sannigrahi et al., 2020). This relationship has been validated in restoring ecosystems through direct comparisons between vegetation indices derived from remote sensing, structural field measurements, and carbon fluxes estimated by eddy covariance, demonstrating that spectral trajectories capture relevant functional changes during recovery (Chasmer et al., 2018). However, numerous studies have pointed out important limitations, especially spectral saturation in dense cover and the difficulty in detecting structural changes in advanced stages of succession (Smith et al., 2007; Liu et al., 2025).

Several case studies illustrate both the potential and limitations of these approaches. In boreal larch forests in Siberia, combining a Forest Recovery Index with the fraction of vegetation cover made it possible to estimate that canopy recovery to prefire conditions may take between 30 and 47 years, with maximum regeneration rates associated with moderate severities (Chu et al., 2016). In this case, the vegetation cover fraction showed greater sensitivity during the early successional stages, while traditional spectral indices tended to stabilize before the forest structure was fully restored. In Amazonian savannahs, on the other hand, NDVI and NBR2 time series based on harmonized Landsat–Sentinel-2 datasets showed rapid recovery, reaching values similar to those prior to the fire after just three rainy seasons (Alves et al., 2025; Bousquet et al., 2022).

In contexts of anthropogenic restoration, such as open-pit mines or energy exploitation areas, vegetation indices have made it possible to identify different patterns of revegetation, stability, and secondary degradation. Multidecadal Landsat series have revealed that grasslands tend to stabilize more quickly than shrub or forest communities, and that re-degradation processes can be detected early through subtle changes in spectral trajectories (Hu et al., 2022). These results underscore the value of remote sensing not only for assessing the initial success of restoration, but also for detecting medium-term failures that require corrective interventions (Wang et al., 2024).

The structural recovery of the ecosystem, expressed in terms of canopy height, woody biomass, and vertical complexity, is not always adequately reflected in optical indices, especially when the leaf area index reaches medium or high values. To overcome these limitations, the combined use of microwave and LiDAR sensors has been intensified. Vegetation optical depth (VOD), especially in the L-band, has shown greater sensitivity to woody biomass and the structural components of trunks and branches, allowing the recovery of biomass after fires in densely vegetated biomes to be tracked on a global scale (Bousquet et al., 2022). In contrast, the C and X bands show a greater response to herbaceous and foliar vegetation, which is particularly useful for characterizing the initial stages of regrowth.

LiDAR data, both airborne and satellite-based, have taken on a central role in the structural validation of recovery. Recent studies have used airborne LiDAR and GEDI data to assess whether areas considered "spectrally recovered" actually meet relevant structural thresholds, such as those defined by the FAO in terms of height and cover (White et al., 2022). At the continental scale, metrics such as Years to Recovery based on NBR have been introduced, which estimate the time needed to reach a certain percentage of predisturbance value, allowing direct comparisons between fires and other forest disturbances, such as logging (White et al., 2022). Post-fire recovery is one of the most extensively documented processes using remote sensing. Recent reviews highlight the combined use of dNDVI, dNBR, moisture indices, fractional vegetation cover, and trend analysis to characterize regeneration across forests, savannas, shrublands, and grasslands (Pérez-Cabello et al., 2021; Kurbanov et al., 2022). A recurring finding is the non-linear relationship between fire severity and recovery rate, often exhibiting convex patterns in which areas affected by moderate severity recover faster than those subjected to low or high severity fires (Meng et al., 2018a; Moni and Avetisyan, 2024). Furthermore, fire recurrence emerges as a critical factor, as repeated fires tend to slow recovery and increase the risk of state transitions, such as shifts from forest to shrubland (Zahabnazouri et al., 2025).

Comparing recovery trajectories between ecoregions poses significant methodological challenges. To address these challenges, the use of relative metrics has been proposed, such as the time needed to reach a percentage of prefire value, which facilitates interregional comparisons and the evaluation of the effectiveness of emergency treatments and active restoration (Pérez-Romero et al., 2025). Studies conducted in the Iberian Peninsula indicate that indices such as NBR, NDVI, and GCI are sensitive in detecting differences between treated and untreated areas, although the response depends on the climatic and edaphic context and the severity of the fire (Pérez-Romero et al., 2025).

Multiscale integration is another fundamental pillar in monitoring recovery. At the local scale, UAVs allow individual trees to be segmented, height and biomass to be estimated, and regeneration to be analyzed at the individual level using photogrammetry and structure-from-motion techniques (Petrov and Medvedev, 2025; Lazzeri et al., 2021). At regional and global scales, Landsat–Sentinel series exploited in cloud computing platforms facilitate systematic trajectory analysis, change point detection, and modeling of recovery metrics that are comparable across regions (Huang et al., 2010).

Finally, the proliferation of multisource, high-frequency data has driven the use of machine learning and deep learning in postfire recovery analysis. Algorithms such as random forest, support vector machines, and deep neural networks are used to integrate spectral indices, radar data, topography, and climate, enabling the classification of recovery states, prediction of return times, and modeling of complex trajectories (Jain et al., 2020; Sun et al., 2025). Recent bibliometric analyses show that, since 2020, artificial intelligence-based applications have become the dominant focus of postfire remote sensing research, with a growing emphasis on long-term monitoring, compound disturbances, and recovery assessment under climate change scenarios (Sun et al., 2025).

6.7 Wildfire and remote sensing—challenges and future directions

Remote sensing has become a pillar of fire management throughout the entire fire cycle: fuels, ignition, spread, emissions, and postfire effects, by providing repeatable and scalable observations that now support operational products and scientific analyses at multiple scales (Chen et al., 2024; huvieco et al., 2020; Szpakowski and Jensen, 2019). However, rapid climate change, the emergence of extreme fires, and the rise of new technologies (AI, commercial constellations, UAV/unmanned aerial systems [UAS], and IoT) are stretching the limits of the current paradigm and opening up new avenues of research with direct implications for safety, planning, and environmental justice (Bowman et al., 2020; Chen et al., 2023; Sun et al., 2025).

(Yang et al., 2024) Firefighting and civil protection services demand information with high spatial and temporal resolution, minute-long latencies, and robust products that work under smoke, cloud cover, and extreme conditions. In contrast, a substantial part of the satellite infrastructure and derived products was designed for large-scale climate and environmental objectives, with moderate resolutions and revisits of days, which introduces inevitable friction in the transition to tactical use (Chen et al., 2024; Chuvieco et al., 2020). Sensors dedicated or widely used for fire (e.g., MODIS/VIIRS in LEO; GEO platforms) have generated series of active hotspots, BA, and emissions proxies of enormous scientific value, but they remain less sensitive to small, short-lived, or low-severity fires, including many prescribed burning scenarios (Chen et al., 2024; Lopresti et al., 2024; Yang et al., 2024). Furthermore, detection and tracking at the urban-forest interface—fragmented landscapes, unrelated thermal sources, and dense smoke occlusion—remain a critical challenge with direct implications for response effectiveness (Allison et al., 2016; Mohapatra and Trinh, 2022; Barmpoutis et al., 2020a).

The historical reliance on passive optical sensors generates systematic gaps in highly cloudy or humid regions and at high latitudes with low illumination, amplifying spatial biases in observation and, by extension, inferences about fire regimes and risks (Chen et al., 2024; Chuvieco et al., 2020). Recent reviews converge on the need for multisensor approaches (optical, thermal, SAR, and LiDAR) as a structural way to overcome these limitations and reduce the fragility of "single-sensor" products in the face of adverse atmospheric conditions (Emilio Chuvieco et al., 2020; Veraverbeke et al., 2018; Ivanovic et al., 2025).

6.7.1 Key challenges throughout the fire cycle and the use of detection

Fuels, hazard, and ignition. Remote estimation of fuel load and structure remains one of the weakest links when it comes to feeding fire behavior models with physically interpretable variables. No single sensor simultaneously resolves fuel mass, horizontal continuity, and vertical structure at the scales required by operational models, especially for surface and fine fuels (Ivanovic et al., 2025; Abdollahi and Yebra, 2025). The moisture content of fine fuels and its relationship to flammability is inferred using optical indices and meteorological variables; although robust continental methodologies have been proposed, biases are amplified in extreme droughts and rapid phenological transitions, precisely when the system enters critical danger thresholds (Abdollahi and Yebra, 2025; Yebra et al., 2018). At the local scale, UAVs and 3D photogrammetry open up opportunities for detailed characterization, but the leap to sustainable regional extrapolation remains uncertain due to ecological heterogeneity, costs, and the absence of consistent scaling frameworks (Keerthinathan et al., 2023; Boroujeni et al., 2024).

In hazard and ignition modeling, many remote sensing-based approaches rely on climatologies and anthropogenic covariates (roads, population density) that are static or low resolution, while social factors (cultural fire practices, governance, land management) are poorly represented. This limits the ability to explain and anticipate ignitions under scenarios of rapid socio-environmental change (Bowman et al., 2020; Chen et al., 2023). The systematic integration of socioeconomic and land-use data with satellite observations is still in its infancy and is particularly relevant for regions that are underrepresented in the literature (Andela et al., 2017; Sun et al., 2025).

Early detection and monitoring of active fires. Satellite detection has advanced through contextual algorithms and near-real-time products, but trade-offs between resolution, revisit time, and latency remain. Geostationary sensors offer minutes of revisit with kilometer-sized pixels; low-orbit sensors provide greater spatial detail but pass every few hours, limiting the tracking of fast fronts and, especially, extreme pyroconvective episodes (Chen et al., 2024; Yang et al., 2024). On the ground, camera and IoT networks and UAS systems provide dense, very high-resolution observations, but face limitations in coverage, energy consumption, connectivity, and algorithmic robustness in real conditions (Barmpoutis et al., 2020b; Chan et al., 2024). Empirical evidence shows that AI-enabled UAVs can improve early detection, fine mapping, and tactical support, but regulatory, autonomy, and onboard processing bottlenecks persist, as do operational risks in smoke, turbulence, and reduced visibility (Allison et al., 2016; Chen et al., 2024; Boroujeni et al., 2024; Keerthinathan et al., 2023).

BA, emissions, and effects. Global BA and emissions products are critical for carbon budgets and climate assessment, but they suffer from systematic underdetection of small fires and discrepancies between products, introducing nontrivial uncertainties in global inferences (Chen et al., 2024; Kurbanov et al., 2022). These uncertainties are particularly relevant in agricultural fires, savannahs, and scrublands, where spatial fragmentation, recurrence, and human practices can bias both BA accounting and its translation into emissions (Andela et al., 2017; Hantson et al., 2015).

In terms of severity and postfire effects, indices such as dNBR and derived metrics remain the standard, but their relationship with specific ecological impacts depends on the ecosystem, species, and regeneration strategy, requiring regional calibrations and caution in the transfer of thresholds (Chuvieco et al., 2020; Kurbanov et al., 2022). Furthermore, effects on understory, roots, and soils, as well as the compound impacts of repeated fires combined with drought or pests, remain poorly captured by standard products, reinforcing the need for structural data (LiDAR) and multisensor integrations (Chen et al., 2023; Veraverbeke et al., 2018).

6.7.2 Cross-cutting methodological challenges

Validation, uncertainty, and scaling. A recurring challenge is the limited or nonindependent validation of many products, based on scarce field data or data not statistically designed for evaluation. Reviews highlight the need for standardized protocols, probabilistic sampling, and explicit quantification of uncertainty, as well as systematic comparisons between products and sensors (Chen et al., 2024; Kurbanov et al., 2022). The absence of open high-resolution references for active fires, fine fuels, or stratified severity limits the objective evaluation of new algorithms, especially AI, and can induce artificial optimism due to overfitting to unrepresentative datasets (Ghali and Akhloufi, 2023; Sun et al., 2025; Yang et al., 2024).

Spatiotemporal scaling is equally critical: linking observations from centimeters (cameras/UAVs), meters (Sentinel-2/Landsat), tens of meters (spatial LiDAR), and kilometers (MODIS/VIIRS/GEO) requires consistent fusion and disaggregation frameworks. Without this bridge, it is difficult to translate information to operational scales for planning and response, especially in extreme fires where dynamics evolve faster than the availability of detailed observations (Chen et al., 2024; Chuvieco et al., 2020).

Big data, AI, and explainability. The explosion of multisensor and multitemporal data overwhelms traditional workflows and pushes toward deep, self-supervised, and multimodal learning. Reviews highlight clear opportunities, greater sensitivity, fine segmentation, and automation, but also risks: dependence on biased training sets, performance drops when extrapolating to new regions/sensors, and lack of interpretability, especially in high-impact applications (Ghali and Akhloufi, 2023; Jain et al., 2020; Yu and Singh, 2025). This fragility compromises operational adoption: a system that "works on paper" but fails in new domains is not acceptable in emergency contexts (Chen et al., 2024; Mohapatra and Trinh, 2022). Hence, the growing relevance of hybrid frameworks in which AI refines or accelerates process-based models (propagation, emissions), rather than replacing them entirely, increasing traceability and trust (Chen et al., 2023; Jain et al., 2020).

6.7.3 New technological directions

Emerging sensors and artificial intelligence are transforming damage mapping from two-dimensional, retrospective approaches to three-dimensional analysis and near-real-time capabilities. In this context, commercial high-resolution optical and SAR constellations, including CubeSats, are beginning to offer near-daily time series with metric resolution, expanding the potential for mapping ignitions, active fronts, and fine effects of fire. However, challenges remain associated with cost, licensing, temporal continuity, management of large data volumes, and the necessary radiometric harmonization for robust multitemporal analysis (Chen et al., 2024; Sun et al., 2025). At the same time, new-generation hyperspectral sensors open up opportunities to characterize fuels, flame temperature, and smoke composition with greater spectral richness, although they require advances in atmospheric corrections and algorithms for more autonomous operation (Thangavel et al., 2023; Veraverbeke et al., 2018).

Spatial LiDAR, for its part, allows 3D structure to be integrated with spectral and thermal products, enabling more realistic fuel and stratified severity models (E. Chuvieco et al., 2020; Ivanovic et al., 2025). Added to these capabilities are UAVs/UASs and their integration with IoT networks, which reinforce multiscale observation in all phases of the fire cycle, albeit with technical challenges in autonomy, onboard computing, connectivity, and cyber-robustness (Chan et al., 2024; Boroujeni et al., 2024; Keerthinathan et al., 2023; Mohapatra and Trinh, 2022). Finally, cloud processing, data cubes, and the development of digital twins of fire landscapes consolidate an ecosystem capable of integrating near-real-time observations, fuel and climate models, propagation simulation, and AI, enabling dynamic severity and risk assessments, and the exploration of scenarios under uncertainty, as summarized by the conceptual progression shown in (Chen et al., 2023; Chen et al., 2024; Jain et al., 2020; Kurbanov et al., 2022).

Recent reviews converge on clear priorities: (1) designing missions and products explicitly aligned with operational and community user requirements, rather than reusing general-purpose sensors (Yu and Singh, 2025; Chen et al., 2024; Chuvieco et al., 2020); (2) institutionalizing open validation frameworks with reference networks and finely annotated datasets that drive comparable improvements in AI (Sun et al., 2025; Yang et al., 2024); (3) consolidating multisensor and multisource fusion, integrating optical/thermal/SAR/LiDAR with UAV, IoT, and socio-economic data (Chan et al., 2024; Ivanovic et al., 2025); (4) moving toward robust and explainable AI, with transparent audits and better performance in regions with little data (Jain et al., 2020; Yu and Singh, 2025); and (5) reducing geographical gaps and capacity inequalities by expanding research and transfer in highly affected but understudied regions (Andela et al., 2017; Sun et al., 2025). Overall, the future is moving toward integrated, intelligent, user-centered systems capable of delivering actionable information from the tree scale to the global climate system scale, in support of safer, more effective, and equitable fire management (Bowman et al., 2020; Chen et al., 2024).

AI disclosure

During the preparation of this work, the authors used DeepL in order to translate the manuscript, as English is not their native language. After using this tool, the authors reviewed and edited the content as needed and take full responsibility for the content of the publication. After using this tool/service, the author(s) reviewed and edited the content as needed and take(s) full responsibility for the content of the publication.

References

Abdollahi, A., Yebra, M., 2025. Challenges and opportunities in remote sensing-based fuel load estimation for wildfire behavior and management: a comprehensive review. Remote Sensing 17 (3), 415. Available at https://doi.org/10.3390/rs17030415.

Albini, F.A., 1976. Estimating wildfire behavior and effects. USDA Forest Service General Technical Report INT-30.

Allison, R., Johnston, J., Craig, G., Jennings, S., 2016. Airborne optical and thermal remote sensing for wildfire detection and monitoring. Sensors 16 (8), 1310. Available at https://doi.org/10.3390/s16081310.

Alonso-Rego, C., Arellano-Pérez, S., Guerra-Hernández, J., Molina-Valero, J.A., Martínez-Calvo, A., Pérez-Cruzado, C., et al., 2021. Estimating stand and fire-related surface and canopy fuel variables in pine stands using low-density airborne and single-scan terrestrial laser scanning data. Remote Sensing 13 (24). Available at https://doi.org/10.3390/rs13245170, https://www.mdpi.com/2072-4292/13/24/5170/pdf.

Alves, D.B., Laffayete Pires da Silveira, A., Cambraia, B.C., Falcão Sobrinho, J., Silva, T.S.F., Pérez-Cabello, F., 2025. Postfire vegetation recovery and spectral separability over amazonian savanna ecosystems using remote sensing time series and fuel loads measurements. Cuadernos de Investigación Geográfica 51 (2), 133–154. Available at https://doi.org/10.18172/cig.6889.

Andela, N., Morton, D.C., Giglio, L., Chen, Y., van der Werf, G.R., Kasibhatla, P.S., et al., 2017. A human-driven decline in global burned area. Science 356 (6345), 1356–1362. Available at https://doi.org/10.1126/science.aal4108.

Aragoneses, E., Chuvieco, E., 2021. Generation and mapping of fuel types for fire risk assessment. Fire 4 (3), 59. Available at https://doi.org/10.3390/fire4030059.

Aragoneses, E., García, M., Ruiz-Benito, P., Chuvieco, E., 2024. Mapping forest canopy fuel parameters at European scale using spaceborne LiDAR and satellite data. Remote Sensing of Environment 303, 114005. Available at https://doi.org/10.1016/j.rse.2024.114005.

Arroyo, L.A., Pascual, C., Manzanera, J.A., 2008. Fire models and methods to map fuel types: the role of remote sensing. Forest Ecology and Management 256 (6), 1239–1252. Available at https://doi.org/10.1016/j.foreco.2008.06.048.

Arumäe, T., Lang, M., 2018. Estimation of canopy cover in dense mixed-species forests using airborne lidar data. European Journal of Remote Sensing 51 (1), 132–141. Available at https://doi.org/10.1080/22797254.2017.1411169.

Barmpoutis, P., Papaioannou, P., Dimitropoulos, K., Grammalidis, N., 2020a. A review on early forest fire detection systems using optical remote sensing. Sensors 20 (22), 6442. Available at https://doi.org/10.3390/s20226442.

Barmpoutis, P., Stathaki, T., Dimitropoulos, K., Grammalidis, N., 2020b. Early fire detection based on aerial 360-degree sensors, deep convolution neural networks and exploitation of fire dynamic textures. Remote Sensing 12 (19), 3177. Available at https://doi.org/10.3390/rs12193177.

Boroujeni, S.P.H., Razi, A., Khoshdel, S., Afghah, F., Coen, J.L., O'Neill, L., et al., 2024. A comprehensive survey of research towards AI-enabled unmanned aerial systems in pre-, active-, and post-wildfire management. Information Fusion 108. Available at https://doi.org/10.1016/j.inffus.2024.102369, https://www.sciencedirect.com/science/journal/15662535.

Bousquet, E., Mialon, A., Rodriguez-Fernandez, N., Mermoz, S., Kerr, Y., 2022. Monitoring post-fire recovery of various vegetation biomes using multi-wavelength satellite remote sensing. Biogeosciences 19 (13), 3317–3336. Available at https://doi.org/10.5194/bg-19-3317-2022, http://www.biogeosciences.net/volumes_and_issues.html.

Bowman, D.M.J.S., Kolden, C.A., Abatzoglou, J.T., Johnston, F.H., van der Werf, G.R., Flannigan, M., 2020. Vegetation fires in the Anthropocene. Nature Reviews Earth and Environment 1 (10), 500–515. Available at https://doi.org/10.1038/s43017-020-0085-3, nature.com/natrevearthenviron/.

Brys, C., La Red Martínez, D.L., Marinelli, M., 2025. Machine learning methods for wildfire risk assessment. Earth Science Informatics 18 (1). Available at https://doi.org/10.1007/s12145-024-01690-z, https://www.springer.com/journal/12145.

Cao, Y., Lei, R., 2024. Progress and prospect of forest fire monitoring based on the multi-source remote sensing data. National Remote Sensing Bulletin 28 (8), 1854–1869. Available at https://doi.org/10.11834/jrs.20242314.

Carvajal-Ramírez, F., da Silva, J.R.M., Agüera-Vega, F., Martínez-Carricondo, P., Serrano, J., Moral, F.J., 2019. Evaluation of fire severity indices based on pre- and post-fire multispectral imagery sensed from UAV. Remote Sensing 11 (9). Available at https://doi.org/10.3390/rs11090993, https://res.mdpi.com/remotesensing/remotesensing-11-00993/article_deploy/remotesensing-11-00993.pdf?filename=&attachment=1.

Chan, C.C., Alvi, S.A., Zhou, X., Durrani, S., Wilson, N., Yebra, M., 2024. A survey on IoT ground sensing systems for early wildfire detection: technologies, challenges, and opportunities. IEEE Access 12, 172785–172819. Available at https://doi.org/10.1109/access.2024.3501336.

Chasmer, L., Baker, T., Carey, S.K., Straker, J., Strilesky, S., Petrone, R., 2018. Monitoring ecosystem reclamation recovery using optical remote sensing: comparison with field measurements and eddy covariance. Science of The Total Environment 642, 436–446. Available at https://doi.org/10.1016/j.scitotenv.2018.06.039.

Chen, X., Hopkins, B., Wang, H., O'Neill, L., Afghah, F., Razi, A., et al., 2022. Wildland fire detection and monitoring using a drone-collected RGB/IR image dataset. IEEE Access 10, 121301–121317. Available at https://doi.org/10.1109/ACCESS.2022.3222805, http://ieeexplore.ieee.org/xpl/RecentIssue.jsp?punumber=6287639.

Chen, C., Xu, T., Sun, F., Zhao, D., 2023. A fire danger index assessment method for short-term pre-warning of wildfires: a case study of Xiangxi, China. Safety Science 167, 106287. Available at https://doi.org/10.1016/j.ssci.2023.106287.

Chen, Y., Morton, D.C., Randerson, J.T., 2024. Remote sensing for wildfire monitoring: insights into burned area, emissions, and fire dynamics. One Earth 7 (6), 1022–1028. Available at https://doi.org/10.1016/j.oneear.2024.05.014, https://www.sciencedirect.com/science/journal/25903322.

Chen, L., Wei, B., Jia, X., Liu, M., Zhao, Y., 2025. Burned area mapping and fire severity assessment of forest–grassland ecosystems using time-series landsat imagery (1985–2023): a case study of daxing'anling region, China. Fire 8 (9), 337. Available at https://doi.org/10.3390/fire8090337.

Chowdhury, E.H., Hassan, Q.K., 2015. Operational perspective of remote sensing-based forest fire danger forecasting systems. ISPRS Journal of Photogrammetry and Remote Sensing 104, 224–236. Available at https://doi.org/10.1016/j.isprsjprs.2014.03.011, www.elsevier.com/inca/publications/store/5/0/3/3/4/0.

Chu, T., Guo, X., 2013. Remote sensing techniques in monitoring post-fire effects and patterns of forest recovery in boreal forest regions: a review. Remote Sensing 6 (1), 470–520. Available at https://doi.org/10.3390/rs6010470Canada, http://www.mdpi.com/2072-4292/6/1/470/pdf.

Chu, T., Guo, X., Takeda, K., 2016. Remote sensing approach to detect post-fire vegetation regrowth in Siberian boreal larch forest. Ecological Indicators 62, 32–46. Available at https://doi.org/10.1016/j.ecolind.2015.11.026.

Chuvieco, E., Mouillot, F., van der Werf, G.R., San Miguel, J., Tanasse, M., Koutsias, N., et al., 2019. Historical background and current developments for mapping burned area from satellite Earth observation. Remote Sensing of Environment 225, 45–64. Available at https://doi.org/10.1016/j.rse.2019.02.013.

Chuvieco, E., Aguado, I., Salas, J., García, M., Yebra, M., Oliva, P., 2020. Satellite remote sensing contributions to wildland fire science and management. Current Forestry Reports 6 (2), 81–96. Available at https://doi.org/10.1007/s40725-020-00116-5.

Chuvieco, E., Pettinari, M.L., Lizundia-Loiola, J., Storm, T., Padilla, M., 2023. Climate-driven changes in fire danger and burned area. Nature Climate Change 13, 1–8. Available at https://doi.org/10.1038/s41558-023-01610-5.

Countryman, C.M., 1972. The fire environment concept. USDA Forest Service General Technical Report PSW-66.

de Almeida, D.R.A., Vedovato, L.B., Fuza, M., Molin, P., Cassol, H., Resende, A.F., et al., 2025. Remote sensing approaches to monitor tropical forest restoration: current methods and future possibilities. Journal of Applied Ecology 62 (2), 188–206. Available at https://doi.org/10.1111/1365-2664.14830, http://onlinelibrary.wiley.com/journal/10.1111/(ISSN)1365-2664.

Dursun, İ., Demir, S., Başayiğit, L., Babalık, A.A., 2025. Post-fire erosion dynamics in the Dim River Basin: a remote sensing and Google Earth Engine approach. Fire Ecology 21 (1). Available at https://doi.org/10.1186/s42408-025-00363-6, https://www.springer.com/journal/42408.

Ejaz, N., Choudhury, S., 2025. A comprehensive survey of the machine learning pipeline for wildfire risk prediction and assessment. Ecological Informatics 90, 103325. Available at https://doi.org/10.1016/j.ecoinf.2025.103325.

Epting, J., Verbyla, D., Sorbel, B., 2005. Evaluation of remotely sensed indices for assessing burn severity in interior Alaska using Landsat TM and ETM+. Remote Sensing of Environment 96 (3-4), 328–339. Available at https://doi.org/10.1016/j.rse.2005.03.002.

Feitosa, F.R., dos Santos, A.M., Alvarado, S.T., da Silva, C.F.A., Nunes, F.G., 2024. Remote sensing applied to the study of fire in savannas: a literature review. Ecological Informatics 79. Available at https://doi.org/10.1016/j.ecoinf.2023.102448.

Fernández-Guisuraga, J.M., Fernandes, P.M., 2024. Prescribed burning mitigates the severity of subsequent wildfires in Mediterranean shrublands. Fire Ecology 20 (1). Available at https://doi.org/10.1186/s42408-023-00233-z, https://www.springer.com/journal/42408.

Fernández-Guisuraga, J.M., Fernandes, P.M., Tárrega, R., Beltrán-Marcos, D., Calvo, L., 2023. Vegetation recovery drivers at short-term after fire are plant community-dependent in mediterranean burned landscapes. Forest Ecology and Management 539. Available at https://doi.org/10.1016/j.foreco.2023.121034, www.elsevier.com/inca/publications/store/5/0/3/3/1/0.

Fernández-Guisuraga, J.M., Fernández-Manso, A., Quintano, C., Fernández-García, V., Cerrillo, A., Marqués, G., et al., 2024. FIREMAP: Cloud-based software to automate the estimation of wildfire-induced ecological impacts and recovery processes using remote sensing techniques. Ecological Informatics 81, 102591. Available at https://doi.org/10.1016/j.ecoinf.2024.102591.

Fernández-Manso, A., Quintano, C., 2020. A synergetic approach to burned area mapping using maximum entropy modeling trained with hyperspectral data and VIIRS hotspots. Remote Sensing 12 (5), 858. Available at https://doi.org/10.3390/rs12050858.

Fernández-Manso, A., Fernández-Manso, O., Quintano, C., 2016. SENTINEL-2A red-edge spectral indices suitability for discriminating burn severity. International Journal of Applied Earth Observation and Geoinformation 50, 170–175. Available at https://doi.org/10.1016/j.jag.2016.03.005.

Ferraz, A., McGaughey, R.J., Reutebuch, S.E., Andersen, H.-E., 2009. Mapping forest canopy fuel characteristics using airborne LiDAR. Remote Sensing of Environment 113 (10), 2173–2184. Available at https://doi.org/10.1016/j.rse.2009.04.014.

Gale, M.G., 2021. Canopy structure and crown fire behavior: a review. Forest Ecology and Management 482, 118873. Available at https://doi.org/10.1016/j.foreco.2020.118873.".

Gale, M.G., Cary, G.J., Yebra, M., Leavesley, A.J., Van Dijk, A.I.J.M., 2022. Comparison of contrasting optical and LiDAR fire severity remote sensing methods in a heterogeneous forested landscape in south-eastern Australia. International Journal of Remote Sensing 43 (7), 2538–2559. Available at https://doi.org/10.1080/01431161.2022.2064197.

Ghali, R., Akhloufi, M.A., 2023. Deep learning approaches for wildland fires using satellite remote sensing data: detection, mapping, and prediction. Fire 6 (5). Available at https://doi.org/10.3390/fire6050192, www.mdpi.com/journal/fire.

Graham, E., Dube, T., Mpakairi, K.S., 2023. Progress in the remote sensing of veld fire occurrence and detection: a review. African Journal of Ecology 61 (3), 541–553. Available at https://doi.org/10.1111/aje.13147.

Guiop-Servan, R.E., Cotrina-Sanchez, A., Puerta-Culqui, J., Oliva-Cruz, M., Barboza, E., 2025. Remote sensing for wildfire mapping: a comprehensive review of advances, platforms, and algorithms. Fire 8 (8). Available at https://doi.org/10.3390/fire8080316, www.mdpi.com/journal/fire.

Han, Y., Zheng, C., Liu, X., Tian, Y., Dong, Z., 2024. Burned area and burn severity mapping with a transformer-based change detection model. IEEE Journal of Selected Topics in Applied Earth Observations and Remote Sensing 17, 13866–13880. Available at https://doi.org/10.1109/jstars.2024.3435857.

Hantson, S., Pueyo, S., Chuvieco, E., 2015. Global fire size distribution is driven by human impact and climate. Global Ecology and Biogeography 24 (1), 77–86. Available at https://doi.org/10.1111/geb.12246.

Hu, X., Ban, Y., Nascetti, A., 2021. Uni-Temporal multispectral imagery for burned area mapping with deep learning. Remote Sensing 13 (8), 1509. Available at https://doi.org/10.3390/rs13081509.

Hu, J., Ye, B., Bai, Z., Feng, Y., 2022. Remote sensing monitoring of vegetation reclamation in the antaibao open-pit mine. Remote Sensing 14 (22), 5634. Available at https://doi.org/10.3390/rs14225634.

Hua, L., Shao, G., 2017. The progress of operational forest fire monitoring with infrared remote sensing. Journal of Forestry Research 28 (2), 215–229. Available at https://doi.org/10.1007/s11676-016-0361-8, https://www.springer.com/journal/11676.

Huang, C., Goward, S.N., Masek, J.G., Thomas, N., Zhu, Z., Vogelmann, J.E., 2010. An automated approach for reconstructing recent forest disturbance history using dense Landsat time series stacks. Remote Sensing of Environment 114 (1), 183–198. Available at https://doi.org/10.1016/j.rse.2009.08.017.

Ivanovic A. Gudelj A.S. Muslim P. Gotovac S. Sanader M.S.E.E.A. 2025 1 Remote Sensing for Assessing Vegetation Fuels in Wildfire Risk and Propagation Modelling 10th International Conference on Smart and Sustainable Technologies, SpliTech 2025 Institute of Electrical and Electronics Engineers Inc. Croatia http://ieeexplore.ieee.org/xpl/mostRecentIssue.jsp?punumber=11091588, https://doi.org/10.23919/SpliTech65624.2025.11091386.

Jain, P., Coogan, S.C.P., Subramanian, S.G., Crowley, M., Taylor, S., Flannigan, M.D., 2020. A review of machine learning applications in wildfire science and management. Environmental Reviews 28 (4), 478–505. Available at https://doi.org/10.1139/er-2020-0019, https://www.nrcresearchpress.com/loi/er.

Janardhan, B. Likhitha, O. Purandeswari, P. Rani, G. Harshini, IoT and AI for Efficient Forest Fire Detection, Monitoring, and Risk Prediction With Real-Time Alerts Proceedings of 5th International Conference on Trends in Material Science and Inventive Materials, ICTMIM 2025. (2025), 1324–1330. https://doi.org/10.1109/ICTMIM65579.2025.10988101.

Keane, R.E., Burgan, R., van Wagtendonk, J., 2005. The role of fuel characterization in fire behavior modeling. Forest Ecology and Management 211 (1–2), 1–18. Available at https://doi.org/10.1016/j.foreco.2005.01.008.

Keerthinathan, P., Amarasingam, N., Hamilton, G., Gonzalez, F., 2023. Exploring unmanned aerial systems operations in wildfire management: data types, processing algorithms and navigation. International Journal of Remote Sensing 44 (18), 5628–5685. Available at https://doi.org/10.1080/01431161.2023.2249604.

Key, C.H., Benson, N.C., 2006. Landscape assessment (LA). In: Lutes, D.C.;, Keane, R.E.;, Caratti, J.F.;, Key, C.H.;, Benson, N.C.;, Sutherland, S.;, Gangi, L.J. (Eds.), FIREMON: fire effects monitoring and inventory system. Gen. Tech. Rep. RMRS-GTR-164-CD. U.S. Department of Agriculture, Forest Service, Rocky Mountain Research Station, Fort Collins, CO, pp. LA-1–55 (2006).

Kim, K.N., Kim, S.H., Won, M.S., Jang, K.C., Choi, W.J., Lee, Y.G., 2019. A feasibility study on the application of tvdi on accessing wildfire danger in the Korean Peninsula. Korean Journal of Remote Sensing 35 (6-3), 1197–1208. Available at https://doi.org/10.7780/kjrs.2019.35.6.3.4, http://www.koreascience.kr/journal/OGCSBN/v35n6_3.view?orgId=ksrs.

Krishna Reddy, K.R., Sharanya, Harini, P.H., Chowdary, K.A., 2025. Forecasting of forest blaze using ensemble learning approach. Proceedings - 3rd International Conference on Self Sustainable Artificial Intelligence Systems, ICSSAS 2025, pp. 300–309. https://doi.org/10.1109/ICSSAS66150.2025.11081260.

Kurbanov, E., Vorobev, O., Lezhnin, S., Sha, J., Wang, J., Li, X., et al., 2022. Remote sensing of forest burnt area, burn severity, and post-fire recovery: a review. Remote Sensing 14 (19), 4714. Available at https://doi.org/10.3390/rs14194714.

Lazzeri, G., Frodella, W., Rossi, G., Moretti, S., 2021. Multitemporal mapping of post-fire land cover using multiplatform prisma hyperspectral and sentinel-uav multispectral data: insights from case studies in portugal and Italy. Sensors 21 (12). Available at https://doi.org/10.3390/s21123982.

Li, B., Ren, H.'e, Tian, J., 2024. Advancements in satellite remote sensing for predicting large-scale wildfire risks: an image processing algorithmic framework. International Journal of Interactive Mobile Technologies (iJIM) 18 (7), 19–33. Available at https://doi.org/10.3991/ijim.v18i07.48601.

Lin, S., Li, L., Liu, S., Gao, G., Zhao, X., Chen, L., et al., 2024. Stratified burn severity assessment by integrating spaceborne spectral and waveform attributes in Great Xing'an Mountain. Remote Sensing of Environment 307, 114152. Available at https://doi.org/10.1016/j.rse.2024.114152.

Liu, P., Zhuang, W., Kou, W., Wang, L., Wang, Q., Deng, Z., 2025. Fire severity outperforms remote sensing indices in exploring post-fire vegetation recovery dynamics in complex plateau mountainous regions. Forests 16 (2), 263. Available at https://doi.org/10.3390/f16020263.

Loboda, T., O'Neal, K.J., Csiszar, I., 2007. Regionally adaptable dNBR-based algorithm for burned area mapping from MODIS data. Remote Sensing of Environment 109 (4), 429–442. Available at https://doi.org/10.1016/j.rse.2007.01.017.

Lopresti, A., Hayden, M.T., Siegel, K., Poulter, B., Stavros, E.N., Dee, L.E., 2024. Remote sensing applications for prescribed burn research. International Journal of Wildland Fire 33 (6). Available at https://doi.org/10.1071/WF23130, http://www.publish.csiro.au/nid/126.htm.

Madkar, S., Sakhare, D.Y., Phutane, K.A., 2022. En Proceedings of the 3rd International Conference on Power, Energy, Control and Transmission Systems (ICPECTS 2022). Video based forest fire and smoke detection using YOLO and CNN.

Makineci, H.B., 2024. Investigation of burned areas with multiplatform remote sensing data on the Rhodes 2023 forest fires. Ain Shams Engineering Journal 15 (2), 102–114. Available at https://doi.org/10.1016/j.asej.2023.102114.

McKenna, P., Erskine, P.D., Lechner, A.M., Phinn, S., 2017. Measuring fire severity using UAV imagery in semi-arid central Queensland, Australia. International Journal of Remote Sensing 38 (14), 4244–4264. Available at https://doi.org/10.1080/01431161.2017.1317942.

McKenna, P.B., Lechner, A.M., Hernandez Santin, L., Phinn, S., Erskine, P.D., 2023. Measuring and monitoring restored ecosystems: can remote sensing be applied to the ecological recovery wheel to inform restoration success? Restoration Ecology 31 (1). Available at https://doi.org/10.1111/rec.13724, http://onlinelibrary.wiley.com/journal/10.1111/(ISSN)1526-100X.

McLauchlan, K.K., Higuera, P.E., Miesel, J., Rogers, B.M., Schweitzer, J., Shuman, J.K., et al., 2020. Fire as a fundamental ecological process: research advances and frontiers. Journal of Ecology 108 (5), 2047–2069. Available at https://doi.org/10.1111/1365-2745.13403, http://onlinelibrary.wiley.com/journal/10.1111/(ISSN)1365-2745.

Meng, R., Wu, J., Zhao, F., Cook, B.D., Hanavan, R.P., Serbin, S.P., 2018a. Measuring short-term post-fire forest recovery across a burn severity gradient in a mixed pine-oak forest using multi-sensor remote sensing techniques. Remote Sensing of Environment 210, 282–296. Available at https://doi.org/10.1016/j.rse.2018.03.019.

Meng, R., Wu, J., Zhao, F., 2018b. Measuring short-term post-fire forest recovery across a burn severity gradient using multi-sensor remote sensing techniques. Remote Sensing of Environment 210, 282–296. Available at https://doi.org/10.1016/j.rse.2018.03.023.

Miller, J.D., Thode, A.E., 2007. Quantifying burn severity in a heterogeneous landscape with a relative version of the delta Normalized Burn Ratio (dNBR). Remote Sensing of Environment 109 (1), 66–80. Available at https://doi.org/10.1016/j.rse.2006.12.006.

Mohapatra, A., Trinh, T., 2022. Early wildfire detection technologies in practice—a review. Sustainability 14 (19), 12270. Available at https://doi.org/10.3390/su141912270.

Moni, D., Avetisyan, D., 2024. Comparative analyses of secondary ecological succession following wildfires in three distinct forest types. a study case from moguer, Spain. Aerospace Research in Bulgaria 36, 94–106. Available at https://doi.org/10.3897/arb.v36.e08.

Muthulakshmi, K., Manimekalai, M.A.P. Gopikrishna, C., 2022. Instant fire detection and toxic fumes monitoring in forests with a remote integrated rover.

Nedkov, R., Velizarova, E., Avetisyan, D., Georgiev, N., 2020. 1 Assessment of forest vegetation state through remote sensing in response to fire impact. Proceedings of SPIE - The International Society for Optical Engineering SPIE Bulgaria 11524. http://spie.org/x1848.xml10.1117/12.2570808.

Pang, Y., Jia, W., Qin, X., Si, L., Liang, X., Lin, X., et al., 2020. Forest fire monitoring using airborne optical full spectrum remote sensing data. Yaogan Xuebao/Journal of Remote Sensing 24 (10), 1280–1292. Available at https://doi.org/10.11834/jrs.20200290, http://www.jors.cn/jrs/ch/index.aspx.

Pausas, J.G., Keeley, J.E., 2021. Wildfires and global change. Frontiers in Ecology and the Environment 19 (7), 387–395. Available at https://doi.org/10.1002/fee.2359, http://onlinelibrary.wiley.com/journal/10.1002/(ISSN)1540-9309.

Pérez-Cabello, F., Montorio, R., Alves, D.B., 2021. Remote sensing techniques to assess post-fire vegetation recovery. Current Opinion in Environmental Science and Health 21. Available at https://doi.org/10.1016/j.coesh.2021.100251, www.journals.elsevier.com/current-opinion-in-environmental-science-and-health.

Pérez-Romero, J., Lucas-Borja, M.E., Zema, D.A., Soria, R., Miralles, I., Blanco-Cano, L., et al., 2025. Early response of post-fire forest treatments across four iberian ecoregions: indicators to maximize its effectiveness by remote sensing. Forests 16 (7). Available at https://doi.org/10.3390/f16071154, http://www.mdpi.com/journal/forests/.

Petrov, O., Medvedev, A., 2025. Analyzing Post-fire vegetation dynamics with ultra-high resolution remote sensing data. Sensing and Spatial Information Sciences XLVIII-G-2025 (G-2025), 1189–1195. Available at https://doi.org/10.5194/isprs-archives-xlviii-g-2025-1189-2025.

Pulvirenti, L., Squicciarino, G., Fiori, E., Fiorucci, P., Ferraris, L., Negro, D., et al., 2020. An automatic processing chain for near real-time mapping of burned forest areas using sentinel-2 Data. Remote Sensing 12 (4), 674. Available at https://doi.org/10.3390/rs12040674.

Quintano, C., Fernández-Manso, A., Calvo, L., Marcos, E., Valbuena, L., 2015. Land surface temperature as potential indicator of burn severity in forest Mediterranean ecosystems. International Journal of Applied Earth Observation and Geoinformation 36, 1–12. Available at https://doi.org/10.1016/j.jag.2014.10.015.

Quintano, C., Fernández-Manso, A., Roberts, D.A., 2020. Enhanced burn severity estimation using fine resolution ET and MESMA fraction images with machine learning algorithm. Remote Sensing of Environment 244, 111815. Available at https://doi.org/10.1016/j.rse.2020.111815.

Quintano, C., Calvo, L., Fernández-Manso, A., Suárez-Seoane, S., Fernandes, P.M., Fernández-Guisuraga, J.M., 2023. First evaluation of fire severity retrieval from PRISMA hyperspectral data. Remote Sensing of Environment 295. Available at https://doi.org/10.1016/j.rse.2023.113670, www.elsevier.com/inca/publications/store/5/0/5/7/3/3.

Quintano, C., Fernández-Manso, A., Fernández-Guisuraga, J.M., Roberts, D.A., 2024. Improving fire severity analysis in mediterranean environments: a comparative study of eemetric and ssebop landsat-based evapotranspiration models. Remote Sensing 16 (2). Available at https://doi.org/10.3390/rs16020361, http://www.mdpi.com/journal/remotesensing.

Ramo, R., García, M., Rodríguez, D., Chuvieco, E., 2018. A data mining approach for global burned area mapping. International Journal of Applied Earth Observation and Geoinformation 73, 39–51. Available at https://doi.org/10.1016/j.jag.2018.05.027.

Ramo, R., Chuvieco, E., 2017. Developing a random forest algorithm for MODIS global burned area classification. Remote Sensing 9 (11), 1193. Available at https://doi.org/10.3390/rs9111193.

Riaño, D., Meier, E., Allgöwer, B., Chuvieco, E., Ustin, S.L., 2003. Modeling airborne laser scanning data for the spatial generation of critical forest parameters in fire behavior modeling. Remote Sensing of Environment 86 (2), 177–186. Available at https://doi.org/10.1016/S0034-4257(03)00098-1.

Rodell, C., Howard, R., Jain, P., Moisseeva, N., Chui, T., Stull, R., 2024. Forecasting hourly wildfire risk: enhancing fire danger assessment using numerical weather prediction. Weather and Forecasting 39 (6), 925–941. Available at https://doi.org/10.1175/waf-d-23-0226.1.

Roteta, E., Bastarrika, A., Padilla, M., Storm, T., Chuvieco, E., 2019. Development of a sentinel-2 burned area algorithm: generation of a small fire database for sub-Saharan Africa. Remote Sensing of Environment 222, 1–17. Available at https://doi.org/10.1016/j.rse.2018.12.011.

Roy, D.P., Boschetti, L., Trigg, S.N., 2006. Remote sensing of fire severity: assessing the performance of the normalized burn ratio. IEEE Geoscience and Remote Sensing Letters 3 (1), 112–116. Available at https://doi.org/10.1109/LGRS.2005.858485.

Ruiz-Benito, P., Vacchiano, G., Lines, E.R., Reyer, C.P.O., Ratcliffe, S., Morin, X., et al., 2020. Forest structure and carbon storage across Europe. Global Ecology and Biogeography 29 (4), 587–601. Available at https://doi.org/10.1111/geb.13049.

Sannigrahi, S., Pilla, F., Basu, B., Sarkar Basu, A., Sarkar, K., Chakraborti, S., et al., 2020. Examining the effects of forest fire on terrestrial carbon emission and ecosystem production in India using remote sensing approaches. Science of the Total Environment 725, 138331. Available at https://doi.org/10.1016/j.scitotenv.2020.138331.

Shahriar, S.A., Choi, Y., Islam, R., 2025. Advanced deep learning approaches for forecasting high-resolution fire weather index (FWI) over CONUS: integration of GNN-LSTM, GNN-TCNN, and GNN-DeepAR. Remote Sensing 17 (3). Available at https://doi.org/10.3390/rs17030515, http://www.mdpi.com/journal/remotesensing.

Shen, S., Seneviratne, S., Wanyan, X., Kirley, M., 2023. FireRisk: A Remote Sensing Dataset for Fire Risk Assessment with Benchmarks Using Supervised and Self-supervised Learning. arXiv, Australia arXiv. Available at https://arxiv.org, https://doi.org/10.48550/arXiv.2303.07035.

Sismanis, M., Chadoulis, R.-T., Manakos, I., Drosou, A., 2023. An unsupervised burned area mapping approach using sentinel-2 images. Land 12 (2), 379. Available at https://doi.org/10.3390/land12020379.

Smith, Lentile, Hudak, Morgan, 2007. Evaluation of linear spectral unmixing and DNBR for predicting postfire recovery in a North American ponderosa pine forest. International Journal of Remote Sensing 28 (22), 5159–5166. Available at https://doi.org/10.1080/01431160701395161.

Stankova, N., 2023. Post-fire recovery monitoring using remote sensing: a review. Aerospace Research in Bulgaria 35, 192–200. Available at https://doi.org/10.3897/arb.v35.e19.

Stefanidou, A., Gitas, I.Z., Korhonen, L., Stavrakoudis, D., Georgopoulos, N., 2020. LiDAR-based estimates of canopy base height for a dense uneven-aged structured forest. Remote Sensing 12 (10). Available at https://doi.org/10.3390/rs12101565.

Sulthana, S.F., Wise, C.T.A., Ravikumar, C.V., Anbazhagan, R., Idayachandran, G., Pau, G., 2023. Review study on recent developments in fire sensing methods. IEEE Access 11, 90269–90282. Available at https://doi.org/10.1109/ACCESS.2023.3306812, http://ieeexplore.ieee.org/xpl/RecentIssue.jsp?punumber=6287639.

Sun, M., Zhang, X., Jin, R., 2025. Visualization of post-fire remote sensing using citespace: a bibliometric analysis. Forests 16 (4), 592. Available at https://doi.org/10.3390/f16040592.

Szpakowski, D.M., Jensen, J.L.R., 2019. A review of the applications of remote sensing in fire ecology. Remote Sensing 11 (22). Available at https://doi.org/10.3390/rs11222638, https://res.mdpi.com/d_attachment/remotesensing/remotesensing-11-02638/article_deploy/remotesensing-11-02638.pdf.

Thangavel, K., Spiller, D., Sabatini, R., Amici, S., Sasidharan, S.T., Fayek, H., et al., 2023. Autonomous satellite wildfire detection using hyperspectral imagery and neural networks: a case study on Australian wildfire. Remote Sensing 15 (3), 720. Available at https://doi.org/10.3390/rs15030720.

Tian, Y., Wu, Z., Li, M., Wang, B., Zhang, X., 2022. Forest fire spread monitoring and vegetation dynamics detection based on multi-source remote sensing images. Remote Sensing 14 (18), 4431. Available at https://doi.org/10.3390/rs14184431.

Tran, B.N., Tanase, M.A., Bennett, L.T., Aponte, C., 2018. Evaluation of spectral indices for assessing fire severity in Australian temperate forests. Remote Sensing 10 (11). Available at https://doi.org/10.3390/rs10111680, https://res.mdpi.com/remotesensing/remotesensing-10-01680/article_deploy/remotesensing-10-01680.pdf?filename=&attachment=1.

Vasconcelos, R.N., Franca Rocha, W.J.S., Costa, D.P., Duverger, S.G., Santana, M.M.Md, Cambui, E.C.B., et al., 2024. Fire detection with deep learning: a comprehensive review. Land 13 (10). Available at https://doi.org/10.3390/land13101696, www.mdpi.com/journal/land/.

Veraverbeke, S., Dennison, P., Gitas, I., Hulley, G., Kalashnikova, O., Katagis, T., et al., 2018. Hyperspectral remote sensing of fire: state-of-the-art and future perspectives. Remote Sensing of Environment 216, 105–121. Available at https://doi.org/10.1016/j.rse.2018.06.020.

Wang, X., Wotton, B.M., Cantin, A.S., Parisien, M.-A., Anderson, K., Moore, B., et al., 2017. cffdrs: an R package for the Canadian forest fire danger rating system. Ecological Processes 6 (1). Available at https://doi.org/10.1186/s13717-017-0070-z.

Wang, R., Sun, Y., Zong, J., Wang, Y., Cao, X., Wang, Y., et al., 2024. Remote sensing application in ecological restoration monitoring: a systematic review. Remote Sensing 16 (12), 2204. Available at https://doi.org/10.3390/rs16122204.

White, J.C., Hermosilla, T., Wulder, M.A., Coops, N.C., 2022. Mapping, validating, and interpreting spatio-temporal trends in post-disturbance forest recovery. Remote Sensing of Environment 271. Available at https://doi.org/10.1016/j.rse.2022.112904, www.elsevier.com/inca/publications/store/5/0/5/7/3/3.

Wooster, M.J., Roberts, G.J., Giglio, L., Roy, D.P., Freeborn, P.H., Boschetti, L., et al., 2021. Satellite remote sensing of active fires: history and current status, applications and future requirements. Remote Sensing of Environment 267, 112694. Available at https://doi.org/10.1016/j.rse.2021.112694.

Wu, M., Huang, Q., Sui, T., Peng, B., Yu, M., 2024. A remote sensing spectral index guided bitemporal residual attention network for wildfire burn severity mapping. IEEE Journal of Selected Topics in Applied Earth Observations and Remote Sensing 17, 17187–17206. Available at https://doi.org/10.1109/JSTARS.2024.3460531, http://ieeexplore.ieee.org/xpl/RecentIssue.jsp?punumber=4609443.

Xie, Z., Song, W., Ba, R., Li, X., Xia, L., 2018. A spatiotemporal contextual model for forest fire detection using Himawari-8 satellite data. Remote Sensing 10 (12), 1992. Available at https://doi.org/10.3390/rs10121992.

Xu, Z., Li, J., Cheng, S., Rui, X., Zhao, Y., He, H., et al., 2025. Deep learning for wildfire risk prediction: integrating remote sensing and environmental data. ISPRS Journal of Photogrammetry and Remote Sensing 227, 632–677. Available at https://doi.org/10.1016/j.isprsjprs.2025.06.002, https://www.sciencedirect.com/science/journal/09242716.

Yang, S., Huang, Q., Yu, M., 2024. Advancements in remote sensing for active fire detection: a review of datasets and methods. Science of the Total Environment 943, 173273. Available at https://doi.org/10.1016/j.scitotenv.2024.173273.

Yebra, M., Quan, X., Riaño, D., Rozas Larraondo, P., van Dijk, A.I.J.M., Cary, G.J., 2018. A fuel moisture content and flammability monitoring methodology for continental Australia based on optical remote sensing. Remote Sensing of Environment 212, 260–272. Available at https://doi.org/10.1016/j.rse.2018.04.053, www.elsevier.com/inca/publications/store/5/0/5/7/3/3.

Yu, S., Singh, M., 2025. Deep learning-based remote sensing image analysis for wildfire risk evaluation and monitoring. Fire 8 (1), 19. Available at https://doi.org/10.3390/fire8010019.

Zahabnazouri, S., Belmont, P., David, S., Wigand, P.E., Elia, M., Capolongo, D., 2025. Detecting burn severity and vegetation recovery after fire using dNBR and dNDVI indices: insight from the bosco difesa grande, Gravina in Southern Italy. Sensors 25 (10), 3097. Available at https://doi.org/10.3390/s25103097.

Zhang, J., Dong, Y., Liu, Y., Kang, Y., Xie, S., Xu, H., 2025. Knowledge domain and emerging trends in monitoring of forest fires using remote sensing: a scientometric review based on citespace analysis. Agricultural and Rural Studies 3 (2). Available at https://doi.org/10.59978/ar03020008, https://sccpress.com/ars.

Fire hazard assessment through remote sensing: burn probability and fire severity predictive modeling

7

Olga Viedma
Department of Environmental Sciences, University of Castilla-La Mancha, Toledo, Spain

Fire is a fundamental ecological process in many ecosystems (Bowman et al., 2009) and represents one of the most recurrent disturbances in the Mediterranean biome (Keeley et al., 2011). In the current context of global change, fire risk factors are intensifying. Meteorological fire danger is already high and is projected to increase under climate change, leading to larger burned areas (BAs) (Dupuy et al., 2020). At the same time, socio-environmental processes such as agricultural abandonment, rural depopulation, and land-use changes have altered ignition patterns and fuel availability, promoting the expansion of shrublands and forested areas (Viedma et al., 2018). Human exposure to fire is also expected to rise in the future, driven by projected population growth in fire-prone regions (Knorr et al., 2016). In this context of combined environmental and social change, the development of improved fire prevention tools is crucial. Fire risk assessment is a cornerstone of prevention, as it provides an objective means to anticipate when and where a fire is most likely to occur, and when it may generate the most severe impacts (Scott et al., 2013). Forest fire risk can be conceptualized as the interaction between hazard (the probability and intensity of fire) and impact (exposure and vulnerability). In this chapter, we specifically address the hazard component of fire risk, namely, burn probability and severity, since understanding how these dimensions evolve across landscapes is essential for guiding effective forest fire management policies and actions.

7.1 Drivers of burn probability and fire severity

Climate-meteorological extremes structure broad-scale fire patterns, but vegetation amount/composition and human influence modulate or even reverse fine-versus coarse-scale relationships (Marlon et al., 2008; Parisien et al., 2011).

7.1.1 Climate and meteorology

Short-term heat and dryness, modulated by antecedent moisture "memory," govern fire activity, and policy/land use can either dampen or amplify these climatic signals (Fréjaville and Curt, 2017). Ignition pressure rises with atmospheric dryness and heat.

Advances in Remote Sensing for Forest Fire Management. DOI: https://doi.org/10.1016/B978-0-443-41692-7.00016-0
© 2027 Elsevier Ltd. All rights are reserved, including those for text and data mining, AI training, and similar technologies.

In the western United States, both vapor-pressure deficit (VPD) and temperature were positively associated with human- and lightning-caused ignitions (Chen et al., 2021), and higher VPD increases the probability that ignitions become large fires (Li et al., 2025). During the extreme 2022 season in SW Europe, record VPD coincided with very low live fuel moisture content (LFMC) and explained substantial variance in BA (Rodrigues et al., 2023). In Europe, higher VPD inflates the upper tail of fire size, with projections indicating much larger maxima this century, particularly in Mediterranean and temperate broadleaf biomes (Grünig et al., 2023). At the regional scale, a pyroregion analysis reveals that VPD is the primary driver of fire severity (Alvarez et al., 2024). In moisture-limited Mediterranean systems, summer aridity and heatwaves elevate BA and large-fire probability (Ruffault et al., 2020). The Peloponnese in 2007 illustrates how a brief, extreme hot–windy window concentrated >190,000 ha into a few events, shifting burning into agro-natural interfaces (Koutsias et al., 2012). At finer scales, heatwave intensity and duration track increases in fire size and, in some regions, stand-replacing outcomes (Elia et al., 2024). Overall, extremely dry dead fuels (≈10%–12%, i.e., very high VPD) mark the onset of widespread burning, although the share of large fires (>500 ha) increases nonlinearly (Resco de Dios et al., 2022). In Euro-Mediterranean settings, an increase in the Fire Weather Index (FWI) from 30 to 60 raised the odds that an ignition became a large fire by ~8× irrespective of season or land use, and seasonal BA scaled with the number of days with FWI >40 (Ager et al., 2014). Moreover, antecedent wetness and fuel buildup condition these meteorological effects. Across Mediterranean regions, wet winters tend to suppress fire activity, whereas hot summers enhance it. Rainfall in late spring can further increase ignition density by stimulating fine-fuel growth that subsequently dries and becomes highly flammable (Sarris and Koutsias, 2014). In NW Spain, wet springs preceded higher subsequent severity, consistent with a fuel-pulse mechanism (García-Llamas et al., 2019b). Globally, one to six month "fuel memory" captured by vegetation indices adds substantial predictive skill beyond instantaneous dryness (Kuhn-Régnier et al., 2021) (Fig. 7.1).

Furthermore, under strong winds and very low humidity, top-down forcing accelerates spread, overrides many fuel controls, and steepens severity responses (Birch et al., 2015; Fernandes et al., 2004; Taylor et al., 2021). Severity rises quickly with fast wind and low relative air humidity in conifer systems (Estes et al., 2017), with strong responses where shrublands dominate (Rodriguez-Jimenez et al., 2023). For example, in the Yeste fire (SE Spain), a faster rate of spread, wind–spread alignment, low RH, and higher temperature were the strongest direct predictors of severity (Viedma et al., 2020b).

7.1.2 Topography

Topography shapes fire regimes through two pathways: where ignitions occur and how far fires spread; and its influence is scale- and season-dependent (Fig. 7.1). Elevation (as a proxy for human accessibility) and slope structure ignition patterns. For example, in temperate Michigan forests, lower elevations (greater access) best predicted human-caused ignitions (Bar Massada et al., 2013); and in west-central

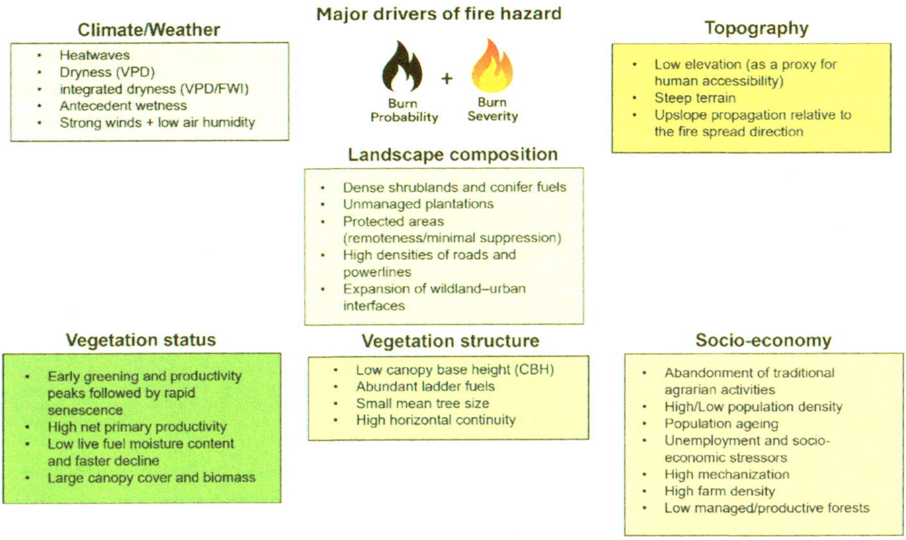

Figure 7.1 Main drivers of fire hazard (burn probability and burn severity). Climate and weather conditions, topography, landscape composition, vegetation status and structure, and socio-economic dynamics interact to shape fire regimes. Collectively, these factors influence both the likelihood of ignition and spread (burn probability) and the ecological and socio-economic impacts (burn severity).

Spain, high elevation and steep slopes strongly correlated with annual fire counts, though their influence declined over time (Viedma et al., 2018). At escalation from ignition to large fire, the slope becomes critical. Across mainland Spain, steeper terrain increased the probability that an ignition exceeded ≥100 ha, especially in the north (Rodrigues et al., 2019). Beyond ignition, slope favors spread and BA when fire weather aligns in Galicia (Fuentes-Santos et al., 2013); it also favors BA, fire frequency, and fire-return interval in Portugal (Bergonse et al., 2022). Fire susceptibility also shifts seasonally along elevation in Liguria—coastal belts in summer, higher inland in winter, illustrating relief–phenology–climate coupling (Tonini et al., 2020). However, topographic effects can be overshadowed by human accessibility or fuel phenology (Bajocco et al., 2010), and predictive skill differs between flat and mountainous municipalities (higher in mountainous areas), indicating that topography conditions both driver signals and model predictability (Scarpa et al., 2024).

Topography also modulates severity through static and directional controls. Severity tends to increase on steeper slopes and on facets with greater solar loading (biome-dependent), and is shaped by elevation, convexity/curvature, terrain ruggedness, and position indices (e.g., concave, moist positions and stream proximity generally dampen severity; Lecina-Diaz et al., 2014; Lentile et al., 2006). Case studies show consistent patterns: higher severity with upslope flow and high-canopy continuity in Colorado subalpine fires (Bigler et al., 2005), and in Galician pine forests where wind–slope alignment and ruggedness amplify crown fire (Fernández-

Alonso et al., 2017). In Spain, severity rose for upslope propagation relative to the fire spread direction and under high solar radiation, with faster head-fire spread ($\gtrsim$26 m minutes^{-1}) and wind alignment linked to high severity (relative dNBR [RdNBR]) (Viedma et al., 2020b, Viedma et al., 2015b). Aspect also interacts with fire behavior. South–north differences are small for head fires but widen for flank/back fires, with higher severity on south aspects (Oliveras et al., 2009). In sum, topography structures ignition via accessibility and exposure (elevation, slope) and governs spread/severity by aligning fuels and flow paths (slope, aspect, curvature), with effects that strengthen where human control is lower and fire-weather windows are favorable (Bergonse et al., 2022; Scarpa et al., 2024; Tonini et al., 2020; Viedma et al., 2018).

7.1.3 Landscape properties

Land-use/land-cover (LULC) and its change, mainly driven by the abandonment of traditional agrarian activities, are central to fire hazard (Moreira et al., 2011) (Fig. 7.1). Pre-1990 agricultural abandonment concentrated in environmentally and socio-economically marginal mountain areas (steep, poor soils, water deficit, shallow rooting, aging/low-skilled farmers, remoteness; Lasanta et al., 2015), triggering succession to shrublands/young forests, increasing fuel loads and connectivity, eroding landscape heterogeneity and natural firebreaks, and raising fire hazard (Quintero et al., 2019; Mantero et al., 2020; Terres et al., 2015; Viedma et al., 2015a). After 1990, abandonment shifted toward warmer, drier, more accessible fringes near settlements, coevolving with urban sprawl, mechanization, market competition, and common agricultural policy (CAP) set-aside/retirement (1988–2008), while aging, low mechanization, and marginality gained weight (Lasanta et al., 2017; Terres et al., 2015; Viedma et al., 2015a). Across Europe, simulations project that 5.6 million hectares of farmland, nearly 3.6% of the agricultural area, will be abandoned between 2015 and 2030, with Spain, Poland, France, Germany, and Italy being the most affected (Perpiña Castillo et al., 2021). Projected scenarios featuring rural abandonment and especially forest expansion indicate that as cropland abandonment progresses, fire frequency will rise abruptly, while fire intensity increases with accumulating fuels (Sil et al., 2019).

Across Mediterranean landscapes, the type of fuel consistently determines where fires start, spread, and burn most intensely (Fig. 7.1). Fire-occurrence models show higher burning probabilities in conifer forests and shrublands, and lower incidence in annual crops and agroforestry mosaics such as dehesas (Foresta et al., 2016; Koutsias et al., 2010; Moreira et al., 2010; Viedma et al., 2018). Resource-selection studies converge on positive selection of shrublands and pine stands, and avoidance of croplands/agroforestry (Bergonse et al., 2022; Fuentes-Santos et al., 2013; Moreno et al., 2011; Ortega et al., 2012). Moreover, plantations are frequently selected, especially when unmanaged or abandoned (Fernandes et al., 2019; Moreira et al., 2023). Severity patterns mirror these occurrence signals. High-severity burning concentrates in Pinus-dominated forests and shrublands, whereas broadleaved stands (e.g., aspen, cork oak) and mixed forests generally dampen severity, as it has been observed in American (Bigler et al., 2005), Asian (Fang et al., 2018), and Iberian case studies (Fernandes et al., 2009; Fernández-García et al., 2023; Rodriguez-Jimenez et al., 2023). Moreover, process-based simulations corroborate and extend these findings. Landscape-scale minimum travel time (MTT) simulations show that agropastoral abandonment (10%–40%) monotonically increased annual burn probability (~+28% at 40%), conditional flame length (~+16%), and potential fire size (~ +31%), roughly doubling the area with probability of flames >2.5 m (Salis et al., 2022). On the contrary, models under 2017 like extremes in Portugal's wildland–urban interface (WUI) showed that converting interface fuels to 100–500 m broadleaf buffers reduced flame length by up to ~59%, rate of spread by ~49%, and fireline intensity by ~5× (Oliveira et al., 2023). Fire simulations in Sardinia (Italy) assessed higher burn probability and higher intensity in dense shrublands, while grass–agriculture mosaics showed lower intensity (Salis et al., 2015).

Phenology and live fuel moisture tightly constrain ignition windows (Fig. 7.1). Early greening and productivity peaks, followed by rapid senescence, load landscapes with dry fine fuels and raise ignition density (Bajocco et al., 2017; Bajocco et al., 2010; Nikonovas et al., 2024). Fuel moisture is the proximate control: lower LFMC in the preceding one to two weeks and faster LFMC decline elevate ignition probability (Jurdao et al., 2012; Sakellariou et al., 2020), and fire occurrence (Luo et al., 2019). Moreover, higher prefire vegetation greenness generally coincides with greater severity, with attenuation at very high greenness (García-Llamas et al., 2019a), and prefire live fuel amount amplifies burn severity. High net primary productivity, used as a proxy for fuel load, supports faster fire spread (Resco de Dios et al., 2025). Larger canopy cover has been associated with higher severity in large forest fires in the United States (Birch et al., 2015), and prefire live biomass (e.g., leaf area index (LAI)/fPAR) consistently predicts greater burn severity in Mediterranean pine systems (Viedma et al., 2020b).

Landscape configuration mediates both where fires start and how far they spread. Fine-scale heterogeneity, through greater contact among fuels and land uses, elevates ignition density (Chas-Amil et al., 2013; Costafreda-Aumedes et al., 2013; Martínez-Fernández et al., 2013; Ortega et al., 2012), whereas homogeneous, well-connected fuels favor propagation and larger BA (Loepfe et al., 2010; Viedma et al., 2009). Moreover, vertical and horizontal structure strongly condition severity (Fig. 7.1). Field-based studies have shown that low canopy base height (CBH), abundant ladder fuels, small mean tree size, many snags, and high within-stand variability in height and diameter at breast height (DBH) increase crown involvement and heat release; by contrast, greater basal area, larger trees, and taller canopies generally moderate severity (Lutze et al., 2025; Taylor et al., 2021). In addition, structure metrics derived from satellite light detection and ranging (LiDAR) (e.g., global ecosystem dynamics investigation [GEDI]) and airborne laser scanning (ALS) converge on these patterns (Fig. 7.1).

Ladder fuels have been identified as the most consistent predictor of severity in California wildfires using GEDI (Hakkenberg et al., 2024), and using ALS in Spain (Fernandez-Manso et al., 2019; Viedma et al., 2020b; Viedma et al., 2015b), the United States (Bishop et al., 2014; Hoe et al., 2018), and Canada (Qi et al., 2022).

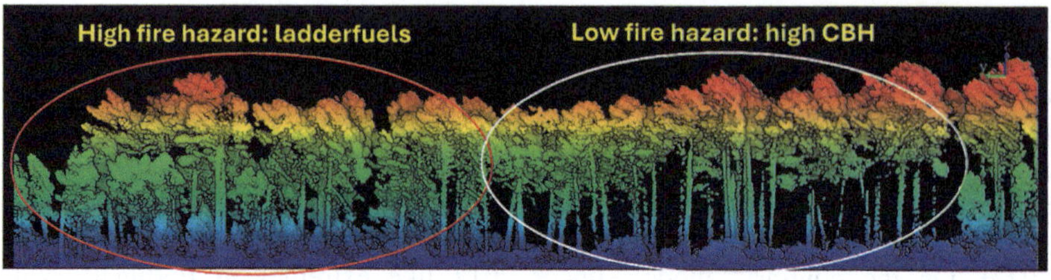

Figure 7.2 Light detection and ranging–derived vertical forest structure illustrating contrasting fire hazard conditions. On the left, dense ladder fuels and low canopy base height (CBH) promote high fire hazard by facilitating vertical continuity and crown ignition. On the right, stands with higher CBH and reduced ladder fuels display lower fire hazard, limiting crown-fire initiation and spread.

Regionally, crown-fire severity increased where CBH was lower, and vertical heterogeneity favored torching (Fernández-Alonso et al., 2017), and high-severity crowns coincided with continuous, high-canopy bulk density (CBD) canopies. Stands with larger crowns and wider spacing burned less severely, whereas dense, regenerating stands with high crown-fuel continuity burned more severely (Qi et al., 2022) (Fig. 7.2. Process-based simulations (FlamMap) using LiDAR show crowning probability rising sharply for CBH $\lesssim$6–7 m, CBD $\gtrsim$0.10 kg m^{-3}, with shrub-dominated surface fuels in Portugal (Botequim et al., 2017); and high crowning probability in Extremadura (Spain) when CBD $\gtrsim$0.08 kg m^{-3} with CBH $\lesssim$1.9 m and tall shrubs are present (Botequim et al., 2019).

Moreover, postfire LiDAR improves discrimination of high-mortality plots compared with imagery alone (Garcia et al., 2017; Viedma et al., 2020a). At the stand scale, high-canopy consumption depresses LiDAR cover metrics, whereas LiDAR height remains robust for recovering prefire tree height (Wulder et al., 2009). High-severity areas exhibit simplified crown and volume structures, declines in crown area, crown LAI, crown volume, and understory/midstory density (Peña-Molina et al., 2024; Viedma et al., 2020a). Overall, unburned and low-severity blocks retain the greatest structural diversity, whereas moderate–high severities converge on simplified configurations (Qi et al., 2022; Viedma et al., 2020a) (Fig. 7.3. These findings show that postfire legacies both reflect and reinforce severity outcomes, and that LiDAR-derived metrics provide mechanistic, tree-level proxies that align with satellite severity indices (e.g., Relativized Burn Ratio [RBR]), discriminate crown versus surface fire, and reveal within-class heterogeneity that optical imagery alone misses (Viedma et al., 2020a) (Fig. 7.3).

Finally, change-explicit LiDAR metrics improve quantification of these losses. A profile-area-change (PAC) metric from matched pre/post ALS correlates far better with plot-level Δbasal area and ΔLAI than satellite-based fire severity indices (e.g., RdNBR), which show spectral saturation at extreme damage (Hu et al., 2019). Multitemporal LiDAR reduces error versus delta Normalized Burn Ratio (dNBR)/RdNBR when predicting basal-area mortality, losses in ladder fuels (2–10 m), and the

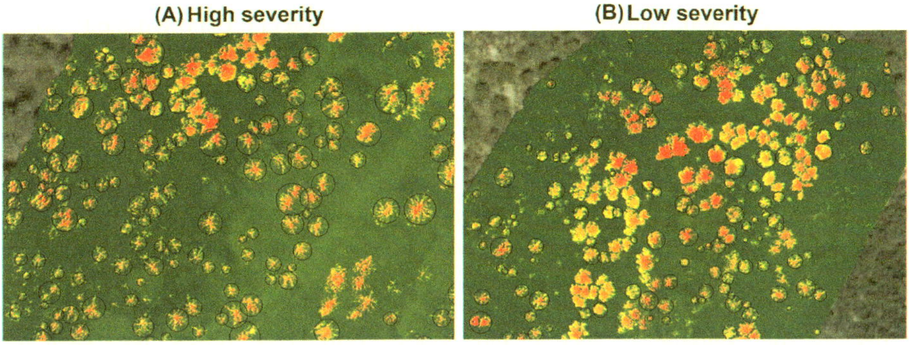

Figure 7.3 High-resolution tree-level map of the Yeste fire, which occurred in SE Spain in August 2017, derived from light detection and ranging onboard an unmanaged aerial vehicle (UAV). Individual crowns are delineated, and vegetation height is represented using a color gradient from green (low height) to red (high height). The maps illustrate spatial heterogeneity in crown damage across stands, highlighting clusters of high severity (A) and low severity (B).

upper canopy. Fusing LiDAR change with spectral fire severity indices further cuts bias (Garcia et al., 2017; Hu et al., 2019). However, LiDAR–satellite coupling is composition-dependent. For example, in Alaska, fractional canopy removal and surface elevation loss (depth of burn) from bi-temporal ALS were tightly related to dNBR in homogeneous black spruce but weakened in mixed stands, cautioning against uniform spectral–structural translations (Alonzo et al., 2017).

7.1.4 Socioeconomic drivers

Demographic and social predictors show robust, scale-dependent signals (Fig. 7.1). High population density generally increases fire occurrence (Ganteaume et al., 2013; Rodrigues and De la Riva, 2014; Vilar et al., 2016) and frequency (Chen et al., 2021; Gelabert et al., 2025; González-Olabarria et al., 2015; Mourão and Martinho, 2014; Nunes, 2012). In addition, low-density localities tend to have larger expected BA per fire because fewer residents reduce early detection and initial attack capacity (Ganteaume et al., 2013; Mourão and Martinho, 2014; Moreira et al., 2023). Moreover, population aging, linked to land abandonment and fuel buildup, is also positively associated with fire frequency (Koutsias et al., 2010) and a larger BA (De Diego et al., 2021). Additionally, unemployment and socioeconomic stressors raise ignition/frequency (Ganteaume et al., 2013; González-Olabarria et al., 2015; Oliveira et al., 2012). Likely, mechanization and high farm density increase ignitions (Martínez-Fernández et al., 2013; Rodrigues and De la Riva, 2014), whereas larger farm size can reduce fire frequency (Ortega et al., 2012), and more agricultural holdings lower the BA (Moreira et al., 2023). Finally, a consistent negative association between grazing and fire occurrence has been observed due to fuel removal (Oliveira et al., 2014).

On the other hand, management, ownership, and protection show mixed patterns (Fig. 7.1). Conflicts around forest policy, declines in forestry-based economies, and suppression-oriented strategies influence fire but are rarely modeled explicitly. Public lands sometimes show more ignitions/frequency (Rodrigues and De la Riva, 2014), yet fire occurrence often concentrates in less managed (often private) forests (Chas-Amil et al., 2015; Martínez-Fernández et al., 2013). In addition, ignition numbers in Galicia rose in common lands when declining productivity aligned with higher shrub cover (Alvarez-Díaz et al., 2015; López-Rodríguez et al., 2021). Protected areas (PAs) can either increase occurrence and BA (Martínez et al., 2009) or reduce them (Foresta et al., 2016; Rodrigues et al., 2016), depending on region and period, reflecting fuel abundance, access, and policy. Temperate forests have seen disproportionate BA inside PAs (Resco de Dios et al., 2025), like in several Canadian PAs, due to remoteness/lightning/minimal suppression (Mansuy et al., 2019). It has been observed that the very large-fire ignition likelihood is highest in PAs (often pine/mixed; Arellano-del-Verbo et al., 2023), and in Galicia, BA increased after the Natura 2000 designation (Rodríguez et al., 2022). Overall, municipal environmental spending and prevention/suppression outlays reduce fire numbers and BA (Mourão and Martinho, 2014).

Accessibility, WUI/WAIs, and networks act as bidirectional drivers (Zambon et al., 2019). Proximity to roads and settlements increases occurrence/frequency (Bar Massada et al., 2013; Fuentes-Santos et al., 2013; Lozano et al., 2008; Zambon et al., 2019), while the BA tends to peak at intermediate/longer distances due to delayed detection and lower fragmentation (Moreno et al., 2011; Zambon et al., 2019). Higher densities of roads and powerlines usually correlate with more ignitions/frequency (Bergonse et al., 2022; Chas-Amil et al., 2015; Oliveira et al., 2012; Pinto et al., 2020; Rodrigues and De la Riva, 2014; Vilar et al., 2016) (Fig. 7.1).

The WUI concentrates fatalities, economic losses, and shows consistently higher probabilities of ignition and burning than adjacent areas (Chappaz and Ganteaume, 2023; Modugno et al., 2016) (Fig. 7.1). Across five Southern European countries, the larger burned surfaces concentrate within ~5 km of WUI (Modugno et al., 2016). In addition, fire hazard is rising in Galician WUIs due to insufficient compliance with fuel-management strip regulations, leading to substantial vertical and horizontal vegetation continuity in proximity to assets (Fernández-álvarez et al., 2019). As WUIs expand, often into landscapes with a recent fire history, both ignition pressure and exposure are increasing (Vacca et al., 2020). WUIs generally raise occurrence and fire counts, especially dense/metropolitan WUIs (Chas-Amil et al., 2015; Chas-Amil et al., 2013; Darques, 2015), although some regions report the highest ignition densities in low-density housing embedded in forests/shrublands that are more vulnerable due to the abandonment and the consequent advancement of shrublands and young trees (Badia et al., 2019). For example, in Sardinia, grassland-dominated WUI shows a high BA (Scarpa et al., 2024). Regarding fire severity, it has been observed that severity tends to be highest around isolated and scattered buildings where shrubland or low-cover forest surrounds structures, whereas very dense WUIs often show lower severity due to fine-scale heterogeneity and fuel breaks (Beltrán-Marcos et al., 2023; Fernández-García et al., 2023).

Finally, temporal diagnostics indicate shifting drivers. In the Iberian Peninsula and adjacent regions, the explanatory power of traditional access/interface proxies has declined, while broader demographic pressure has gained importance for occurrence; signs for PAs have flipped over time (less fire activity; Rodrigues et al., 2016). Moreover, in Spain, it has been assessed that fire activity has migrated from rugged conifer landscapes toward lower-elevation, high-radiation agro-mosaics, and near settlements and access networks (Viedma et al., 2018; Vilar et al., 2016).

7.2 Statistical approaches

7.2.1 Ignitions and fire occurrences

Ignition points (x, y coordinates) are a primary descriptor for assessing fire patterns and modeling fire risk. However, due to the uncertainties regarding fire ignition locations, and several times, the lack of such spatially explicit information, secondary fire descriptors are used (i.e., n° of fires and/or fire density; Morgan et al., 2001). The simplest way to derive fire density is to divide the number of fires by the unit area in km^2 (Oliveira et al., 2014). However, more sophisticated methods have been applied using the Kernel density method (González-Olabarria et al., 2015; Koutsias et al., 2012; Oliveira et al., 2012; Vilar del Hoyo et al., 2011). In addition, to improve fire density calculation at any aggregation unit, "the fire domain extent" correction was applied. In this case, only the reliably BA within each spatial unit was considered as the forested areas (Oliveira et al., 2012; Oliveira et al., 2014; Vilar del Hoyo et al., 2011), or only WUI-non-WUI areas at the municipality level (Chas-Amil et al., 2013; Modugno et al., 2016) (Fig. 7.4.

Usually, continuous fire density values have been reclassified into a binary variable (presence/absence; Costafreda-Aumedes et al., 2013), whereas in others, fire density values have been reclassified into intervals (very low, low, medium, high, and very high) (González-Olabarria et al., 2015) to later be transformed into a binary variable (0 [low] and 1 [high]) (Vilar del Hoyo et al., 2011) using only the lowest and the highest ranges; although the removal of the central part may be controversial from a statistical point of view. Accordingly, ignition modeling and fire occurrence ($\geq$1 fire per unit period) are operationalized as a binary response, making logistic generalized linear models (GLM) or generalized additive models (GAM) the canonical framework. Moreover, global and regionally stratified models (province/municipality/grid) have addressed heterogeneity and nonstationarity, including Geographically Weighted Linear Regressions (GWLR) for local coefficients (Koutsias et al., 2010; Rodrigues et al., 2016; Rodrigues et al., 2019) (Fig. 7.4).

Where absences are unreliable, presence-only Maximum Entropy models (MaxEnt) are used (Arellano-del-Verbo et al., 2023; Chen et al., 2021; Foresta et al., 2016; Mansuy et al., 2019; Vilar et al., 2016). On the other hand, if occurrence is counted (frequency), Poisson/NegBin GLMs (with overdispersion control) and zero-inflated/hurdle or hierarchical space-time models are adopted (Chas-Amil et al., 2015; Rodrigues et al., 2016; Viedma et al., 2018). Finally, machine learning (Classification and Regression Trees [CART], Random Forests [RFs], or Support

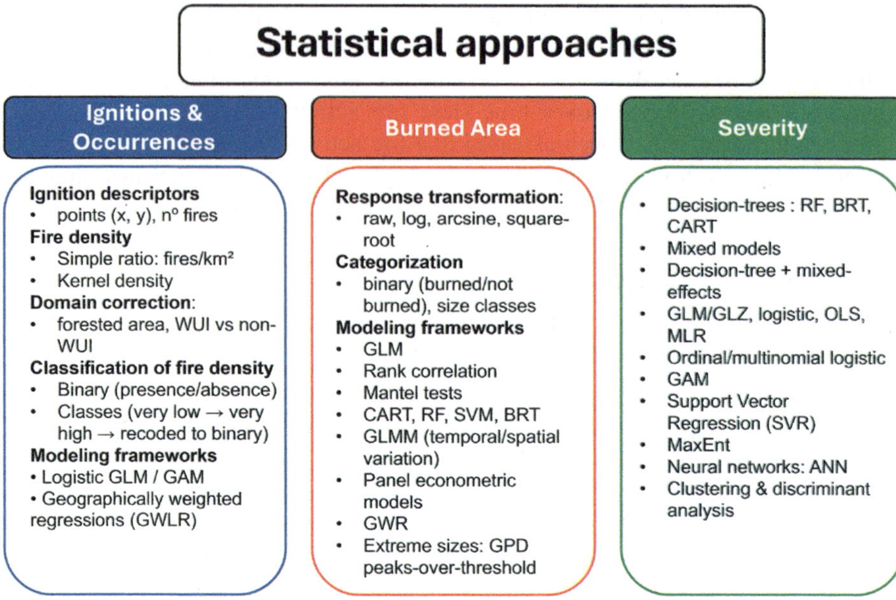

Figure 7.4 Overview of statistical approaches commonly applied to assess fire hazard: ignitions and occurrences (left), burned area analysis (center), and fire severity (right). Together, these approaches provide complementary insights into different components of the fire regime.

Vector Machine [SVM]) complements GLM/GAM for ignitions probability and fire occurrence when interactions are strong (Bajocco et al., 2010; Bergonse et al., 2022; Gelabert et al., 2025; Lozano et al., 2008; Mourão and Martinho, 2014; Oliveira et al., 2012; Tonini et al., 2020); and finally, joint occurrence–size frameworks combine spatial logistic models with extreme-value tails (integrated nested laplace approximation–stochastic partial differential equation [INLA–SPDE]) for occurrence; generalized Pareto distribution [GPD] for size; Pinto et al., 2020) (Fig. 7.4).

7.2.2 *Burned area*

Continuous BA is analyzed untransformed or after log, arcsine (proportions), or square-root transforms, but BA is also binarized (burned/not burned by % threshold) or grouped into size classes (Loepfe et al., 2010; Viedma et al., 2009). Continuous/proportional BA uses GLM (Viedma et al., 2009); rank correlations (Nunes, 2012); partial Mantel test (Loepfe et al., 2010); CART (Mourão and Martinho, 2014); RFs (Bergonse et al., 2022; Tonini et al., 2020); SVM (Scarpa et al., 2024); and Boosted Regression Trees (BRT; Parisien et al., 2011). Moreover, generalized linear mixed models (GLMM) decompose temporal and spatial variation in municipal/parish data, including BA or fire indices derived from principal component analysis (PCA)-(Moreira et al., 2023; Resco de Dios et al., 2025); panel data econometric models (De

Diego et al., 2021); and geographically weighted regressions (GWR; Fuentes-Santos et al., 2013) have been used. Finally, size extremes are modeled via peaks-overthreshold (GPD), isolating tail risk from mean behavior (Pinto et al., 2020) (Fig. 7.4).

7.2.3 Fire severity

Tree-ensemble models, especially RFs, have been widely used for severity prediction. RF regressors/classifiers underpin mapping or probability models (Birch et al., 2015; Fernández-García et al., 2023; García-Llamas et al., 2019b). BRTs are also prominent, with applications to continuous severity and class probabilities (Fang et al., 2018; Viedma et al., 2020b; Viedma et al., 2015b). Moreover, decision-tree models, CART, and related regression trees have been used either alone or alongside other methods to predict severity classes or continuous responses (Bishop et al., 2014; Fernandes et al., 2009; Lentile et al., 2006). Finally, mixed-effects modeling has been paired with decision-tree models to analyze severity regimes (Beltrán-Marcos et al., 2023; Fernandes et al., 2009) or used mixed-effects modeling alone (Grünig et al., 2023; Hakkenberg et al., 2024) (Fig. 7.4).

On the other hand, generalized linear modeling (GLM/GLZ), logistic regression, and ordinary least squares (OLS) appear across multiple scales and datasets. Ordinal logistic (proportional odds) models (Bigler et al., 2005), multinomial logistic (Montealegre et al., 2014), OLS, and related linear forms (Oliveras et al., 2009), as well as multiple linear regression (Alvarez et al., 2024), were applied. In addition, GAM has been used to model severity responses with smooth, nonparametric terms (Estes et al., 2017; Hoe et al., 2018). Furthermore, support-vector methods appear in postfire severity/burn-impact pipelines via support vector regression (SVR; Garcia et al., 2017). Likely, MaxEnt approach was used to fit separate one-class models for low, moderate, and high severity, and to generate per-pixel suitability (probability of presence) surfaces (Fernandez-Manso et al., 2019), as well as artificial neural networks (ANNs; Peña-Molina et al., 2024). Finally, clustering and discriminant-analysis approaches were used to delineate severity-related spectral groups and derive discriminant rules for mapping (Alonzo et al., 2017; Viedma et al., 2020a) (Fig. 7.4).

7.3 Concluding remarks

Results across climate-based studies are robust: (1) compound hot–dry conditions (heatwaves and droughts) most strongly elevate large-fire risk; (2) antecedent wetness/fuel memory shapes seasonal susceptibility; (3) integrated dryness (VPD/FWI) tracks escalation from ignition to large fire and underpins rising risk under warming. Key differences hinge on regime and governance: fuel-limited contexts can show weaker or even negative climate–area links without fuel buildup, and strong suppression can decouple observed trends from climatic propensity.

Collectively, the empirical and modeling evidence converges. Landscape composition and management strongly modulate fire behavior. Dense shrublands and conifer fuels with high connectivity elevate occurrence and severity, whereas broadleaf and mixed forests dampen extremes. In addition, landscape configuration is pivotal. Strategic fragmentation and targeted treatments (higher CBH, lower CBD, shrub-understory control) reduce flame lengths, spread rates, and BA, while retaining large, fire-resistant trees (Stephens et al., 2010) (Fig. 7.5. At the stand scale, fuel-centric "fire-smart" management, that is, fuel isolation, structural modification, and type conversion, remains the primary climate adaptation in Mediterranean forests (Fernandes, 2013). Fuel controls gain importance with aridity and where shrub encroachment dominates (Pausas and Fernández-Muñoz, 2012). Prescribed burning (PB) has consistent empirical support: it mitigates subsequent wildfire behavior/severity and aids suppression, with effectiveness governed by time-since-treatment, fuel productivity, and fire weather (Fernandes et al., 2013). In Mediterranean shrublands, PB reduced severity ~24% (RBR) at first encounter and remained influential for approximately five years, though diminishing under extreme weather; repeated PB helps cap fuel buildup (Fernández-Guisuraga and Fernandes, 2024). In maritime pine, PB effects on crown scorch/char persist approximately two to six years, suggesting approximately four-year intervals (Espinosa et al., 2019) (Fig. 7.5).

Likely agricultural belts and fine-grained mosaics around WUI both buffer damage and create containment opportunities. Managing abandoned farmland, for example, promoting High Nature Value farmlands (HNVf), reverses homogenization, with cobenefits for biodiversity and ecosystem services (Lasanta et al., 2015; Sil et al., 2019). Scenario modeling confirms that expanding HNVf, especially combined with fire-smart forestry, markedly reduces BA and economic losses while sustaining services (Lecina-Diaz et al., 2023). Governance should adopt the "Fire Smart Territory" socio-ecological framework—risk-based planning, prior zoning of responses (from aggressive attack to monitored/no-suppression within prescription), systematic cost-effectiveness

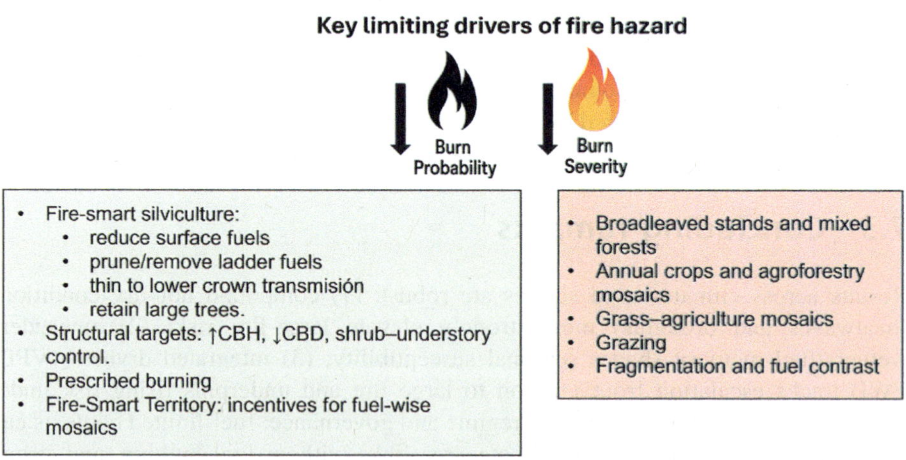

Figure 7.5 Management strategies and land-use configurations that reduce fire hazard. Left: fire-smart silviculture and structural fuel treatments. Right: landscape-scale approaches that enhance fuel heterogeneity.

assessment, and bottom-up integration of traditional fire use (Fernandes, 2013; Tedim et al., 2016). Climate-smart agendas centered on large-scale (re)forestation must be redesigned to be fire-smart: avoid dense, even-aged monocultures that raise megafire risk and carbon reversal; and favor heterogeneous mosaics, vertical discontinuities, low-flammability (often native resprouting) species, and lower densities over stand lifetimes (Regos et al., 2023) (Fig. 7.5). Rewilding requires explicit treatment of disturbance, trophic complexity, and social trade-offs through participatory design (Canibe-Iglesias et al., 2022). Fire hazard mitigation is even more challenging in PAs because legislation regulates and limits management, which, in concert with land-tenure constraints, complicates fuel treatment and land-use allocation (Lecina-Diaz et al., 2023). In protected "living laboratories" (UNESCO Biosphere Reserves), aligning climate- and fire-smart policies may require rezoning (e.g., expanded core areas), strategic fuel management in high-hazard biodiversity hotspots, and spatially explicit planning tools (Campos et al., 2022). Policy instruments (e.g., CAP reform, payments for ecosystem services) should reward farmers/landowners for prevention functions and for conversions to more fire-resistant woodlands (Regos et al., 2023) (Fig. 7.5).

AI disclosure

During the preparation of this work, the author(s) used ChatGPT in order to enhance the clarity of the text and check for grammatical accuracy. After using this tool/service, the author(s) reviewed and edited the content as needed and take(s) full responsibility for the content of the publication.

References

Ager, A.A., Preisler, H.K., Arca, B., Spano, D., Salis, M., 2014. Wildfire risk estimation in the Mediterranean area. Environmetrics (London, Ont.) 25 (6), 384–396. Available from. Available at https://doi.org/10.1002/env.2269, http://doi.wiley.com/10.1002/env.2269.

Alonzo, M., Morton, D.C., Cook, B.D., Andersen, H.E., Babcock, C., Pattison, R., 2017. Patterns of canopy and surface layer consumption in a boreal forest fire from repeat airborne lidar. Environmental Research Letters 12 (6). Available at https://doi.org/10.1088/1748-9326/aa6ade.

Alvarez, A., Lecina-Diaz, J., Batllori, E., Duane, A., Brotons, L., Retana, J., 2024. Spatiotemporal patterns and drivers of extreme fire severity in Spain for the period 1985–2018. Agricultural and Forest Meteorology 358. Available at https://doi.org/10.1016/j.agrformet.2024.110185.

Alvarez-Díaz, M., González-Gómez, M., Otero-Giraldez, M.S., 2015. Detecting the socio-economic driving forces of the fire catastrophe in NW Spain. European Journal of Forest Research 134 (6), 1087–1094. Available from. Available at https://doi.org/10.1007/s10342-015-0911-1, http://www.scopus.com/inward/record.url?eid=2-s2.0-84943815715&partnerID=40&md5=451b38291f54939e177c58199a3a9652.

Arellano-del-Verbo, G., Urbieta, I.R., Moreno, J.M., 2023. Large-fire ignitions are higher in protected areas than outside them in West-Central Spain. Fire 6 (1). Available at https://doi.org/10.3390/fire6010028.

Badia, A., Pallares-Barbera, M., Valldeperas, N., Gisbert, M., 2019. Wildfires in the wildland-urban interface in Catalonia: vulnerability analysis based on land use and land cover change. Science of the Total Environment 673, 184–196. Available at https://doi.org/10.1016/j.scitotenv.2019.04.012.

Bajocco, S., Rosati, L., Ricotta, C., 2010. Knowing fire incidence through fuel phenology: a remotely sensed approach. Ecological Modelling 221 (1), 59–66. Available at https://doi.org/10.1016/j.ecolmodel.2008.12.024.

Bajocco, S., Koutsias, N., Ricotta, C., 2017. Linking fire ignitions hotspots and fuel phenology: the importance of being seasonal. Ecological Indicators 82, 433–440. Available at https://doi.org/10.1016/j.ecolind.2017.07.027.

Bar Massada, A., Syphard, A.D., Stewart, S.I., Radeloff, V.C., 2013. Wildfire ignition-distribution modelling: a comparative study in the Huron-Manistee National Forest, Michigan, USA. International Journal of Wildland Fire 22 (2), 174–183. Available at https://doi.org/10.1071/WF11178.

Beltrán-Marcos, D., Calvo, L., Fernández-Guisuraga, J.M., Fernández-García, V., Suárez-Seoane, S., 2023. Wildland-urban interface typologies prone to high severity fires in Spain. Science of the Total Environment 894. Available at https://doi.org/10.1016/j.scitotenv.2023.165000.

Bergonse, R., Oliveira, S., Zêzere, J.L., Moreira, F., Ribeiro, P.F., Leal, M., et al., 2022. Biophysical controls over fire regime properties in Central Portugal. Science of the Total Environment 810. Available at https://doi.org/10.1016/j.scitotenv.2021.152314.

Bigler, C., Kulakowski, D., Veblen, T.T., 2005. Multiple disturbance interactions and drought influence fire severity in rocky mountain subalpine forests. Ecology 86 (11), 3018–3029. Available from. Available at https://doi.org/10.1890/05-0011, http://www.esajournals.org/doi/abs/10.1890/05-0011.

Birch, D.S., Morgan, P., Kolden, C.A., Abatzoglou, J.T., Dillon, G.K., Hudak, A.T., et al., 2015. Vegetation, topography and daily weather influenced burn severity in central Idaho and western Montana forests. Ecosphere 6 (1). Available at https://doi.org/10.1890/ES14-00213.1.

Bishop, B.D., Dietterick, B.C., White, R.A., Mastin, T.B., 2014. Classification of plot-level fire-caused tree mortality in a redwood forest using digital orthophotography and LiDAR. Remote Sensing 6 (3), 1954–1972. Available at https://doi.org/10.3390/rs6031954.

Botequim, B., Arias-Rodil, M., Garcia-Gonzalo, J., Silva, A., Marques, S., Borges, J.G., et al., 2017. Modeling post-fire mortality in pure and mixed forest stands in Portugal-A forest planning-oriented model. Sustainability (Switzerland) 9 (3). Available at https://doi.org/10.3390/su9030390.

Botequim, B., Fernandes, P.M., Borges, J.G., González-Ferreiro, E., Guerra-Hernández, J., 2019. Improving silvicultural practices for Mediterranean forests through fire behaviour modelling using LiDAR-derived canopy fuel characteristics. International Journal of Wildland Fire 28 (11), 823–839. Available at https://doi.org/10.1071/WF19001.

Bowman, D.M.J.S., Balch, J.K., Artaxo, P., Bond, W.J., Carlson, J.M., Cochrane, M.A., et al., 2009. Fire in the earth system. Science (New York, N.Y.) 324 (5926), 481–484. Available from. Available at https://doi.org/10.1126/science.1163886, http://www.sciencemag.org/content/324/5926/481.abstract.

Campos, J.C., Rodrigues, S., Sil, Â., Hermoso, V., Freitas, T.R., Santos, J.A., et al., 2022. Climate regulation ecosystem services and biodiversity conservation are enhanced differently by climate-and fire-smart landscape management. Environmental Research Letters 17 (5). Available at https://doi.org/10.1088/1748-9326/ac64b5.

Canibe-Iglesias, M.C., Hermoso, V., Campos, J.C., Carvalho-Santos, C., Fernandes, P.M., Freitas, T.R., et al., 2022. Climate- and fire-smart landscape scenarios call for redesigning protection regimes to achieve multiple management goals. Journal of Environmental Management 322. Available at https://doi.org/10.1016/j.jenvman.2022.116045.

Chappaz, F., Ganteaume, A., 2023. Role of land-cover and WUI types on spatio-temporal dynamics of fires in the French Mediterranean area. Risk Analysis 43 (5), 1032–1057. Available at https://doi.org/10.1111/risa.13979.

Chas-Amil, M.L., Prestemon, J.P., McClean, C.J., Touza, J., 2015. Human-ignited wildfire patterns and responses to policy shifts. Applied Geography 56. pp. 164–176. Available at http://www.scopus.com/inward/record.url?eid=2-s2.0-84919797522&partnerID=40&md5=70a6ac942389535a913d7e7f413e4bf1 Department of Quantitative Economics, University of Santiago de Compostela, Baixada Burgo das Nacións s/nSantiago de Compostela, Spain Southern Research Station, USDA Forest ServiceResearch Triangle Park, NC, United States Environment Department, Universi.

Chas-Amil, M.L., Touza, J., García-Martínez, E., 2013. Forest fires in the wildland-urban interface: a spatial analysis of forest fragmentation and human impacts. Applied Geography 43, 127–137. Available at https://doi.org/10.1016/j.apgeog.2013.06.010.

Chen, B., Jin, Y., Scaduto, E., Moritz, M., Goulden, M.L., Randerson, J.T., 2021. Climate, fuel, and land use shaped the spatial pattern of wildfire in California's Sierra Nevada. Journal of Geophysical Research: Biogeosciences, e2020JG005786 Available from. Available at https://doi.org/10.1029/2020JG005786, https://onlinelibrary.wiley.com/doi/10.1029/2020JG005786.

Costafreda-Aumedes, S., Garcia-Martin, A., Vega-Garcia, C., 2013. The relationship between landscape patterns and human-caused fire occurrence in Spain. Forest Systems 22 (1), 71–81. Available at https://doi.org/10.5424/fs/2013221-02685, https://www.scopus.com/inward/record.url?eid=2-s2.0-84875731318&partnerID Departamento de Ingeniería Agroforestal, Universidad de Lleida, Alcalde Rovira Roure, 191, 25198 Lleida, Spain Centro Universitario de la Defensa, Academia General Militar, Ctra. de Huesca, s/n, 50090 Zaragoza, Spain.

Darques, R., 2015. Mediterranean cities under fire. A critical approach to the wildland-urban interface. Applied Geography 59, 10–21. Available at https://doi.org/10.1016/j.apgeog.2015.02.008.

De Diego, J., Rúa, A., Fernández, M., 2021. Vulnerability variables and their effect on wildfires in Galicia (Spain). A panel data analysis. Land 10 (1004). Available at https://doi.org/10.3390/land10101004, https://doi.org/10.3390/land Available from.

Dupuy, J.-luc, Fargeon, H., Martin-StPaul, N., Pimont, F., Ruffault, J., Guijarro, M., et al., 2020. Climate change impact on future wildfire danger and activity in southern Europe: a review. Annals of Forest Science 77 (35). Available at https://doi.org/10.1007/s13595-020-00933-5/Published, https://doi.org/10.1007/s13595-020-00933-5 Available from.

Elia, M., Lafortezza, R., Onofrio, C., Costa-Saura, J.M., Bacciu, V., Giannico, V., et al., 2024. Coupling heat wave and wildfire occurrence across multiple ecoregions within a Eurasia longitudinal gradient. Science of the Total Environment 912. Available at https://doi.org/10.1016/j.scitotenv.2023.169269.

Espinosa, J., Palheiro, P., Loureiro, C., Ascoli, D., Esposito, A., Fernandes, P.M., 2019. Fire-severity mitigation by prescribed burning assessed from fire-treatment encounters in maritime pine stands. Canadian Journal of Forest Research 49 (2), 205–211. Available at https://doi.org/10.1139/cjfr-2018-0263.

Estes, B.L., Knapp, E.E., Skinner, C.N., Miller, J.D., Preisler, H.K., 2017. Factors influencing fire severity under moderate burning conditions in the Klamath Mountains, northern California, USA. Ecosphere 8 (5). Available at https://doi.org/10.1002/ecs2.1794.

Fang, L., Yang, J., White, M., Liu, Z., 2018. Predicting potential fire severity using vegetation, topography and surface moisture availability in a Eurasian boreal forest landscape. Forests 9 (3). Available at https://doi.org/10.3390/f9030130.

Fernandes, P.M., 2013. Fire-smart management of forest landscapes in the Mediterranean basin under global change. Landscape and Urban Planning 110 (1), 175–182. Available at https://doi.org/10.1016/j.landurbplan.2012.10.014.

Fernandes, P.A.M., Loureiro, C.A., Botelho, H.S., 2004. Fire behaviour and severity in a maritime pine stand under differing fuel conditions. Annals of Forest Science 61 (6), 537–544. Available at https://doi.org/10.1051/forest:2004048.

Fernandes, P.M., Botelho, H.S., Rego, F.C., Loureiro, C., 2009. Empirical modelling of surface fire behaviour in maritime pine stands. International Journal of Wildland Fire 18 (6), 698–710.

Fernandes, P.M., Davies, G.M., Ascoli, D., Fernández, C., Moreira, F., Rigolot, E., et al., 2013. Prescribed burning in southern Europe: developing fire management in a dynamic landscape. Frontiers in Ecology and the Environment 11 (SUPPL. 1). Available at https://doi.org/10.1890/120298.

Fernandes, P.M., Delogu, G.M., Leone, V., Ascoli, D., 2019. Wildfire Policies Contribution to Foster Extreme Wildfires Extreme Wildfire Events and Disasters: Root Causes and New Management Strategies. Elsevier, pp. 187–200. Available at 10.1016/B978-0-12-815721-3.00010-2.

Fernandez-Manso, A., Quintano, C., Roberts, D.A., 2019. Burn severity analysis in Mediterranean forests using maximum entropy model trained with EO-1 Hyperion and LiDAR data. ISPRS Journal of Photogrammetry and Remote Sensing 155, 102–118. Available at https://doi.org/10.1016/j.isprsjprs.2019.07.003.

Fernández-Alonso, J.M., Vega, J.A., Jiménez, E., Ruiz-González, A.D., Álvarez-González, J.G., 2017. Spatially modeling wildland fire severity in pine forests of Galicia, Spain. European Journal of Forest Research 136 (1), 105–121. Available at https://doi.org/10.1007/s10342-016-1012-5.

Fernández-García, V., Beltrán-Marcos, D., Calvo, L., 2023. Building patterns and fuel features drive wildfire severity in wildland-urban interfaces in Southern Europe. Landscape and Urban Planning 231. Available at https://doi.org/10.1016/j.landurbplan.2022.104646.

Fernández-Guisuraga, J.M., Fernandes, P.M., 2024. Prescribed burning mitigates the severity of subsequent wildfires in Mediterranean shrublands. Fire Ecology 20 (1). Available at https://doi.org/10.1186/s42408-023-00233-z.

Fernández-álvarez, M., Armesto, J., Picos, J., 2019. LiDAR-based wildfire prevention in WUI: the automatic detection, measurement and evaluation of forest fuels. Forests 10 (2). Available at https://doi.org/10.3390/f10020148.

Foresta, M., Carranza, M.L., Garfì, V., Di Febbraro, M., Marchetti, M., Loy, A., 2016. A systematic conservation planning approach to fire risk management in Natura 2000 sites. Journal of Environmental Management 181, 574–581.

Fréjaville, T., Curt, T., 2017. Seasonal changes in the human alteration of fire regimes beyond the climate forcing. Environmental Research Letters 12 (3). Available at https://doi.org/10.1088/1748-9326/aa5d23.

Fuentes-Santos, I., Marey-Pérez, M.F., González-Manteiga, W., 2013. Forest fire spatial pattern analysis in Galicia (NW Spain). Journal of Environmental Management 128 (0),

30–42. Available from. Available at http://www.sciencedirect.com/science/article/pii/S0301479713002648.

Ganteaume, A., Camia, A., Jappiot, M., San-Miguel-Ayanz, J., Long-Fournel, M., Lampin, C., 2013. A review of the main driving factors of forest fire ignition over Europe. Environmental Management 51 (3), 651–662. Available at https://doi.org/10.1007/s00267-012-9961-z.

Garcia, M., Saatchi, S., Casas, A., Koltunov, A., Ustin, S., Ramirez, C., et al., 2017. Quantifying biomass consumption and carbon release from the California Rim fire by integrating airborne LiDAR and Landsat OLI data. Journal of Geophysical Research: Biogeosciences 122 (2), 340–353. Available at https://doi.org/10.1002/2015JG003315.

García-Llamas, P., Suárez-Seoane, S., Taboada, A., Fernández-García, V., Fernández-Guisuraga, J.M., Fernández-Manso, A., et al., 2019a. Assessment of the influence of biophysical properties related to fuel conditions on fire severity using remote sensing techniques: a case study on a large fire in NW Spain. International Journal of Wildland Fire 28 (7), 512–520.

García-Llamas, P., Suárez-Seoane, S., Taboada, A., Fernández-Manso, A., Quintano, C., Fernández-García, V., et al., 2019b. Environmental drivers of fire severity in extreme fire events that affect Mediterranean pine forest ecosystems. Forest Ecology and Management 433, 24–32. Available at https://doi.org/10.1016/j.foreco.2018.10.051.

Gelabert, P.J., Jiménez-Ruano, A., Ochoa, C., Alcasena, F., Sjöström, J., Marrs, C., et al., 2025. Assessing human-caused wildfire ignition likelihood across Europe. Egusphere 143 Available from. Available at https://doi.org/10.5194/egusphere-2025-143, https://egusphere.copernicus.org/preprints/2025/egusphere-2025-143/.

González-Olabarria, J.R., Mola-Yudego, B., Coll, L., 2015. Different factors for different causes: analysis of the spatial aggregations of fire ignitions in Catalonia (Spain). Risk Analysis 35 (7), 1197–1209. Available at https://doi.org/10.1111/risa.12339, http://www.scopus.com/inward/record.url?eid=2 Forest Science Centre of Catalonia (CEMFOR-CTFC), E-25280, Ctra. de St. Llorenç de Morunys, Solsona, Spain CREAF, Cerdanyola del Vallès, Spain Norwegian Forest and Landscape Institute, PO Box 115, Ås, Norway School of Forest Sciences, University of Easter.

Grünig, M., Seidl, R., Senf, C., 2023. Increasing aridity causes larger and more severe forest fires across Europe. Global Change Biology 29 (6), 1648–1659. Available at https://doi.org/10.1111/gcb.16547.

Hakkenberg, C.R., Clark, M.L., Bailey, T., Burns, P., Goetz, S.J., 2024. Ladder fuels rather than canopy volumes consistently predict wildfire severity even in extreme topographic-weather conditions. Communications Earth and Environment 5 (1). Available at https://doi.org/10.1038/s43247-024-01893-8.

Hoe, M.S., Dunn, C.J., Temesgen, H., 2018. Multitemporal LiDAR improves estimates of fire severity in forested landscapes. International Journal of Wildland Fire 27 (9), 581–594. Available at https://doi.org/10.1071/WF17141.

Hu, T., Ma, Q., Su, Y., Battles, J.J., Collins, B.M., Stephens, S.L., et al., 2019. A simple and integrated approach for fire severity assessment using bi-temporal airborne LiDAR data. International Journal of Applied Earth Observation and Geoinformation 78, 25–38. Available at https://doi.org/10.1016/j.jag.2019.01.007.

Jurdao, S., Chuvieco, E., Arevalillo, J.M., 2012. Modelling fire ignition probability froM satellite estiMates of live fuel Moisture content. Fire Ecology 8 (1). Available at https://doi.org/10.4996/fireecology.0801077.

Keeley, J.E., Pausas, J.G., Rundel, P.W., Bond, W.J., Bradstock, R.A., 2011. Fire as an evolutionary pressure shaping plant traits. Trends in Plant Science 16 (8), 406–411.

Knorr, W., Arneth, A., Jiang, L., 2016. Demographic controls of future global fire risk. Nature Climate Change 6 (8), 781–785. Available at https://doi.org/10.1038/nclimate2999.

Koutsias, N., Martinez-Fernandez, J., Allgower, B., 2010. Do factors causing wildfires vary in space? Evidence from geographically weighted regression. Giscience & Remote Sensing 47 (2), 221–240.

Koutsias, N., Arianoutsou, M., Kallimanis, A.S., Mallinis, G., Halley, J.M., Dimopoulos, P., 2012. Where did the fires burn in Peloponnisos, Greece the summer of 2007? Evidence for a synergy of fuel and weather. Agricultural and Forest Meteorology 156 (0), 41–53. Available from. Available at http://www.sciencedirect.com/science/article/pii/S0168192311003431.

Kuhn-Régnier, A., Voulgarakis, A., Nowack, P., Forkel, M., Prentice, I.C., Harrison, S.P., 2021. The importance of antecedent vegetation and drought conditions as global drivers of burnt area. Biogeosciences 18 (12), 3861–3879. Available at https://doi.org/10.5194/bg-18-3861-2021.

Lasanta, T., Nadal-Romero, E., Arnáez, J., 2015. Managing abandoned farmland to control the impact of re-vegetation on the environment. The state of the art in Europe. Environmental Science and Policy 52, 99–109. Available at https://doi.org/10.1016/j.envsci.2015.05.012.

Lasanta, T., Arnáez, J., Pascual, N., Ruiz-Flaño, P., Errea, M.P., Lana-Renault, N., 2017. Space–time process and drivers of land abandonment in Europe. Catena 149, 810–823. Available at https://doi.org/10.1016/j.catena.2016.02.024.

Lecina-Diaz, J., Alvarez, A., Retana, J., 2014. Extreme fire severity patterns in topographic, convective and wind-driven historical wildfires of mediterranean pine forests. PLoS One 9 (1), e85127. Available at https://doi.org/10.1371/journal.pone.0085127 Available from.

Lecina-Diaz, J., Chas-Amil, M.L., Aquilué, N., Sil, Â., Brotons, L., Regos, A., et al., 2023. Incorporating fire-smartness into agricultural policies reduces suppression costs and ecosystem services damages from wildfires. Journal of Environmental Management 337. Available at https://doi.org/10.1016/j.jenvman.2023.117707.

Lentile, L.B., Smith, F.W., Shepperd, W.D., 2006. Influence of topography and forest structure on patterns of mixed severity fire in ponderosa pine forests of the South Dakota Black Hills, USA. International Journal of Wildland Fire 15 (4), 557–566. Available at https://doi.org/10.1071/WF05096.

Li, F., Zhu, Q., Yuan, K., Huang, H., Radeloff, V.C., Chen, M., 2025. Exacerbating risk in human-ignited large fires over western United States due to lower flammability thresholds and greenhouse gas emissions. PNAS Nexus 4 (2). Available at https://doi.org/10.1093/pnasnexus/pgaf012.

Loepfe, L., Martinez-Vilalta, J., Oliveres, J., Piñol, J., Lloret, F., 2010. Feedbacks between fuel reduction and landscape homogenisation determine fire regimes in three Mediterranean areas. Forest Ecology and Management 259 (12), 2366–2374. Available from. Available at https://doi.org/10.1016/j.foreco.2010.03.009, http://www.scopus.com/inward/record.url?eid=2-s2.0-77952111804&partnerID=40&md5=d6d4eaf5bb0a71cd4 Center for Ecological Research and Forestry Applications (CREAF), Autonomous University of Barcelona, E-08193 Bellaterra, Spain.

Lozano, F.J., Suárez-Seoane, S., Kelly, M., Luis, E., 2008. A multi-scale approach for modeling fire occurrence probability using satellite data and classification trees: a case study in a mountainous Mediterranean region. Remote Sensing of Environment 112 (3), 708–719. Available at https://doi.org/10.1016/j.rse.2007.06.006, http://www.scopus.com/inward/record.url?eid=2-s2.0-39749164604&partnerID=40&md5=
72458be5f236436d467861f Área de Ecología, Fac. CC. Biológicas y Ambientales, Universidad de León, 24071 Campus de Vegazana, León, Spain Department of

Environmental Sciences, Policy and Management, University of California, Berkeley, 137 Mulford Hall #3114, Berkeley, CA 94720-311.

Luo, K., Quan, X., He, B., Yebra, M., 2019. Effects of live fuel moisture content on wildfire occurrence in fire-prone regions over southwest China. Forests 10 (10). Available at https://doi.org/10.3390/f10100887.

Lutze, M.T., Trouvé, R.ël, Baker, P.J., Nitschke, C.R., 2025. The impact of a severe wildfire on canopy structure and composition in a lowland mixed-eucalypt forest in southeastern Australia. Fire Ecology 21 (1). Available at https://doi.org/10.1186/s42408-025-00366-3.

López-Rodríguez, G., Rodríguez-Vicente, V., Marey-Pérez, M.F., 2021. Study of forest productivity in the occurrence of forest fires in galicia (Spain). Sustainability (Switzerland) 13 (15). Available at https://doi.org/10.3390/su13158472.

Mansuy, N., Miller, C., Parisien, M.A., Parks, S.A., Batllori, E., Moritz, M.A., 2019. Contrasting human influences and macro-environmental factors on fire activity inside and outside protected areas of North America. Environmental Research Letters 14 (6). Available at https://doi.org/10.1088/1748-9326/ab1bc5.

Mantero, G., Morresi, D., Marzano, R., Motta, R., Mladenoff, D.J., Garbarino, M., 2020. The influence of land abandonment on forest disturbance regimes: a global review. Landscape Ecology 35 (12), 2723–2744. Available at https://doi.org/10.1007/s10980-020-01147-w.

Marlon, J.R., Bartlein, P.J., Carcaillet, C., Gavin, D.G., Harrison, S.P., Higuera, P.E., et al., 2008. Climate and human influences on global biomass burning over the past two millennia. Nature Geoscience 1 (10), 697–702. Available at https://doi.org/10.1038/ngeo313.

Martínez, J., Vega-Garcia, C., Chuvieco, E., 2009. Human-caused wildfire risk rating for prevention planning in Spain. Journal of Environmental Management 90 (2), 1241–1252. Available at https://doi.org/10.1016/j.jenvman.2008.07.005.

Martínez-Fernández, J., Chuvieco, E., Koutsias, N., 2013. Modelling long-term fire occurrence factors in Spain by accounting for local variations with geographically weighted regression. Natural Hazards and Earth System Sciences 13 (2), 311–327. Available from. Available at https://doi.org/10.5194/nhess-13-311-2013, http://www.nat-hazards-earth-syst-sci.net/13/311/2013/.

Modugno, S., Balzter, H., Cole, B., Borrelli, P., 2016. Mapping regional patterns of large forest fires in Wildland-Urban Interface areas in Europe. Journal of Environmental Management 172, 112–126. Available at https://doi.org/10.1016/j.jenvman.2016.02.013.

Montealegre, A.L., Lamelas, M.T., Tanase, M.A., De la Riva, J., 2014. Forest fire severity assessment using ALS data in a mediterranean environment. Remote Sensing 6 (5), 4240–4265. Available at https://doi.org/10.3390/rs6054240.

Moreira, F., Catry, F.X., Rego, F., Bacao, F., 2010. Size-dependent pattern of wildfire ignitions in Portugal: when do ignitions turn into big fires? Landscape Ecology 25 (9), 1405–1417.

Moreira, F., Viedma, O., Arianoutsou, M., Curt, T., Koutsias, N., Rigolot, E., et al., 2011. Landscape - wildfire interactions in southern Europe: Implications for landscape management. Journal of Environmental Management 92 (10). Available at https://doi.org/10.1016/j.jenvman.2011.06.028.

Moreira, F., Leal, M., Bergonse, R., Canadas, M.J., Novais, A., Oliveira, S., et al., 2023. Recent trends in fire regimes and associated territorial features in a fire-prone mediterranean region. Fire 6 (2). Available at https://doi.org/10.3390/fire6020060.

Moreno, J.M., Viedma, O., Zavala, G., Beĺn, L., 2011. Landscape variables influencing forest fires in central Spain. International Journal of Wildland Fire 20 (5), 678–689. Available at https://doi.org/10.1071/WF10005.

Morgan, P., Hardy, C.C., Swetnam, T.W., Rollins, M.G., Long, D.G., 2001. Mapping fire regimes across time and space: understanding coarse and fine-scale fire patterns. International Journal of Wildland Fire 10 (4), 329–342.

Mourão, P.R., Martinho, V.D., 2014. The choices of the fire - Debating socioeconomic determinants of the fires observed at Portuguese municipalities. Forest Policy and Economics 43, 29–40. Available at https://doi.org/10.1016/j.forpol.2014.01.007.

Nikonovas, T., Santín, C., Belcher, C.M., Clay, G.D., Kettridge, N., Smith, T.E.L., et al., 2024. Vegetation phenology as a key driver for fire occurrence in the UK and comparable humid temperate regions. International Journal of Wildland Fire 33 (10). Available at https://doi.org/10.1071/WF23205.

Nunes, A.N., 2012. Regional variability and driving forces behind forest fires in Portugal an overview of the last three decades (1980–2009). Applied Geography 34, 576–586. Available at https://doi.org/10.1016/j.apgeog.2012.03.002.

Oliveira, S., Oehler, F., San-Miguel-Ayanz, J., Camia, A., Pereira, J.M.C., 2012. Modeling spatial patterns of fire occurrence in Mediterranean Europe using multiple regression and random forest. Forest Ecology and Management 275, 117–129. Available from. Available at https://doi.org/10.1016/j.foreco.2012.03.003, http://www.sciencedirect.com/science/article/pii/S0378112712001272.

Oliveira, S., Pereira, J.M.C., San-Miguel-Ayanz, J., Lourenço, L., 2014. Exploring the spatial patterns of fire density in Southern Europe using geographically weighted regression. Applied Geography 51, 143–157. Available from. Available at https://doi.org/10.1016/j.apgeog.2014.04.002, http://www.sciencedirect.com/science/article/pii/S0143622814000630.

Oliveira, A.S., Silva, J.S., Guiomar, N., Fernandes, P., Nereu, M., Gaspar, J., et al., 2023. The effect of broadleaf forests in wildfire mitigation in the WUI – a simulation study. International Journal of Disaster Risk Reduction 93. Available at https://doi.org/10.1016/j.ijdrr.2023.103788.

Oliveras, I., Gracia, M., Moŕ, G., Retana, J., 2009. Factors influencing the pattern of fire severities in a large wildfire under extreme meteorological conditions in the Mediterranean basin. International Journal of Wildland Fire 18 (7), 755–764. Available at https://doi.org/10.1071/WF08070.

Ortega, M., Saura, S., González-Avila, S., Gómez-Sanz, V., Elena-Rosselló, R., 2012. Landscape vulnerability to wildfires at the forest-agriculture interface: half-century patterns in Spain assessed through the SISPARES monitoring framework. Agroforestry Systems 85 (3), 331–349. Available at https://doi.org/10.1007/s10457-011-9423-2, http://www.scopus.com/inward/record.url?eid=2-s2.0-84864324621&partnerID=40&md5=111ea51f6 Unit of Forest Fires, Department of Silviculture and Forest Management, Centre of Forest Research CIFOR, Instituto National de Investigación y Tecnología Agraria y Alimentaria, INIA, Carretera de La Coruña, km 7.5, 28040 Madrid, Spain Department of Forest.

Parisien, M.A., Parks, S.A., Krawchuk, M.A., Flannigan, M.D., Bowman, L.M., Moritz, M.A., 2011. Scale-dependent controls on the area burned in the boreal forest of Canada, 1980-2005. Ecological Applications 21 (3), 789–805.

Pausas, J.G., Fernández-Muñoz, S., 2012. Fire regime changes in the Western Mediterranean Basin: from fuel-limited to drought-driven fire regime. Climatic Change 110 (1-2), 215–226. Available at https://doi.org/10.1007/s10584-011-0060-6.

Perpiña Castillo, C., Jacobs-Crisioni, C., Diogo, V., Lavalle, C., 2021. Modelling agricultural land abandonment in a fine spatial resolution multi-level land-use model: an application for the EU. Environmental Modelling and Software 136. Available at https://doi.org/10.1016/j.envsoft.2020.104946.

Peña-Molina, E., Moya, D., Tomé, J.L., Marino, E., Fajardo-Cantos, Á., González-Romero, J., et al., 2024. Postfire damage zoning with open low-density LiDAR data sources in semi-arid forests of the Iberian Peninsula. Remote Sensing Applications: Society and Environment 33. Available at https://doi.org/10.1016/j.rsase.2023.101114.

Pinto, G.A.S.J., Rousseu, F., Niklasson, M., Drobyshev, I., 2020. Effects of human-related and biotic landscape features on the occurrence and size of modern forest fires in Sweden. Agricultural and Forest Meteorology 291. Available at https://doi.org/10.1016/j.agrformet.2020.108084.

Qi, Y., Coops, N.C., Daniels, L.D., Butson, C.R., 2022. Assessing the effects of burn severity on post-fire tree structures using the fused drone and mobile laser scanning point clouds. Frontiers in Environmental Science 10. Available at https://doi.org/10.3389/fenvs.2022.949442.

Quintero, N., Viedma, O., Urbieta, I.R., Moreno, J.M., 2019. Assessing landscape fire hazard by multitemporal automatic classification of landsat time series using the google earth engine in west-central Spain. Forests 10 (6), 518 518.

Regos, A., Pais, S., Campos, J.C., Lecina-Diaz, J., 2023. Nature-based solutions to wildfires in rural landscapes of Southern Europe: let's be fire-smart!. International Journal of Wildland Fire 32 (6), 942–950. Available at https://doi.org/10.1071/WF22094.

Resco de Dios, V., Cunill Camprubí, À., Pérez-Zanón, N., Peña, J.C., Martínez del Castillo, E., Rodrigues, M., et al., 2022. Convergence in critical fuel moisture and fire weather thresholds associated with fire activity in the pyroregions of Mediterranean Europe. Science of The Total Environment 806, 151462. Available at https://doi.org/10.1016/J.SCITOTENV.2021.151462 151462.

Resco de Dios, V., Schütze, S.J., Cunill Camprubí, À., Balaguer-Romano, R., Boer, M.M., Fernandes, P.M., 2025. Protected areas as hotspots of wildfire activity in fire-prone temperate and Mediterranen biomes. Journal of Environmental Management 385. Available at https://doi.org/10.1016/j.jenvman.2025.125669.

Rodrigues, M., De la Riva, J., 2014. An insight into machine-learning algorithms to model human-caused wildfire occurrence. Environmental Modelling and Software 57, 192–201. Available from. Available at https://doi.org/10.1016/j.envsoft.2014.03.003GEOFOREST, https://www.scopus.com/inward/record.url?eid=2-s2.0-84901505388&partnerID=40&md5=db47a3645bbc11d2 Group, IUCA, Department of Geography and Land Management, University of Zaragoza, Pedro Cerbuna 12, 50009 Zaragoza, Spain.

Rodrigues, M., Jiménez, A., de la Riva, J., 2016. Analysis of recent spatial–temporal evolution of human driving factors of wildfires in Spain. Natural Hazards 84 (3), 2049–2070. Available at https://doi.org/10.1007/s11069-016-2533-4.

Rodrigues, M., Costafreda-Aumedes, S., Comas, C., Vega-García, C., 2019. Spatial stratification of wildfire drivers towards enhanced definition of large-fire regime zoning and fire seasons. Science of the Total Environment 689, 634–644. Available at https://doi.org/10.1016/j.scitotenv.2019.06.467.

Rodrigues, M., Cunill Camprubí, À., Balaguer-Romano, R., Coco Megía, C.J., Castañares, F., Ruffault, J., et al., 2023. Drivers and implications of the extreme 2022 wildfire season in Southwest Europe. Science of the Total Environment 859. Available at https://doi.org/10.1016/j.scitotenv.2022.160320.

Rodriguez-Jimenez, F., Fernandes, P.M., Fernández-Guisuraga, J.M., Alvarez, X., Lorenzo, H., 2023. Drivers and Trends in the Size and Severity of Forest Fires Endangering WUI

Areas: a Regional Case Study. Forests 14 (12). Available at https://doi.org/10.3390/f14122366.

Rodríguez, G.L., Rodríguez Vicente, V., Pérez, M.F.M., 2022. Influence of the declaration of protected natural areas on the evolution of forest fires in collective lands in Galicia (Spain). Forests 13 (8). Available at https://doi.org/10.3390/f13081161.

Ruffault, J., Curt, T., Moron, V., Trigo, R.M., Mouillot, F., Koutsias, N., et al., 2020. Increased likelihood of heat-induced large wildfires in the Mediterranean Basin. Scientific Reports 10 (1). Available at https://doi.org/10.1038/s41598-020-70069-z.

Sakellariou, S., Cabral, P., Caetano, M., Pla, F., Painho, M., Christopoulou, O., et al., 2020. Remotely sensed data fusion for spatiotemporal geostatistical analysis of forest fire hazard. Sensors (Switzerland) 20 (17), 1–20. Available at https://doi.org/10.3390/s20175014.

Salis, M., Ager, A.A., Alcasena, F.J., Arca, B., Finney, M.A., Pellizzaro, G., et al., 2015. Analyzing seasonal patterns of wildfire exposure factors in Sardinia, Italy. Environmental Monitoring and Assessment 187 (1). Available at https://doi.org/10.1007/s10661-014-4175-x.

Salis, M., Del Giudice, L., Jahdi, R., Alcasena-Urdiroz, F., Scarpa, C., Pellizzaro, G., et al., 2022. Spatial patterns and intensity of land abandonment drive wildfire hazard and likelihood in mediterranean agropastoral areas. Land 11 (11). Available at https://doi.org/10.3390/land11111942.

Sarris, D., Koutsias, N., 2014. Ecological adaptations of plants to drought influencing the recent fire regime in the Mediterranean. Agricultural and Forest Meteorology 184, 158–169. Available at https://doi.org/10.1016/j.agrformet.2013.09.002.

Scarpa, C., Elia, M., D'Este, M., Salis, M., Rodrigues, M., Arca, B., et al., 2024. Modelling wildfire activity in wildland–urban interface (WUI) areas of Sardinia, Italy. International Journal of Wildland Fire 33 (12). Available at https://doi.org/10.1071/WF24109.

Scott, J.H. Thompson, M.P. Calkin, D.E., 2013. A wildfire risk assessment framework for land and resource management. USDA Forest Service / UNL Faculty Publications. 328.

Sil, Â., Fernandes, P.M., Rodrigues, A.P., Alonso, J.M., Honrado, J.P., Perera, A., et al., 2019. Farmland abandonment decreases the fire regulation capacity and the fire protection ecosystem service in mountain landscapes. Ecosystem Services 36. Available at https://doi.org/10.1016/j.ecoser.2019.100908.

Stephens, S.L., Millar, C.I., Collins, B.M., 2010. Operational approaches to managing forests of the future in Mediterranean regions within a context of changing climates. Environmental Research Letters 5 (2). Available at https://doi.org/10.1088/1748-9326/5/2/024003.

Taylor, C., Blanchard, W., Lindenmayer, D.B., 2021. What are the associations between thinning and fire severity? Austral Ecology aec.13096, Available from. Available at https://doi.org/10.1111/AEC.13096, https://onlinelibrary.wiley.com/doi/10.1111/aec.13096.

Tedim, F., Leone, V., Xanthopoulos, G., 2016. A wildfire risk management concept based on a social-ecological approach in the European Union: fire Smart Territory. International Journal of Disaster Risk Reduction 18, 138–153. Available at https://doi.org/10.1016/j.ijdrr.2016.06.005.

Terres, J.M., Scacchiafichi, L.N., Wania, A., Ambar, M., Anguiano, E., Buckwell, A., et al., 2015. Farmland abandonment in Europe: identification of drivers and indicators, and development of a composite indicator of risk. Land use Policy 49, 20–34. Available at https://doi.org/10.1016/j.landusepol.2015.06.009.

Tonini, M., D'andrea, M., Biondi, G., Degli Esposti, S., Trucchia, A., Fiorucci, P., 2020. A machine learning-based approach for wildfire susceptibility mapping. the case study of the liguria region in italy. Geosciences (Switzerland) 10 (3). Available at https://doi.org/10.3390/geosciences10030105.

Vacca, P., Caballero, D., Pastor, E., Planas, E., 2020. WUI fire risk mitigation in Europe: a performance-based design approach at home-owner level. Journal of Safety Science and Resilience 1 (2), 97–105. Available at https://doi.org/10.1016/j.jnlssr.2020.08.001.

Viedma, O., Angeler, D.G., Moreno, J.M., 2009. Landscape structural features control fire size in a Mediterranean forested area of central Spain. International Journal of Wildland Fire 18 (5), 575–583. Available at https://doi.org/10.1071/WF08030.

Viedma, O., Moity, N., Moreno, J.M., 2015a. Changes in landscape fire-hazard during the second half of the 20th century: agriculture abandonment and the changing role of driving factors. Agriculture, Ecosystems and Environment 207, 126–140. Available at https://doi.org/10.1016/j.agee.2015.04.011.

Viedma, O., Quesada, J., Torres, I., De Santis, A., Moreno, J.M., 2015b. Fire severity in a large fire in a *Pinus pinaster* forest is highly predictable from burning conditions, stand structure, and topography. Ecosystems 18 (2). Available at https://doi.org/10.1007/s10021-014-9824-y.

Viedma, O., Urbieta, I.R., Moreno, J.M., 2018. Wildfires and the role of their drivers are changing over time in a large rural area of west-central Spain. Scientific Reports 8 (1). Available at https://doi.org/10.1038/s41598-018-36134-4.

Viedma, O., Almeida, D.R.A., Moreno, J.M., 2020a. Postfire tree structure from high-resolution LiDAR and RBR Sentinel 2A fire severity metrics in a pinus halepensis-dominated burned stand. Remote Sensing 12, 3554. Available at https://doi.org/10.3390/rs12213554 Advances in Forest Fire Behaviour Modelling Using Remote Sensing).

Viedma, O., Chico, F., Fernández, J.J., Madrigal, C., Safford, H.D., Moreno, J.M., 2020b. Disentangling the role of prefire vegetation vs. burning conditions on fire severity in a large forest fire in SE Spain. Remote Sensing of Environment 247, 111891 111891.

Vilar, L., Gómez, I., Martínez-Vega, J., Echavarría, P., Riaño, D., Martín, M.P., 2016. Multitemporal modelling of socio-economic wildfire drivers in central Spain between the 1980s and the 2000s: comparing generalized linear models to machine learning algorithms. PLoS One 11 (8). Available at https://doi.org/10.1371/journal.pone.0161344.

Vilar del Hoyo, L., Isabel, M.P.M., Vega, F.J.M., 2011. Logistic regression models for human-caused wildfire risk estimation: analysing the effect of the spatial accuracy in fire occurrence data. European Journal of Forest Research 130 (6), 983–996. Available at https://doi.org/10.1007/s10342-011-0488-2.

Wulder, M.A., White, J.C., Alvarez, F., Han, T., Rogan, J., Hawkes, B., 2009. Characterizing boreal forest wildfire with multi-temporal Landsat and LIDAR data. Remote Sensing of Environment 113 (7), 1540–1555. Available at https://doi.org/10.1016/j.rse.2009.03.004.

Zambon, I., Cerdà, A., Cudlin, P., Serra, P., Pili, S., Salvati, L., 2019. Road network and the spatial distribution of wildfires in the valencian community (1993–2015). Agriculture (Switzerland) 9 (5). Available at https://doi.org/10.3390/agriculture9050100.

Integrating empirical and stochastic models towards wildfire risk mitigation

8

Marcos Rodrigues[1,2]
[1]Department of Geography and Land Management, University of Zaragoza, Zaragoza, Spain, [2]GEOFOREST-IUCA group, University Institute for Environmental Sciences, University of Zaragoza, Zaragoza, Spain

8.1 Case study/activity

8.1.1 Objective

The primary objective of this study was to develop and implement a methodology that bridges the gap between empirical fire history and predictive stochastic modeling. By utilizing the SCENFIRE framework (A flexible scenario-based algorithm for burn probability-based estimates of exposure to wildfires), the research sought to move beyond static risk indices and instead simulate the complete "risk life cycle"—from the point of ignition and the failure of initial suppression to the eventual landscape-scale spread. Central to this objective was the coupling of empirical models (detailing human-caused, natural, and suppression-escape probabilities) with stochastic spread algorithms to generate a library of thousands of plausible, yet theoretical, wildfire perimeters.

Furthermore, the study aimed to streamline the modeling process by introducing an automated selection algorithm. This method was designed to ensure that simulated fire perimeters accurately mirror historical fire-size distributions without the need for labor-intensive, iterative parameter fine-tuning. Ultimately, the goal was to provide a robust, scenario-based tool for evaluating burn probability (BP), allowing risk managers to identify key exposure hot spots that help in prioritizing strategic mitigation efforts.

8.1.2 Scope

The study was conducted in the county of Calatayud in eastern Spain (Autonomous Region of Aragon), a Mediterranean transition zone where wildfire activity is driven by a complex interplay of extreme aridity, fuel continuity from rural abandonment, and high-velocity wind corridors (Fig. 8.1). The region experienced one of the most devastating fires in the summer of 2022, when a single fire event spread over 12,000 hectares, the second largest in Aragon since 1968 (Rodrigues and de la Riva, 2024). Ecologically, the area is a mosaic of diverse fuel types and protected values; it includes numerous urban settlements interspersed with protected areas such as Special Protection Areas for Birds (SPA) and Special Areas of Conservation (SAC). According to the Scott and Burgan classification system, the fuel arrangement

Advances in Remote Sensing for Forest Fire Management. DOI: https://doi.org/10.1016/B978-0-443-41692-7.00019-6
© 2027 Elsevier Ltd. All rights are reserved, including those for text and data mining, AI training, and similar technologies.

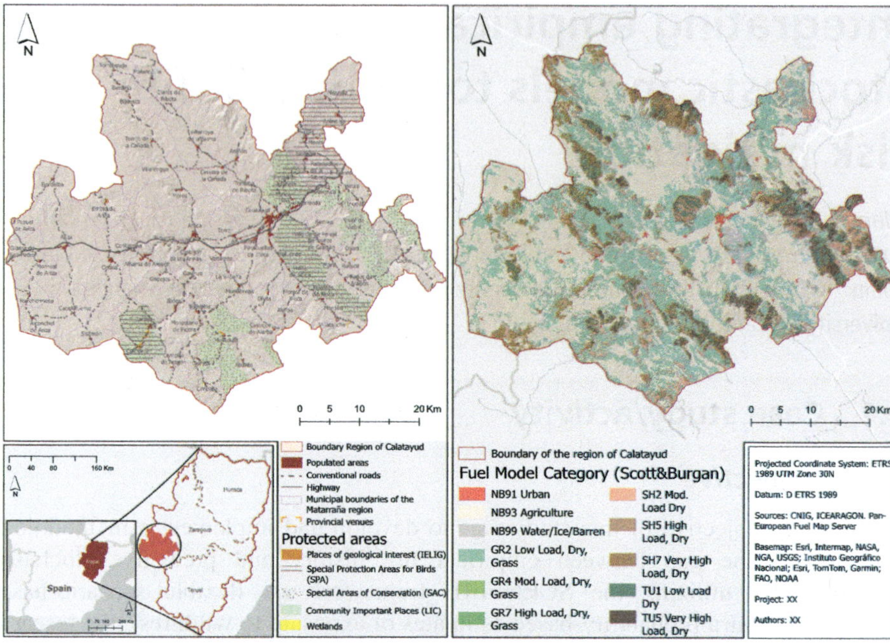

Figure 8.1 Description of the study area. The left panel shows the administrative division, urban settlements, and protected areas. The right panel displays the arrangement of fuel models according to Scott and Burgan's classification. Map lines delineate study areas and do not necessarily depict accepted national boundaries.

consists of a high-load mix of dry grasses, shrubs, and timber-understory models that facilitate rapid fire propagation. The topography further modulates fire behavior, making the region an ideal yet challenging domain for testing integrated empirical and stochastic risk assessment frameworks.

The methodological workflow sought to reproduce the "risk life cycle" (Fig. 8.2), meaning a fire that starts (ignition) and escapes control (suppression capacity) spreads through the landscape (spread and behavior), damaging exposed assets (exposure and risk). Such an approach mimics the sequence of events of a fire (from inception to impact). Instead of synthesizing risk components into an index, it supports complementary and deeper insights for risk assessment, such as developing novel metrics of exposure, transmission of risk, and other related information supporting risk management (Rodrigues et al., 2022). The working hypothesis was that combining empirical (i.e., based on historical data) and stochastic (e.g., spread simulation algorithms) fire modeling facilitates the assessment of wildfire exposure of natural values and human communities at risk, given its ability to reduce uncertainty and capture the variability in fire ignition location and changing weather conditions modulating fire behavior.

However, this modeling approach is very time-consuming and data-demanding. It requires a comprehensive set of information about landscape configuration that includes details about the shape of the relief (elevation, slope, and aspect), the arrangement and

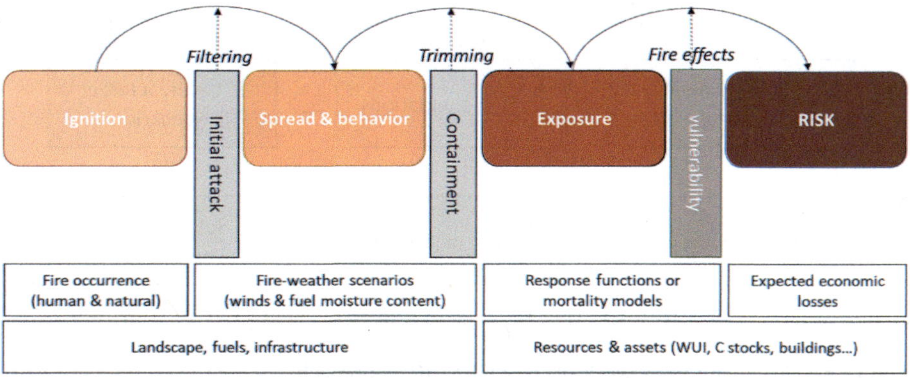

Figure 8.2 Risk assessment framework based on the wildfire life cycle.

type of fuels (surface fuel models), and canopy features (canopy height, density, base height, and bulk density). In turn, simulations must reproduce the spatial pattern of fire occurrence and the historical fire regime, which demands spatial depictions of the potential for fire ignition (both natural and human-caused) and accurate records of past fire events. Accordingly, analyses were constrained to eastern Spain since, in this particular domain, all information was readily available from several datasets of ignition (Rodrigues et al., 2023; Gelabert et al., 2025), suppression potential (Rodrigues et al., 2019; Alawode et al., 2025), fuels, and landscape (Kutchartt et al., 2024). The latter represents one of the most significant efforts to produce wall-to-wall input data for wildfire spread simulations in Europe, developed by integrating state-of-the-art remote sensing products (e.g., global ecosystem dynamics investigation [GEDI] data) with specialized machine learning models. This collection of data, coupled with fire-weather data from ERA5-Land, was assembled into BP estimates using dedicated Python batch processing scripts (to retrieve input data and run spread simulations) and the SCENFIRE R package (to ensemble simulations; Fig. 8.3) into several scenarios according to fire occurrence (Rodrigues, 2026).

8.1.3 Audience

The study is designed to engage two primary audiences. The methodology will be particularly valuable to wildfire modelers, especially those specializing in stochastic fire spread simulation, as it establishes a robust foundation for integrating models into scenario-specific analyses. The study's outcomes—a collection of detailed maps illustrating exposure levels across various scenarios and conditions—will be highly relevant to risk managers, offering actionable insights for strategic planning and decision-making.

8.1.4 Rationale

Methods for evaluating wildfire likelihood and impact have become increasingly accurate, effectively bridging the gap between research and practice (Chuvieco et al., 2023). Within

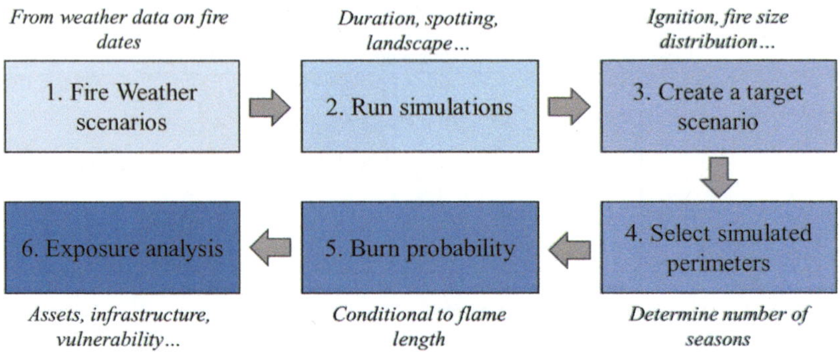

Figure 8.3 Modeling framework of the SCENFIRE approach.

the current landscape of risk assessment, two primary frameworks have emerged as industry standards (Finney, 2005; Chuvieco et al., 2014). The most common approach utilizes short-term assessments driven by fire risk indices, which integrate immediate danger levels with local vulnerabilities. This model is currently a cornerstone of the European Union's framework, utilized by the European Forest Fire Information System (EFFIS; CEU. JRC., 2022) to provide insights into short-term risk potential. In contrast, a second approach relies on quantitative simulation techniques designed for long-term strategic planning. This method, famously adopted by the United States' cohesive strategy, provides a data-rich baseline for high-level decision support and risk mitigation (Haas et al., 2013; Alcasena et al., 2021). This proposal aligns with the latter simulation-based philosophy, employing cross-boundary methods to evaluate wildfire exposure under the shifting pressures of a changing climate.

However, global and climate change have made quantitative risk assessments more complex due to the broad range of potential future scenarios (Fargeon et al., 2020; Jones et al., 2022; Galizia et al., 2023; Pérez-Invernón et al., 2023). Contemporary approaches typically leverage ignition models to identify starting points and use spread algorithms to grow theoretical fire perimeters (Parisien et al., 2019). Parameters such as wind speed (WS), direction, fuel moisture, and duration are iteratively calibrated to replicate historical fire regimes, a process that is often labor-intensive and time-consuming (Aparício et al., 2023). Within this context, integrating empirical ignition models with stochastic spread simulations offers a robust framework for assessing wildfire exposure to natural resources and communities. This approach reduces uncertainty by capturing variability in ignition locations and the dynamic weather conditions that drive fire behavior. By combining these methods, we can generate realistic long-term exposure patterns that account for the interplay between local factors (topography and landscape) and regional factors (climate variability). Our approach aims to offer an alternative method for integrating simulations into risk and exposure metrics. It simplifies the calibration process, reducing complexity and time investment, while ensuring seamless adaptation to a range of realistic scenarios, thereby enhancing both efficiency and practicality in wildfire risk assessment.

The method unfolds across three steps aimed (1) at characterizing the pattern of occurrence (understood as the probability that a fire starts and escapes initial suppression) using machine learning techniques, (2) to simulate thousands of theoretical fire events spreading through the landscape using minimum travel time (MTT)–based algorithms (MTT; Finney, 2002), and (3) finally assembling simulations according to specific scenarios. This leads to comprehensive assessments of long-term patterns of fire exposure, adequately weighting the footprint of fire occurrence and suppression while accounting for the local (topography and landscape) and regional (climate variability) factors (Finney, 2005). The outputs of the simulations are eventually integrated into straightforward metrics of exposure that inform about the likelihood of fire affecting every location inside the simulated domain, the so-called BP. The present experiment serves as full-testing case study of the proposed methods.

8.1.5 Expected results and deliverables

This case study showcases the proposed procedure for producing meaningful depictions of wildfire exposure. Specifically, we tested several combinations of fire occurrence based on kernel density estimates (KDE), human and natural ignition potential, and the capability to control fires during the initial attack stage. Simulated perimeters were assembled into BP (i.e., the likelihood that a fire will affect a given location) by integrating these ignition-related products using data from 2000 to 2016 (Santalla et al., 2017). These results were subsequently compared to the most recent large fire events in the region.

To explore these estimates, we produced a set of raster layers representing four ignition scenarios: stand-alone kernel estimates, a full combination of ignition factors, human-caused ignition potential only, and a coupling of kernel estimates with full ignition potential. These combinations resulted in four distinct depictions of BP. The resulting spatial patterns were subsequently compared to illustrate the sensitivity of exposure metrics to the ignition footprint and to highlight the critical role that landscape configuration plays in determining BP estimates.

The proposed study is expected to demonstrate that the integration of empirical and stochastic methods significantly enhances the precision and utility of landscape-level wildfire risk evaluations. By synthesizing models of ignition and spread into the SCENFIRE framework, we anticipate a more streamlined and computationally efficient workflow that eliminates the traditional need for labor-intensive, iterative parameter fine-tuning. Central to these expectations is the systematic comparison of the ignition frameworks. The study will reveal how specific ignition "footprints" influence the final BP. Ultimately, the study aims to deliver actionable geospatial products for risk managers and wildfire modelers.

8.1.6 Safety considerations

The methodologies employed in this study are inherently data-intensive, requiring the integration of historical wildfire records, geospatial land-cover

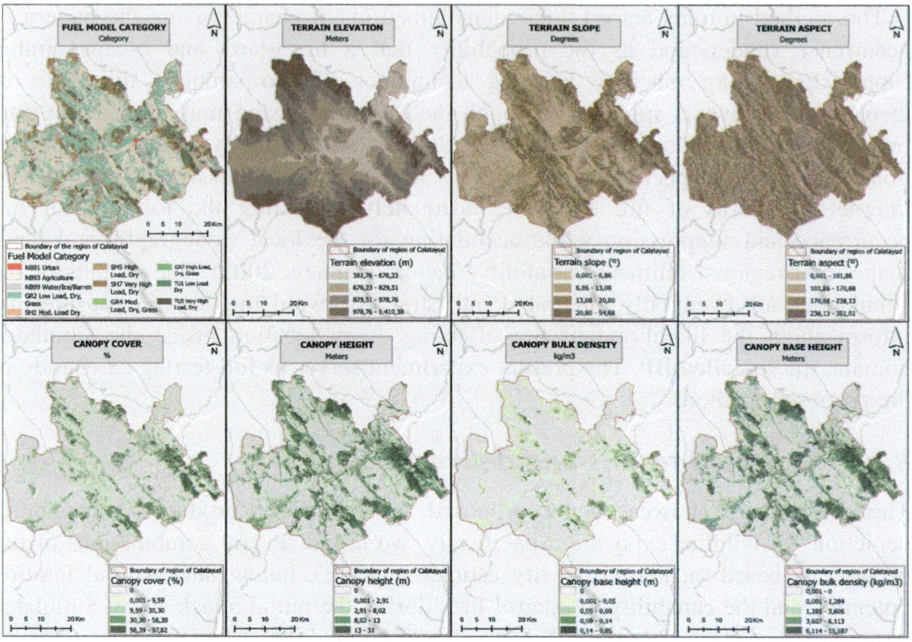

Figure 8.4 Landscape file used in wildfire spread simulations. Spatial distribution of variables required by the minimum travel time algorithm. Map lines delineate study areas and do not necessarily depict accepted national boundaries.

data, accessibility metrics, and fire-weather variables. Specifically, wildfire spread simulations necessitate high-resolution data on fuel complexes, topography, and canopy structure parameters (Fig. 8.4). Beyond spread modeling, the subcomponents of the risk framework demand specialized inputs. Empirical models of human-caused ignition were developed from a variety of variables, such as distance to the road network, the arrangement of the wildland-urban interface, fuel types, and dead fuel moisture content (Gelabert et al., 2025). Furthermore, modeling naturally caused fires accounted for the characteristics of lightning flashes (polarity and density) coupled with topography and fuel moisture content (Rodrigues et al., 2023). In turn, modeling the potential for escaping fires also necessitated information regarding aerial suppression means and fire simultaneity (Rodrigues et al., 2019).

These data were retrieved from official repositories like the Spanish wildfire database (Estadística General de Incendios Forestales [EGIF]), geographic information systems (GIS) vector topographic maps (IGN, 2018), and dedicated datasets for supporting fire risk assessment (Kutchartt et al., 2024). Data on ignition and suppression were retrieved from the Zenodo repository of project FireCycle (CNS2023–144228).

8.2 Actions taken/workflow/tools used/simulations and analyses

8.2.1 Overall workflow

The backbone of our approach rests in producing a vast library of plausible simulated fires under a wide array of conditions, assuming a random pattern of ignition. Then we assemble individual perimeters accounting for the specific probability of occurrence based on the likelihood of fire occurrence and the probability of a given fire-weather scenario of WS and direction. Furthermore, we consider different fire durations and fuel moisture content scenarios. A dedicated selection algorithm was applied to integrate these simulated perimeters, prioritizing alignment with a given frequency distribution of fire sizes. By weighing each perimeter according to its probability of occurrence, the algorithm ensures realistic spatial patterns of wildfire incidence and associated fire-weather conditions.

The workflow begins by characterizing weather forcing via historical reanalysis into synthetic scenarios. Subsequently, stochastic simulations are executed using randomly allocated ignitions, weighted by historical frequency or burned area. From the resulting fire library, a selection algorithm extracts perimeters to populate a target fire-size distribution and spatial probability of fire occurrence surface derived empirically. This iterative sampling continues until the discrepancy between the simulated and target distributions is minimized. Finally, the optimized perimeters are aggregated to calculate BP and secondary fire metrics. This approach offers several key advantages. First, it eliminates the need for fine-tuning simulation parameters by generating a comprehensive pool of scenarios, which can be automated using scripting tools (e.g., FConstMTT batch processing). Second, it enables seamless adaptation to different fire-size distributions without requiring recalibration of the simulation process (Fig. 8.3).

8.2.2 Defining the probability of occurrence

We used fire occurrence probability (Po) to decide where selected simulated fires start. This probability can be modeled through multiple frameworks, such as the likelihood of anthropogenic versus lightning-driven ignitions, the risk of fires breaching initial containment, or through spatial patterns derived from kernel density estimation. Ideally, this value represents the chance that a fire—started by either people or lightning—successfully escapes initial control efforts and begins to spread across the landscape (Alcasena et al., 2022b). To calculate this, we used data at a 200 m resolution for three products: the likelihood of human-caused ignitions (Ph), the likelihood of lightning-caused ignitions (Pn), and the probability that a fire will escape suppression (Pe). Operationally, Po was calculated as follows:

$$P_o = (P_h \cup P_n) \cap P_e \tag{8.1}$$

where *Ph* is the probability of anthropogenic ignition; *Pn* the probability of natural ignition; and *Pe* is the probability of escape from initial attack.

Nonetheless, *Po* was also depicted following standard practice, which included accounting for human-caused ignitions via empirical models (Alcasena et al., 2019), KDE (Amatulli et al., 2007), and their combination (Alcasena et al., 2022b).

8.2.3 Defining weather scenarios

Simulations were proportionately distributed across synoptic weather patterns retrieved from observed wildfires in the region (Aparício et al., 2023). K-means clustering was performed on fire records using meteorological variables (including temperature, relative humidity, WS, and wind direction) sourced from ERA5-Land reanalysis data at noon on the day of ignition. This analysis yielded four distinct clusters, each representing a unique weather pattern associated with a specific frequency of wildfire occurrence (Table 8.1).

8.2.4 Calculating burn probability

To simulate fire propagation, we utilized the FConstMTT command-line application (Missoula Fire Sciences Laboratory, 2024). This tool implements the core MTT algorithm to model spatial fire growth (Finney, 2002). We generated a library of 270,000 wildfire events simulated under extreme fire-weather conditions (fifth percentile dead fuel moisture content, 95th percentile WS, and 90%–60% of live woody-herbaceous moisture content). Simulations were proportionately distributed across the four synoptic weather patterns (Table 8.1). From the collection of plausible fires, we generated four ensembles of BP, one for each Po scenario. Each scenario featured a total of 30,000 fire seasons (calibrated against historical fire-size distributions; Parisien et al., 2019; Aparício et al., 2023), which roughly meant selecting 63,000 perimeters to account for a yearly average burned area of 411 hectares across the 17 years of data records. Annual burn probability (BPa) was calculated as:

$$BP_a = (N_p/y) \tag{8.2}$$

where BP_a denotes the annual burn probability; N is the number of simulated fire perimeters intersecting grid cell p; and y represents the effective number of simulated years.

8.2.5 Challenges and solutions

Two main challenges were addressed in this work. The first was finding a suitable way to implement multiple scenarios of BP while minimizing the computational effort. The standard wildfire simulation workflow is a cyclical process that begins with the spatial allocation of ignition seed points and the definition of meteorological scenarios, including wind profiles and fuel moisture. These inputs drive a spread algorithm to simulate fire growth, which is then statistically compared against historical fire-size distributions. If the simulated outputs do not align with historical records, the process enters an iterative calibration loop where parameters are adjusted; otherwise, the final BP maps and fire behavior metrics are derived. This

Table 8.1 Summary of weather type (WT) conditions used in wildfire spread simulation. WT stands for the dominant weather types observed in the region obtained by using a K-means cluster procedure from historical wildfires (minutes k=4); wind speed (WS) is the 95th percentile of WS (km hours^{-1}); relative humidity (RH) is the 5th percentile of RH (%); T is the 95th percentile of temperature (°C); Freq. WT is the relative frequency of observation of the WT; wind direction (WD) is the dominant WD; Freq. WD is the relative frequency of WD within the WT; and Freq. Simulation is the proportion of burned area across all WT and WD.

WT	WS	RH	T	Freq. WT	WD	Freq. WD	Freq. simulation
1	30.60	64.16	12.95	0.07	N	0.46	0.03
					NE	0.04	0.00
					E	0.21	0.02
					SE	0.13	0.01
					S	0.04	0.00
					SW	0.08	0.01
					W	0.04	0.00
					NW	0.00	0.00
2	22.96	25.16	26.50	0.41	N	0.11	0.04
					NE	0.18	0.07
					E	0.18	0.07
					SE	0.11	0.05
					S	0.11	0.05
					SW	0.12	0.05
					W	0.06	0.02
					NW	0.12	0.05
3	29.53	12.77	34.70	0.27	N	0.10	0.03
					NE	0.08	0.02
					E	0.10	0.03
					SE	0.05	0.01
					S	0.10	0.03
					SW	0.29	0.08
					W	0.20	0.05
					NW	0.08	0.02
4	35.34	31.19	20.92	0.24	N	0.56	0.14
					NE	0.01	0.00
					E	0.13	0.03
					SE	0.10	0.02
					S	0.10	0.02
					SW	0.09	0.02
					W	0.00	0.00
					NW	0.00	0.00

sequence complicates exploring different configurations of ignition footprints and/or fire-size distributions. We sought to overcome this limitation by following the SCENFIRE's protocol. A second challenge involved determining whether BP estimates derived from limited temporal data could effectively capture emerging fire risks. This was evaluated by examining Calatayud County and its surroundings—an area where the largest fire between 2000 and 2016 reached approximately 4800 hectares, but which was subsequently impacted by a massive 12,000 hectare wildfire in the summer of 2022 (San-Miguel-Ayanz et al., 2012).

8.3 Results

8.3.1 Spatial footprints of fire occurrence

Four products were considered to address the probability of fire occurrence, namely, KDE calculated from historical wildfires, empirical models of human-caused ignition and lightning-caused ignition probability, and probability of fires escaping suppression. These were combined to produce four distinct realizations of the probability of occurrence, encompassing the most common approaches to account for fire ignition patterns (Fig. 8.5).

8.3.2 Spatial patterns of burn probability

The spatial distribution of BP was evaluated across four distinct ignition frameworks to assess the sensitivity of the model to different ignition drivers. The scenarios

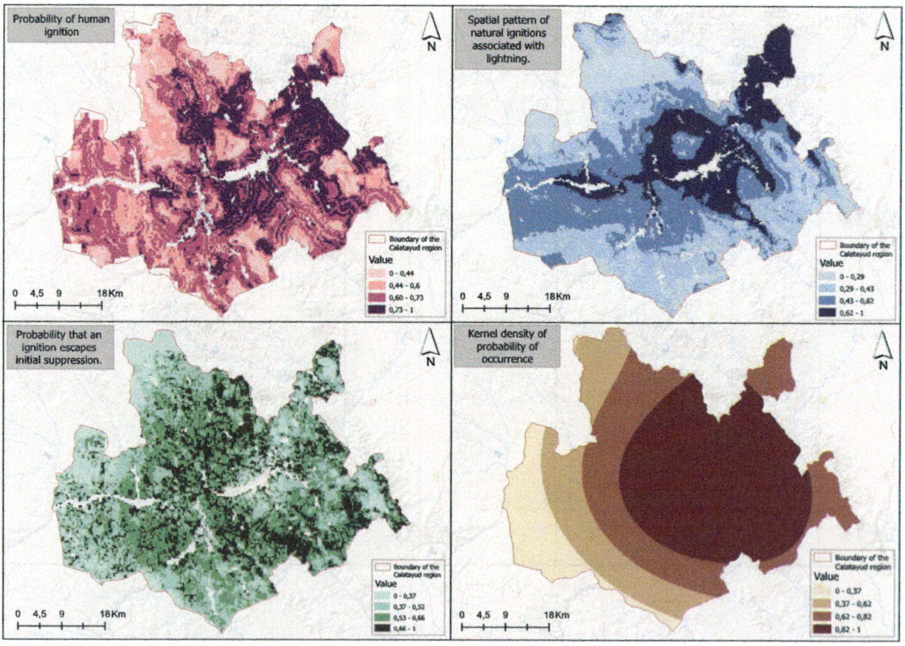

Figure 8.5 Spatial distribution of fire occurrence. Spatial distribution of human-caused (top-left), lightning-caused (top-right), escaping fires (bottom-left), and kernel ignition density (bottom-right). Map lines delineate study areas and do not necessarily depict accepted national boundaries.

compared baseline human-caused ignition models (IG-H) and KDE, both integrated with natural ignitions and the probability of fires escaping initial suppression efforts (IG-FULL and KDE-IG-FULL). These scenarios aggregated approximately 68,000 theoretical fires each to represent 30,000 plausible fire seasons according to historical fire data. The four subsets of selected perimeters reproduced the historical fire distribution satisfactorily, with an average percent discrepancy below 1.5% (Fig. 8.6). Majority of fires group close to the 10–15 hectares, decreasing in frequency up to the largest size recorded (4800 hectares).

BP maps were created from the four subset realizations, showing the potential for fire exposure (Fig. 8.7). To facilitate interpretation, BP was classified into six interval classes at key breakpoints of probability, representing fire return intervals equivalent to 10,000 (BP=0.0001), 1000 (0.001), 500 (0.002), 300 (0.0033), 200 (0.009), and 150 (0.0066) years. Across all four simulations, the study area exhibits a consistent core spatial structure. High-risk areas (BP>0.005) are persistently identified in the central-western and eastern corridors, close to Ateca and the main city of Calatayud. The spatial footprint does not fully align with the distribution of ignition probabilities (Fig. 8.5). This suggests that the spatial footprint of fire risk is strongly governed by landscape connectivity and fuel continuity, which facilitate fire propagation regardless of the initial ignition source. Conversely, the westernmost and southern sectors maintain consistently lower probabilities (BP<0.001), likely functioning as fire refugia due to topographic shielding or fragmented fuel loads.

The choice of ignition model introduces more subtle shifts in the risk profile than expected, although it remains relevant. When only human ignitions are considered (IG-H), BP patterns seem more fragmented and spatially heterogeneous. High-probability areas appear as small, discontinuous patches, reflecting the clustered and localized nature of human-caused ignitions. This configuration emphasizes fine-scale variability and highlights potential ignition hotspots, but it also produces sharper gradients and a noisier spatial pattern, which may overrepresent localized recurrence rather than sustained landscape-scale burning.

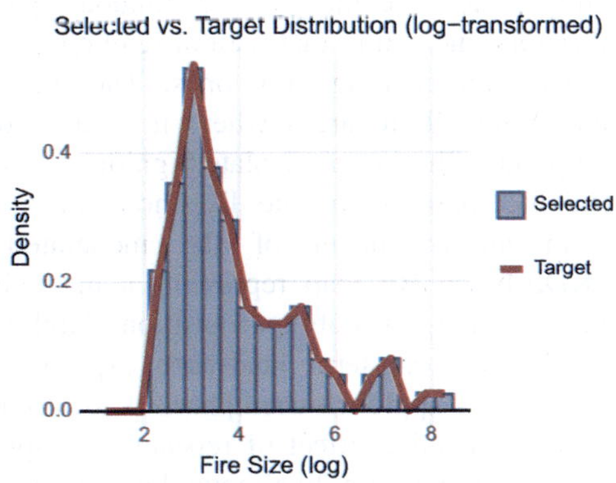

Figure 8.6 Fire-size distribution of selected simulated perimeters. Discrepancy between the historical fire-size distribution (red line) and the selected fire perimeters used to calculate burn probability. Distributions were log-transformed to facilitate comparison.

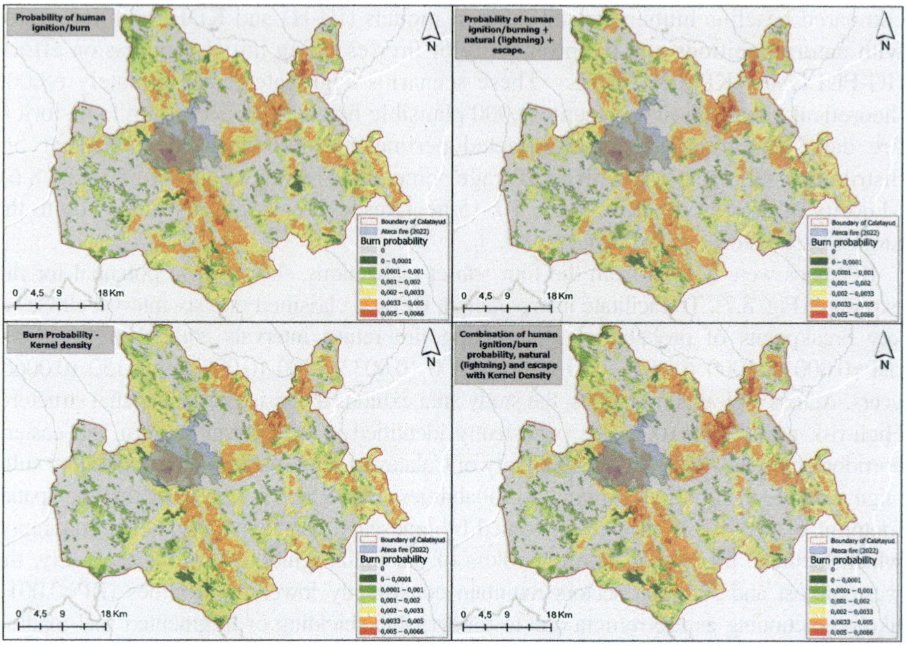

Figure 8.7 Spatial distribution of burn probability (BP) under different scenarios of fire occurrence. Spatial distribution of BP. Top-left, human-caused ignition; Top-right, human and lightning-caused ignition plus likelihood of escaping fires; Bottom-left, kernel density estimate from historical wildfires; Bottom-right, combination of kernel density, ignition, and escaping probabilities. The blue polygon outlines the fire scar of the Ateca fire obtained from the European Forest Fire Information System. Map lines delineate study areas and do not necessarily depict accepted national boundaries.

Including both human and natural ignitions (IG-FULL) increased spatial coherence in BP. High-probability zones become more continuous, particularly along the southern and eastern sectors, indicating that natural ignitions reinforce existing fire-prone landscapes rather than creating new ones. The incorporation of escape probability further constrains BP to areas where fires are more likely to spread beyond initial ignition points, filtering out isolated ignitions that rarely develop into significant burns. The incorporation of the likelihood of fires breaching initial suppression results in an expansion of the moderate-to-high-risk classes (0.002–0.005). The KDE-based BP map represents a markedly smoother spatial pattern, with broad gradients and fewer abrupt transitions. High BP areas expand into continuous regions, reflecting long-term landscape exposure to fire rather than individual ignition events. This approach deemphasizes localized ignition clustering and instead captures the cumulative effect of repeated fire spread, making it well suited for characterizing regional fire likelihood but less sensitive to fine-scale ignition processes. The combined KDE-IG-FULL map integrates these perspectives, producing a BP pattern that is both smooth and spatially structured. High-probability areas align with zones of frequent ignitions and high escape likelihood, while low-probability areas remain sharply defined. This hybrid representation most effectively reflects BPa as the outcome of interacting ignition sources and fire spread processes, providing a balanced and ecologically realistic depiction of fire risk across the landscape.

The overlay of the 2022 Ateca wildfire perimeter (approx. 12,000 ha) served as reference for the simulated BP estimates. This extreme event, which more than doubled the maximum fire size recorded during the 2000–16 calibration period, provides a rigorous test for the model's ability to anticipate emerging fire conditions. Across all four ignition scenarios, the highest BP classes (BP>0.005) show a strong spatial correlation with the actual footprint of the Ateca fire. The model successfully identified the Calatayud district as a high-vulnerability zone, with the core of the 2022 perimeter aligning precisely with the central-western high-risk corridor identified in the simulations. While the baseline IG-H and KDE models captured the central core, the alternatives incorporating the escape factor slightly expanded the high-risk (orange and red) zones far enough north to encompass the full extent of the 2022 perimeter. The spatial footprint of the observed fire in the central part of the study area was most accurately reproduced by the KDE-IG-FULL formulation. This model clearly delineates the central burned area as a coherent low-probability zone, closely matching the geometry and continuity of the recorded fire occurrence. In contrast, ignition-based maps relying primarily on human ignitions produce a more fragmented representation of this area, reflecting localized ignition patterns rather than realized fire spread. Incorporating natural ignitions and escape probability improves spatial coherence but still underrepresents the integrity of the observed fire footprint. Overall, the hybrid KDE formulation provides the most faithful depiction of BPa as constrained by both empirical fire occurrence and the combined effects of human and natural ignitions and fire-escape processes.

8.4 Learning and knowledge outcomes

8.4.1 Main findings

Based on the integrated model that combines KDE with natural and human ignition potential and the probability of escaping fires (KDE-IG-FULL), mitigation strategies should prioritize the "risk skeleton" formed by high landscape connectivity and fuel continuity. The spatial alignment between this scenario and the Ateca fire perimeter evidenced that once an ignition occurs, it is the landscape's physical configuration that dictates fire propagation far more than the initial ignition source did. Consequently, the most critical mitigation strategy is the implementation of large-scale, strategic fuel treatments (such as fuel breaks and selective thinning) within the central-western and eastern corridors identified as persistent high-risk zones. These interventions must be designed to disrupt the connectivity of fuels that would allow fires to overcome initial suppression efforts, effectively turning these identified high-probability paths into manageable suppression lines. Furthermore, the inclusion of the "escape" factor highlighted a need for enhanced suppression infrastructure in remote areas that would otherwise be considered lower risk when using baseline human ignition or KDE models. Mitigation initiatives may include the strategic placement of water points, and strict prevention measures along escaping hot spots.

8.4.2 Lessons learnt

- The integration of empirical fire occurrence models with the MTT algorithm allowed for the identification of emerging fire risks, even when events exceed historical maximums recorded in training data.
- Exposure estimates based on empirical models unravel key pressure points more accurately than traditional KDE.
- Incorporating the probability of fires escaping initial suppression improved the final footprint of extreme wildfire hot spots.
- The strong spatial correlation between high BP corridors across modeled scenarios and the actual 2022 Ateca perimeter suggests that fuel continuity and topography are the fundamental drivers of fire propagation once ignition occurs.
- The hybrid formulation of KDE, plus fire ignition and escaping potential, was the best alternative when comparing BP with the Ateca fire scar.

8.4.3 Research opportunities

Despite the satisfactory outcomes, further research is still required. Two distinct lines can be drawn from our results. The first relates to the incorporation of burning intensity in the depiction of BP. As is, our BP estimates reflect the annual likelihood of a fire affecting each location in the region. However, to better link BP with potential damage, environmental loss, or smoke emissions, BP must be constrained by the level of burning intensity, for instance, by calculating BPa conditional on a given flame length threshold (Thompson et al., 2011). A second work path must explore future conditions in fire regimes. That is, accounting for the expected change in overall burned area and fire size due to climate warming, together with potential declines in fuel moisture content and changes in the spatial arrangement of ignition likelihood (Galizia et al., 2023). This would likely improve the capability of the approach to detect potential hot spots of BP. Complementarily, our findings enable the calculation of devoted exposure metrics such as risk transmission hot spots or the overall exposure of buildings, roads, and other key infrastructure (Alcasena et al., 2022a).

8.5 Funding

This work was financed by projects FireCycle (CNS2023–144228), funded by the Spanish Ministry of Science and Innovation (MCIN/AEI/10.13039/501100011033); project "Understanding Wildfire Risk: A Local-Scale Assessment Framework in Chile," financed by the AXA Research Fund; and HIDROGIF (PROY_T21_24), funded by the Government of Aragon.

References

Alawode, G.L., Gelabert, P.J., Rodrigues, M., 2025. A spatially explicit containment modelling approach for escaped wildfires in a Mediterranean climate using machine learning. Geomatics, Natural Hazards and Risk 16 (1). Available at https://doi.org/10.1080/19475705.2024.2447514.

Alcasena, F.J., Ager, A.A., Bailey, J.D., Pineda, N., Vega-García, C., 2019. Towards a comprehensive wildfire management strategy for Mediterranean areas: framework development and implementation in Catalonia, Spain. Journal of Environmental Management 231, 303–320. Available at https://doi.org/10.1016/j.jenvman.2018.10.027.

Alcasena, F., Ager, A., Le Page, Y., Bessa, P., Loureiro, C., Oliveira, T., 2021. Assessing wildfire exposure to communities and protected areas in Portugal. Fire 4 (4). Available at https://doi.org/10.3390/fire4040082.

Alcasena, F., Rodrigues, M., Gelabert, P., Vega-García, C., 2022a. Advancing new methods for creating fire-resilient communities in mediterranean areas. Environmental Sciences Proceedings 17 (1). Available at https://doi.org/10.3390/environsciproc2022017070, https://www.mdpi.com/2673-4931/17/1/70.

Alcasena, F., Rodrigues, M., Salis, M., Resco, V., Palaiologou, P., Ager, A., et al., 2022b. Implementing a probabilistic fire modeling system at the pan-European level. Advances in Forest Fire Research 2022, 1.ª Edição. Imprensa da Universidade de Coimbra, pp. 1149–1152. Available at http://books.uc.pt/chapter?chapter=9789892622989174.

Amatulli, G., Peréz-Cabello, F., de la Riva, J., 2007. Mapping lightning/human-caused wildfires occurrence under ignition point location uncertainty. Ecological Modelling 200 (3-4), 321–333. Available at https://doi.org/10.1016/j.ecolmodel.2006.08.001.

Aparício, B.A., Benali, A., Pereira, J.M.C., Sá, A.C.L., 2023. MTTfireCAL package for R—an innovative, comprehensive, and fast procedure to calibrate the MTT fire spread modelling system. Fire 6 (6). Available at https://doi.org/10.3390/fire6060219.

CEU. JRC., 2022. Pan-European Wildfire Risk Assessment. Publications Office, LU. Available at https://data.europa.eu/doi/10.2760/9429.

Chuvieco, Aguado, Jurdao, Pettinari, Yebra, Salas, et al., 2014. Integrating geospatial information into fire risk assessment. International Journal of Wildland Fire 23 (5), 606–619. Available at https://doi.org/10.1071/WF12052.

Chuvieco, E., Yebra, M., Martino, S., Thonicke, K., Gómez-Giménez, M., San-Miguel, J., et al., 2023. Towards an integrated approach to wildfire risk assessment: when, where, what and how may the landscapes burn. Fire 6 (5). Available at https://doi.org/10.3390/fire6050215.

Fargeon, Pimont, Martin-StPaul, De Caceres, Ruffault, Barbero, et al., 2020. Projections of fire danger under climate change over France: where do the greatest uncertainties lie? Climatic Change 160 (3), 479–493. Available at https://doi.org/10.1007/s10584-019-02629-w.

Finney, M.A., 2002. Fire growth using minimum travel time methods. Canadian Journal of Forest Research 32 (8), 1420–1424. Available at https://doi.org/10.1139/x02-068.

Finney, M.A., 2005. The challenge of quantitative risk analysis for wildland fire. Forest Ecology and Management 211 (1-2), 97–108. Available at https://doi.org/10.1016/j.foreco.2005.02.010.

Galizia, L.F., Barbero, R., Rodrigues, M., Ruffault, J., Pimont, F., Curt, T., 2023. Global warming reshapes european pyroregions. Earth's Future 11 (5). Available at https://doi.org/10.1029/2022EF003182.

Gelabert, P., Jiménez-Ruano, A., Ribalaygua, J., Torres, L., Rodrigues, M., 2025. Human-caused ignition pathways under climate change scenarios in Eastern Spain. Geomatics, Natural Hazards and Risk 16 (1). Available at https://doi.org/10.1080/19475705.2025.2472864.

Haas, J.R., Calkin, D.E., Thompson, M.P., 2013. A national approach for integrating wildfire simulation modeling into Wildland Urban Interface risk assessments within the United States. Landscape and Urban Planning 119, 44–53. Available at https://doi.org/10.1016/j.landurbplan.2013.06.011.

IGN, 2018. Mapas Vectoriales y Bases Cartográficas y Topográficas a escala 1:25,000 Para Explotación y Consulta Mediante Sistemas de Información Geográfica. Instituto Geográfico Nacional, Ministerio de Fomento, Madrid. Available at https://www.ign.es/web/ign/portal.

Jones, M.W., Abatzoglou, J.T., Veraverbeke, S., Andela, N., Lasslop, G., Forkel, M., et al., 2022. Global and regional trends and drivers of fire under climate change. Reviews of Geophysics 60 (3). Available at https://doi.org/10.1029/2020RG000726.

Kutchartt, E., González-Olabarria, J.R., Aquilué, N., Garcia-Gonzalo, J., Trasobares, A., Botequim, B., et al., 2024. Pan-European fuel map server: an open-geodata portal for supporting fire risk assessment. Geomatica 76 (2). Available at https://doi.org/10.1016/j.geomat.2024.100036.

Missoula Fire Sciences Laboratory, 2024. Command Line Applications for Fire Behavior and First Order Fire Effects Calculations https://www.firelab.org/ (or specific software URL if available).

Parisien, M.-A., Dawe, D.A., Miller, C., Stockdale, C.A., Armitage, O.B., 2019. Applications of simulation-based burn probability modelling: a review. International Journal of Wildland Fire 28 (12), 913–926. Available at https://doi.org/10.1071/WF19069.

Pérez-Invernón, F.J., Gordillo-Vázquez, F.J., Huntrieser, H., Jöckel, P., 2023. Variation of lightning-ignited wildfire patterns under climate change. Nature Communications 14 (1). Available at https://doi.org/10.1038/s41467-023-36500-5.

Rodrigues, M., 2026. SCENFIRE: A flexible scenario-based algorithm for burn probability-based estimates of exposure to wildfires. Journal of Computational Science, 99, 102937. https://doi.org/10.1016/j.jocs.2026.102937.

Rodrigues, M., de la Riva, J., 2024. The large forest fires in Aragón: brief analysis and retrospective in the context of the 2022 summer LOS GRANDES INCENDIOS FORESTALES EN ARAGÓN: BREVE ANÁLISIS Y RETROSPECTIVA EN EL CONTEXTO DE LA TEMPORADA DE VERANO DE 2022. Pirineos 179. Available at https://doi.org/10.3989/pirineos.2024.179002.

Rodrigues, M., Alcasena, F., Vega-García, C., 2019. Modeling initial attack success of wildfire suppression in Catalonia, Spain. Science of the Total Environment 666, 915–927. Available at https://doi.org/10.1016/j.scitotenv.2019.02.323.

Rodrigues, M., Zúñiga-Antón, M., Alcasena, F., Gelabert, P., Vega-Garcia, C., 2022. Integrating geospatial wildfire models to delineate landscape management zones and inform decision-making in Mediterranean areas. Safety Science 147. Available at https://doi.org/10.1016/j.ssci.2021.105616.

Rodrigues, M., Jiménez-Ruano, A., Gelabert, P.J., De Dios, V.R., Torres, L., Ribalaygua, J., et al., 2023. Modelling the daily probability of lightning-caused ignition in the Iberian Peninsula. International Journal of Wildland Fire 32 (3), 351–362. Available at https://doi.org/10.1071/WF22123.

San-Miguel-Ayanz, J., Schulte, E., Schmuck, G., Camia, A., Strobl, P., Liberta, G., et al., 2012. Comprehensive monitoring of wildfires in europe: the European Forest Fire Information System (EFFIS). Approaches to Managing Disaster - Assessing Hazards, Emergencies and Disaster Impacts. InTech, pp. 87–105.

Santalla A.L., Correal A.M., María J.C., Roberti I.B., Ramiro M.M.F., Cuzzi M.L. 2017. EGIFWEB, una nueva aplicación informática para la gestión de la Base de Datos Nacional de Incendios Forestales. Sociedad Española de Ciencias Forestales. 1-14.

Thompson, M.P., Calkin, D.E., Finney, M.A., Ager, A.A., Gilbertson-Day, J.W., 2011. Integrated national-scale assessment of wildfire risk to human and ecological values. Stochastic Environmental Research and Risk Assessment 25 (6), 761–780. Available at https://doi.org/10.1007/s00477-011-0461-0.

Postfire damage assessment: burned area and fire severity

9

Carmen Quintano[1] *and Alfonso Fernández-Manso*[2]
[1]Electronic Technology Department, IuFOR, University of Valladolid, Valladolid, Spain,
[2]Agrarian Science and Engineering Department, University of León, Ponferrada, Spain

9.1 Introduction

Wildfires represent one of the most destructive forms of natural disturbance, posing significant threats to ecosystems, economic stability, and human well-being. In recent decades, the frequency and intensity of extreme wildfire events have increased markedly (Jones et al., 2024), exerting widespread impacts on both human societies and natural systems around the globe (Cunningham et al., 2024).

In this context, remote sensing (RS) has become an indispensable tool, offering distinct advantages such as the ability to monitor extensive geographical areas, acquire information noninvasively, access remote and inaccessible regions, and deliver high-resolution, consistent data (Nelson et al., 2024). These capabilities make RS particularly effective for the detection and assessment of fire-impacted forest landscapes (Yang et al., 2024; Zagalikis, 2023).

Research dedicated to mapping fire-affected regions typically focuses on two primary objectives: the delineation of burned areas (BA) and the evaluation of burn severity (Key, 2006). BA mapping involves binary classification of regions as either affected or unaffected by fire. In contrast, burn severity analysis involves classifying affected zones into multiple categories, ranging from low to high severity, based on the extent of ecological alteration (Wu et al., 2024). These classification products are essential for quantifying environmental damage, guiding ecological restoration, supporting emergency response planning, and informing management and policy decisions (Fernández-Guisuraga et al., 2021; Leblon et al., 2012).

This chapter provides a comprehensive review of RS data and methodological approaches employed in the estimation of BA and assessment of burn severity. Emphasis is placed on satellite-based observations; studies relying on UAV-acquired data are not considered. Section 9.2 outlines the principal sensors mounted on satellite platforms relevant to wildfire monitoring. Section 9.3 discusses the key computational algorithms employed, while Section 9.4 reviews available operational products. Finally, Section 9.5 presents the main conclusions and outlines prospective directions for future research.

9.2 Sensors for mapping burned area/burn severity

RS is based on the detection and analysis of electromagnetic radiation that is reflected, emitted, or scattered by objects or areas on Earth's surface (Chuvieco et al.,

Advances in Remote Sensing for Forest Fire Management. DOI: https://doi.org/10.1016/B978-0-443-41692-7.00006-8
© 2027 Elsevier Ltd. All rights are reserved, including those for text and data mining, AI training, and similar technologies.

2019). Among the most commonly used instruments for assessing the surface changes following a fire are passive optical sensors, which capture radiative energy within the solar spectrum, including the visible (VIS; 0.4–0.7 μm), near-infrared (NIR; 0.8–1.4 μm), and shortwave infrared (SWIR) regions. These spectral bands are particularly relevant for evaluating surface alterations caused by fire (Lentile et al., 2006). Postfire transformations such as vegetation charring, increased soil exposure, and ash deposition significantly modify surface reflectance characteristics. Typically, NIR reflectance tends to decrease in BA, while reflectance in the red (0.6–0.7 μm) and SWIR bands generally increases (Chuvieco et al., 2019; Lentile et al., 2006). Additionally, red-edge wavelengths have been demonstrated to be effective indicators for assessing burn severity (Fernández-Manso et al., 2016). Beyond optical data, thermal sensors contribute valuable information by capturing heat signatures and providing estimates of surface temperature (Chen et al., 2024). Radar sensors, which actively emit and receive signals, offer the advantage of being able to penetrate smoke, cloud cover, and darkness, thereby producing high-resolution imagery under various conditions, including nighttime (Ghali and Akhloufi, 2023).

The selection of the appropriate RS platform is a key determinant of the accuracy and applicability of wildfire assessments. Platform characteristics such as orbital altitude, inclination, overpass time, and revisit frequency play crucial roles in data utility (Chen et al., 2024). While aerial and UAV-based sensors offer high-resolution capabilities, their use is often constrained by cost, limited spatial coverage, and sensitivity to meteorological conditions. Consequently, satellite platforms in low Earth orbit (LEO, below 1000 km) are most commonly used for wildfire monitoring (Chuvieco et al., 2020). These satellites provide imagery with a range of spatial resolutions, from coarse (hundreds of meters) to fine (submeter), which influences their suitability for various scales of analysis (Nelson et al., 2024).

At the global scale, commonly used LEO sensors for wildfire applications include the Moderate-Resolution Imaging Spectroradiometer (MODIS) (Terra, Aqua), the Visible Infrared Imaging Radiometer Suite (VIIRS) (Suomi NPP [suomi national polar-orbiting partnership (SNPP)], the National Oceanic and Atmospheric Administration [NOAA-20], NOAA-21), the Sea and Land Surface Temperature Radiometer (SLSTR) (Sentinel-3), the Advanced Very High Resolution Radiometer (AVHRR) (NOAA and MetOp series), and the vegetation (VGT) Satellite Pour l'Observation de la Terre (SPOT, Proba-V). At the regional scale, instruments such as the Enhanced Thematic Mapper (ETM+) (Landsat-7), the Operational Land Image (OLI) and thermal infra-red sensor (TIRS) (Landsat-8 and -9), the Multi-Spectral Instrument (MSI) (Sentinel-2), and advanced spaceborne thermal emission and reflection radiometer (ASTER) (Terra) are frequently employed (Nelson et al., 2024). Due to their similar spectral configurations and shared coordinate systems, Sentinel-2 and Landsat-8 imagery are often combined in BA mapping studies (e.g., Quintano et al., 2018).

For high-resolution, local-scale analysis, commercial constellations such as PlanetScope, with over 430 Dove and SuperDove satellites (Planet Labs), and the WorldView series (DigitalGlobe), provide detailed spatial imagery suitable for individual fire event studies (Chuvieco et al., 2019).

Alternatively, geostationary Earth orbit (GEO) satellites, located approximately 35,786 km above Earth's surface, offer continuous regional monitoring. However, these platforms are characterized by lower spatial resolution and diminished performance at high latitudes (Nelson et al., 2024). Examples of sensors on GEO satellites include advanced baseline imager (ABI) (on GOES-16, -17, and -18), the Flexible Combined Imager (FCI) (on Meteosat Third Generation), and SEVIRI (on Meteosat Second Generation).

For further technical specifications on satellite platforms and sensors commonly used in wildfire research, such as satellite names, launch dates, spectral characteristics, spatial and temporal resolutions, comprehensive tabular reviews are available in sources like (Nelson et al., 2024; Chuvieco et al., 2019, 2020).

9.3 Operational burned area/burn severity mapping methods

The methodologies employed for BA estimation vary according to both the spatial scale of analysis (local, regional, or global) and the specific objectives of the assessment (Szpakowski and Jensen, 2019). At global and regional scales, diverse methodological approaches have led to the development of multiple BA products (Kurbanov et al., 2022), which will be discussed in the section on operational products.

At the local scale, BA mapping is typically conducted using sensors with high to moderate spatial resolution (less than 100 m). These sensors facilitate detailed change detection by means of spectral indices (SIs) and image differencing techniques (Han et al., 2024). Such methods are especially effective in capturing the fine-scale heterogeneity of fire-affected landscapes. At this resolution, the estimation of BA often overlaps with burn severity assessment, as both rely on similar data sources and analytical techniques (Meng and Zhao, 2017), many of which will be explored in the following section.

Burn severity estimation methods can be broadly categorized into four main groups: (1) traditional approaches that do not involve machine learning (ML) or deep learning (DL); (2) ML-based approaches; (3) DL-based approaches; and (4) cloud computing–based methodologies (Meng and Zhao, 2017). Each category encompasses a range of techniques with varying degrees of complexity and applicability, depending on data availability, computational resources, and the desired spatial and temporal scales of analysis (Wu et al., 2024).

9.3.1 Traditional approaches (not based on machine learning or deep learning algorithms)

Approaches for estimating BA and burn severity encompass a wide array of techniques, ranging from traditional spectral index classifications to more advanced physical and data-driven models (Meng and Zhao, 2017). A foundational category includes methods based on the classification of SIs, which highlight spectral reflectance changes caused by fire events. Widely used indices such as the Normalized Difference Vegetation Index (NDVI), Normalized Burn Ratio (NBR), and its derivatives, Soil-Adjusted Vegetation Index (SAVI), Global Environment Monitoring Index (GEMI), and red-edge-based indices are integral to this approach. The classification of delta NBR (dNBR) using thresholds derived from the Composite Burn Index (CBI) has become a standard practice for burn severity estimation (Key and Benson, 2006). Comprehensive reviews by Meng and Zhao (2017) and Chuvieco et al. (2019, 2020) provide detailed information on SI-based methods.

Another approach involves subpixel classifiers, which address the mixed nature of postfire landscapes, typically composed of vegetation remnants, exposed soil, and ash (Quintano et al., 2013). This mixture poses a subpixel problem for which Linear Spectral Mixture Analysis (LSMA) (Shimabukuro and Smith, 1991) has proven effective (Kokaly et al., 2007). LSMA models pixel reflectance as a linear combination of endmember reflectances (Quintano et al., 2012), outperforming traditional SI-based methods in many cases (Lentile et al., 2006, 2009; Quintano et al., 2006; Smith et al., 2007; Veraverbeke and Hook, 2013). However, LSMA assumes a single spectrum per endmember, limiting its flexibility. The Multiple Endmember Spectral Mixture Analysis (MESMA) (Roberts et al., 1998) overcomes this by allowing a variable number of endmembers per pixel. MESMA has been validated for burn severity mapping in multiple studies (Meng and Zhao, 2017; Quintano et al., 2013, 2017; Veraverbeke and Hook, 2013). Hyperspectral data further enhance MESMA's performance, leveraging higher spectral resolution (Lewis et al., 2017; Quintano et al., 2023; Somers et al., 2016; Tane et al., 2018; Veraverbeke et al., 2014, 2018).

A third methodological group involves Radiative Transfer Models (RTMs), which simulate radiative processes within vegetation canopies based on physical principles under varying vegetation conditions (e.g., leaf type and condition, leaf area index [LAI], and vegetation fractional cover), making them broadly applicable across ecosystems (Fernández-Guisuraga et al., 2021; Meng and Zhao, 2017). RTMs are used in both forward and inverse modes: the former simulates spectral signatures under defined scenarios, while the latter retrieves burn severity (e.g., CBI values) from postfire imagery (Chuvieco et al., 2006; Chuvieco and Kasischke, 2007; De Santis and Chuvieco, 2007). Enhancements to RTMs include geometric models (De Santis et al., 2009; Yin et al., 2020) and 3D formulations (Disney et al., 2011).

While these methods primarily rely on optical data, alternative data sources, such as thermal infrared, radar, and light detection and ranging (LiDAR), have also proven valuable. Thermal imagery, though historically underused in burn severity analysis, captures changes in land surface temperature (LST), which can indicate reduced evapotranspiration due to vegetation loss (Lentile et al., 2006). LST has emerged as a robust indicator of postfire severity (Ge, 2011; Quintano et al., 2015; Veraverbeke et al., 2012; Vlassova et al., 2014). Combining thermal and spectral data improves estimation accuracy; for instance, LST has been integrated with MESMA-derived fractions (Quintano et al., 2017, 2019) and used to enhance SI-based assessments (Harris et al., 2011; Veraverbeke et al., 2011; Zheng et al., 2016).

Wildfires also affect evapotranspiration (ET), making it a valuable indicator for understanding fire dynamics (Clark et al., 2012; Dore et al., 2010; Rocha and Shaver, 2011). ET is a key hydrological process linking vegetation condition and water availability (Atchley et al., 2018). Changes in ET postfire have been explored using RS models (Fang et al., 2018; Lentile et al., 2006; Montes-Helu et al., 2009). In Mediterranean ecosystems, ET derived from the mapping evapo transpiration at high resolution with internalized calibration (METRIC) model has shown utility in assessing fire severity (Fernández-Manso and Quintano, 2020). More recently, Quintano et al. (2024) demonstrated that simplified fine-resolution ET models can outperform traditional dNBR-based assessments.

In terms of active sensors, synthetic aperture radar (SAR) and LiDAR have seen extensive application in BA detection (Abdikan et al., 2022; Addison and Oommen, 2018; Belenguer-Plomer et al., 2019; Lasaponara et al., 2023; Zhang et al., 2024). SAR techniques rely on differences in surface moisture content between burned and unburned areas (Stroppiana et al., 2015), whereas LiDAR excels at capturing vertical forest structure and canopy changes (Kane et al., 2014). When combined with passive optical data, LiDAR contributes to significantly improved assessments of fire impact (Abdikan et al., 2022; Addison and Oommen, 2018; Belenguer-Plomer et al., 2019; Lasaponara et al., 2023; Zhang et al., 2024).

9.3.2 Machine learning–based approaches

The application of ML techniques in fire mapping research is a relatively recent development compared to their use in other scientific fields (Alkhatib et al., 2023). Nonetheless, a wide range of ML methods has already been adopted (see Table 9.1), encompassing both tree-based algorithms, such as random forest (RF), classification and regression trees (CART), and extreme gradient boosting (XGBoost), and nontree-based approaches, including support vector machines (SVM) and maximum entropy (MaxEnt) (Wu et al., 2024).

Among these, RF (Breiman, 2001) is particularly notable for its high accuracy, especially in identifying low to moderate fire severity categories (Wu et al., 2024), and for its robust performance across diverse environments (Quintano et al., 2020). RF is among the most widely utilized ML models in studies assessing wildfire damage (Fernández-Guisuraga et al., 2024a; Gibson et al., 2020; Hultquist et al., 2014; Quintano et al., 2020). Its advantages include fast computational processing, broad applicability, the ability to assess variable importance, and the provision of clear classification accuracy metrics (Collins et al., 2018; Rodriguez-Galiano et al., 2012).

CART models (Breiman et al., 1984) are appreciated for their intuitive, flowchart-like structure, which facilitates interpretability. However, they are prone to over-fitting, particularly when they are deep or not pruned appropriately (Hastie et al., 2009). Examples of CART applied to burn severity estimation can be found in the works of Tuffly (2017) and Liu et al. (2025). Meanwhile, XGBoost, originally developed for regression tasks, has proven highly effective in predicting wildfire

Table 9.1 Summary of machine learning–based approaches.

Algorithm	Advantages	Limitations	Remote sensing applications	Some examples of real-world case studies
Random forest (RF)	• High accuracy • Robust to overfitting • Handles large datasets well • Can deal with correlated variables	• Black-box model • High computational demand with large datasets	• Burned area (BA) classification • Fire severity estimation • Multitemporal image fusion	• Hultquist et al. (2014) • Gibson et al. (2020) • Quintano et al. (2020, 2023) • Fernández-Guisuraga et al. (2024a)
Classification and regression trees	• Easy to interpret • Fast training • Useful for basic rule-based classification	• Lower accuracy than RF and XGBoost • Tends to overfit if not pruned	• Burned versus unburned area classification • Binary change detection	• Tuffly (2017) • Liu et al. (2025)
Extreme gradient boosting (XGBoost)	• Excellent predictive performance • Overfitting control • Handles missing data effectively	• Requires fine hyperparameter tuning • Lower interpretability compared to simpler trees	• Burn severity mapping with multispectral imagery • Time-series analysis postfire	• Irawan et al. (2024) • Militino et al. (2024)
Support vector machines	• Performs well with small datasets • Suitable for high-dimensional spaces • Effective for nonlinear boundaries	• Sensitive to kernel and parameter selection • Not easily scalable to big data	• High-resolution fire classification • Separation of spectrally similar burned/ unburned pixels	• Hultquist et al. (2014) • Smith et al. (2021) • Ling et al. (2025)

Maximum entropy (MaxEnt)	• Ideal for presence-only or presence/absence data • Requires minimal data • Outputs interpretable probability maps	• Assumes variable independence • Less effective in multiclass or highly structured spatial tasks	• BA probability mapping • Wildfire occurrence modeling	• Fernández-Manso and Quintano (2020) • Fernandez-Manso et al. (2019) • Quintano et al. (2020)

probability and extent (Irawan et al., 2024; Militino et al., 2024), while also helping to identify critical variables for fire risk management (Wu et al., 2024).

In the domain of nontree-based methods, SVM (Hastie et al., 2004) relies on hyperplanes to define class boundaries by maximizing margins (Hastie et al., 2009). SVMs have been effectively used to classify burn severity based on postfire satellite imagery, showing competitive performance with XGBoost in regression tasks and enabling the generation of detailed risk maps with multiple susceptibility categories (Ling et al., 2025; Smith et al., 2021).

Lastly, MaxEnt modeling (Jaynes, 1957; Phillips et al., 2006), traditionally applied in species distribution studies, is increasingly utilized in fire severity assessments with promising results (Fernandez-Manso et al., 2019; Fernández-Manso and Quintano, 2020; Quintano et al., 2020). As a one-class classification algorithm that requires presence-only data, MaxEnt presents an efficient alternative to more data-demanding classifiers (Li and Guo, 2010).

9.3.3 Deep learning–based approaches

DL models have shown remarkable effectiveness in detecting and mapping wildfires using satellite-based RS data, often outperforming classical ML approaches (Ghali and Akhloufi, 2023). These models are particularly adept at identifying intricate patterns in imagery, such as delineating fire perimeters and assessing fire intensity levels (Wu et al., 2024). Moreover, they are capable of processing complex, high-dimensional datasets that include multispectral, multitemporal, and multimodal satellite imagery (Garcia-Garcia et al., 2017). Among the most notable architectures used in these tasks are convolutional neural networks (CNNs) and U-Net models.

CNNs, as described by LeCun et al. (2015), are a category of DL algorithms that learn hierarchical feature representations from data via successive layers of abstraction. These networks apply convolution operations through layered neurons, often combined with nonlinear activation functions and pooling layers that help in dimensionality reduction. In wildfire mapping scenarios, CNNs have been used successfully to integrate radar and optical data, for example, (Belenguer-Plomer et al., 2021) utilized data from Sentinel-1 and Sentinel-2 satellites to identify BA with high precision. Other studies have also leveraged CNNs to accurately estimate burn severity (Dixon et al., 2023; Hu et al., 2023; Zikiou et al., 2024).

In contrast, the U-Net architecture, widely employed in RS, has its roots in biomedical image segmentation due to its encoder-decoder design (Rogan and Franklin, 2001). This architecture enhances the foundational fully convolutional network (FCN) by enabling better localization through symmetric contracting and expanding paths (Lee et al., 2022). The encoder module captures high-level features through convolutional layers, max pooling, and concatenation, while the decoder reconstructs spatially detailed segmentation maps via upsampling operations (Garcia-Garcia et al., 2017). U-Net-based models have proven effective in identifying BA (Luft et al., 2022; Martins et al., 2022; Shirvani et al., 2023; Zhang et al., 2023) and in classifying wildfire burn severity (Farasin et al., 2020; Monaco et al., 2021; Widya and Lee, 2024; Wu et al., 2024), reinforcing the robustness and adaptability of DL techniques in postfire analysis and RS applications.

9.3.4 Cloud-based algorithms

Recent developments in cloud-based geospatial processing platforms have significantly enhanced the efficiency of handling large-scale geospatial datasets (Nelson et al., 2024). One of the most widely adopted solutions is Google Earth Engine (GEE), a cloud-based geospatial analysis tool launched in December 2010, which offers interfaces in both JavaScript and Python (Gorelick et al., 2017). Since its inception, GEE has continually evolved, expanding its capabilities (Zhao et al., 2021). It is freely accessible to researchers, while commercial and governmental users can access it through paid options (Garima et al., 2024). Notably, other robust cloud-based platforms, such as Amazon Web Services (AWS), Microsoft Azure (including the Planetary Computer), and Sentinel Hub, also provide competitive services in this field (Nelson et al., 2024).

At its core, GEE combines a massive repository of satellite imagery with high-performance cloud computing infrastructure (Sangeetham et al., 2024). The platform hosts an extensive collection of both raw and processed datasets (Kumar and Mutanga, 2018), which can be used at various scales from global to local applications (Pérez-Cutillas et al., 2023). This includes over four decades of satellite observations sourced from major Earth observation missions like Landsat, MODIS, NOAA, Sentinel, and ALOS (Gorelick et al., 2017). In addition, GEE integrates datasets related to climate, weather, and geophysics, which are updated daily and readily accessible for analytical use (Velastegui-Montoya et al., 2023). Users can not only utilize the platform's built-in data but also upload and analyze their own datasets or combine both sources in their workflows (Amani et al., 2020).

Functionally, GEE supports a wide range of spectral and spatial analysis operations, enabling batch processing of satellite images (Pérez-Cutillas et al., 2023). These operations range from basic arithmetic calculations to sophisticated image analysis techniques and ML applications. GEE provides application programming interfaces (APIs) that support standard programming languages like JavaScript and Python, facilitating flexible development environments (Tamiminia et al., 2020). The platform includes a variety of classification algorithms, such as RF (Badda et al., 2023; dos Santos et al., 2023; Magidi et al., 2021), SVM (Awad, 2021; Ebadati et al., 2024), and CART (Chen et al., 2022). It also offers multiple supervised classification approaches (Amani et al., 2019; Hasan et al., 2022), along with clustering techniques like K-Means, Cobweb, and Simple Non-Iterative Clustering (SNIC) (Gorelick et al., 2017).

For temporal change detection, GEE integrates algorithms such as LandTrendR, which has been widely used to identify BA (Liu et al., 2023). Additionally, the Spectral Mixture Analysis (SMA) method was incorporated into GEE by Daldegan et al. (2019) for similar purposes. Another notable tool related to wildfires, FIREMAP, built on GEE, is openly available for mapping both BA and fire severity levels (Fernández-Guisuraga et al., 2024b).

Despite these extensive capabilities, one major limitation of GEE is its lack of support for DL frameworks or neural network-based models (Nelson et al., 2024). As a result, researchers who require such methods must export their processed data to local systems or alternative cloud environments that are compatible with DL technologies (Arruda et al., 2021; Pinto et al., 2021).

9.4 Burned area products

BA products are satellite-derived datasets that provide vital information on the global spatial and temporal occurrence of wildfires (Zagalikis, 2023; Nelson et al., 2024). These datasets are typically generated from moderate to coarse resolution satellite imagery, ranging from 250 m to 1 km, using automated threshold-based algorithms that rely on burn indices to detect fire-affected regions (Chen et al., 2024). Historically, most BA products have depended on data from MODIS due to its consistent and long-term imagery archive since 1999. However, with the emergence of newer satellite constellations such as VIIRS and Sentinel-3, the production of widely used fire products is transitioning to these platforms in preparation for the end of the MODIS mission, expected in 2026 (Lizundia-Loiola et al., 2022).

Numerous studies have examined the accuracy and limitations of these products, revealing considerable discrepancies between different BA datasets (Humber et al., 2019; Katagis and Gitas, 2022). These variations are largely linked to the coarse spatial resolution and the nature of automated detection algorithms (Vanderhoof et al., 2017). To address these limitations, higher-resolution BA products have been developed using Landsat imagery, offering improved accuracy over MODIS, VIIRS, and Sentinel-3 outputs (Vanderhoof et al., 2021). Nevertheless, these higher-resolution sensors face their own challenges; Landsat 8 and Sentinel-2 have longer revisit cycles compared to VIIRS and Sentinel-3, and Sentinel-2 lacks thermal infrared bands, which are valuable for fire detection and monitoring (Vanderhoof et al., 2021). For a comprehensive overview of the most widely used BA products, readers can refer to the review by Chuvieco et al. (2019).

More recently, cloud-based geospatial analysis platforms such as GEE have enabled global-scale BA assessments by integrating data from multiple satellite missions at varying resolutions. Prominent examples include the Global Annual Burned Area Mapping (GABAM 2015) project, which leverages Landsat data (Long et al., 2019), and Sentinel-2 BAM, which combines imagery from Sentinel-2 with active fire detections from VIIRS and MODIS (Bastarrika et al., 2024).

9.5 Conclusion and future perspectives

Over the past decade, the use of RS combined with ML algorithms has emerged as a powerful approach for monitoring wildfires, particularly for mapping BA and assessing burn severity. Numerous studies have demonstrated that ML classifiers such as RF, CART, and XGBoost, as well as nontree-based methods like SVM and MaxEnt, offer high levels of accuracy and flexibility in fire-related RS tasks. These algorithms have been successfully applied to multispectral satellite data, often leveraging SIs like the NBR and its derivative dNBR, which are sensitive to vegetation loss and fire impact.

The success of these models is closely linked to the spatial and temporal resolution of the input satellite imagery. While widely used products such as MODIS provide consistent time-series data, their coarse resolution limits the precision of BA delineation. In contrast, higher-resolution imagery from sensors such as Sentinel-2 and Landsat 8 enables more detailed mapping. However, these platforms are limited by their revisit frequency, and in the case of Sentinel-2, the absence of thermal bands reduces their effectiveness in some fire detection scenarios. The integration of data from multiple sensors, combining optical, thermal, and radar observations, has proven to improve classification outcomes. Cloud-based platforms like GEE have further facilitated the fusion and processing of these datasets, allowing for large-scale BA analysis across diverse geographic regions.

Despite these advances, several limitations persist. Discrepancies between different BA products remain significant, often due to differences in algorithm design, data resolution, and thresholding techniques. Moreover, although platforms like GEE provide access to massive data archives and computational infrastructure, they do not natively support DL frameworks. As a result, researchers must frequently export processed datasets to external environments for DL-based analyses, which complicates workflows and limits automation potential.

Looking ahead, the integration of DL into fire-related RS workflows represents a promising direction. Architectures such as CNNs and U-Net have already demonstrated exceptional performance in pixel-level classification tasks in other domains and are increasingly being tested for BA mapping. These models are capable of capturing complex spatial patterns and temporal dynamics, but they require high computational resources and substantial training data. Hybrid models that combine ML and DL approaches may help balance interpretability and predictive power, allowing for more robust and flexible fire severity modeling.

Another critical avenue for future work lies in the real-time and near real-time application of these models. With the increasing availability of satellite data and the rise of cloud computing environments like AWS, Microsoft Azure, and the Planetary Computer, there is growing potential for automated fire monitoring systems that can support emergency response operations. However, the success of such systems also depends on rigorous validation procedures. Field data, including observations from unmanned aerial vehicles (UAVs), should be more widely used to validate and calibrate RS-derived outputs, particularly in regions where satellite data may be limited by cloud cover or vegetation complexity.

Finally, there is a growing recognition of the need for regional customization in model development. Fire behavior varies significantly across ecosystems, and models trained in one geographic or climatic context may perform poorly in another. Tailoring models to specific vegetation types, topographies, and climate regimes can enhance classification accuracy and relevance. In addition to detection and classification, future research could also focus on probabilistic modeling to predict fire risk, using approaches such as MaxEnt or Bayesian inference to inform fire prevention and management strategies.

In conclusion, the synergy between RS technologies and ML offers a transformative pathway for improving how we monitor, understand, and respond to wildfires. As climate change continues to intensify fire regimes globally, advancing these analytical tools will be essential for both scientific research and practical fire management.

AI disclosure

During the preparation of this work, the author(s) used ChatGPT in order to write the chapter in English correctly, as they are not native speakers. After using this tool/service, the author(s) reviewed and edited the content as needed and take(s) full responsibility for the content of the publication.

References

Abdikan, S., Bayik, C., Sekertekin, A., Bektas Balcik, F., Karimzadeh, S., Matsuoka, M., et al., 2022. Burned area detection using multi-sensor SAR, optical, and thermal data in mediterranean pine forest. Forests 13 (2). Available at https://doi.org/10.3390/f13020347, https://www.mdpi.com/1999-4907/13/2/347/pdf.

Addison, P., Oommen, T., 2018. Utilizing satellite radar remote sensing for burn severity estimation. International Journal of Applied Earth Observation and Geoinformation 73, 292–299. Available at https://doi.org/10.1016/j.jag.2018.07.002.

Alkhatib, R., Sahwan, W., Alkhatieb, A., Schütt, B., 2023. A brief review of machine learning algorithms in forest fires science. Applied Sciences (Switzerlan. (2023).

Amani, M., Brisco, B., Afshar, M., Mirmazloumi, S.M., Mahdavi, S., Mirzadeh, S.M.J., et al., 2019. A generalized supervised classification scheme to produce provincial wetland inventory maps: an application of Google Earth Engine for big geo data processing. Big Earth Data 3 (4), 378–394. Available at https://doi.org/10.1080/20964471.2019.1690404, https://www.tandfonline.com/action/journalInformation?journalCode=tbed20.

Amani, M., Ghorbanian, A., Ahmadi, S.A., Kakooei, M., Moghimi, A., Mirmazloumi, S.M., et al., 2020. Google earth engine cloud computing platform for remote sensing big data applications: a comprehensive review. IEEE Journal of Selected Topics in Applied Earth Observations and Remote Sensing 13, 5326–5350. Available at https://doi.org/10.1109/JSTARS.2020.3021052, http://ieeexplore.ieee.org/xpl/RecentIssue.jsp?punumber=4609443.

Arruda, V.L.S., Piontekowski, V.J., Alencar, A., Pereira, R.S., Matricardi, E.A.T., 2021. An alternative approach for mapping burn scars using Landsat imagery, google earth engine, and deep learning in the Brazilian Savanna. Remote Sensing Applications: Society and Environment 22. Available at https://doi.org/10.1016/j.rsase.2021.100472, http://www.journals.elsevier.com/remote-sensing-applications-society-and-environment/.

Atchley, A.L., Kinoshita, A.M., Lopez, S.R., Trader, L., Middleton, R., 2018. Simulating surface and subsurface water balance changes due to burn severity. Vadose Zone Journal 17 (1). Available at https://doi.org/10.2136/vzj2018.05.0099, https://dl.sciencesocieties.org/publications/vzj/pdfs/17/1/180099.

Awad, M., 2021. Google Earth Engine (GEE) cloud computing based crop classification using radar, optical images and Support Vector Machine Algorithm (SVM). 2021 IEEE 3rd International Multidisciplinary Conference on Engineering Technology, IMCET 2021. 9781665405874. pp. 71-76. Institute of Electrical and Electronics Engineers Inc. Lebanon. http://ieeexplore.ieee.org/xpl/mostRecentIssue.jsp?punumber=9665416. 10.1109/IMCET53404.2021.9665519.

Badda, H., Cherif, E.K., Boulaassal, H., Wahbi, M., Yazidi Alaoui, O., Maatouk, M., et al., 2023. Improving the accuracy of random forest classifier for identifying burned areas in the tangier-tetouan-Al hoceima region using google earth engine. Remote Sensing 15 (17), 4226. Available at https://doi.org/10.3390/rs15174226.

Bastarrika, A., Rodriguez-Montellano, A., Roteta, E., Hantson, S., Franquesa, M., Torre, L., et al., 2024. An automatic procedure for mapping burned areas globally using Sentinel-2 and VIIRS/MODIS active fires in Google Earth Engine. ISPRS Journal of Photogrammetry and Remote Sensing 218, 232–245. Available at https://doi.org/10.1016/j.isprsjprs.2024.08.019, https://www.sciencedirect.com/science/journal/09242716.

Belenguer-Plomer, M.A., Tanase, M.A., Fernandez-Carrillo, A., Chuvieco, E., 2019. Burned area detection and mapping using Sentinel-1 backscatter coefficient and thermal anomalies. Remote Sensing of Environment 233. Available at https://doi.org/10.1016/j.rse.2019.111345, www.elsevier.com/inca/publications/store/5/0/5/7/3/3.

Belenguer-Plomer, M.A., Tanase, M.A., Chuvieco, E., Bovolo, F., 2021. CNN-based burned area mapping using radar and optical data. Remote Sensing of Environment 260. Available at https://doi.org/10.1016/j.rse.2021.112468, www.elsevier.com/inca/publications/store/5/0/5/7/3/3.

Breiman, L., 2001. Random forests. Machine Learning 45 (1), 5–32. Available at https://doi.org/10.1023/a:1010933404324.

Breiman, L., Friedman, J.H., Olshen, R.A., Stone, C.J., 1984. Classification and Regression Trees.

Chen, H., Yunus, A.P., Nukapothula, S., Avtar, R., 2022. Modelling arctic coastal plain lake depths using machine learning and google earth engine. Physics and Chemistry of the Earth 126. Available at https://doi.org/10.1016/j.pce.2022.103138, http://www.journals.elsevier.com/physics-and-chemistry-of-the-earth/.

Chen, Y., Morton, D.C., Randerson, J.T., 2024. Remote sensing for wildfire monitoring: Insights into burned area, emissions, and fire dynamics. One Earth 7 (6), 1022–1028. Available at https://doi.org/10.1016/j.oneear.2024.05.014, https://www.sciencedirect.com/science/journal/25903322.

Chuvieco, E., Kasischke, E.S., 2007. Remote sensing information for fire management and fire effects assessment. Journal of Geophysical Research: Biogeosciences 112 (1). Available at https://doi.org/10.1029/2006JG000230, http://onlinelibrary.wiley.com/journal/10.1002/(ISSN)2169-8961.

Chuvieco, E., Riaño, D., Danson, F.M., Martin, P., 2006. Use of a radiative transfer model to simulate the postfire spectral response to burn severity. Journal of Geophysical Research: Biogeosciences 111 (G4). Available at https://doi.org/10.1029/2005jg000143.

Chuvieco, E., Mouillot, F., van der Werf, G.R., San Miguel, J., Tanasse, M., Koutsias, N., et al., 2019. Historical background and current developments for mapping burned area from satellite Earth observation. Remote Sensing of Environment 225, 45–64. Available at https://doi.org/10.1016/j.rse.2019.02.013, www.elsevier.com/inca/publications/store/5/0/5/7/3/3.

Chuvieco, E., Aguado, I., Salas, J., García, M., Yebra, M., Oliva, P., 2020. Satellite remote sensing contributions to wildland fire science and management. Current Forestry Reports 6 (2), 81–96. Available at https://doi.org/10.1007/s40725-020-00116-5.

Clark, K.L., Skowronski, N., Gallagher, M., Renninger, H., Schäfer, K., 2012. Effects of invasive insects and fire on forest energy exchange and evapotranspiration in the New Jersey pinelands. Agricultural and Forest Meteorology 166-167, 50–61. Available at https://doi.org/10.1016/j.agrformet.2012.07.007.

Collins, L., Griffioen, P., Newell, G., Mellor, A., 2018. The utility of random forests for wildfire severity mapping. Remote Sensing of Environment 216, 374–384. Available at https://doi.org/10.1016/j.rse.2018.07.005.

Cunningham, C.X., Williamson, G.J., Bowman, D.M.J.S., 2024. Increasing frequency and intensity of the most extreme wildfires on earth. Nature Ecology and Evolution 8 (8), 1420–1425. Available at https://doi.org/10.1038/s41559-024-02452-2, https://www.nature.com/natecolevol/.

Daldegan, G.A., Roberts, D.A., Ribeiro, Fd.F., 2019. Spectral mixture analysis in Google Earth Engine to model and delineate fire scars over a large extent and a long time-series in a rainforest-savanna transition zone. Remote Sensing of Environment 232. Available at https://doi.org/10.1016/j.rse.2019.111340, www.elsevier.com/inca/publications/store/5/0/5/7/3/3.

De Santis, A., Chuvieco, E., 2007. Burn severity estimation from remotely sensed data: performance of simulation versus empirical models. Remote Sensing of Environment 108 (4), 422–435. Available at https://doi.org/10.1016/j.rse.2006.11.022.

De Santis, A., Chuvieco, E., Vaughan, P.J., 2009. Short-term assessment of burn severity using the inversion of PROSPECT and GeoSail models. Remote Sensing of Environment 113 (1), 126–136. Available at https://doi.org/10.1016/j.rse.2008.08.008.

Disney, M.I., Lewis, P., Gomez-Dans, J., Roy, D., Wooster, M.J., Lajas, D., 2011. 3D radiative transfer modelling of fire impacts on a two-layer savanna system. Remote Sensing of Environment 115 (8), 1866–1881. Available at https://doi.org/10.1016/j.rse.2011.03.010.

Dixon, D.J., Zhu, Y., Brown, C.F., Jin, Y., 2023. Satellite detection of canopy-scale tree mortality and survival from California wildfires with spatio-temporal deep learning. Remote Sensing of Environment 298. Available at https://doi.org/10.1016/j.rse.2023.113842, www.elsevier.com/inca/publications/store/5/0/5/7/3/3.

Dore, S., Kolb, T.E., Montes-Helu, M., Eckert, S.E., Sullivan, B.W., Hungate, B.A., et al., 2010. Carbon and water fluxes from ponderosa pine forests disturbed by wildfire and thinning. Ecological Applications 20 (3), 663–683. Available at https://doi.org/10.1890/09-0934.1.

dos Santos, S.M.B., Duverger, S.G., Bento-Gonçalves, A., Franca-Rocha, W., Vieira, A., Teixeira, G., 2023. Remote sensing applications for mapping large wildfires based on machine learning and time series in northwestern Portugal. Fire 6 (2). Available at https://doi.org/10.3390/fire6020043, www.mdpi.com/journal/fire.

Ebadati, B., Attarzadeh, R., Alikhani, M., Youssefi, F., Pirasteh, S., 2024. Rapid post-wildfire burned vegetation assessment with google earth engine (Case Study: 2023 Canada Wildfires). International Archives of the Photogrammetry, Remote Sensing and Spatial Information Sciences - ISPRS Archives 48 (3), 45–53. Available at https://doi.org/10.5194/isprs-archives-XLVIII-3-W3-2024-45-2024, http://www.isprs.org/proceedings/XXXVIII/4-W15/.

Fang, L., Yang, J., White, M., Liu, Z., 2018. Predicting potential fire severity using vegetation, topography and surface moisture availability in a eurasian boreal forest landscape. Forests 9 (3), 130. Available at https://doi.org/10.3390/f9030130.

Farasin, A., Colomba, L., Garza, P., 2020. Double-step U-net: a deep learning-based approach for the estimation of wildfire damage severity through sentinel-2 satellite data. Applied Sciences 10 (12), 4332. Available at https://doi.org/10.3390/app10124332.

Fernández-Guisuraga, J.M., Suárez-Seoane, S., Calvo, L., 2021. Radiative transfer modeling to measure fire impact and forest engineering resilience at short-term. ISPRS Journal of Photogrammetry and Remote Sensing 176, 30–41. Available at https://doi.org/10.1016/j.isprsjprs.2021.04.002, www.elsevier.com/inca/publications/store/5/0/3/3/4/0.

Fernández-Guisuraga, J.M., Calvo, L., Quintano, C., Fernández-Manso, A., Fernandes, P.M., 2024a. Linking crown fire likelihood with post-fire spectral variability in Mediterranean fire-prone ecosystems. International Journal of Wildland Fire 33 (4). Available at https://doi.org/10.1071/WF23174, http://www.publish.csiro.au/nid/126.htm.

Fernández-Guisuraga, J.M., Fernández-Manso, A., Quintano, C., Fernández-García, V., Cerrillo, A., Marqués, G., et al., 2024b. FIREMAP: cloud-based software to automate the estimation of wildfire-induced ecological impacts and recovery processes using remote sensing techniques. Ecological Informatics 81, 102591. Available at https://doi.org/10.1016/j.ecoinf.2024.102591.

Fernández-Manso, A., Quintano, C., 2020. A synergetic approach to burned area mapping using maximum entropy modeling trained with hyperspectral data and VIIRS hotspots. Remote Sensing 12 (5), 858. Available at https://doi.org/10.3390/rs12050858.

Fernández-Manso, A., Fernández-Manso, O., Quintano, C., 2016. SENTINEL-2A red-edge spectral indices suitability for discriminating burn severity. International Journal of Applied Earth Observation and Geoinformation 50, 170–175. Available at https://doi.org/10.1016/j.jag.2016.03.005.

Fernandez-Manso, A., Quintano, C., Roberts, D.A., 2019. Burn severity analysis in Mediterranean forests using maximum entropy model trained with EO-1 Hyperion and LiDAR data. ISPRS Journal of Photogrammetry and Remote Sensing 155, 102–118. Available at https://doi.org/10.1016/j.isprsjprs.2019.07.003, www.elsevier.com/inca/publications/store/5/0/3/3/4/0.

Garcia-Garcia, A., Orts-Escolano, S., Oprea, S.O., Villena-Martinez, V., Garcia-Rodriguez, J., 2017. A review on deep learning techniques applied to semantic segmentation. arXiv. Available at https://doi.org/10.48550/arxiv.1704.06857, https://arxiv.org.

Garima, Acharjya, K., Karthikeyan, M.P., 2024. GOOGLE EARTH ENGINE FOR REMOTE SENSING BIG DATA: A COMPREHENSIVE REVIEW. Proceedings of International Conference on Sustainable Computing and Integrated Communication in Changing Landscape of AI, ICSCAI 2024. 9798350355338. Institute of Electrical and Electronics Engineers Inc. India. http://ieeexplore.ieee.org/xpl/mostRecentIssue.jsp?punumber=10865984. 10.1109/ICSCAI61790.2024.10867007.

Ge, J., 2011. Satellite-observed surface temperature changes after the 2004 taylor complex fire in Alaska. Earth Interactions 15 (11), 1–14. Available at https://doi.org/10.1175/2010ei361.1.

Ghali, R., Akhloufi, M.A., 2023. Deep learning approaches for wildland fires remote sensing: classification, detection, and segmentation. Remote Sensing 15 (7), 1821. Available at https://doi.org/10.3390/rs15071821.

Gibson, R., Danaher, T., Hehir, W., Collins, L., 2020. A remote sensing approach to mapping fire severity in south-eastern Australia using sentinel 2 and random forest. Remote Sensing of Environment 240, 111702. Available at https://doi.org/10.1016/j.rse.2020.111702.

Gorelick, N., Hancher, M., Dixon, M., Ilyushchenko, S., Thau, D., Moore, R., 2017. Google earth engine: planetary-scale geospatial analysis for everyone. Remote Sensing of Environment 202, 18–27. Available at https://doi.org/10.1016/j.rse.2017.06.031.

Han, Y., Zheng, C., Liu, X., Tian, Y., Dong, Z., 2024. Burned area and burn severity mapping with a transformer-based change detection model. IEEE Journal of Selected Topics in Applied Earth Observations and Remote Sensing 17, 13866–13880. Available at https://doi.org/10.1109/JSTARS.2024.3435857, http://ieeexplore.ieee.org/xpl/RecentIssue.jsp?punumber=4609443.

Harris, S., Veraverbeke, S., Hook, S., 2011. Evaluating spectral indices for assessing fire severity in chaparral ecosystems (Southern California) Using MODIS/ASTER (MASTER) airborne simulator data. Remote Sensing 3 (11), 2403–2419. Available at https://doi.org/10.3390/rs3112403.

Hasan, S.H., Al-Hameedawi, A.N.M., Ismael, H.S., 2022. Supervised classification model using google earth engine development environment for wasit governorate. IOP Conference Series: Earth and Environmental Science 961 (1). Available at https://doi.org/10.1088/1755-1315/961/1/012051, https://iopscience.iop.org/journal/1755-1315.

Hastie, T., Rosset, S., Tibshirani, R., Zhu, J., 2004. The entire regularization path for the support vector machine. Journal of Machine Learning Research 5, 1391–1415. Available at http://www.jmlr.org/papers/volume5/hastie04a/hastie04a.pdf.

Hastie, T., Tibshirani, J., 2009. Friedman, The Elements of Statistical Learning: Data Mining, Inference, and Prediction. Available at https://doi.org/10.1007/978-0-387-84858-7.

Hu, X., Zhang, P., Ban, Y., 2023. Large-scale burn severity mapping in multispectral imagery using deep semantic segmentation models. ISPRS Journal of Photogrammetry and Remote Sensing 196, 228–240. Available at https://doi.org/10.1016/j.isprsjprs.2022.12.026.

Hultquist, C., Chen, G., Zhao, K., 2014. A comparison of Gaussian process regression, random forests and support vector regression for burn severity assessment in diseased forests. Remote Sensing Letters 5 (8), 723–732. Available at https://doi.org/10.1080/2150704X.2014.963733.

Humber, M.L., Boschetti, L., Giglio, L., Justice, C.O., 2019. Spatial and temporal intercomparison of four global burned area products. International Journal of Digital Earth 12 (4), 460–484. Available at https://doi.org/10.1080/17538947.2018.1433727, http://www.tandfonline.com/toc/tjde20/current.

Irawan, A.M., Vall-llossera, M., López-Martínez, C., Camps, A., Chaparro, D., Portal, G., et al., 2024. Land, jet stream, and other atmospheric effects on burned area estimation during the South Asian heatwave of 2022. International Journal of Applied Earth Observation and Geoinformation 128, 103720. Available at https://doi.org/10.1016/j.jag.2024.103720.

Jaynes, E.T., 1957. Information theory and statistical mechanics. Physical Review 106 (4), 620–630. Available at https://doi.org/10.1103/physrev.106.620.

Jones, M.W., Kelley, D.I., Burton, C.A., Di Giuseppe, F., Barbosa, M.L.F., Brambleby, E., et al., 2024. State of Wildfires 2023-2024. Earth System Science Data 16 (8), 3601–3685. Available at https://doi.org/10.5194/essd-16-3601-2024, www.earth-system-science-data.net.

Kane, V.R., North, M.P., Lutz, J.A., Churchill, D.J., Roberts, S.L., Smith, D.F., et al., 2014. Assessing fire effects on forest spatial structure using a fusion of landsat and airborne LiDAR data in Yosemite national park. Remote Sensing of Environment 151, 89–101. Available at https://doi.org/10.1016/j.rse.2013.07.041, www.elsevier.com/inca/publications/store/5/0/5/7/3/3.

Katagis, T., Gitas, I.Z., 2022. Assessing the accuracy of MODIS MCD64A1 C6 and FireCCI51 burned area products in mediterranean ecosystems. Remote Sensing 14 (3), 602. Available at https://doi.org/10.3390/rs14030602.

Key, C.H., 2006. Ecological and sampling constraints on defining landscape fire severity. Fire Ecology 2 (2), 34–59. Available at https://doi.org/10.4996/fireecology.0202034.

Key, C.H., Benson, N.C., 2006. Landscape assessment: ground measure of severity, the composite burn index; and remote sensing of severity, the normalized burn ratio FIREMON: Fire Effects Monitoring and Inventory System, pp. 1–51.

Kokaly, R.F., Rockwell, B.W., Haire, S.L., King, T.V.V., 2007. Characterization of post-fire surface cover, soils, and burn severity at the Cerro Grande Fire, New Mexico, using hyperspectral and multispectral remote sensing. Remote Sensing of Environment 106 (3), 305–325. Available at https://doi.org/10.1016/j.rse.2006.08.006.

Kumar, L., Mutanga, O., 2018. Google earth engine applications since inception: usage, trends, and potential. Remote Sensing 10 (10), 1509. Available at https://doi.org/10.3390/rs10101509.

Kurbanov, E., Vorobev, O., Lezhnin, S., Sha, J., Wang, J., Li, X., et al., 2022. Remote sensing of forest burnt area, burn severity, and post-fire recovery: a review. Remote Sensing 14 (19), 4714. Available at https://doi.org/10.3390/rs14194714.

Lasaponara, R., Fattore, C., Modica, G., 2023. Imaging burned areas and fire severity in mediterranean fragmented ecosystems using sentinel-1 and sentinel-2: the case study of tortoli–ogliastra fire (Sardinia). IEEE Geoscience and Remote Sensing Letters 20, 1–5. Available at https://doi.org/10.1109/lgrs.2023.3324945.

Leblon, B., Bourgeau-Chavez, L., San-Miguel-Ayanz, J., 2012. Use of Remote Sensing in Wildfire Management. InTech. Available at https://doi.org/10.5772/45829.

LeCun, Y., Bengio, Y., Hinton, G., 2015. Deep learning. Nature 521 (7553), 436–444. Available at https://doi.org/10.1038/nature14539.

Lee, C., Park, S., Kim, T., Liu, S., Md Reba, M.N., Oh, J., et al., 2022. Machine learning-based forest burned area detection with various input variables: a case study of South Korea. Applied Sciences (Switzerland) 12 (19). Available at https://doi.org/10.3390/app121910077, www.mdpi.com/journal/applsci/.

Lentile, L.B., Holden, Z.A., Smith, A.M.S., Falkowski, M.J., Hudak, A.T., Morgan, P., et al., 2006. Remote sensing techniques to assess active fire characteristics and post-fire effects. International Journal of Wildland Fire 15 (3), 319–345. Available at https://doi.org/10.1071/WF05097.

Lentile, L.B., Smith, A.M.S., Hudak, A.T., Morgan, P., Bobbitt, M.J., Lewis, S.A., et al., 2009. Remote sensing for prediction of 1-year post-fire ecosystem condition. International Journal of Wildland Fire 18 (5), 594–608. Available at https://doi.org/10.1071/WF07091.

Lewis, S.A., Hudak, A.T., Robichaud, P.R., Morgan, P., Satterberg, K.L., Strand, E.K., et al., 2017. Indicators of burn severity at extended temporal scales: a decade of ecosystem response in mixed-conifer forests of western Montana. International Journal of Wildland Fire 26 (9), 755–771. Available at https://doi.org/10.1071/WF17019, http://www.publish.csiro.au/nid/126.htm.

Li, W., Guo, Q., 2010. A maximum entropy approach to one-class classification of remote sensing imagery. International Journal of Remote Sensing 31 (8), 2227–2235. Available at https://doi.org/10.1080/01431161003702245.

Ling, X., Zhang, G., Zheng, Y., Xiao, H., Yang, Y., Zhou, F., et al., 2025. Research on the simulation model of dynamic shape for forest fire burned area based on grid paths from satellite remote sensing images. Remote Sensing 17 (1), 140. Available at https://doi.org/10.3390/rs17010140.

Liu, W., Guan, H., Hesp, P.A., Batelaan, O., 2023. Remote sensing delineation of wildfire spatial extents and post-fire recovery along a semi-arid climate gradient. Ecological Informatics 78, 102304. Available at https://doi.org/10.1016/j.ecoinf.2023.102304.

Liu, P., Zhuang, W., Kou, W., Wang, L., Wang, Q., Deng, Z., 2025. Fire severity outperforms remote sensing indices in exploring post-fire vegetation recovery dynamics in complex plateau mountainous regions. Forests 16 (2), 263. Available at https://doi.org/10.3390/f16020263.

Lizundia-Loiola, J., Franquesa, M., Khairoun, A., Chuvieco, E., 2022. Global burned area mapping from Sentinel-3 Synergy and VIIRS active fires. Remote Sensing of Environment 282, 113298. Available at https://doi.org/10.1016/j.rse.2022.113298.

Long, T., Zhang, Z., He, G., Jiao, W., Tang, C., Wu, B., et al., 2019. 30 m resolution global annual burned area mapping based on landsat images and google earth engine. Remote Sensing 11 (5), 489. Available at https://doi.org/10.3390/rs11050489.

Luft, H., Schillaci, C., Ceccherini, G., Vieira, D., Lipani, A., 2022. Deep learning based burnt area mapping using sentinel 1 for the santa cruz mountains lightning complex (CZU) and Creek Fires 2020. Fire 5 (5), 163. Available at https://doi.org/10.3390/fire5050163.

Magidi, J., Nhamo, L., Mpandeli, S., Mabhaudhi, T., 2021. Application of the random forest classifier to map irrigated areas using google earth engine. Remote Sensing 13 (5), 1–15. Available at https://doi.org/10.3390/rs13050876, https://www.mdpi.com/2072-4292/13/5/876/pdf.

Martins, V.S., Roy, D.P., Huang, H., Boschetti, L., Zhang, H.K., Yan, L., 2022. Deep learning high resolution burned area mapping by transfer learning from Landsat-8 to PlanetScope. Remote Sensing of Environment 280, 113203. Available at https://doi.org/10.1016/j.rse.2022.113203.

Meng, R., Zhao, F., 2017. Remote sensing of fire effects: A Review for Recent Advances In Burned Area and Burn Severity Mapping Remote Sensing of Hydrometeorological Hazards. CRC Press, United States, pp. 261–284. Available at https://doi.org/10.1201/b20993, http://www.tandfebooks.com/doi/book/10.1201/b20993.

Militino, A.F., Goyena, H., Pérez-Goya, U., Ugarte, M.D., 2024. Logistic regression versus XGBoost for detecting burned areas using satellite images. Environmental and Ecological Statistics 31 (1), 57–77. Available at https://doi.org/10.1007/s10651-023-00590-7.

Monaco, S., Greco, S., Farasin, A., Colomba, L., Apiletti, D., Garza, P., et al., 2021. Attention to fires: multi-channel deep learning models for wildfire severity prediction. Applied Sciences 11 (22), 11060. Available at https://doi.org/10.3390/app112211060.

Montes-Helu, M.C., Kolb, T., Dore, S., Sullivan, B., Hart, S.C., Koch, G., et al., 2009. Persistent effects of fire-induced vegetation change on energy partitioning and evapotranspiration in ponderosa pine forests. Agricultural and Forest Meteorology 149 (3-4), 491–500. Available at https://doi.org/10.1016/j.agrformet.2008.09.011.

Nelson, D.M., He, Y., Moore, G.W.K., 2024. Trends and applications in wildfire burned area mapping: Remote sensing data, cloud geoprocessing platforms, and emerging algorithms. Geomatica 76 (1), 100008. Available at https://doi.org/10.1016/j.geomat.2024.100008.

Pérez-Cutillas, P., Pérez-Navarro, A., Conesa-García, C., Zema, D.A., Amado-Álvarez, J.P., 2023. What is going on within google earth engine? A systematic review and meta-analysis. Remote Sensing Applications: Society and Environment 29, 100907. Available at https://doi.org/10.1016/j.rsase.2022.100907.

Phillips, S.J., Anderson, R.P., Schapire, R.E., 2006. Maximum entropy modeling of species geographic distributions. Ecological Modelling 190 (3-4), 231–259. Available at https://doi.org/10.1016/j.ecolmodel.2005.03.026.

Pinto, M.M., Trigo, R.M., Trigo, I.F., Dacamara, C.C., 2021. A practical method for high-resolution burned area monitoring using sentinel-2 and viirs. Remote Sensing 13 (9). Available at https://doi.org/10.3390/rs13091608, https://www.mdpi.com/2072-4292/13/9/1608/pdf.

Quintano, C., Fernández-Manso, A., Fernández-Manso, O., Shimabukuro, Y.E., 2006. Mapping burned areas in Mediterranean countries using spectral mixture analysis from a uni-temporal perspective. International Journal of Remote Sensing 27 (4), 645–662. Available at https://doi.org/10.1080/01431160500212195, https://www.tandfonline.com/loi/tres20.

Quintano, C., Fernández-Manso, A., Shimabukuro, Y.E., Pereira, G., 2012. Spectral unmixing. International Journal of Remote Sensing 33 (17), 5307–5340. Available at https://doi.org/10.1080/01431161.2012.661095, https://www.tandfonline.com/loi/tres20.

Quintano, C., Fernández-Manso, A., Roberts, D.A., 2013. Multiple endmember spectral mixture analysis (MESMA) to map burn severity levels from Landsat images in Mediterranean countries. Remote Sensing of Environment 136, 76–88. Available at https://doi.org/10.1016/j.rse.2013.04.017.

Quintano, C., Fernández-Manso, A., Calvo, L., Marcos, E., Valbuena, L., 2015. Land surface temperature as potential indicator of burn severity in forest Mediterranean ecosystems. International Journal of Applied Earth Observation and Geoinformation 36, 1–12. Available at https://doi.org/10.1016/j.jag.2014.10.015.

Quintano, C., Fernandez-Manso, A., Roberts, D.A., 2017. Burn severity mapping from Landsat MESMA fraction images and land surface temperature. Remote Sensing of Environment 190, 83–95. Available at https://doi.org/10.1016/j.rse.2016.12.009, www.elsevier.com/inca/publications/store/5/0/5/7/3/3.

Quintano, C., Fernández-Manso, A., Fernández-Manso, O., 2018. Combination of landsat and sentinel-2 MSI data for initial assessing of burn severity. International Journal of Applied Earth Observation and Geoinformation 64, 221–225. Available at https://doi.org/10.1016/j.jag.2017.09.014.

Quintano, C., Fernández-Manso, A., Calvo, L., Roberts, D.A., 2019. Vegetation and soil fire damage analysis based on species distribution modeling trained with multispectral satellite data. Remote Sensing 11 (15), 1832. Available at https://doi.org/10.3390/rs11151832.

Quintano, C., Fernández-Manso, A., Roberts, D.A., 2020. Enhanced burn severity estimation using fine resolution ET and MESMA fraction images with machine learning algorithm. Remote Sensing of Environment 244, 111815. Available at https://doi.org/10.1016/j.rse.2020.111815.

Quintano, C., Calvo, L., Fernández-Manso, A., Suárez-Seoane, S., Fernandes, P.M., Fernández-Guisuraga, J.M., 2023. First evaluation of fire severity retrieval from PRISMA hyperspectral data. Remote Sensing of Environment 295, 113670. Available at https://doi.org/10.1016/j.rse.2023.113670.

Quintano, C., Fernández-Manso, A., Fernández-Guisuraga, J.M., Roberts, D.A., 2024. Improving fire severity analysis in mediterranean environments: a comparative study of eeMETRIC and SSEBop landsat-based evapotranspiration models. Remote Sensing 16 (2). Available at https://doi.org/10.3390/rs16020361, http://www.mdpi.com/journal/remotesensing.

Roberts, D.A., Gardner, M., Church, R., Ustin, S., Scheer, G., Green, R.O., 1998. Mapping chaparral in the Santa Monica Mountains using multiple endmember spectral mixture models. Remote Sensing of Environment 65 (3), 267–279. Available at https://doi.org/10.1016/s0034-4257(98)00037-6.

Rocha, A.V., Shaver, G.R., 2011. Postfire energy exchange in arctic tundra: the importance and climatic implications of burn severity. Global Change Biology 17 (9), 2831–2841. Available at https://doi.org/10.1111/j.1365-2486.2011.02441.x.

Rodriguez-Galiano, V.F., Ghimire, B., Rogan, J., Chica-Olmo, M., Rigol-Sanchez, J.P., 2012. An assessment of the effectiveness of a random forest classifier for land-cover classification. ISPRS Journal of Photogrammetry and Remote Sensing 67 (1), 93–104. Available at https://doi.org/10.1016/j.isprsjprs.2011.11.002.

Rogan, J., Franklin, J., 2001. Mapping wildfire burn severity in southern california forests and shrublands using enhanced thematic mapper imagery. Geocarto International 16 (4), 91–106. Available at https://doi.org/10.1080/10106040108542218.

Sangeetham, R., Reddy S, N., Navven, K.P., Guru Jawahar, J., Ramya, K., Pragathi, B., 2024. A review on google earth engine: an open access cloud analysis platform for planetary scale satellite data. E3S Web of Conferences 591, 09011. Available at https://doi.org/10.1051/e3sconf/202459109011.

Shimabukuro, Y.E., Smith, J.A., 1991. The least-squares mixing models to generate fraction images derived from remote sensing multispectral data. IEEE Transactions on Geoscience and Remote Sensing 29 (1), 16–20. Available at https://doi.org/10.1109/36.103288.

Shirvani, Z., Abdi, O., Goodman, R.C., 2023. High-resolution semantic segmentation of woodland fires using residual attention unet and time series of sentinel-2. Remote Sensing 15 (5), 1342. Available at https://doi.org/10.3390/rs15051342.

Smith, A.M.S., Drake, N.A., Wooster, M.J., Hudak, A.T., Holden, Z.A., Gibbons, C.J., 2007. Production of Landsat ETM+ reference imagery of burned areas within Southern African savannahs: comparison of methods and application to MODIS. International Journal of Remote Sensing 28 (12), 2753–2775. Available at https://doi.org/10.1080/01431160600954704.

Smith, C.W., Panda, S.K., Bhatt, U.S., Meyer, F.J., Badola, A., Hrobak, J.L., 2021. Assessing wildfire burn severity and its relationship with environmental factors: a case study in interior alaska boreal forest. Remote Sensing 13 (10). Available at https://doi.org/10.3390/rs13101966, https://www.mdpi.com/2072-4292/13/10/1966/pdf.

Somers, B., Tits, L., Roberts, D., Wetherley, E., 2016. Endmember library approaches to resolve spectral mixing problems in remotely sensed data: potential, challenges, and applications. Data Handling in Science and Technology 30, 551–577. Available at https://doi.org/10.1016/B978-0-444-63638-6.00017-6, http://www.elsevier.com/wps/find/bookdescription.cws_home/BS_DHST/description#description.

Stroppiana, D., Azar, R., Calò, F., Pepe, A., Imperatore, P., Boschetti, M., et al., 2015. Integration of optical and SAR data for burned area mapping in Mediterranean regions. Remote Sensing 7 (2), 1320–1345. Available at https://doi.org/10.3390/rs70201320, http://www.mdpi.com/2072-4292/7/2/1320/pdf.

Szpakowski, D.M., Jensen, J.L.R., 2019. A review of the applications of remote sensing in fire ecology. Remote Sensing 11 (22). Available at https://doi.org/10.3390/rs11222638, https://res.mdpi.com/d_attachment/remotesensing/remotesensing-11-02638/article_deploy/remotesensing-11-02638.pdf.

Tamiminia, H., Salehi, B., Mahdianpari, M., Quackenbush, L., Adeli, S., Brisco, B., 2020. Google Earth Engine for geo-big data applications: a meta-analysis and systematic review. ISPRS Journal of Photogrammetry and Remote Sensing 164, 152–170. Available at https://doi.org/10.1016/j.isprsjprs.2020.04.001.

Tane, Z., Roberts, D., Veraverbeke, S., Casas, Á., Ramirez, C., Ustin, S., 2018. Evaluating endmember and band selection techniques for multiple endmember spectral mixture analysis using post-fire imaging spectroscopy. Remote Sensing 10 (3), 389. Available at https://doi.org/10.3390/rs10030389.

Tuffly, M., 2017. Creating a burn severity map using classification and regression tree (CART). Available from: http://www.eriaconsultants.Com/documents.

Vanderhoof, M.K., Brunner, N., Beal, Y.J.G., Hawbaker, T.J., 2017. Evaluation of the U.S. Geological survey landsat burned area essential climate variable across the conterminous U.S. Using commercial high-resolution imagery. Remote Sensing 9 (7). Available at https://doi.org/10.3390/rs9070743, http://www.mdpi.com/journal/remotesensing.

Vanderhoof, M.K., Hawbaker, T.J., Teske, C., Ku, A., Noble, J., Picotte, J., 2021. Mapping wetland burned area from sentinel-2 across the southeastern united states and its contributions relative to landsat-8 (2016–2019). Fire 4 (3). Available at https://doi.org/10.3390/fire4030052, https://www.mdpi.com/2571-6255/4/3/52/pdf.

Velastegui-Montoya, A., Montalván-Burbano, N., Carrión-Mero, P., Rivera-Torres, H., Sadeck, L., Adami, M., 2023. Google earth engine: a global analysis and future trends. Remote Sensing 15 (14), 3675. Available at https://doi.org/10.3390/rs15143675.

Veraverbeke, S., Hook, S.J., 2013. Evaluating spectral indices and spectral mixture analysis for assessing fire severity, combustion completeness and carbon emissions. International Journal of Wildland Fire 22 (5), 707–720. Available at https://doi.org/10.1071/WF12168.

Veraverbeke, S., Harris, S., Hook, S., 2011. Evaluating spectral indices for burned area discrimination using MODIS/ASTER (MASTER) airborne simulator data. Remote Sensing of Environment 115 (10), 2702–2709. Available at https://doi.org/10.1016/j.rse.2011.06.010.

Veraverbeke, S., Verstraeten, W.W., Lhermitte, S., Van De Kerchove, R., Goossens, R., 2012. Assessment of post-fire changes in land surface temperature and surface albedo, and their relation with fireburn severity using multitemporal MODIS imagery. International Journal of Wildland Fire 21 (3), 243–256. Available at https://doi.org/10.1071/WF10075.

Veraverbeke, S., Stavros, E.N., Hook, S.J., 2014. Assessing fire severity using imaging spectroscopy data from the airborne visible/infrared imaging spectrometer (AVIRIS) and comparison with multispectral capabilities. Remote Sensing of Environment 154, 153–163. Available at https://doi.org/10.1016/j.rse.2014.08.019, www.elsevier.com/inca/publications/store/5/0/5/7/3/3.

Veraverbeke, S., Dennison, P., Gitas, I., Hulley, G., Kalashnikova, O., Katagis, T., et al., 2018. Hyperspectral remote sensing of fire: State-of-the-art and future perspectives. Remote Sensing of Environment 216, 105–121. Available at https://doi.org/10.1016/j.rse.2018.06.020.

Vlassova, L., Pérez-Cabello, F., Mimbrero, M.R., Llovería, R.M., García-Martín, A., 2014. Analysis of the relationship between land surface temperature and wildfire severity in a series of landsat images. Remote Sensing 6 (7), 6136–6162. Available at https://doi.org/10.3390/rs6076136, http://www.mdpi.com/2072-4292/6/7/6136/pdf.

Widya, L.K., Lee, C.W., 2024. Geospatial technologies for estimating post-wildfire severity through satellite imagery and vegetation types: a case study of the Gangneung Wildfire, South Korea. Geosciences Journal 28 (2), 247–260. Available at https://doi.org/10.1007/s12303-023-0045-2, https://www.springer.com/journal/12303.

Wu, M., Huang, Q., Sui, T., Peng, B., Yu, M., 2024. A remote sensing spectral index guided bitemporal residual attention network for wildfire burn severity mapping. IEEE Journal of Selected Topics in Applied Earth Observations and Remote Sensing 17, 17187–17206. Available at https://doi.org/10.1109/jstars.2024.3460531.

Yang, S., Huang, Q., Yu, M., 2024. Advancements in remote sensing for active fire detection: a review of datasets and methods. Science of the Total Environment 943, 173273. Available at https://doi.org/10.1016/j.scitotenv.2024.173273.

Yin, C., He, B., Yebra, M., Quan, X., Edwards, A.C., Liu, X., et al., 2020. Improving burn severity retrieval by integrating tree canopy cover into radiative transfer model simulation. Remote Sensing of Environment 236, 111454. Available at https://doi.org/10.1016/j.rse.2019.111454.

Zagalikis, G., 2023. Remote Sensing and GIS Applications in Wildfires. IntechOpen. Available at https://doi.org/10.5772/intechopen.111616.

Zhang, P., Ban, Y., Nascetti, A., 2023. Total-variation regularized U-Net for wildfire burned area mapping based on Sentinel-1 C-Band SAR backscattering data. ISPRS Journal of Photogrammetry and Remote Sensing 203, 301–313. Available at https://doi.org/10.1016/j.isprsjprs.2023.07.024.

Zhang, P., Hu, X., Ban, Y., Nascetti, A., Gong, M., 2024. Assessing sentinel-2, sentinel-1, and ALOS-2 PALSAR-2 Data for large-scale wildfire-burned area mapping: insights from the 2017–2019 Canada Wildfires. Remote Sensing 16 (3), 556. Available at https://doi.org/10.3390/rs16030556.

Zhao, Q., Yu, L., Li, X., Peng, D., Zhang, Y., Gong, P., 2021. Progress and trends in the application of google earth and google earth engine. Remote Sensing 13 (18), 3778. Available at https://doi.org/10.3390/rs13183778.

Zheng, Z., Zeng, Y., Li, S., Huang, W., 2016. A new burn severity index based on land surface temperature and enhanced vegetation index. International Journal of Applied Earth Observation and Geoinformation 45, 84–94. Available at https://doi.org/10.1016/j.jag.2015.11.002.

Zikiou, N., Rushmeier, H., Capel, M.I., Kandakji, T., Rios, N., Lahdir, M., 2024. Remote sensing and machine learning for accurate fire severity mapping in Northern Algeria. Remote Sensing 16 (9), 1517. Available at https://doi.org/10.3390/rs16091517.

Remote sensing techniques to assess post-fire regeneration and ecosystem resilience

10

José Manuel Fernández-Guisuraga[1,2]
[1]Area of Ecology, Department of Biodiversity and Environmental Management, Faculty of Biological and Environmental Sciences, University of León, León, Spain, [2]Institute for the Environment and Global Change (IMACG), University of León, León, Spain

10.1 Ecological resilience in fire-prone ecosystems

Wildfires are a ubiquitous ecological driver of change in terrestrial ecosystems worldwide (Bennett et al., 2016; Pausas and Millán, 2019; Smith-Tripp et al., 2024). In the western Mediterranean Basin, wildfires affect approximately half a million hectares of forest each year (Fernández-Guisuraga et al., 2024b), exerting notable impacts on terrestrial carbon dynamics, surface energy budgets, and broader climate system processes (Ward et al., 2012). At more local scales, many fire-prone ecosystems in this region have demonstrated a notable capacity to recover their prefire structure and function under historical fire regimes, exhibiting strong ecological resilience to disturbance (Seidl et al., 2014; Johnstone et al., 2016).

Nevertheless, ongoing and projected increases in wildfire frequency and severity (Duane et al., 2021; Pausas and Keeley, 2021), driven by current aridification trends (Fréjaville and Curt, 2015) and widespread rural land abandonment (Fernandes, 2013) in the western Mediterranean Basin, pose significant risks to ecosystem resilience (Doblas-Miranda et al., 2017). In particular, repeated high-severity fire events may disrupt key regenerative mechanisms, such as resprouting and seedling recruitment (González-De Vega et al., 2016; Johnstone et al., 2016), along with soil functions and processes (Fernández-Guisuraga et al., 2023c), thereby compromising the capacity of plant communities to recover after fire disturbance (Fernández-Guisuraga and Calvo, 2023).

The increasing frequency and severity of wildfire activity in the western Mediterranean Basin is reshaping forest recovery trajectories by modifying plant species composition and structural attributes of forest ecosystems (Calvo et al., 2008; Doblas-Miranda et al., 2017; Jimeno-Llorente et al., 2023). According to the framework proposed by Seidl and Turner (2022), ecosystem responses to wildfire disturbance can be categorized into four main trajectories: (1) maintenance of both plant community composition and structure, indicating forest resilience; (2) structural change without compositional shift, referred to as restructuring; (3) compositional change with structural stability, or reassembly; and (4) concurrent changes in both composition and structure, leading to forest replacement by an alternative ecosystem state.

Advances in Remote Sensing for Forest Fire Management. DOI: https://doi.org/10.1016/B978-0-443-41692-7.00010-X
© 2027 Elsevier Ltd. All rights are reserved, including those for text and data mining, AI training, and similar technologies.

These divergent forest recovery trajectories may be shaped by fine-scale ecological processes, such as postfire shifts in spatial and temporal resource allocation, which can influence local vegetation dynamics by mediating competitive and facilitative interactions among plant species in the community (Fernández-Guisuraga et al., 2022). At broader spatial scales, fire severity and prefire plant community composition stand out among the most relevant factors shaping early-stage plant community trajectories in Mediterranean regions (Chergui et al., 2018; Viana-Soto et al., 2020; Fernández-Guisuraga et al., 2023a). Other factors may include soil type and quality (Torres et al., 2017), climatic conditions, and postfire weather patterns (Bright et al., 2019), and the surrounding landscape configuration and topography (Christopoulou et al., 2019). Remarkably, these factors often interact in complex ways, influencing vegetation recovery across spatial and temporal scales (Bastos et al., 2011).

Several studies have suggested that it is uncommon for high-severity, stand-replacing wildfire events to induce alternative and stable ecosystem states in fire-adapted Mediterranean ecosystems (Neary and Leonard, 2015; Nolè et al., 2022). However, recent evidence has demonstrated that fire-vegetation feedbacks that promote changes in fuel composition, load, and flammability-related traits after high-severity wildfires may trigger permanent nonforest states when such events recur in reburning areas (Fernández-Guisuraga and Calvo, 2023). Consequently, gaining insights into the temporal dynamics and environmental controls of postfire vegetation recovery is essential for designing forest management strategies that mitigate the ecological and social impacts of wildfires (Mitri and Gitas, 2012; Bartels et al., 2016).

The most desirable recovery pathway for a given ecosystem depends on the ecosystem services it supports (Millar and Stephenson, 2015) and on how postfire recovery is conceptualized and assessed (Bartels et al., 2016; Smith-Tripp et al., 2024). Therefore, accurately identifying the trajectory an ecosystem is following becomes essential for guiding postfire decision-making, particularly in the context of global change (Fernández-Guisuraga et al., 2020). In this context, field-based inventories provide the necessary detail to assess the recovery pathways described by Seidl and Turner (2022). However, fieldwork alone is often constrained by the generally low accessibility of burned areas, the limited spatial exhaustiveness achievable across extensive burned landscapes, and the high financial costs typically associated with large-scale field sampling campaigns (Bergen et al., 2009; Fernández-Guisuraga et al., 2024b).

Nowadays, synoptic remote sensing Earth observations represent the most viable and cost-effective alternative for assessing postfire recovery dynamics across large spatial and/or temporal scales (Bright et al., 2019; Nolè et al., 2022; Smith-Tripp et al., 2024), especially when combined with accurate field data for calibration or validation purposes (Kibler et al., 2019). The following sections will detail the remote sensing products and techniques implemented in the assessment of postfire ecosystem recovery, as well as future perspectives aimed at overcoming certain limitations currently observed in the obtention of wall-to-wall recovery estimates.

10.2 Optical remote sensing data in the assessment of postfire ecosystem recovery and ecosystem resilience

Most studies assessing the resilience of forest ecosystems to fire through remote sensing techniques have relied on broadband optical data with low to moderate spatial resolution, primarily acquired from open satellite missions of the European Space Agency (ESA) (e.g., Sentinel-2) or the National Aeronautics and Space Administration (NASA) (e.g., Moderate Resolution Imaging Spectroradiometer [MODIS], Landsat) (Chuvieco et al., 2020; Pérez-Cabello et al., 2021).

Conventionally, spectral vegetation indices (SVI), such as the normalized difference vegetation index (NDVI) or the normalized burn ratio (NBR), have been computed from broadband optical satellite data to describe the spectral recovery with respect to the prefire scenario in terms of vegetation greenness (Ireland and Petropoulos, 2015; Storey et al., 2016; Hislop et al., 2018; Gibson et al., 2022; Koutsias et al., 2024), since SVIs are not strict physical quantities (Hudak et al., 2007). Other studies have implemented SVIs within change detection algorithms, such as the Vegetation Change Tracker and the Composite2Change (Huang et al., 2010; Hermosilla et al., 2015), applied to pre and postfire time series of optical satellite data (Zhao et al., 2016; White et al., 2017). In addition, relationships between field plot data and SVIs or derived textural metrics have been established using empirical, statistical models to produce wall-to-wall postfire recovery maps at the wildfire scale (Cuevas-González et al., 2009; Fernández-Guisuraga et al., 2019a,b).

A primary limitation of these approaches to assess postfire ecosystem recovery lies in the limited generalization capacity of SVI in heterogeneous landscapes encompassing multiple plant communities, due to their context-dependent nature and lack of a physically based foundation (De Santis and Chuvieco, 2007; Fernández-Guisuraga et al., 2021a,b). Additionally, the performance of SVI may be compromised by (1) confounding background signals from charred legacies and exposed soils shortly after fire events (Solans Vila and Barbosa, 2010; Meng et al., 2017), (2) inconsistencies between vegetation greenness as represented by SVI and the biophysical traits that actually determine canopy reflectance (Veraverbeke et al., 2012), and (3) spectral signal saturation under dense canopy cover conditions (Mutanga et al., 2023).

To address these limitations, several studies have proposed the use of physically based remote sensing techniques, such as advanced spectral mixture models and radiative transfer models (RTMs), to assess the resilience of forest ecosystems to fire by retrieving ecologically meaningful vegetation biophysical variables beyond vegetation greenness. For instance, Fernandez-Manso et al. (2016) implemented multiple endmember spectral mixture analysis (MESMA) (Roberts et al., 1998) to estimate postfire vegetation recovery in Mediterranean pine forests in northwest Spain. By retrieving the shade-normalized, subpixel green vegetation fraction (i.e., fractional vegetation cover -FCOVER-) from a 13-year Landsat TM/ETM+ time series, the authors quantified ecosystem recovery trajectories across different fire severity levels. The results showed strong agreement with field-based studies from the same region, supporting the ecological validity of the MESMA-derived recovery metrics. The use of MESMA was compared to spectral mixture analysis based on SVI by (Fernández-Guisuraga et al., 2020) in the same study region, in order to evaluate the accuracy of postfire vegetation recovery estimates against field data. The MESMA approach yielded higher accuracy and reduced underestimation of vegetation recovery compared to SVI-based methods. The subpixel capabilities of spectral mixture analysis, including the MESMA algorithm, have even been leveraged to assess ecosystem resilience to fire by evaluating species-level recovery trajectories in chamise chaparral communities in California and, in the case of conifer forests, by analyzing species-level recovery at the canopy level (Kibler et al., 2019).

RTMs had received limited attention in the field of fire ecology and were primarily used in forward mode to simulate the reflectance signal of fire-induced ecological effects in Mediterranean ecosystems and retrieve fire severity at the plot level from the RTM inversion using optical satellite data (Chuvieco et al., 2006; De Santis and Chuvieco, 2007; De Santis et al., 2009). However, more recently, Fernández-Guisuraga et al. (2021b) proposed a hybrid RTM inversion approach to retrieve FCOVER from optical satellite data within a heterogeneous postfire landscape in the western Mediterranean Basin, characterized by a mosaic of forest and shrubland ecosystems. In a companion study, Fernández-Guisuraga et al. (2021a) applied the hybrid RTM inversion framework to estimate short-term resilience to fire across different ecosystem types affected by a mixed-severity wildfire. By retrieving FCOVER from a time series of pre and postfire high-resolution Deimos-2 optical data, the authors evidenced that vegetation recovery rates were significantly influenced by fire severity and the regenerative strategies of dominant species. The ecosystems dominated by facultative or obligate seeders exhibited delayed recovery compared to those dominated by resprouter species, which is consistent with previous field-based observations worldwide (Calvo et al., 2003; Chergui et al., 2018; Kibler et al., 2019).

The advent of high-resolution optical satellite data (e.g., QuickBird, WorldView-2) in recent years has increasingly been leveraged to assess postfire vegetation recovery across diverse biomes worldwide (Mitri and Gitas, 2012; Vanderhoof et al., 2018; Christopoulou et al., 2019; Fang et al., 2019). Most of these studies have adopted a geographic object-based image analysis (GEOBIA) approach (Blaschke, 2010), as it frequently yields better results than conventional pixel-based approaches when applied to high-resolution optical data by enabling more accurate detection of postfire vegetation legacies in classification workflows (Fernández-Guisuraga et al., 2022). First, GEOBIA involves the scene segmentation into homogeneous objects that incorporate not only spectral but also spatial and contextual information (Jobin et al., 2008; Blaschke, 2010; Myint et al., 2011). Second, this object-based framework reduces within-class spectral variability and minimizes the effects of spatial noise inherent in high-resolution datasets (Whiteside et al., 2011). Despite the potential of GEOBIA approaches, most studies analyzing very high spatial resolution data acquired by optical sensors onboard unmanned aerial vehicles (UAVs) in burned landscapes still rely on pixel-based approaches to describe the ecosystem spectral recovery (Fernández-Guisuraga et al., 2018; Larrinaga and Brotons, 2019; Pádua et al., 2020; Talucci et al., 2020; van Blerk et al., 2022; Gillespie et al., 2024). This trend contrasts with the method followed by Fernández-Guisuraga et al. (2022), who implemented multiresolution segmentation and machine learning classification algorithms to produce wall-to-wall estimates of pine sapling and shrub competitor abundance at the species level across multiple wildfires spanning a climatic gradient in Spain. Their findings indicated that intra-specific competition among pine saplings was negligible in the short term after fire, whereas increasing cover and height of neighboring shrub species had a significant nonlinear effect on sapling growth.

10.3 Active remote sensing data in the assessment of postfire ecosystem recovery and ecosystem resilience

Although previous studies have demonstrated a certain linkage between ecosystem spectral recovery and postfire forest structural development assessed using active remote sensing data such as airborne light detection and ranging (LiDAR) (Wulder et al., 2009; Bishop et al., 2014; McCarley et al., 2017; White et al., 2023), these relationships may be secondary and context-dependent, varying according to differences in species composition and vegetation structure (Boucher et al., 2023; Eastburn et al., 2024), as well as the characteristics of the associated disturbance regimes (Thapa et al., 2025). Indeed, optical sensors typically depict spectral recovery of the uppermost strata in forest ecosystems due to canopy occlusion (Hislop et al., 2018; Hoe et al., 2018; Fernández-Guisuraga et al., 2021a), but the remotely sensed signal may actually reflect a mixture of canopy and recovering subcanopy vegetation strata in burned landscapes, without the ability to distinguish between them (Gibson et al., 2022). Active sensors such as LiDAR or synthetic aperture radar (SAR) are usually required to quantify postfire ecosystem recovery in terms of structural characteristics (Gordon et al., 2017).

Unlike optical remote sensing data, airborne or spaceborne LiDAR and SAR data can accurately characterize vegetation structure and biophysical traits through the vertical vegetation profile, owing to their sensitivity to the amount and spatial arrangement of photosynthetic and nonphotosynthetic elements (Bergen et al., 2009). Airborne LiDAR data, either alone or fused with optical data to spatially and temporally augment the inherently limited coverage of airborne surveys (García et al., 2017; Fisher et al., 2024), have been used to accurately quantify fire-induced changes in forest structure and to monitor postfire vegetation recovery in Mediterranean forests (Martín-Alcón et al., 2015; Viana-Soto et al., 2022) as well as in a wide range of biomes worldwide (Bolton et al., 2015; Gordon et al., 2017; Matasci et al., 2018; Meng et al., 2018). In addition to its limited spatial coverage, airborne LiDAR acquisitions are constrained by high operational costs, complex flight logistics, and limited temporal revisit frequency (Wulder et al., 2012). In recent years, LiDAR data from the Global Ecosystem Dynamics Investigation (GEDI), the first spaceborne mission explicitly designed for monitoring terrestrial ecosystems, have also been extensively used for this purpose (Huettermann et al., 2023; Lin et al., 2023; Guerra-

Hernández et al., 2024; Percival et al., 2024; Clark et al., 2025), owing to their open-access nature, broad spatial coverage, and ability to provide repeat observations over time under resonance orbit conditions (Guerra-Hernández et al., 2024). For instance, Clark et al. (2025) concluded that bitemporal GEDI waveforms acquired across several large wildfires in California provided consistent and ecologically meaningful forest structure metrics, offering novel insights into postfire structural dynamics at both the canopy and subcanopy levels.

In addition to airborne and spaceborne LiDAR applications, SAR data acquisition holds strong potential for monitoring postfire vegetation recovery and ecosystem resilience due to the sensitivity of backscattered microwave energy to the distribution of scattering elements throughout the vertical profile of plant communities (Tanase et al., 2015; Chuvieco et al., 2020). Moreover, the ESA Copernicus program currently provides open and systematic global coverage of C-band SAR data through the Sentinel-1 mission. Several studies have qualitatively analyzed the evolution of C- and/or L-band backscatter signatures, including indices such as the SAR-based vegetation structural perpendicular index (RADAR-VSPI), using pre and postfire time series or fire-related chronosequences in subtropical, Mediterranean, and boreal regions worldwide (Tanase et al., 2011; Zhou et al., 2019; Chhabra et al., 2022; De Luca et al., 2022; Singh et al., 2025). These studies concluded that L-band cross-polarized backscatter was more sensitive than C-band data, because of signal attenuation of the latter, to fire-induced changes in ecosystem structure, allowing for a more accurate identification of recovery patterns as a function of fire severity and vegetation type. Finally, other studies have leveraged C-band SAR and field data to produce wall-to-wall estimates of ecosystem resilience to fire based on the recovery of vertical structural diversity (VSD) at the wildfire scale in northwestern Spain (Fernández-Guisuraga et al., 2023b; Jimeno-Llorente et al., 2023). The authors found that C-band SAR data accurately predicted VSD four years after fire and that plant communities dominated by resprouter species experienced lower fire severity and exhibited significantly greater VSD recovery than those dominated by facultative or obligate seeders.

10.4 Future perspectives

Despite the substantial progress made in recent years with the deployment of new sensors and Earth observation missions, as well as the development of novel remote sensing-based techniques, particularly those grounded in physically based principles, several limitations still persist in the assessment of ecosystem resilience to fire. A key challenge lies in the difficulty of simultaneously resolving spectral and structural postfire recovery trajectories across the entire vertical profile of plant communities, particularly in ecosystems with complex structures, heterogeneous compositions, and marked variability in ecosystem recovery processes (White et al., 2023; Smith-Tripp et al., 2025). In such contexts, mixing signals from the canopy and regenerating understory, together with the challenges associated with accurately estimating fine-scale subcanopy structures, can confound the ecological interpretation of remotely sensed data (Bao et al., 2024; Mandl et al., 2024; Post et al., 2025). Overcoming this limitation would allow for a more accurate assessment of ecosystem recovery pathways and improve our ability to anticipate long-term ecosystem states. Moreover, the integration of structural and biophysical information into unified and ecologically meaningful indicators of ecological resilience remains limited. Importantly, no clear consensus currently exists on which remotely sensed metrics best reflect ecological resilience, complicating cross-site comparisons and long-term assessments (Runge et al., 2025). Lastly, the lack of integration between remote sensing observations and mechanistic ecological models (Dronova and Taddeo, 2022; Gibson et al., 2025) hampers a deeper understanding of the processes driving postfire recovery and limits the predictive power of current approaches under future disturbance scenarios.

To address these challenges, at least to some extent, future research should prioritize the implementation of novel remote sensing techniques aimed at improving the estimation of ecosystem structural characteristics and their dynamics following fire disturbance. For instance, recent advances in airborne LiDAR processing techniques enable the accurate derivation of leaf area density profiles and associated metrics, thereby facilitating the identification of changes in vegetation density and vertical discontinuities in ecosystem strata throughout the entire vertical profile, even at the individual tree level (Kamoske et al., 2019; Viedma et al., 2024). These methodologies have recently been applied to estimate the fuel complex for fire risk assessment (Fernández-Guisuraga and Calvo, 2025; Fernández-Guisuraga et al., 2025), yet they remain largely underexplored in the assessment of vegetation recovery dynamics during postfire succession.

In connection with the above, the increasing deployment of LiDAR sensors onboard UAVs offers a solution to many of the limitations associated with conventional airborne or spaceborne platforms, as UAV-LiDAR scans can achieve much higher point densities (often exceeding 100 pulses per m^2) and enhanced ranging accuracy (Cai et al., 2024). As a result, low-altitude UAV-LiDAR acquisitions are well suited to capturing fine-scale features such as subcanopy vegetation structure and microtopography, even under complex terrain conditions (Wallace et al., 2016; Fernández-Guisuraga et al., 2024a). This is particularly relevant, as the structural characteristics of early-successional vegetation, often undetected by conventional airborne or spaceborne sensors, provide crucial insights into recruitment dynamics and short-term ecosystem resilience to fire (Lindenmayer et al., 2019; Falk et al., 2022). In addition, UAV-based surveys are less logistically demanding than traditional airborne LiDAR campaigns, requiring fewer planning resources and infrastructure (Kellner et al., 2019), which makes them particularly advantageous for rapid deployment in postfire monitoring contexts. Moreover, the integration (Sankey et al., 2017) of hyperspectral data acquired by spectroscopic sensors onboard UAVs holds great promise for disentangling fine-scale spatial variability associated with spectrally homogeneous recovering vegetation and heterogeneous species composition in postfire landscapes (Hwang et al., 2023), particularly when fused with UAV-LiDAR data.

Finally, the adoption of multispectral LiDAR systems can further enhance the performance of conventional single-channel LiDAR and optical data fusion approaches in postfire ecosystem recovery studies by simultaneously retrieving structural parameters and biophysical properties throughout the vertical canopy profile (Wallace et al., 2012; Wallace et al., 2014). This includes an improved characterization of lower canopy and understory traits, which are often poorly captured by optical sensors alone under dense canopy conditions (Georgopoulos et al., 2021), with the potential to provide an integrated perspective on spectral and structural recovery processes, enhancing the detection of subtle ecological changes across postfire time series.

AI disclosure

During the preparation of this work, the author(s) used ChatGPT in order to improve the readability and linguistic clarity of the text. The final content was reviewed and validated by the authors to ensure its accuracy and scientific integrity. After using this tool/service, the author(s) reviewed and edited the content as needed and take(s) full responsibility for the content of the publication.

References

Bao, S.G., Wang, W.J., Liu, Z., Zhang, H.K., Wang, L., Ma, J., et al., 2024. Revealing post-megafire spectral and compositional recovery in the Siberian boreal forest using Landsat time series and regression-based unmixing approach. Remote Sensing of Environment 311. Available at https://doi.org/10.1016/j.rse.2024.114307, https://www.sciencedirect.com/science/journal/00344257.

Bartels, S.F., Chen, H.Y.H., Wulder, M.A., White, J.C., 2016. Trends in post-disturbance recovery rates of Canada's forests following wildfire and harvest. Forest Ecology and Management 361, 194–207. Available at https://doi.org/10.1016/j.foreco.2015.11.015, www.elsevier.com/inca/publications/store/5/0/3/3/1/0.

Bastos, A., Gouveia, C.M., DaCamara, C.C., Trigo, R.M., 2011. Modelling post-fire vegetation recovery in Portugal. Biogeosciences 8 (12), 3593–3607. Available at https://doi.org/10.5194/bg-8-3593-2011.

Bennett, L.T., Bruce, M.J., MacHunter, J., Kohout, M., Tanase, M.A., Aponte, C., 2016. Mortality and recruitment of fire-tolerant eucalypts as influenced by wildfire severity and recent prescribed fire. Forest Ecology and Management 380, 107–117. Available at https://doi.org/10.1016/j.foreco.2016.08.047, www.elsevier.com/inca/publications/store/5/0/3/3/1/0.

Bergen, K.M., Goetz, S.J., Dubayah, R.O., Henebry, G.M., Hunsaker, C.T., Imhoff, M.L., et al., 2009. Remote sensing of vegetation 3-D structure for biodiversity and habitat: review and implications for lidar and radar spaceborne missions. Journal of Geophysical Research: Biogeosciences 114 (G2). Available at https://doi.org/10.1029/2008jg000883.

Bishop, B.D., Dietterick, B.C., White, R.A., Mastin, T.B., 2014. Classification of plot-level fire-caused tree mortality in a redwood forest using digital orthophotography and LiDAR. Remote Sensing 6 (3), 1954–1972. Available at https://doi.org/10.3390/rs6031954, http://www.mdpi.com/2072-4292/6/3/1954/pdf.

Blaschke, T., 2010. Object based image analysis for remote sensing. ISPRS Journal of Photogrammetry and Remote Sensing 65 (1), 2–16. Available at https://doi.org/10.1016/j.isprsjprs.2009.06.004.

van Blerk, J.J., West, A.G., Smit, J., Altwegg, R., Hoffman, M.T., 2022. UAVs improve detection of seasonal growth responses during post-fire shrubland recovery. Landscape Ecology 37 (12), 3179–3199. Available at https://doi.org/10.1007/s10980-022-01535-4.

Bolton, D.K., Coops, N.C., Wulder, M.A., 2015. Characterizing residual structure and forest recovery following high-severity fire in the western boreal of Canada using Landsat time-series and airborne lidar data. Remote Sensing of Environment 163, 48–60. Available at https://doi.org/10.1016/j.rse.2015.03.004, www.elsevier.com/inca/publications/store/5/0/5/7/3/3.

Boucher, P.B., Hockridge, E.G., Singh, J., Davies, A.B., 2023. Flying high: sampling savanna vegetation with UAV-lidar. Methods in Ecology and Evolution 14 (7), 1668–1686. Available at https://doi.org/10.1111/2041-210X.14081, http://onlinelibrary.wiley.com/journal/10.1111/(ISSN)2041-210X.

Bright, B.C., Hudak, A.T., Kennedy, R.E., Braaten, J.D., Henareh Khalyani, A., 2019. Examining post-fire vegetation recovery with Landsat time series analysis in three western North American forest types. Fire Ecology 15 (1). Available at https://doi.org/10.1186/s42408-018-0021-9, http://fireecology.org/journal/journal-archive/.

Cai, S., Zhang, W., Zhang, S., Yu, S., Liang, X., 2024. Branch architecture quantification of large-scale coniferous forest plots using UAV-LiDAR data. Remote Sensing of Environment 306, 114121. Available at https://doi.org/10.1016/j.rse.2024.114121.

Calvo, L., Santalla, S., Marcos, E., Valbuena, L., Tárrega, R., Luis, E., 2003. Regeneration after wildfire in communities dominated by *Pinus pinaster*, an obligate seeder, and in others dominated by *Quercus pyrenaica*, a typical resprouter. Forest Ecology and Management 184 (1-3), 209–223. Available at https://doi.org/10.1016/s0378-1127(03)00207-x.

Calvo, L., Santalla, S., Valbuena, L., Marcos, E., Tárrega, R., Luis-Calabuig, E., 2008. Post-fire natural regeneration of a *Pinus pinaster* forest in NW Spain. Plant Ecology 197 (1), 81–90. Available at https://doi.org/10.1007/s11258-007-9362-1.

Chergui, B., Fahd, S., Santos, X., 2018. Quercus suber forest and Pinus plantations show different post-fire resilience in Mediterranean north-western Africa. Annals of Forest Science 75 (2). Available at https://doi.org/10.1007/s13595-018-0742-6.

Chhabra, A., Rüdiger, C., Yebra, M., Jagdhuber, T., Hilton, J., 2022. RADAR-vegetation structural perpendicular index (R-VSPI) for the quantification of wildfire impact and post-fire vegetation recovery. Remote Sensing 14 (13), 3132. Available at https://doi.org/10.3390/rs14133132.

Christopoulou, A., Mallinis, G., Vassilakis, E., Farangitakis, G.P., Fyllas, N.M., Kokkoris, G.D., et al., 2019. Assessing the impact of different landscape features on post-fire forest recovery with multitemporal remote sensing data: the case of mount taygetos (southern Greece). International Journal of Wildland Fire 28 (7), 521–532. Available at https://doi.org/10.1071/WF18153, http://www.publish.csiro.au/nid/126.htm.

Chuvieco, E., Riaño, D., Danson, F.M., Martin, P., 2006. Use of a radiative transfer model to simulate the postfire spectral response to burn severity. Journal of Geophysical Research: Biogeosciences 111 (G4). Available at https://doi.org/10.1029/2005jg000143.

Chuvieco, E., Aguado, I., Salas, J., García, M., Yebra, M., Oliva, P., 2020. Satellite remote sensing contributions to wildland fire science and management. Current Forestry Reports 6 (2), 81–96. Available at https://doi.org/10.1007/s40725-020-00116-5.

Clark, M.L., Hakkenberg, C.R., Bailey, T., Burns, P., Goetz, S.J., 2025. Changes in GEDI-based measures of forest structure after large California wildfires relative to pre-fire conditions. Remote Sensing of Environment 323. Available at https://doi.org/10.1016/j.rse.2025.114718, https://www.sciencedirect.com/science/journal/00344257.

Cuevas-González, M., Gerard, F., Balzter, H., Riaño, D., 2009. Analysing forest recovery after wildfire disturbance in boreal Siberia using remotely sensed vegetation indices. Global Change Biology 15 (3), 561–577. Available at https://doi.org/10.1111/j.1365-2486.2008.01784.x.

Doblas-Miranda, E., Alonso, R., Arnan, X., Bermejo, V., Brotons, L., de las Heras, J., et al., 2017. A review of the combination among global change factors in forests, shrublands and pastures of the Mediterranean Region: beyond drought effects. Global and Planetary Change 148, 42–54. Available at https://doi.org/10.1016/j.gloplacha.2016.11.012.

Dronova, I., Taddeo, S., 2022. Remote sensing of phenology: towards the comprehensive indicators of plant community dynamics from species to regional scales. Journal of Ecology 110 (7), 1460–1484. Available at https://doi.org/10.1111/1365-2745.13897.

Duane, A., Castellnou, M., Brotons, L., 2021. Towards a comprehensive look at global drivers of novel extreme wildfire events. Climatic Change 165 (3-4). Available at https://doi.org/10.1007/s10584-021-03066-4.

Eastburn, J.F., Campbell, M.J., Dennison, P.E., Anderegg, W.R., Barrett, K.J., Fekety, P.A., et al., 2024. Ecological and climatic transferability of airborne lidar-driven aboveground biomass models in Piñon-Juniper woodlands. GIScience and Remote Sensing 61 (1). Available at https://doi.org/10.1080/15481603.2024.2363577, http://www.tandfonline.com/loi/tgrs20?open=50&repitition=0#vol_50.

Falk, D.A., van Mantgem, P.J., Keeley, J.E., Gregg, R.M., Guiterman, C.H., Tepley, A.J., et al., 2022. Mechanisms of forest resilience. Forest Ecology and Management 512. Available at https://doi.org/10.1016/j.foreco.2022.120129, www.elsevier.com/inca/publications/store/5/0/3/3/1/0.

Fang, L., Crocker, E.V., Yang, J., Yan, Y., Yang, Y., Liu, Z., 2019. Competition and burn severity determine post-fire sapling recovery in a nationally protected boreal forest of china: an analysis from very high-resolution satellite imagery. Remote Sensing 11 (6). Available at https://doi.org/10.3390/RS11060603, https://www.mdpi.com/2072-4292/11/6/603.

Fernandes, P.M., 2013. Fire-smart management of forest landscapes in the Mediterranean basin under global change. Landscape and Urban Planning 110 (1), 175–182. Available at https://doi.org/10.1016/j.landurbplan.2012.10.014, www.elsevier.com/inca/publications/store/5/0/3/3/4/7.

Fernandez-Manso, A., Quintano, C., Roberts, D.A., 2016. Burn severity influence on post-fire vegetation cover resilience from Landsat MESMA fraction images time series in Mediterranean forest ecosystems. Remote Sensing of Environment 184, 112–123. Available at https://doi.org/10.1016/j.rse.2016.06.015, www.elsevier.com/inca/publications/store/5/0/5/7/3/3.

Fernández-Guisuraga, J.M., Calvo, L., 2023. Fuel build-up promotes an increase in fire severity of reburned areas in fire-prone ecosystems of the western Mediterranean Basin. Fire Ecology 19 (1). Available at https://doi.org/10.1186/s42408-023-00232-0.

Fernández-Guisuraga, J.M., Calvo, L., 2025. Fire severity shows limited dependence on fuel structure under adverse fire weather conditions: a case study of two extreme wildfire events. Fire Ecology 21 (1). Available at https://doi.org/10.1186/s42408-025-00371-6, https://www.springer.com/journal/42408.

Fernández-Guisuraga, J., Sanz-Ablanedo, E., Suárez-Seoane, S., Calvo, L., 2018. Using unmanned aerial vehicles in postfire vegetation survey campaigns through large and heterogeneous areas: opportunities and challenges. Sensors 18 (2), 586. Available at https://doi.org/10.3390/s18020586.

Fernández-Guisuraga, J.M., Calvo, L., Fernández-García, V., Marcos-Porras, E., Taboada, Á., Suárez-Seoane, S., 2019a. Efficiency of remote sensing tools for post-fire management along a climatic gradient. Forest Ecology and Management 433, 553–562. Available at https://doi.org/10.1016/j.foreco.2018.11.045, www.elsevier.com/inca/publications/store/5/0/3/3/1/0.

Fernández-Guisuraga, J.M., Suárez-Seoane, S., Calvo, L., 2019b. Modeling *Pinus pinaster* forest structure after a large wildfire using remote sensing data at high spatial resolution. Forest Ecology and Management 446, 257–271. Available at https://doi.org/10.1016/j.foreco.2019.05.028, www.elsevier.com/inca/publications/store/5/0/3/3/1/0.

Fernández-Guisuraga, J.M., Calvo, L., Suárez-Seoane, S., 2020. Comparison of pixel unmixing models in the evaluation of post-fire forest resilience based on temporal series of satellite imagery at moderate and very high spatial resolution. ISPRS Journal of Photogrammetry and Remote Sensing 164, 217–228. Available at https://doi.org/10.1016/j.isprsjprs.2020.05.004, www.elsevier.com/inca/publications/store/5/0/3/3/4/0.

Fernández-Guisuraga, J.M., Suárez-Seoane, S., Calvo, L., 2021a. Radiative transfer modeling to measure fire impact and forest engineering resilience at short-term. ISPRS Journal of Photogrammetry and Remote Sensing 176, 30–41. Available at https://doi.org/10.1016/j.isprsjprs.2021.04.002, www.elsevier.com/inca/publications/store/5/0/3/3/4/0.

Fernández-Guisuraga, J.M., Verrelst, J., Calvo, L., Suárez-Seoane, S., 2021b. Hybrid inversion of radiative transfer models based on high spatial resolution satellite reflectance data improves fractional vegetation cover retrieval in heterogeneous ecological systems after fire. Remote Sensing of Environment 255, 112304. Available at https://doi.org/10.1016/j.rse.2021.112304.

Fernández-Guisuraga, J.M., Calvo, L., Suárez-Seoane, S., 2022. Monitoring post-fire neighborhood competition effects on pine saplings under different environmental conditions by means of UAV multispectral data and structure-from-motion photogrammetry. Journal of Environmental Management 305, 114373. Available at https://doi.org/10.1016/j.jenvman.2021.114373.

Fernández-Guisuraga, J.M., Fernandes, P.M., Tárrega, R., Beltrán-Marcos, D., Calvo, L., 2023a. Vegetation recovery drivers at short-term after fire are plant community-dependent in mediterranean burned landscapes. Forest Ecology and Management 539. Available at https://doi.org/10.1016/j.foreco.2023.121034, www.elsevier.com/inca/publications/store/5/0/3/3/1/0.

Fernández-Guisuraga, J.M., Suárez-Seoane, S., Calvo, L., 2023b. Radar and multispectral remote sensing data accurately estimate vegetation vertical structure diversity as a fire resilience indicator. Remote Sensing in Ecology and Conservation 9 (1), 117–132. Available at https://doi.org/10.1002/rse2.299, http://onlinelibrary.wiley.com/journal/10.1002/(ISSN)2056-3485.

Fernández-Guisuraga, J.M., Calvo, L., Huerta, S., Marcos, E., 2023c. Burnt wood management enhances soil multifunctionality at the medium term after a large wildfire in north-west Spain. International Journal of Wildland Fire 32 (1), 99–110. Available at https://doi.org/10.1071/wf22091.

Fernández-Guisuraga, J.M., Calvo, L., Enterkine, J., Price, W.J., Dinkins, J.B., Jensen, K.S., et al., 2024a. Estimating vegetation and litter biomass fractions in rangelands using structure-from-motion and LiDAR datasets from unmanned aerial vehicles. Landscape

Ecology 39 (10). Available at https://doi.org/10.1007/s10980-024-01979-w, https://www.springer.com/journal/10980.

Fernández-Guisuraga, J.M., Calvo, L., Pérez-Rodríguez, L.A., Suárez-Seoane, S., 2024b. Integrating physical-based models and structure-from-motion photogrammetry to retrieve fire severity by ecosystem strata from very high resolution UAV Imagery. Fire 7 (9). Available at https://doi.org/10.3390/fire7090304, www.mdpi.com/journal/fire.

Fernández-Guisuraga, J.M., Monzón-González, A., Fernández-García, V., Peña-Pérez, S.A., Calvo, L., 2025. Estimation of prometheus fuel types using physically based remote sensing techniques. Fire Ecology 21 (1). Available at https://doi.org/10.1186/s42408-025-00373-4, https://www.springer.com/journal/42408.

Fisher, G.B., Elmore, A.J., Fitzpatrick, M.C., McNeil, D.J., Atkins, J.W., Larkin, J.L., 2024. Mapping recent timber harvest activity in a temperate forest using single date airborne LiDAR surveys and machine learning: lessons for conservation planning. GIScience and Remote Sensing 61 (1). Available at https://doi.org/10.1080/15481603.2024.2379198, http://www.tandfonline.com/loi/tgrs20?open=50&repitition=0#vol_50.

Fréjaville, T., Curt, T., 2015. Spatiotemporal patterns of changes in fire regime and climate: defining the pyroclimates of south-eastern France (Mediterranean Basin). Climatic Change 129 (1-2), 239–251. Available at https://doi.org/10.1007/s10584-015-1332-3.

García, M., Saatchi, S., Casas, A., Koltunov, A., Ustin, S.L., Ramirez, C., et al., 2017. Extrapolating forest canopy fuel properties in the California Rim fire by combining airborne LiDAR and landsat OLI data. Remote Sensing 9 (4). Available at https://doi.org/10.3390/rs9040394, http://www.mdpi.com/journal/remotesensing.

Georgopoulos, N., Gitas, I.Z., Stefanidou, A., Korhonen, L., Stavrakoudis, D., 2021. Estimation of individual tree stem biomass in an uneven-aged structured coniferous forest using multispectral LiDAR Data. Remote Sensing 13 (23), 4827. Available at https://doi.org/10.3390/rs13234827.

Gibson, R.K., White, L.A., Hislop, S., Nolan, R.H., Dorrough, J., 2022. The post-fire stability index; a new approach to monitoring post-fire recovery by satellite imagery. Remote Sensing of Environment 280. Available at https://doi.org/10.1016/j.rse.2022.113151, www.elsevier.com/inca/publications/store/5/0/5/7/3/3.

Gibson, R.K., Driscoll, D.A., Macdonald, K.J., Williamson, G.J., Nolan, R.H., Doherty, T.S., et al., 2025. Remotely sensed fire heterogeneity and biomass recovery predicts empirical biodiversity responses. Global Ecology and Biogeography 34 (4). Available at https://doi.org/10.1111/geb.70040, http://onlinelibrary.wiley.com/journal/10.1111/(ISSN)1466-8238.

Gillespie, M., Okin, G.S., Meyer, T., Ochoa, F., 2024. Evaluating burn severity and post-fire woody vegetation regrowth in the kalahari using uav imagery and random forest algorithms. Remote Sensing 16 (21), 3943. Available at https://doi.org/10.3390/rs16213943.

González-De Vega, S., De las Heras, J., Moya, D., 2016. Resilience of Mediterranean terrestrial ecosystems and fire severity in semiarid areas: responses of Aleppo pine forests in the short, mid and long term. Science of The Total Environment 573, 1171–1177. Available at https://doi.org/10.1016/j.scitotenv.2016.03.115.

Gordon, C.E., Price, O.F., Tasker, E.M., 2017. Mapping and exploring variation in post-fire vegetation recovery following mixed severity wildfire using airborne LiDAR: Ecological Applications 27 (5), 1618–1632. Available at https://doi.org/10.1002/eap.1555, www.esajournals.org.

Guerra-Hernández, J., Pereira, J.M.C., Stovall, A., Pascual, A., 2024. Impact of fire severity on forest structure and biomass stocks using NASA GEDI data. Insights from the 2020 and

2021 wildfire season in Spain and Portugal. Science of Remote Sensing 9. Available at https://doi.org/10.1016/j.srs.2024.100134, https://www.sciencedirect.com/science/journal/26660172.

Hermosilla, T., Wulder, M.A., White, J.C., Coops, N.C., Hobart, G.W., 2015. Regional detection, characterization, and attribution of annual forest change from 1984 to 2012 using Landsat-derived time-series metrics. Remote Sensing of Environment 170, 121–132. Available at https://doi.org/10.1016/j.rse.2015.09.004, www.elsevier.com/inca/publications/store/5/0/5/7/3/3.

Hislop, S., Jones, S., Soto-Berelov, M., Skidmore, A., Haywood, A., Nguyen, T., 2018. Using landsat spectral indices in time-series to assess wildfire disturbance and recovery. Remote Sensing 10 (3), 460. Available at https://doi.org/10.3390/rs10030460.

Hoe, M.S., Dunn, C.J., Temesgen, H., 2018. Multitemporal LiDAR improves estimates of fire severity in forested landscapes. International Journal of Wildland Fire 27 (9), 581–594. Available at https://doi.org/10.1071/WF17141, http://www.publish.csiro.au/nid/126.htm.

Huang, C., Goward, S.N., Masek, J.G., Thomas, N., Zhu, Z., Vogelmann, J.E., 2010. An automated approach for reconstructing recent forest disturbance history using dense Landsat time series stacks. Remote Sensing of Environment 114 (1), 183–198. Available at https://doi.org/10.1016/j.rse.2009.08.017.

Hudak, A.T., Morgan, P., Bobbitt, M.J., Smith, A.M.S., Lewis, S.A., Lentile, L.B., et al., 2007. The relationship of multispectral satellite imagery to immediate fire effects. Fire Ecology 3 (1), 64–90. Available at https://doi.org/10.4996/fireecology.0301064.

Huettermann, S., Jones, S., Soto-Berelov, M., Hislop, S., 2023. Using Landsat time series and bi-temporal GEDI to compare spectral and structural vegetation responses after fire. International Journal of Applied Earth Observation and Geoinformation 122, 103403. Available at https://doi.org/10.1016/j.jag.2023.103403.

Hwang, K., Harpold, A.A., Tague, C.L., Lowman, L., Boisramé, G.F.S., Lininger, K.B., et al., 2023. Seeing the disturbed forest for the trees: remote sensing is underutilized to quantify critical zone response to unprecedented disturbance. Earth's Future 11 (8). Available at https://doi.org/10.1029/2022EF003314, http://onlinelibrary.wiley.com/journal/10.1002/(ISSN)2328-4277.

Ireland, G., Petropoulos, G.P., 2015. Exploring the relationships between post-fire vegetation regeneration dynamics, topography and burn severity: a case study from the Montane Cordillera Ecozones of Western Canada. Applied Geography 56, 232–248. Available at https://doi.org/10.1016/j.apgeog.2014.11.016, http://www.elsevier.com/inca/publications/store/3/0/3/9/0/index.htt.

Jimeno-Llorente, L., Marcos, E., Fernández-Guisuraga, J.M., 2023. The effects of fire severity on vegetation structural complexity assessed using SAR data are modulated by plant community types in Mediterranean fire-prone ecosystems. Fire 6 (12), 450. Available at https://doi.org/10.3390/fire6120450.

Jobin, B., Labrecque, S., Grenier, M., Falardeau, G., 2008. Object-based classification as an alternative approach to the traditional pixel-based classification to identify potential habitat of the grasshopper sparrow. Environmental Management 41 (1), 20–31. Available at https://doi.org/10.1007/s00267-007-9031-0.

Johnstone, J.F., Allen, C.D., Franklin, J.F., Frelich, L.E., Harvey, B.J., Higuera, P.E., et al., 2016. Changing disturbance regimes, ecological memory, and forest resilience. Frontiers in Ecology and the Environment 14 (7), 369–378. Available at https://doi.org/10.1002/fee.1311, http://onlinelibrary.wiley.com/journal/10.1002/(ISSN)1540-9309.

Kamoske, A.G., Dahlin, K.M., Stark, S.C., Serbin, S.P., 2019. Leaf area density from airborne LiDAR: comparing sensors and resolutions in a temperate broadleaf forest ecosystem.

Forest Ecology and Management 433, 364–375. Available at https://doi.org/10.1016/j.foreco.2018.11.017, www.elsevier.com/inca/publications/store/5/0/3/3/1/0.

Kellner, J.R., Armston, J., Birrer, M., Cushman, K.C., Duncanson, L., Eck, C., et al., 2019. New opportunities for forest remote sensing through ultra-high-density drone lidar. Surveys in Geophysics 40 (4), 959–977. Available at https://doi.org/10.1007/s10712-019-09529-9, www.wkap.nl/journalhome.htm/0169-3298.

Kibler, C.L., Parkinson, A.M.L., Peterson, S.H., Roberts, D.A., D'Antonio, C.M., Meerdink, S.K., et al., 2019. Monitoring post-fire recovery of chaparral and conifer species using field surveys and landsat time series. Remote Sensing 11 (24). Available at https://doi.org/10.3390/rs11242963, https://res.mdpi.com/d_attachment/remotesensing/remotesensing-11-02963/article_deploy/remotesensing-11-02963.pdf.

Koutsias, N., Panourgia, K., Nakas, G., Petanidou, T., 2024. The importance of landscape and fire-history as factors explaining post-fire vegetation recovery in a Mediterranean island using Sentinel-2 satellite data. Science of The Total Environment 957, 177443. Available at https://doi.org/10.1016/j.scitotenv.2024.177443.

Larrinaga, A.R., Brotons, L., 2019. Greenness indices from a low-cost UAV imagery as tools for monitoring post-fire forest recovery. Drones 3 (1), 1–16. Available at https://doi.org/10.3390/drones3010006, https://www.mdpi.com/2504-446X/3/1/6/pdf.

Lin, S., Zhang, H., Liu, S., Gao, G., Li, L., Huang, H., 2023. Characterizing post-fire forest structure recovery in the great xing'an mountain using gedi and time series landsat data. Remote Sensing 15 (12), 3107. Available at https://doi.org/10.3390/rs15123107.

Lindenmayer, D.B., Westgate, M.J., Scheele, B.C., Foster, C.N., Blair, D.P., 2019. Key perspectives on early successional forests subject to stand-replacing disturbances. Forest Ecology and Management 454. Available at https://doi.org/10.1016/j.foreco.2019.117656, www.elsevier.com/inca/publications/store/5/0/3/3/1/0.

De Luca, G., Silva, J.M.N., Modica, G., 2022. Short-term temporal and spatial analysis for post-fire vegetation regrowth characterization and mapping in a Mediterranean ecosystem using optical and SAR image time-series. Geocarto International 37 (27), 15428–15462. Available at https://doi.org/10.1080/10106049.2022.2097482, http://www.tandfonline.com/toc/tgei20/current.

Mandl, L., Viana-Soto, A., Seidl, R., Stritih, A., Senf, C., 2024. Unmixing-based forest recovery indicators for predicting long-term recovery success. Remote Sensing of Environment 308, 114194. Available at https://doi.org/10.1016/j.rse.2024.114194.

Martín-Alcón, S., Coll, L., De Cáceres, M., Guitart, L., Cabré, M., Just, A., et al., 2015. Combining aerial LiDAR and multispectral imagery to assess postfire regeneration types in a Mediterranean forest. Canadian Journal of Forest Research 45 (7), 856–866. Available at https://doi.org/10.1139/cjfr-2014-0430, http://pubs.nrc-cnrc.gc.ca/rp-ps/inDetail.jsp?jcode=cjfr&lang=eng&vol=39&is=1.

Matasci, G., Hermosilla, T., Wulder, M.A., White, J.C., Coops, N.C., Hobart, G.W., et al., 2018. Three decades of forest structural dynamics over Canada's forested ecosystems using Landsat time-series and lidar plots. Remote Sensing of Environment 216, 697–714. Available at https://doi.org/10.1016/j.rse.2018.07.024, www.elsevier.com/inca/publications/store/5/0/5/7/3/3.

McCarley, T.R., Kolden, C.A., Vaillant, N.M., Hudak, A.T., Smith, A.M.S., Wing, B.M., et al., 2017. Multi-temporal LiDAR and Landsat quantification of fire-induced changes to forest structure. Remote Sensing of Environment 191, 419–432. Available at https://doi.org/10.1016/j.rse.2016.12.022, www.elsevier.com/inca/publications/store/5/0/5/7/3/3.

Meng, R., Wu, J., Schwager, K.L., Zhao, F., Dennison, P.E., Cook, B.D., et al., 2017. Using high spatial resolution satellite imagery to map forest burn severity across spatial scales in a Pine Barrens ecosystem. Remote Sensing of Environment 191, 95–109. Available at https://doi.org/10.1016/j.rse.2017.01.016, www.elsevier.com/inca/publications/store/5/0/5/7/3/3.

Meng, R., Wu, J., Zhao, F., Cook, B.D., Hanavan, R.P., Serbin, S.P., 2018. Measuring short-term post-fire forest recovery across a burn severity gradient in a mixed pine-oak forest using multi-sensor remote sensing techniques. Remote Sensing of Environment 210, 282–296. Available at https://doi.org/10.1016/j.rse.2018.03.019, www.elsevier.com/inca/publications/store/5/0/5/7/3/3.

Millar, C.I., Stephenson, N.L., 2015. Temperate forest health in an era of emerging megadisturbance. Science (New York, N.Y.) 349 (6250), 823–826. Available at https://doi.org/10.1126/science.aaa9933, http://www.sciencemag.org/content/349/6250/823.full.pdf.

Mitri, G.H., Gitas, I.Z., 2012. Mapping post-fire forest regeneration and vegetation recovery using a combination of very high spatial resolution and hyperspectral satellite imagery. International Journal of Applied Earth Observation and Geoinformation 20 (1), 60–66. Available at https://doi.org/10.1016/j.jag.2011.09.001, http://www.elsevier.com/locate/jag.

Mutanga, O., Masenyama, A., Sibanda, M., 2023. Spectral saturation in the remote sensing of high-density vegetation traits: a systematic review of progress, challenges, and prospects. ISPRS Journal of Photogrammetry and Remote Sensing 198, 297–309. Available at https://doi.org/10.1016/j.isprsjprs.2023.03.010.

Myint, S.W., Gober, P., Brazel, A., Grossman-Clarke, S., Weng, Q., 2011. Per-pixel vs. object-based classification of urban land cover extraction using high spatial resolution imagery. Remote Sensing of Environment 115 (5), 1145–1161. Available at https://doi.org/10.1016/j.rse.2010.12.017.

Neary, D.G., Leonard, J., 2015. Wildland fire: impacts on forest, woodland, and grassland ecological processes. Wildland Fires: A Worldwide Reality. Nova Science Publishers, Inc, United States, pp. 35–112. Available at https://www.novapublishers.com/catalog/product_info.php?products_id=55623.

Nolè, A., Rita, A., Spatola, M.F., Borghetti, M., 2022. Biogeographic variability in wildfire severity and post-fire vegetation recovery across the European forests via remote sensing-derived spectral metrics. Science of the Total Environment 823, 153807. Available at https://doi.org/10.1016/j.scitotenv.2022.153807.

Pausas, J.G., Keeley, J.E., 2021. Wildfires and global change. Frontiers in Ecology and the Environment 19 (7), 387–395. Available at https://doi.org/10.1002/fee.2359, http://onlinelibrary.wiley.com/journal/10.1002/(ISSN)1540-9309.

Pausas, J.G., Millán, M.M., 2019. Greening and browning in a climate change hotspot: the Mediterranean Basin. Bioscience 69 (2), 143–151. Available at https://doi.org/10.1093/biosci/biy157, http://bioscience.oxfordjournals.org/.

Percival, J.E.H., Sato, H., Razanaparany, T.P., Rakotomamonjy, A.H., Razafiarison, Z.L., Kitajima, K., 2024. Non fire-adapted dry forest of northwestern madagascar: escalating and devastating trends revealed by Landsat timeseries and GEDI lidar data. PLoS One 19 (2). Available at https://doi.org/10.1371/journal.pone.0290203, https://journals.plos.org//plosone/article?id=10.1371/journal.pone.0290203.

Post, A.J., Forbes, B., Cooper, Z., Faro, K., Seel, C., Clark, M., et al., 2025. Using handheld mobile laser scanning to quantify fine-scale surface fuels and detect changes post-disturbance in northern California forests. Ecological Indicators 172. Available at https://doi.org/10.1016/j.ecolind.2025.113276, https://www.sciencedirect.com/science/journal/1470160X.

Pádua, L., Guimarães, N., Adão, T., Sousa, A., Peres, E., Sousa, J.J., 2020. Effectiveness of sentinel-2 in multi-temporal post-fire monitoring when compared with UAV imagery. ISPRS International Journal of Geo-Information 9 (4), 225. Available at https://doi.org/10.3390/ijgi9040225.

Pérez-Cabello, F., Montorio, R., Alves, D.B., 2021. Remote sensing techniques to assess post-fire vegetation recovery. Current Opinion in Environmental Science & Health 21, 100251. Available at https://doi.org/10.1016/j.coesh.2021.100251.

Roberts, D.A., Gardner, M., Church, R., Ustin, S., Scheer, G., Green, R.O., 1998. Mapping Chaparral in the Santa Monica Mountains using multiple endmember spectral mixture models. Remote Sensing of Environment 65 (3), 267–279. Available at https://doi.org/10.1016/s0034-4257(98)00037-6.

Runge, K., Tucker, M., Crowther, T.W., Fournier de Laurière, C., Guirado, E., Bialic-Murphy, L., et al., 2025. Monitoring terrestrial ecosystem resilience using earth observation data: identifying consensus and limitations across metrics. Global Change Biology 31 (3). Available at https://doi.org/10.1111/gcb.70115, http://onlinelibrary.wiley.com/journal/10.1111/(ISSN)1365-2486.

Sankey, T., Donager, J., McVay, J., Sankey, J.B., 2017. UAV lidar and hyperspectral fusion for forest monitoring in the southwestern USA. Remote Sensing of Environment 195, 30–43. Available at https://doi.org/10.1016/j.rse.2017.04.007, www.elsevier.com/inca/publications/store/5/0/5/7/3/3.

De Santis, A., Chuvieco, E., Vaughan, P.J., 2009. Short-term assessment of burn severity using the inversion of PROSPECT and GeoSail models. Remote Sensing of Environment 113 (1), 126–136. Available at https://doi.org/10.1016/j.rse.2008.08.008.

De Santis, A., Chuvieco, E., 2007. Burn severity estimation from remotely sensed data: performance of simulation versus empirical models. Remote Sensing of Environment 108 (4), 422–435. Available at https://doi.org/10.1016/j.rse.2006.11.022.

Seidl, R., Turner, M.G., 2022. Post-disturbance reorganization of forest ecosystems in a changing world. Proceedings of the National Academy of Sciences 119 (28). Available at https://doi.org/10.1073/pnas.2202190119.

Seidl, R., Rammer, W., Spies, T.A., 2014. Disturbance legacies increase the resilience of forest ecosystem structure, composition, and functioning. Ecological Applications 24 (8), 2063–2077. Available at https://doi.org/10.1890/14-0255.1, http://www.esajournals.org/doi/pdf/10.1890/14-0255.1.

Singh, H., Srivastava, P.K., Prasad, R., Srivastava, S.K., 2025. Tracking post-fire vegetation regrowth and burned areas using bitemporal sentinel-1 SAR data: a google earth engine approach in heath vegetation of mooloolah river national park, queensland, Australia. Remote Sensing 17 (12). Available at https://doi.org/10.3390/rs17122031, http://www.mdpi.com/journal/remotesensing.

Smith-Tripp, S.M., Coops, N.C., Mulverhill, C., White, J.C., Axelson, J., 2024. Landsat assessment of variable spectral recovery linked to post-fire forest structure in dry sub-boreal forests. ISPRS Journal of Photogrammetry and Remote Sensing 208, 121–135. Available at https://doi.org/10.1016/j.isprsjprs.2024.01.008, https://www.sciencedirect.com/science/journal/09242716.

Smith-Tripp, S., Coops, N.C., Mulverhill, C., White, J.C., Gergel, S., 2025. Early spectral dynamics are indicative of distinct growth patterns in post-wildfire forests. Remote Sensing in Ecology and Conservation 11 (2), 221–238. Available at https://doi.org/10.1002/rse2.420, http://bit.ly/1lUx0ZG.

Solans Vila, J.P., Barbosa, P., 2010. Post-fire vegetation regrowth detection in the Deiva Marina region (Liguria-Italy) using Landsat TM and ETM+ data. Ecological Modelling 221 (1), 75–84. Available at https://doi.org/10.1016/j.ecolmodel.2009.03.011.

Storey, E.A., Stow, D.A., O'Leary, J.F., 2016. Assessing postfire recovery of chamise chaparral using multi-temporal spectral vegetation index trajectories derived from Landsat imagery. Remote Sensing of Environment 183, 53–64. Available at https://doi.org/10.1016/j.rse.2016.05.018, www.elsevier.com/inca/publications/store/5/0/5/7/3/3.

Talucci, A.C., Forbath, E., Kropp, H., Alexander, H.D., DeMarco, J., Paulson, A.K., et al., 2020. Evaluating post-fire vegetation recovery in Cajander Larch Forests in Northeastern Siberia using UAV derived vegetation indices. Remote Sensing 12 (18). Available at https://doi.org/10.3390/RS12182970, https://res.mdpi.com/d_attachment/remotesensing/remotesensing-12-02970/article_deploy/remotesensing-12-02970.pdf.

Tanase, M., de la Riva, J., Santoro, M., Pérez-Cabello, F., Kasischke, E., 2011. Sensitivity of SAR data to post-fire forest regrowth in Mediterranean and boreal forests. Remote Sensing of Environment 115 (8), 2075–2085. Available at https://doi.org/10.1016/j.rse.2011.04.009.

Tanase, M.A., Kennedy, R., Aponte, C., 2015. Radar burn ratio for fire severity estimation at canopy level: an example for temperate forests. Remote Sensing of Environment 170, 14–31. Available at https://doi.org/10.1016/j.rse.2015.08.025.

Thapa, N., Narine, L.L., Wilson, A.E., 2025. Forest aboveground biomass estimation using airborne LiDAR: a systematic review and meta-analysis. Journal of Forestry. Available at https://doi.org/10.1007/s44392-025-00029-w, https://link.springer.com/journal/44392.

Torres, J., Marques, J., Alves, P., Costa, H., Honrado, J., 2017. Local lithological drivers of post-fire vegetation recovery and implications for fire-prone regions. Ecological Research 32 (1), 37–49. Available at https://doi.org/10.1007/s11284-016-1415-2.

Vanderhoof, M.K., Burt, C., Hawbaker, T.J., 2018. Time series of high-resolution images enhances efforts to monitor post-fire condition and recovery, Waldo Canyon fire, Colorado, USA. International Journal of Wildland Fire 27 (10), 699–713. Available at https://doi.org/10.1071/WF17177, http://www.publish.csiro.au/nid/126.htm.

Veraverbeke, S., Gitas, I., Katagis, T., Polychronaki, A., Somers, B., Goossens, R., 2012. Assessing post-fire vegetation recovery using red–near infrared vegetation indices: accounting for background and vegetation variability. ISPRS Journal of Photogrammetry and Remote Sensing 68 (1), 28–39. Available at https://doi.org/10.1016/j.isprsjprs.2011.12.007

Viana-Soto, A., Aguado, I., Salas, J., García, M., 2020. Identifying post-fire recovery trajectories and driving factors using landsat time series in fire-prone mediterranean pine forests. Remote Sensing 12 (9), 1499. Available at https://doi.org/10.3390/rs12091499.

Viana-Soto, A., García, M., Aguado, I., Salas, J., 2022. Assessing post-fire forest structure recovery by combining LiDAR data and Landsat time series in Mediterranean pine forests. International Journal of Applied Earth Observation and Geoinformation 108, 102754. Available at https://doi.org/10.1016/j.jag.2022.102754.

Viedma, O., Silva, C.A., Moreno, J.M., Hudak, A.T., 2024. LadderFuelsR: a new automated tool for vertical fuel continuity analysis and crown base height detection using light detection and ranging. Methods in Ecology and Evolution 15 (11), 1958–1967. Available at https://doi.org/10.1111/2041-210x.14427.

Wallace, A., Nichol, C., Woodhouse, I., 2012. Recovery of forest canopy parameters by inversion of multispectral LiDAR Data. Remote Sensing 4 (2), 509–531. Available at https://doi.org/10.3390/rs4020509.

Wallace, A.M., McCarthy, A., Nichol, C.J., Ren, X., Morak, S., Martinez-Ramirez, D., et al., 2014. Design and evaluation of multispectral LiDAR for the recovery of arboreal parameters. IEEE Transactions on Geoscience and Remote Sensing 52 (8), 4942–4954. Available at https://doi.org/10.1109/TGRS.2013.2285942.

Wallace, L., Lucieer, A., Malenovský, Z., Turner, D., Vopěnka, P., 2016. Assessment of forest structure using two uav techniques: a comparison of airborne laser scanning and structure from motion (SfM) point clouds. Forests 7 (3), 62. Available at https://doi.org/10.3390/f7030062.

Ward, D.S., Kloster, S., Mahowald, N.M., Rogers, B.M., Randerson, J.T., Hess, P.G., 2012. The changing radiative forcing of fires: global model estimates for past, present and future. Atmospheric Chemistry and Physics 12 (22), 10857–10886. Available at https://doi.org/10.5194/acp-12-10857-2012.

White, J.C., Wulder, M.A., Hermosilla, T., Coops, N.C., Hobart, G.W., 2017. A nationwide annual characterization of 25 years of forest disturbance and recovery for Canada using Landsat time series. Remote Sensing of Environment 194, 303–321. Available at https://doi.org/10.1016/j.rse.2017.03.035, www.elsevier.com/inca/publications/store/5/0/5/7/3/3.

White, J.C., Hermosilla, T., Wulder, M.A., 2023. Pre-fire measures of boreal forest structure and composition inform interpretation of post-fire spectral recovery rates. Forest Ecology and Management 537. Available at https://doi.org/10.1016/j.foreco.2023.120948, www.elsevier.com/inca/publications/store/5/0/3/3/1/0.

Whiteside, T.G., Boggs, G.S., Maier, S.W., 2011. Comparing object-based and pixel-based classifications for mapping savannas. International Journal of Applied Earth Observation and Geoinformation 13 (6), 884–893. Available at https://doi.org/10.1016/j.jag.2011.06.008, http://www.elsevier.com/locate/jag.

Wulder, M.A., White, J.C., Alvarez, F., Han, T., Rogan, J., Hawkes, B., 2009. Characterizing boreal forest wildfire with multi-temporal Landsat and LIDAR data. Remote Sensing of Environment 113 (7), 1540–1555. Available at https://doi.org/10.1016/j.rse.2009.03.004.

Wulder, M.A., White, J.C., Nelson, R.F., Næsset, E., Ørka, H.O., Coops, N.C., et al., 2012. Lidar sampling for large-area forest characterization: a review. Remote Sensing of Environment 121, 196–209. Available at https://doi.org/10.1016/j.rse.2012.02.001.

Zhao, F.R., Meng, R., Huang, C., Zhao, M., Zhao, F.A., Gong, P., et al., 2016. Long-term post-disturbance forest recovery in the greater yellowstone ecosystem analyzed using Landsat time series stack. Remote Sensing 8 (11). Available at https://doi.org/10.3390/rs8110898, http://www.mdpi.com/2072-4292/8/11/898/pdf.

Zhou, Z., Liu, L., Jiang, L., Feng, W., Samsonov, S.V., 2019. Using long-term SAR backscatter data to monitor post-fire vegetation recovery in tundra environment. Remote Sensing 11 (19), 2230. Available at https://doi.org/10.3390/rs11192230.

Wildland-urban interface: dynamics, planning strategies, and remote sensing for fire risk mitigation

11

David Beltrán-Marcos[1] *and Leonor Calvo*[1,2]
[1]Area of Ecology, Department of Biodiversity and Environmental Management, Faculty of Biological and Environmental Sciences, University of León, León, Spain, [2]Institute for the Environment and Global Change (IMACG), University of León, León, Spain

11.1 Introduction

The increased risk of sixth-generation wildfires (i.e., large wildfires of extreme fire behavior) involves areas of high socio-economic value (Molina-Terrén et al., 2019), such as wildland–urban interfaces (WUI), where humans, goods, and ecosystem services are highly vulnerable (Modugno et al., 2016; Badia et al., 2019). WUI areas are geographical zones where human development, such as settlements and constructions, meets or intermixes with wildland areas (Bento-Gonçalves and Vieira, 2020; Ganteaume et al., 2021), creating a new landscape context formed by conjunction of buildings and wildland fuels (Lampin-Maillet et al., 2009). In WUI areas, vegetation is not merely perceived as an aesthetic component, but rather as a fundamental element of the urban environment that provides a wide array of ecosystem services (Depietri and Orenstein, 2020), making these locations particularly attractive places to live (Schug et al., 2023). These services include climate regulation, air and water quality enhancement, noise attenuation, biodiversity promotion, and significant benefits to the physical and mental well-being of residents (Depietri and Orenstein, 2020). Nonetheless, the combination of expanding residential areas into forested landscapes, the infrastructure development without adequate land-use planning, the loss of traditional land use derived from rural abandonment, the proliferation of rapid-growing forest plantations, and the increasing frequency of extreme climatic events has substantially heightened the vulnerability of WUI areas (Radeloff et al., 2018; Rosenthal et al., 2021). As a result, these zones have become critical hotspots for both wildfire ignition and propagation (Calviño-Cancela et al., 2017; Syphard et al., 2019; Vacca et al., 2020). These drivers have also significantly altered historical fire regimes, leading to increased wildfire frequency and severity (Pausas and Fernández-Muñoz, 2012; Beltrán-Marcos et al., 2024), and consequently reducing the resilience of WUI areas (Molina-Terrén et al., 2019; Pastor et al., 2020).

Advances in Remote Sensing for Forest Fire Management. DOI: https://doi.org/10.1016/B978-0-443-41692-7.00011-1
© 2027 Elsevier Ltd. All rights are reserved, including those for text and data mining, AI training, and similar technologies.

These emerging fire regime patterns are causing unprecedented damage to society (Roces-Díaz et al., 2022), both directly (such as the loss of human lives and the infrastructure damage) and indirectly through cascading effects on ecosystem services (Rosenthal et al., 2021). These latter include disruptions to supporting services (e.g., soil fertility), provisioning services (e.g., forage availability), regulating services (e.g., carbon sequestration), and cultural services (e.g., landscape aesthetics). In this context, effective fire risk mitigation in the WUI requires a comprehensive understanding of its spatial dynamics, the underlying drivers that increase their vulnerability, and the spatial planning strategies that can reduce the impacts of extreme wildfires (Caggiano et al., 2020).

In recent years, advances in remote sensing (RS) have significantly enhanced monitoring capabilities in WUI areas, providing effective tools for risk assessment and the planning of mitigation strategies (Bento-Gonçalves and Vieira, 2020). The integration of empirical models with geospatial data derived from multispectral satellite imagery, such as NASA/USGS Landsat and the European Copernicus Sentinel-2 missions, has enabled the detection of vegetation cover changes across the globe (Lampin-Maillet et al., 2011; Badia et al., 2019; Li et al., 2022), as well as the characterization of fuel loads prone to high-severity wildfires, including large biomass accumulations in WUI areas (Beltrán-Marcos et al., 2023; Fernández-García, Beltrán-Marcos and Calvo, 2023). Furthermore, these satellite products have also been extensively used to map urban expansion in WUI landscapes (Hammer et al., 2007; Schug et al., 2023). The most common approach to exploit this type of optical imagery is by computing spectral indices, such as the Normalized Difference Vegetation Index (NDVI) or the differenced Normalized Burn Ratio (dNBR), which are, respectively, essential for monitoring seasonal vegetation productivity and assessing postfire ecological impacts, both crucial factors for identifying highly vulnerable WUI areas (Beltrán-Marcos et al., 2023; Fernández-Manso et al., 2024). Nevertheless, more robust and generalizable approaches like Radiative Transfer Models (RTMs) (Fernández-Guisuraga et al., 2023) or Multiple Endmember Spectral Mixture Analysis (MESMA) (Fernández-Manso et al., 2024) have improved the accuracy in estimating key biophysical variables within WUI areas, including the Fractional Vegetation Cover (FCOVER), the Fraction of Absorbed Photosynthetically Active Radiation (FAPAR), and Canopy Water Content (CWC). In parallel, the use of active RS technologies, such as light detection and ranging (LiDAR) and Synthetic Aperture Radar (SAR) from Sentinel-1, has provided detailed information on vegetation vertical structure, fuel load density, and terrain morphology, which are critical components for modeling fire spread dynamics and assessing infrastructure exposure levels (Collins et al., 2024; Sankey et al., 2024).

The integration of these datasets into geospatial platforms such as Google Earth Engine (GEE) or other multiscale analytical software has enabled the development of high-resolution spatiotemporal fire risk maps, which support the identification of critical hotspots and the prioritization of targeted prevention measures (Sarricolea et al., 2020). This anticipatory capability offered by RS technologies is particularly vital in WUI contexts, where wildfires threaten not only natural ecosystems but also entire communities and human safety.

11.2 Wildland–urban interface dynamics and vulnerability drivers in the context of extreme wildfires

Globally, WUI areas have increased by 12.56% between 1985 and 2020 (Schug et al., 2023; Chen et al., 2024). In fact, burned area mapping obtained from active fire data collected by the Moderate Resolution Imaging Spectroradiometer (MODIS) between 2001 and 2020, and global human settlement population datasets (GHS-POP) (Schiavina et al., 2022), revealed that 12.54% of all WUI zones worldwide, encompassing an estimated 10.11 million people, lie within a 2 km buffer zone identified as highly threatened by severe wildfire activity (Chen et al., 2024). At the regional scale, the largest concentrations of vulnerable WUI areas were observed in the United States, Africa, Brazil, China, Indonesia, and across various areas of Europe (Fig. 11.1). In particular, in regions such as Mediterranean Basin and the western coast of the United States, the unregulated expansion of WUI areas has been a major contributing factor to the escalation of large-scale and extreme behavior wildfires, such as those that occurred in California in 2024 and across the Iberian Peninsula (Spain and Portugal) in 2025. For instance, in northwestern (NW) Spain, where WUI areas account for over 8% of the total land surface, the frequency and fire severity of Level 2 wildfires (those that represent immediate threats to population clusters) have increased significantly over the past decade. This trend has been driven by unplanned urban expansion, agricultural abandonment, and extreme meteorological conditions (Rodriguez-Jimenez et al., 2023).

In the context of potential fire risk, WUIs are complex areas where infrastructures and settlements are located within or adjacent to vegetated landscapes (Modugno et al., 2016). Under the more simplistic characterization, WUIs can be distinguished between interface areas (built-up zones with sparse vegetation located near continuous patches of flammable wildland vegetation) and intermix areas, where flammable vegetation is fully intermingled with anthropogenic infrastructures (Radeloff et al., 2005). Therefore, the fire vulnerability of WUI areas is determined by a combination of (1) environmental factors, such as fire regime, fuel characteristics, and topography; (2) anthropogenic factors, such as population density or risk perception; and (3) institutional factors, such as the development of prevention plans or fire defense infrastructure (Keeley and Syphard, 2019) (Fig. 11.2). The interaction of these variables produces diverse risk profiles that demand context-specific mitigation strategies tailored to the characteristics of each territory. Additionally, dynamic variables related to human behavior and evacuation logistics have emerged as critical components in assessing WUI risk. In this sense, Gwynne et al. (2019) introduce the concept of "dynamic vulnerability" through a simulation framework that integrates fire development with the mobility patterns of affected people and vehicular traffic. This approach allows for a more accurate analysis of a community's response based on its infrastructure, evacuation capabilities, and the evolution of the fire.

The WUI fire vulnerability associated with anthropogenic factors is also intrinsically linked to risk perception and the level of social awareness among

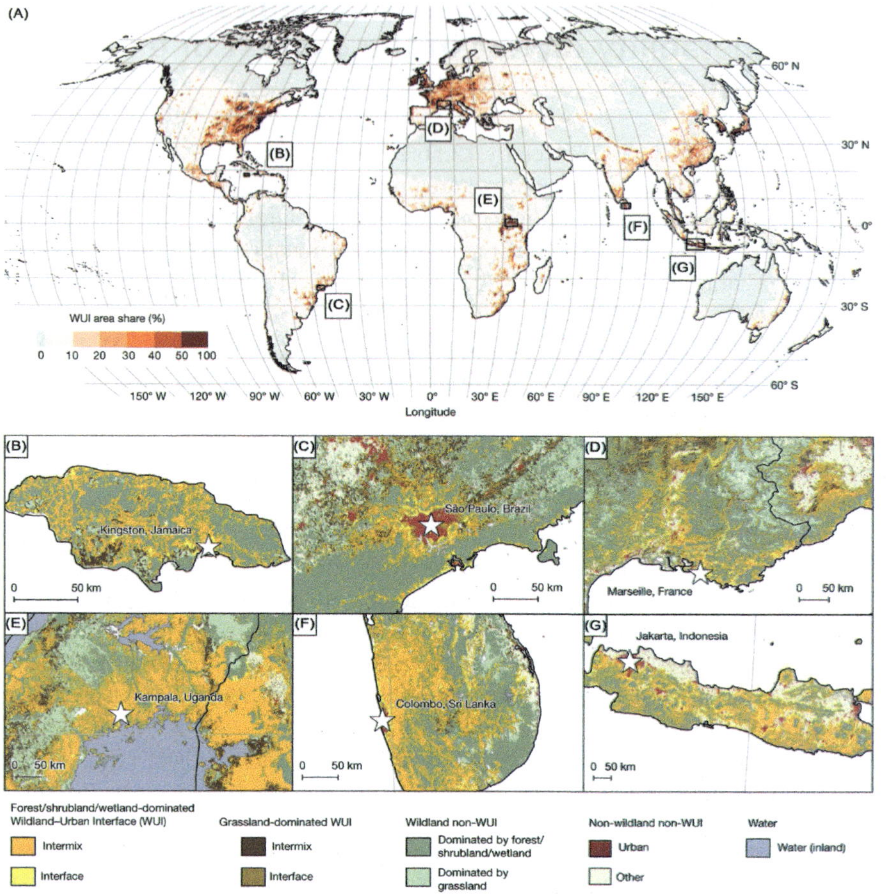

Figure 11.1 (A) Percentage of the Earth's surface dominated by wildland–urban interface (WUI) areas in 2020, and (B–G), map of WUI classified as intermix, interface, and non-WUI areas for Jamaica (B), Brazil (C), France (D), Uganda (E), Sri Lanka (F), and Indonesia (G). The interactive global map of this classification can be consulted at https://silvis.forest.wisc.edu/data/globalwui. Map lines delineate study areas and do not necessarily depict accepted national boundaries.
Source: This figure originates from Schug, F., Bar-Massada, A., Carlson, A.R., Cox, H., Hawbaker, T.J., Helmers, D., Hostert, P., Kaim, D., Kasraee, N.K., Martinuzzi, S., Mockrin, M.H., Pfoch, K.A. & Radeloff, V.C. (2023). The global wildland–urban interface. Nature, 621(7977), 94–99. https://doi.org/10.1038/s41586-023-06320-0.

communities living in WUI areas. According to authors such as Chas-Amil et al. (2022), the actions adopted by the population to protect themselves against wildfires can significantly reduce suppression costs and fire-related damages. Consequently, the development of self-protection plans, such as clearing defensible spaces around anthropogenic structures, as well as the implementation of educational and training

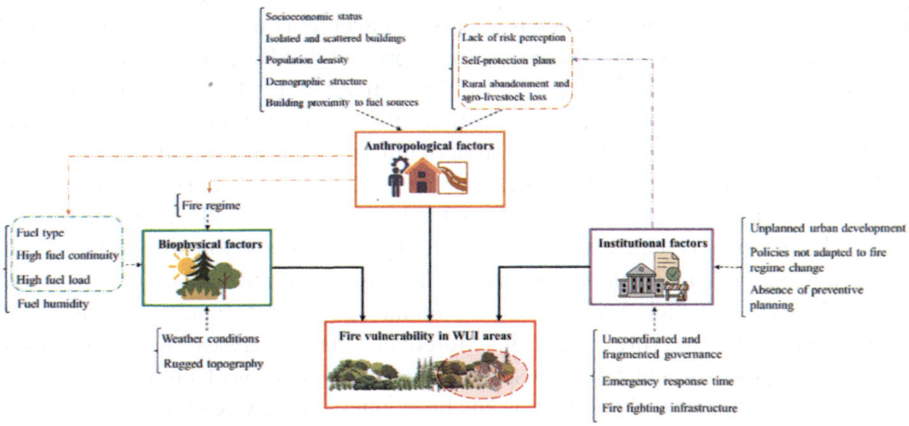

Figure 11.2 Structural diagram of the main factors related to the fire vulnerability in the wildland–urban interface (WUI).

programs, is essential in socially vulnerable municipalities. These communities are often characterized by low population density, remote locations, and a relatively high proportion of elderly residents (Oliveira et al., 2020). Therefore, the concept of vulnerability must be expanded to encompass social dimensions, as proposed by Bar-Massada et al. (2023), integrating physical wildfire exposure with variables such as age, socioeconomic status, response time, and access to defense infrastructures. The combination of these factors shapes risk maps of highly localized communities, which require differentiated mitigation strategies. This perspective is particularly relevant in WUI contexts that present additional challenges, such as rugged topography, limited road infrastructure, or a high degree of population dispersion, as is the case in isolated WUIs (Sturtevant and Cleland, 2007; Beltrán-Marcos et al., 2023).

On the other hand, one of the main fire vulnerability drivers in WUIs related to biophysical factors is the accumulation of vegetative fuel within urban environments, which increases both the likelihood of ignition and the rate of fire spread (Calviño-Cancela et al., 2017). In areas where urban expansion has fragmented natural landscapes without the implementation of adequate fuel-reduction measures, exposure to fire is significantly higher (Bar-Massada et al., 2023). In this sense, the absence of land-use planning strategies that include defensible buffer zones and a smart landscape fragmentation, the widespread use of flammable building materials, and the close proximity of residential structures to dense vegetation have all been identified as critical drivers of extreme wildfire events in WUI areas (Bento-Gonçalves and Vieira, 2020). This accumulation of biomass is especially critical in territories where rural abandonment and land-use change have fostered the proliferation of highly flammable species and horizontal vegetation continuity (Beltrán-Marcos et al., 2023; Fernández-García et al., 2023). Authors such as Calviño-Cancela et al. (2017) and Rodriguez-Jimenez et al. (2023) report that in regions of NW Spain, homogeneous stands of eucalyptus and pine, combined with the scattered expansion of housing across forested landscapes, create conditions conducive to high-intensity wildfires in WUI areas, as evidenced by the wave of large-scale wildfires in 2025. In addition, the configuration of the landscape, especially with the abandonment of the agricultural-forest mosaic structure, directly influences fire spread. Beltrán-Marcos et al. (2023) and Fernández-García et al. (2023) demonstrated that large wildfires tend to propagate with high fire severity across homogeneous landscapes dominated by continuous forest vegetation and characterized by low building density. The loss of traditional agricultural and livestock uses, and their replacement by unmanaged forest stands, increases both fuel connectivity and fire intensity. Moreover, their analyses in the European Mediterranean Basin and Spain, respectively, suggest that most high-severity WUI wildfires occur in areas where there is an abrupt interface between built environments and dense vegetation with unmanaged transition zones dominated by shrub communities, such as those located across the NW Iberian Peninsula, facilitating fire ignition and reducing, or even preventing, the effectiveness of fire suppression efforts.

In addition to the aforementioned factors, weak institutional coordination and fragmented multilevel governance significantly hinder the effective implementation of wildfire prevention policies in WUI areas, thereby increasing their vulnerability. Oliveira et al. (2020) emphasize that the lack of cooperation across levels of government, coupled with fragmented governance between urban planning, forestry, and civil protection sectors, has obstructed the development and execution of integrated strategies in WUI territories, particularly in Southern Europe. This fragmented institutional framework has also altered the continuity of previously established landscape management plans and the implementation of territory-based mitigation measures (Pastor et al., 2020), such as strategic vegetation mosaics and low-combustibility strips, which could substantially reduce WUI vulnerability in the face of emerging sixth-generation wildfires.

11.3 Urban planning and prevention strategies

The residential development expanding from peri-urban and rural areas into forested zones with high fuel loads, without the implementation of preventive measures such as fuel management strips or restrictions on vegetation types near structures, involves a significant risk to both human life and infrastructure (Caggiano et al., 2020). Moreover, the continuity of vegetation between built structures increases the likelihood of ignition and complicates fire suppression operations (Sarricolea et al., 2020). In light of this situation, mitigation strategies must incorporate preventive planning approaches based on landscape structuration, fuel management, and the implementation of self-protection plans at the local scale.

Among the most effective approaches to this preventive planning is the establishment of buffer zones, defined as transition areas between wild vegetation and buildings, designed to reduce fuel continuity and provide time and space for emergency services to act (Vacca et al., 2020; Ganteaume et al., 2021). These zones not only decrease fire severity by reducing biomass loads but also serve as access corridors and strategic defense lines for firefighting crews. However, the effective implementation of such buffers requires detailed knowledge of landscape structure and urban settlement patterns, information that is not always readily available. For instance, in the Mediterranean Basin, the distances established to delimit WUI buffers around populated areas and enforce fuel-reduction measures range from 50 to 400 m, depending on specific local regulations (Modugno et al., 2016; Bento-Gonçalves and Vieira, 2020; Pastor et al., 2020). In countries like France, significant progress has been made in legislating mandatory perimeter defense zones for new developments located in forested environments. For example, owners are required to maintain a vegetation-free strip of at least 50 m around their properties (Lampin-Maillet et al., 2011). These strips act as self-protection spaces, limiting fire spread and reducing structural vulnerability. Nevertheless, their effectiveness depends largely on compliance by residents and ongoing maintenance, two aspects that remain insufficient in many Mediterranean regions. In the case of the United States, the National Fire Plan (NFP) has promoted fuel-reduction treatments since 2000, with a specific focus on WUI areas. Despite this, recent studies show that only 3% of the treated area is located within the WUI, with an additional 8% falling within a 2.5-km buffer surrounding these zones, reflecting a gap between mitigation policies and the actual spatial distribution of wildfire risk (Radeloff et al., 2018).

On the other hand, urban planning in WUI areas should be strategically designed to reduce population exposure through both spatial planning and fuel management. For instance, reducing homogeneous fuel types, particularly shrub-dominated fuels, surrounding isolated and dispersed WUIs has been identified as a critical intervention to mitigate fire severity (Beltrán-Marcos et al., 2023). Furthermore, the implementation of different silvicultural practices, such as pruning and removing dead and lower branches to increase canopy base height and eliminate ladder fuels in forested WUI systems, as well as breaking vertical fuel continuity by reducing understory fuel accumulation, is essential to decreasing wildfire risk (Fernandes, 2013). From a land management perspective, promoting structurally diverse landscape mosaics in WUI areas is key to enhancing both fire resistance and postfire resilience (Fernandes, 2013). Landscapes composed of resilient vegetation mosaics, especially those maintained in regions with long fire-free intervals (>25 years), not only function as effective wildfire mitigation tools but also enhance the delivery of ecosystem services (Moghli et al., 2022). In this regard, fuel-type conversion strategies within postwildfire reforestation policies, favoring less flammable vegetation types, can significantly contribute to building more fire-resilient WUIs (Dimitrakopoulos and Papaioannou, 2001). This resilience can also be achieved by supporting traditional and sustainable land-use practices, such as extensive livestock grazing or silvicultural work by local forestry crews. These practices are crucial for retaining populations in territories affected by severe rural depopulation, where landscape fragmentation can hinder the spread and intensity of large wildfires (Azevedo et al., 2011).

Finally, self-protection plans represent a critical complement to the aforementioned wildfire prevention strategies. These plans, typically developed by residents with technical support from public authorities and emergency services, enable the designation of evacuation routes, safe zones, and response protocols in the event of an extreme wildfire. Moreover, they reinforce risk awareness and foster a culture of prevention among communities. For instance, experience from Spanish regions such as Catalonia has demonstrated that when self-protection plans are integrated with well-designed buffer zones, the resilience of WUI areas to extreme wildfire events significantly increases (Badia et al., 2019).

11.4 Application of remote sensing for wildland–urban interface management

In the context of changing fire regimes, related to the increasing occurrence of high-severity wildfires, the ability to model fire spread dynamics and identify the key drivers of vulnerability in WUI areas have become essential (Ganteaume et al., 2021). RS approaches have emerged as a critical tool for both pre and postfire management in WUI zones, providing up-to-date, multiscale, and comparable data that enables reliable risk characterization, fire impact assessment, and landscape mapping.

One of the most significant recent advances has been the fine-scale spatial identification of the relationships between fuel characteristics, building patterns, and fire severity in large wildfire events (Beltrán-Marcos et al., 2023; Fernández-García et al., 2023). This understanding is essential for defining forest management guidelines and policies aimed at reducing the ecological and socio-economic impacts of wildfires in WUI areas. To achieve this, (Fernández-García, Beltrán-Marcos and Calvo, 2023) mapped different prefire fuel metrics such as fractional vegetation cover (FCOV), FAPAR, the leaf area index (LAI), and CWC, as well as their homogeneity in the landscape using multispectral and radar satellite data from missions like Sentinel-2 and Sentinel-1. These RS sources provide a very suitable temporal (<1 week) and spatial (10–20 m) resolutions for effective landscape management in WUI areas. Moreover, their estimation relied on increasingly adopted physically based methodologies, such as RTM models, which simulate the interaction between electromagnetic radiation, land surfaces, and the atmosphere. This approach enables more meaningful and generalizable measurements of critical biophysical variables (Fernández-Guisuraga et al., 2023). Traditionally, prefire vegetation amount quantification using satellite data has relied on empirical approaches based on statistical relationships between field measurements at the plot level and spectral vegetation indices (Michael et al., 2018; Syphard, Rustigian-Romsos and Keeley, 2021). However, the lack of physical meaning associated with these spectral indices in heterogeneous burned landscapes (Fernández-Guisuraga et al., 2023) results in limited generalizability across different vegetation types, thereby introducing uncertainty into the estimation of fuel loads in specific and complex environments such as the WUI. In addition to these biophysical variables, the incorporation of building patterns as a covariate in wildfire behavior modeling has improved the understanding of WUI typologies that are most vulnerable to fire impacts, particularly in the Iberian Peninsula (Beltrán-Marcos et al., 2023). In this context, the integration of geospatial information allowed the identification of landscape situations more prone to high-severity wildfires. Specifically, WUI typologies composed of isolated, scattered, or sparsely clustered buildings enclosed in a dense shrub matrix were found to exhibit the highest risk of severe wildfire impacts. Nonetheless, significant differences were observed along a bioclimatic gradient, with isolated WUIs affected consistently by high-severity wildfires across Atlantic, Transitional, and Mediterranean areas (Figs. 11.3 and 11.4). These lines of research are complemented by the findings of (Fernández-Manso et al., 2024), which represent a significant methodological advancement in the generation of fire risk maps in WUI environments. By integrating hyperspectral data from the PRISMA satellite, MESMA calibrated with field-based Composite Burn Index (CBI) measurements, and machine learning algorithms such as random forest (RF), this study developed a highly accurate predictive model at the regional scale. This model identifies areas with a high

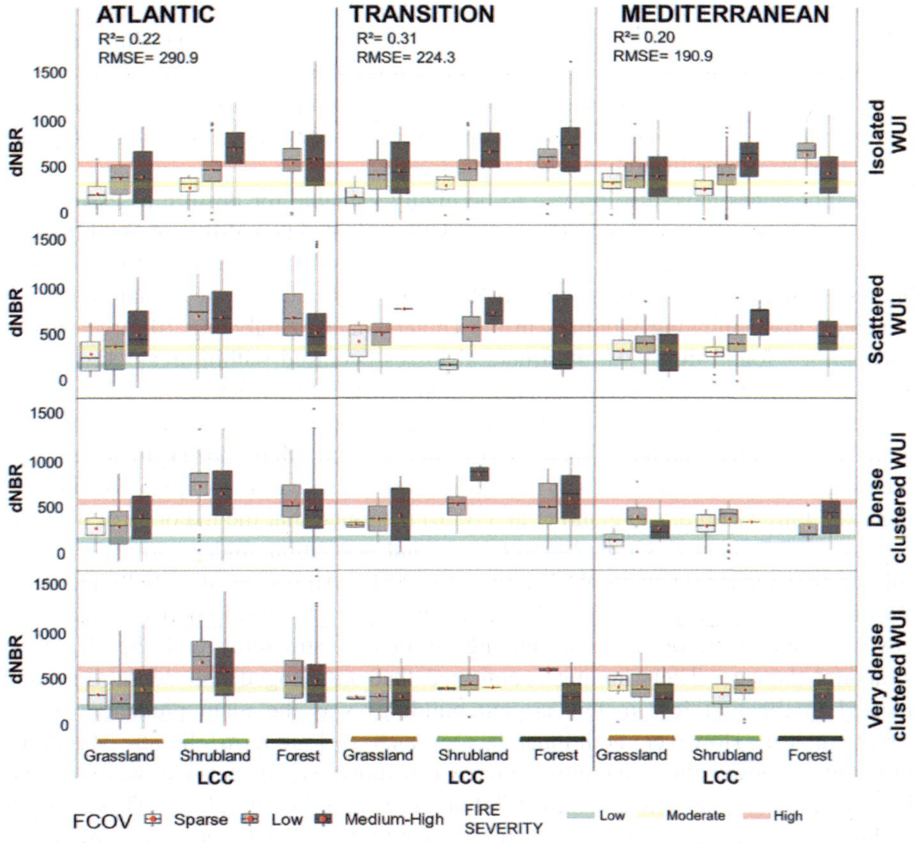

Figure 11.3 Boxplot displaying fire severity differenced Normalized Burn Ratio (dNBR) values along the climatic gradient and within each WUI type, grouped by building density pattern, Land Cover Class (LCC), and fractional vegetation cover (FCOV). The thresholds for fire severity categories were established according to the values set by the European Forest Fire Information Service.

Very dense clustered Atlantic WUI High-FCOV shrubland | *Dense clustered Transition WUI High-FCOV forest* | *Isolated Mediterranean WUI High-FCOV shrubland*

Figure 11.4 Example of wildland–urban interface (WUI) situations prone to high-severity wildfires in different climatic contexts (Atlantic, Transition, and Mediterranean) of the Iberian Peninsula.

probability of experiencing high-severity wildfires based on topo-climatic and fuel-related variables. A similar RS approach was employed by Rodriguez-Jimenez et al. (2023), who analyzed the driving factors and temporal trends in fire size and severity, focusing on wildfires that threaten WUI areas. Their work reinforces the relevance of combining satellite-derived indicators and fire behavior metrics to inform targeted mitigation strategies in high-risk WUI landscapes.

RS approaches have also proven to be highly valuable for postfire assessments in WUI areas. For instance, Caggiano et al. (2020) evaluated how different WUI building mapping methods (e.g., based on georeferenced census data versus object-based image analysis (OBIA) using high-resolution RS techniques) capture the actual patterns of structural loss following wildfires in the United States. Analyzing 70 large wildfire events, they found that 97% of structural losses occurred within areas classified as WUI when using OBIA-derived maps, compared to 86% when using census-based delineations. These object-based classification techniques have also been successfully employed by other authors, such as (McNamara and Mell, 2022), to assess both building damage and destruction and to inform defensive action planning in WUI contexts. Furthermore, Caggiano et al. (2020) identified that the majority of damaged structures were located within 100 m of wildland vegetation, underscoring the critical role of fuel management in WUI areas. Similarly, Ahmed et al. (2018) integrated WorldView-2 imagery with field data to analyze the relationship between vegetation cover within 30-m buffers surrounding residential structures and the degree of structural damage. Their findings confirmed that unmanaged and dense vegetation located in the first few meters around buildings was the main predictor of total structural loss. However, additional factors such as long-term climatic variability, proximity to power lines, and altitude have also shown significant relative importance in explaining structural loss within WUI areas (Syphard et al., 2021). These insights highlight the multifactorial nature of fire vulnerability in WUI areas and reinforce the necessity of combining high-resolution spatial data with detailed environmental and infrastructure variables to enhance postfire impact assessment and planning.

From an operational perspective, multiple RS analyses have been conducted in recent years to assess the expansion of WUI areas by integrating multitemporal data to detect changes in land cover and building density. For example, Tang et al. (2024)

developed a unified global WUI database at a spatial resolution of 9 km using RF models and MODIS satellite products. This enabled the authors to analyze global WUI expansion and fire occurrences within those areas for the period 2001–20. Their results showed a 24% relative global increase in WUI extent during that time frame, accompanied by a rise in both wildfire frequency and burned area in WUI landscapes. In the same line, Bar-Massada et al. (2023) combined high-resolution vegetation maps derived from Sentinel-2 imagery with building footprint data from the 2019 World Settlement Footprint dataset (Marconcini et al., 2020) to produce the first detailed map of WUI areas across Europe. Their analysis revealed that 7.4% of the European land surface is composed of WUI settlements that could present a substantial wildfire risk, although with considerable variability both between and within countries. In the case of the United States, the contributions of Carlson et al. (2022) and Li et al. (2022) showed that the spatial extent of WUI areas has grown significantly over recent decades, intensifying risks related to wildfires, biodiversity loss, and socio-environmental conflicts. Both studies utilized Landsat imagery (30 m resolution), land cover classifications, and building footprint data derived from Microsoft Bing to generate retrospective and highly detailed maps of WUI areas from 1990 to 2020. This allowed for an in-depth analysis of WUI evolution and the associated increase in wildfire exposure and risk across the United States.

In general, all the aforementioned RS analyses highlight the critical need for high-resolution and up-to-date spatial data to accurately assess postfire damages and feed predictive models. These data sources play a fundamental role in supporting strategic decision-making in WUI environments, both in the short and long term. In this context, RS through optical and radar imagery, as well as georeferenced building data, has become a key tool for generating high-resolution WUI maps that integrate both social (e.g., building density or demographic characteristics) and biophysical components (e.g., vegetation types or fuel moisture). Nonetheless, several challenges remain, particularly regarding spatial resolution limitations, model validation, and scalability across different geographic contexts. Additionally, there is an urgent need to better integrate these tools into decision-making processes to enhance the effectiveness of wildfire risk mitigation and land-use planning strategies in fire-prone WUI regions.

11.5 Conclusions and final remarks

The increase in fuel load and connectivity across natural landscapes, combined with complex socioeconomic challenges related to urban expansion, rural depopulation, and land abandonment, has transformed WUI areas into critical hotspots for extreme wildfire impacts (Chen et al., 2024). However, WUI fire risk is not only determined by fuel accumulation or land cover changes near urban settlements but also by spatial planning, community preparedness, and the effectiveness of institutional governance (Bar-Massada et al., 2023).

The widespread use of RS and geospatial modeling has significantly enhanced our ability to map and monitor WUI risk dynamics. These RS tools have made it possible to identify, with high accuracy, key biophysical drivers such as prefire fuel type, vegetation structure, and fuel amount, which have emerged as strong predictors of fire severity, particularly in Southern European contexts. In this region, WUI typologies composed of isolated or scattered buildings embedded within dense shrublands have consistently proven to be especially vulnerable to high-severity wildfire events. Nonetheless, the WUI is not a globally uniform concept (Tang et al., 2024), as it encompasses a wide range of building patterns, vegetation structures, and socio-political dynamics, especially in developing regions such as parts of Africa, where WUI expansion is particularly pronounced (Schug et al., 2023). This WUI heterogeneity underscores the need for context-specific definitions and typologies to support accurate risk assessments and effective prevention strategies. Therefore, given the current alteration of fire regimes, international institutional frameworks should prioritize WUI areas in land-use planning, population management, and protection policies that are essential for addressing the challenges raised by the new generation of larger and more severe wildfires.

References

Ahmed, M., Rahaman, K., Hassan, Q., 2018. Remote sensing of wildland fire-induced risk assessment at the community level. Sensors 18 (5), 1570. Available at https://doi.org/10.3390/s18051570.

Azevedo, J.C., Moreira, C., Castro, J.P., Loureiro, C., 2011. Agriculture Abandonment, Land-use Change and Fire Hazard in Mountain Landscapes in Northeastern Portugal. Springer Science and Business Media LLC, pp. 329–351. Available at 10.1007/978-3-642-12754-0_14.

Badia, A., Pallares-Barbera, M., Valldeperas, N., Gisbert, M., 2019. Wildfires in the wildland-urban interface in Catalonia: Vulnerability analysis based on land use and land cover change. Science of the Total Environment 673, 184–196. Available at https://doi.org/10.1016/j.scitotenv.2019.04.012.

Bar-Massada, A., Alcasena, F., Schug, F., Radeloff, V.C., 2023. The wildland – urban interface in Europe: spatial patterns and associations with socioeconomic and demographic variables. Landscape and Urban Planning 235, 104759. Available at https://doi.org/10.1016/j.landurbplan.2023.104759.

Beltrán-Marcos, D., Calvo, L., Fernández-Guisuraga, J.M., Fernández-García, V., Suárez-Seoane, S., 2023. Wildland-urban interface typologies prone to high severity fires in Spain. Science of the Total Environment 894, 165000. Available at https://doi.org/10.1016/j.scitotenv.2023.165000.

Beltrán-Marcos, D., Suárez-Seoane, S., Fernández-Guisuraga, J.M., Azevedo, J.C., Calvo, L., 2024. Fire regime attributes shape pre-fire vegetation characteristics controlling extreme fire behavior under different bioregions in Spain. Fire Ecology 20 (1). Available at https://doi.org/10.1186/s42408-024-00276-w, https://www.springer.com/journal/42408.

Bento-Gonçalves, A., Vieira, A., 2020. Wildfires in the wildland-urban interface: key concepts and evaluation methodologies. Science of the Total Environment 707, 135592. Available at https://doi.org/10.1016/j.scitotenv.2019.135592.

Caggiano, M.D., Hawbaker, T.J., Gannon, B.M., Hoffman, C.M., 2020. Building loss in wui disasters: evaluating the core components of the wildland–urban interface definition. Fire

3 (4), 1–17. Available at https://doi.org/10.3390/fire3040073, https://www.mdpi.com/2571-6255/3/4/73.

Calviño-Cancela, M., Chas-Amil, M.L., García-Martínez, E.D., Touza, J., 2017. Interacting effects of topography, vegetation, human activities and wildland-urban interfaces on wildfire ignition risk. Forest Ecology and Management 397, 10–17. Available at https://doi.org/10.1016/j.foreco.2017.04.033, www.elsevier.com/inca/publications/store/5/0/3/3/1/0.

Carlson, A.R., Helmers, D.P., Hawbaker, T.J., Mockrin, M.H., Radeloff, V.C., 2022. The wildland–urban interface in the United States based on 125 million building locations. Ecological Applications 32 (5). Available at https://doi.org/10.1002/eap.2597, http://onlinelibrary.wiley.com/journal/10.1002/(ISSN).

Chas-Amil, M.L., Nogueira-Moure, E., Prestemon, J.P., Touza, J., 2022. Spatial patterns of social vulnerability in relation to wildfire risk and wildland-urban interface presence. Landscape and Urban Planning 228. Available at https://doi.org/10.1016/j.landurbplan.2022.104577, www.elsevier.com/inca/publications/store/5/0/3/3/4/7.

Chen, B., Wu, S., Jin, Y., Song, Y., Wu, C., Venevsky, S., et al., 2024. Wildfire risk for global wildland–urban interface areas. Nature Sustainability 7 (4), 474–484. Available at https://doi.org/10.1038/s41893-024-01291-0.

Collins, L., Guindon, L., Lloyd, C., Taylor, S.W., White, S., 2024. Fractional cover mapping of wildland-urban interface fuels using Landsat, Sentinel 1 and PALSAR imagery. Remote Sensing of Environment 308, 114189. Available at https://doi.org/10.1016/j.rse.2024.114189.

Depietri, Y., Orenstein, D.E., 2020. Managing fire risk at the wildland-urban interface requires reconciliation of tradeoffs between regulating and cultural ecosystem services. Ecosystem Services 44, 101108. Available at https://doi.org/10.1016/j.ecoser.2020.101108.

Dimitrakopoulos, A.P., Papaioannou, K.K., 2001. Flammability assessment of Mediterranean forest fuels. Fire Technology 37 (2), 143–152. Available at https://doi.org/10.1023/A:1011641601076.

Fernandes, P.M., 2013. Fire-smart management of forest landscapes in the Mediterranean basin under global change. Landscape and Urban Planning 110 (1), 175–182. Available at https://doi.org/10.1016/j.landurbplan.2012.10.014, www.elsevier.com/inca/publications/store/5/0/3/3/4/7.

Fernández-García, V., Beltrán-Marcos, D., Calvo, L., 2023. Building patterns and fuel features drive wildfire severity in wildland-urban interfaces in Southern Europe. Landscape and Urban Planning 231, 104646. Available at https://doi.org/10.1016/j.landurbplan.2022.104646.

Fernández-Guisuraga, J.M., Calvo, L., Quintano, C., Fernández-Manso, A., Fernandes, P.M., 2023. Fractional vegetation cover ratio estimated from radiative transfer modeling outperforms spectral indices to assess fire severity in several Mediterranean plant communities. Remote Sensing of Environment 290. Available at https://doi.org/10.1016/j.rse.2023.113542, www.elsevier.com/inca/publications/store/5/0/5/7/3/3.

Fernández-Manso, A., Quintano, C., Fernández-Guisuraga, J.M., Roberts, D., 2024. Next-gen regional fire risk mapping: integrating hyperspectral imagery and National Forest Inventory data to identify hot-spot wildland-urban interfaces. Science of the Total Environment 940, 173568. Available at https://doi.org/10.1016/j.scitotenv.2024.173568.

Ganteaume, A., Barbero, R., Jappiot, M., Maillé, E., 2021. Understanding future changes to fires in southern Europe and their impacts on the wildland-urban interface. Journal of Safety Science and Resilience 2 (1), 20–29. Available at https://doi.org/10.1016/j.jnlssr.2021.01.001.

Gwynne, S., Ronchi, E., Bénichou, N., Kinateder, M., Kuligowski, E., Gomaa, I., et al., 2019. Modeling and mapping dynamic vulnerability to better assess WUI evacuation performance. Fire and Materials 43 (6), 644–660. Available at https://doi.org/10.1002/fam.2708.

Hammer, R.B., Radeloff, V.C., Fried, J.S., Stewart, S.I., 2007. Wildlandurban interface housing growth during the 1990s in California, Oregon, and Washington. International Journal of Wildland Fire 16 (3), 255–265. Available at https://doi.org/10.1071/WF05077.

Keeley, J.E., Syphard, A.D., 2019. Twenty-first century California, USA, wildfires: fuel-dominated vs. wind-dominated fires. Fire Ecology 15 (1). Available at https://doi.org/10.1186/s42408-019-0041-0, http://fireecology.org/journal/journal-archive/.

Lampin-Maillet, C., Jappiot, M., Long, M., Morge, D., Ferrier, J.P., 2009. Characterization and mapping of dwelling types for forest fire prevention. Computers, Environment and Urban Systems 33 (3), 224–232. Available at https://doi.org/10.1016/j.compenvurbsys.2008.07.003.

Lampin-Maillet, C., Long-Fournel, M., Ganteaume, A., Jappiot, M., Ferrier, J.P., 2011. Land cover analysis in wildland–urban interfaces according to wildfire risk: a case study in the South of France. Forest Ecology and Management 261 (12), 2200–2213. Available at https://doi.org/10.1016/j.foreco.2010.11.022.

Li, S., Dao, V., Kumar, M., Nguyen, P., Banerjee, T., 2022. Mapping the wildland-urban interface in California using remote sensing data. Scientific Reports 12 (1). Available at https://doi.org/10.1038/s41598-022-09707-7, www.nature.com/srep/index.html.

Marconcini, M., Metz-Marconcini, A., Üreyen, S., Palacios-Lopez, D., Hanke, W., Bachofer, F., et al., 2020. Outlining where humans live, the World Settlement Footprint 2015. Scientific Data 7 (1). Available at https://doi.org/10.1038/s41597-020-00580-5.

McNamara, D., Mell, W., 2022. Towards the use of remote sensing for identification of building damage, destruction, and defensive actions at wildland-urban interface fires. Fire Technology 58 (1), 641–672. Available at https://doi.org/10.1007/s10694-021-01170-6.

Michael, Y., Lensky, I.M., Brenner, S., Tchetchik, A., Tessler, N., Helman, D., 2018. Economic assessment of fire damage to urban forest in the wildland-urban interface using planet satellites constellation images. Remote Sensing 10 (9). Available at https://doi.org/10.3390/rs10091479, https://res.mdpi.com/remotesensing/remotesensing-10-01479/article_deploy/remotesensing-10-01479-v3.pdf?filename=&attachment=1.

Modugno, S., Balzter, H., Cole, B., Borrelli, P., 2016. Mapping regional patterns of large forest fires in Wildland–Urban Interface areas in Europe. Journal of Environmental Management 172, 112–126. Available at https://doi.org/10.1016/j.jenvman.2016.02.013.

Moghli, A., Santana, V.M., Baeza, M.J., Pastor, E., Soliveres, S., 2022. Fire recurrence and time since last fire interact to determine the supply of multiple ecosystem services by mediterranean forests. Ecosystems 25 (6), 1358–1370. Available at https://doi.org/10.1007/s10021-021-00720-x, http://www.springerlink.com/content/.

Molina-Terrén, D.M., Xanthopoulos, G., Diakakis, M., Ribeiro, L., Caballero, D., Delogu, G.M., et al., 2019. Analysis of forest fire fatalities in Southern Europe: Spain, Portugal, Greece and Sardinia (Italy). International Journal of Wildland Fire 28 (2), 85–98. Available at https://doi.org/10.1071/WF18004, http://www.publish.csiro.au/nid/126.htm.

Oliveira, S., Gonçalves, A., Benali, A., Sá, A., Zêzere, J.L., Pereira, J.M., 2020. Assessing risk and prioritizing safety interventions in human settlements affected by large wildfires. Forests 11 (8). Available at https://doi.org/10.3390/F11080859, https://res.mdpi.com/d_attachment/forests/forests-11-00859/article_deploy/forests-11-00859-v2.pdf.

Pastor, E., Muñoz, J.A., Caballero, D., Àgueda, A., Dalmau, F., Planas, E., 2020. Wildland–urban interface fires in Spain: summary of the policy framework and recommendations for improvement. Fire Technology 56 (5), 1831–1851. Available at https://doi.org/10.1007/s10694-019-00883-z, http://www.springerlink.com/content/.

Pausas, J.G., Fernández-Muñoz, S., 2012. Fire regime changes in the Western Mediterranean Basin: from fuel-limited to drought-driven fire regime. Climatic Change 110 (1-2), 215–226. Available at https://doi.org/10.1007/s10584-011-0060-6.

Radeloff, V.C., Hammer, R.B., Stewart, S.I., Fried, J.S., Holcomb, S.S., McKeefry, J.F., 2005. The wildland–urban interface in the United States. Ecological Applications 15 (3), 799–805. Available at https://doi.org/10.1890/04-1413.

Radeloff, V.C., Helmers, D.P., Kramer, H.A., Mockrin, M.H., Alexandre, P.M., Bar-Massada, A., et al., 2018. Rapid growth of the US wildland-urban interface raises wildfire risk. Proceedings of the National Academy of Sciences of the United States of America 115 (13), 3314–3319. Available at https://doi.org/10.1073/pnas.1718850115, http://www.pnas.org/content/pnas/115/13/3314.full.pdf.

Roces-Díaz, J.V., Santín, C., Martínez-Vilalta, J., Doerr, S.H., 2022. A global synthesis of fire effects on ecosystem services of forests and woodlands. Frontiers in Ecology and the Environment 20 (3), 170–178. Available at https://doi.org/10.1002/fee.2349, http://onlinelibrary.wiley.com/journal/10.1002/(ISSN).

Rodriguez-Jimenez, F., Fernandes, P.M., Fernández-Guisuraga, J.M., Alvarez, X., Lorenzo, H., 2023. Drivers and trends in the size and severity of forest fires endangering WUI areas: a regional case study. Forests 14 (12). Available at https://doi.org/10.3390/f14122366, http://www.mdpi.com/journal/forests/.

Rosenthal, A., Stover, E., Haar, R.J., Reid, C., 2021. Health and social impacts of California wildfires and the deficiencies in current recovery resources: an exploratory qualitative study of systems-level issues. PLoS One 16 (3), e0248617. Available at https://doi.org/10.1371/journal.pone.0248617.

Sankey, T.T., Tango, L., Tatum, J., Sankey, J.B., 2024. Forest fire, thinning, and flood in wildland-urban interface: UAV and lidar-based estimate of natural disaster impacts. Landscape Ecology 39 (3). Available at https://doi.org/10.1007/s10980-024-01811-5, https://www.springer.com/journal/10980.

Sarricolea, P., Serrano-Notivoli, R., Fuentealba, M., Hernández-Mora, M., de la Barrera, F., Smith, P., et al., 2020. Recent wildfires in central chile: detecting links between burned areas and population exposure in the wildland urban interface. Science of The Total Environment 706, 135894. Available at https://doi.org/10.1016/j.scitotenv.2019.135894.

Schiavina, M., Melchiorri, M., Corbane, C., Freire, S., Batista e Silva, F., 2022. Built-up areas are expanding faster than population growth: regional patterns and trajectories in Europe. Journal of Land Use Science 17 (1), 591–608. Available at https://doi.org/10.1080/1747423x.2022.2055184.

Schug, F., Bar-Massada, A., Carlson, A.R., Cox, H., Hawbaker, T.J., Helmers, D., et al., 2023. The global wildland–urban interface. Nature 621 (7977), 94–99. Available at https://doi.org/10.1038/s41586-023-06320-0, https://www.nature.com/nature/.

Sturtevant, B.R., Cleland, D.T., 2007. Human and biophysical factors influencing modern fire disturbance in northern Wisconsin. International Journal of Wildland Fire 16 (4), 398–413. Available at https://doi.org/10.1071/WF06023.

Syphard, A.D., Rustigian-Romsos, H., Mann, M., Conlisk, E., Moritz, M.A., Ackerly, D., 2019. The relative influence of climate and housing development on current and projected future fire patterns and structure loss across three California landscapes. Global Environmental Change 56, 41–55. Available at https://doi.org/10.1016/j.gloenvcha.2019.03.007, www.elsevier.com/inca/publications/store/3/0/4/2/5.

Syphard, A.D., Rustigian-Romsos, H., Keeley, J.E., 2021. Multiple-scale relationships between vegetation, the wildland–urban interface, and structure loss to wildfire in California. Fire 4 (1). Available at https://doi.org/10.3390/fire4010012, https://www.mdpi.com/2571-6255/4/1/12/pdf.

Tang, W., He, C., Emmons, L., Zhang, J., 2024. Global expansion of wildland-urban interface (WUI) and WUI fires: insights from a multiyear worldwide unified database (WUWUI). Environmental Research Letters 19 (4), 044028. Available at https://doi.org/10.1088/1748-9326/ad31da.

Vacca, P., Caballero, D., Pastor, E., Planas, E., 2020. WUI fire risk mitigation in Europe: a performance-based design approach at home-owner level. Journal of Safety Science and Resilience 1 (2), 97–105. Available at https://doi.org/10.1016/j.jnlssr.2020.08.001.

Assessing the impact of wildfires on ecosystem services and biodiversity in Mediterranean landscapes

12

José Manuel Fernández-Guisuraga[1,2]
[1]Area of Ecology, Department of Biodiversity and Environmental Management, Faculty of Biological and Environmental Sciences, University of León, León, Spain, [2]Institute for the Environment and Global Change (IMACG), University of León, León, Spain

12.1 Impact of wildfires on the delivery of ecosystem services in Mediterranean fire-prone landscapes

Human societies obtain a broad spectrum of ecological, socio-cultural, and economic benefits from multiple ecosystem functions and processes, which are collectively referred to as ecosystem services (Garland et al., 2021). In many biomes worldwide, wildfires are a natural ecological process that not only shape ecosystem structure, function, and dynamics, but also act as a selective force on species adapted to recurrent fire disturbance (Pausas and Keeley, 2014). When occurring within their historical frequency and intensity ranges, wildfires can contribute to the conservation of multiple ecosystem functions and services (Thom and Seidl, 2016; Pausas and Keeley, 2019). For example, the reproductive success of certain plant species depends on the occurrence of wildfire events (Supriya, 2017), while the landscape heterogeneity promoted by low-intensity wildfires can enhance pollination, habitat diversity, and recreational value of ecosystems (Lee et al., 2015; Pausas and Keeley, 2019). However, in Mediterranean ecosystems, fire regime shifts have been shown to hinder the provision of ecosystem services by up to 70% compared to unburned counterparts (Taboada et al., 2021; Moghli et al., 2022; García-Llamas et al., 2024).

In this context, the rural transformation that took place across southern European countries during the 20th century led to the widespread abandonment of traditional land-use systems based on agrosilvopastoral practices, including grazing, prescribed burning, and manual vegetation clearance (Morán-Ordóñez et al., 2013; García-Llamas et al., 2019). Indeed, rural land abandonment has resulted in significant changes in landscape structure and connectivity, leading to a marked increase in fuel amounts and horizontal continuity across the landscape (Moreira et al., 2011). Such transformations, together with anthropogenic climate change (Tejedor et al., 2024), have led to profound modifications in fire regimes (Chergui et al., 2018; Pausas and Keeley, 2021), fostering the expansion of flammable vegetation types and promoting wildfires of greater extent, intensity, severity, and recurrence (Raftoyannis et al., 2014;

Advances in Remote Sensing for Forest Fire Management. DOI: https://doi.org/10.1016/B978-0-443-41692-7.00014-7
© 2027 Elsevier Ltd. All rights are reserved, including those for text and data mining, AI training, and similar technologies.

Duane et al., 2021; Lecina-Diaz et al., 2023). These dynamics are exemplified by the extreme fire season recorded in southern Europe during the summer of 2022, when prolonged heatwaves and drought conditions contributed to the burning of approximately 470,000 hectares across Portugal, Spain, and France, with 21 extreme wildfire events exceeding 5000 ha each (Rodrigues et al., 2023).

Changes in fire regime attributes are challenging the persistence of plant communities worldwide (Enright et al., 2015), a process largely driven by ecosystem conversions to alternative stable states (Seidl et al., 2014; Fernández-Guisuraga and Calvo, 2023), which depends on two principal mechanisms as described by Karavani et al. (2018): (1) the capacity of individual plants to survive fire, which is strongly influenced by fire severity and intensity, and (2) the resilience of plant communities, determined by their ability to regenerate after fire and strongly tied to fire return intervals. For species that regenerate from canopy or soil seed banks, the time between consecutive wildfires becomes critical, as it must allow individuals to reach reproductive maturity and produce viable seeds in sufficient quantity (Calvo et al., 2008; Fernández-García et al., 2019). Consequently, if wildfires recur too frequently, they may hinder postfire recruitment due to immaturity risk (interval squeeze syndrome) (Enright et al., 2015) and increase ecosystem vulnerability (Baeza et al., 2007; Moghli et al., 2022). For its part, the ability to resprout is a fundamental trait that supports the resistance and resilience of Mediterranean forests to wildfire disturbance (Enright et al., 2015). Repeated high-severity wildfires have been reported to impair resprouting capacity due to damage to bud-forming tissues, depletion of nonstructural carbohydrate reserves, and reduced availability of water and nutrients (Keeley et al., 2008; Paula and Ojeda, 2009; Pausas et al., 2016; Fairman et al., 2019). Moreover, under altered fire disturbance regimes, wildfires can produce substantial heat-driven alterations in the physicochemical (Mataix-Solera et al., 2011; Romeo et al., 2020) and biological (Rincón and Pueyo, 2010; Fernández-García et al., 2020; Sáenz de Miera et al., 2020; Fernández-Guisuraga et al., 2023b; Fernández-Guisuraga et al., 2023a) properties of soils, particularly within the uppermost centimeters (Badía et al., 2017). Nonetheless, the magnitude and persistence of these changes are highly variable and depend largely on two main factors: (1) the elapsed time since the fire, given the transient character of many heat-induced effects (Hart et al., 2005), and (2) the role of biotic feedbacks mediated by interactions among above and belowground communities (Adkins et al., 2020).

In this emerging scenario, fire-induced ecological impacts on vegetation and soils can severely disrupt ecosystem functionality and service delivery (Fernández-Guisuraga et al., 2023b; García-Llamas et al., 2024). Such impacts may be long-lasting or even irreversible, and frequently result in trade-offs between different ecosystem functions and services (Lee et al., 2015; Schmerbeck et al., 2015; Taboada et al., 2021; Fernández-Guisuraga et al., 2023c). Indeed, the type and magnitude of services that ecosystems can deliver are closely linked to the composition, structure, and functional traits of plant communities (Yapp et al., 2010; Shi et al., 2023), as well as to soil functional characteristics (Huerta et al., 2022). For instance, repeated high-severity wildfires have been associated with vegetation shifts from forest to shrubland (Fernández-Guisuraga and Calvo, 2023), leading to a reduction in regulating services such as carbon sequestration and the loss of cultural services such as recreational value (Starbuck et al., 2006; Hurteau and Brooks, 2011). Likewise, wildfires can profoundly affect soil fungal communities (Salo and Kouki, 2018), diminishing the availability of edible fungi and thus impairing food production services (Franco-Manchón et al., 2019). Long-term research in Mediterranean fire-prone ecosystems dominated by obligate seeder species has further documented the detrimental effects of wildfires on a number of regulating, provisioning, and supporting ecosystem services, including wood and timber supply, mushroom harvesting, disturbance regulation, and biodiversity support (Taboada et al., 2021; Moghli et al., 2022). Similarly, García-Llamas et al. (2024) found that in seeder-dominated ecosystems, increased fire recurrence markedly reduced the potential supply of services linked to mature forest stands. In contrast, resprouter-dominated ecosystems showed no substantial changes in service provision regardless of fire recurrence. Notwithstanding, Taboada et al. (2021) and García-Llamas et al. (2024) evidenced new opportunities for the supply of provisioning ecosystem services associated with wildfire occurrence, including livestock grazing and honey production.

12.2 Ecosystem services and wildfires: a remote sensing perspective

The decline in the provision of ecosystem services can have significant adverse consequences for both rural communities and broader society in Mediterranean regions worldwide (Pereira et al., 2021). In response, there is growing interest among researchers and policymakers in monitoring and assessing fire-induced impacts on the flow of ecosystem services to society (Lucas-Borja et al., 2019; Fernández-Guisuraga et al., 2023). A wide range of methodologies exists for monitoring ecosystem services based on field data collection of the ecosystem functions that sustain the supply of each ecosystem service (Meyer et al., 2015). However, these approaches alone are not feasible for achieving spatially exhaustive coverage at frequent temporal scales (Pettorelli et al., 2018), particularly in extensive burned landscapes covering a wide range of vegetation assemblages and environmental settings (Fernández-Guisuraga et al., 2022a).

Remote sensing products and techniques have become a valuable tool for examining the intricate relationships between environmental and societal systems (De Araujo Barbosa et al., 2015). This rapidly evolving technology has been widely implemented to retrieve wall-to-wall products depicting several ecosystem functions, processes, and services by integrating multiple sensors and data sources in grasslands, shrublands, forests, and agroforestry systems worldwide (De Araujo Barbosa et al., 2015; Pettorelli et al., 2018; Masenyama et al., 2022; Paula et al., 2023; Schirpke et al., 2023). One of the major strengths of remote sensing technology lies in its ability to provide synoptic and temporally frequent observations through comprehensive and extensive datasets (Pettorelli et al., 2014). This makes it particularly valuable for detecting and monitoring changes in ecosystem functions and services at multiple spatial and temporal scales (Cabello et al., 2012; Fernández-Guisuraga et al., 2022b).

In addition, as more advanced remote sensing Earth observation products continue to emerge, their contribution to the mapping, modeling, and valuation of ecosystem services and goods is expected to expand significantly (De Araujo Barbosa et al., 2015).

Although remote sensing data offer valuable information for monitoring ecosystem services at regional to global scales, conventional remote sensing techniques implemented to date, such as land cover classifications or spectral vegetation indices (SVI) calculated from optical satellite data, often deliver only indirect estimates or surrogates of key indicators related to ecosystem functioning (De Araujo Barbosa et al., 2015; Dawson et al., 2016; Masenyama et al., 2022; Paula et al., 2023). In this context, remote sensing-derived proxies frequently require integration with field-based observations in empirical models to reliably capture the targeted ecosystem functions and services (Pettorelli et al., 2018), and they typically face generalization and applicability issues in complex landscapes that encompass multiple plant communities and environmental settings due to their context-dependent nature and lack of a physically based foundation (Fernández-Guisuraga et al., 2021).

In the fields of fire ecology and remote sensing, ecosystem functions and services are often overlooked in most postfire assessments that rely on remote sensing data (Marcos et al., 2023), in contrast to the more common focus on evaluating postfire vegetation recovery dynamics alone (Bright et al., 2019; Fernández-Guisuraga et al., 2021; Nolè et al., 2022; Smith-Tripp et al., 2024). Despite this general trend, a few recent studies have highlighted the potential of remote sensing-derived proxies, computed from both optical and active sensors, for quantifying the impacts of wildfires on the provision of ecosystem functions and services, including (1) gross primary productivity and photosynthetic capacity (Hemes et al., 2023; Marcos et al., 2023), (2) soil fertility (Rosero-Vlasova et al., 2018; Fernández-Guisuraga et al., 2022; Beltrán-Marcos et al., 2023), (3) water quality (Chafer et al., 2016; Pinardi et al., 2023), (4) air quality (Cruz-López et al., 2019; Griffin et al., 2024), and (5) carbon sequestration (Reddy et al., 2015; Fernández-Guisuraga et al., 2022c; Fernández-Guisuraga et al., 2022d; Liu and Popescu, 2022; Pascual and Guerra-Hernández, 2023; Feng et al., 2024).

Most of these studies are related to proxy- or process-based monitoring of structural and compositional features linked to the spectral signal or the structure of recovering vegetation in ecosystems. However, a framework for assessing key functional indicators of soil functioning, which holds important implications for the provision of ecosystem services in postfire scenarios (Pereira et al., 2021; Huerta et al., 2022; Fernández-Guisuraga et al., 2023), is typically underrepresented in the literature due to the challenges associated with evaluating fire-induced impacts on the soil compartment (Beltrán-Marcos et al., 2023; Quintano et al., 2023). In this context, the European Union soil strategy has emphasized the need to intensify efforts to assess soil properties and functions following fire disturbance events. In recent years, several studies have advanced the integration of remote sensing techniques for monitoring postfire soil functioning in Mediterranean landscapes. Among these, synthetic aperture radar (SAR) data have shown promising potential, despite being scarcely explored in this context to date. Recent findings highlight the effectiveness of L-band SAR backscatter data for retrieving soil organic carbon stocks and indicators of soil fertility, outperforming C-band SAR and optical data both in accuracy and temporal transferability, from immediate postfire conditions to one year after disturbance (Fernández-Guisuraga et al., 2022d). Similarly, Beltrán-Marcos et al. (2023) proposed a data fusion technique combining Sentinel-2 imagery (moderate spatial but high spectral resolution) with optical data acquired from unmanned aerial vehicles (UAVs) (high spatial but low spectral resolution) to generate spatially and spectrally enhanced scenes through image sharpening procedures. This data fusion approach substantially improved the estimation of soil organic carbon stocks compared to the use of Sentinel-2 or UAV datasets individually.

On the other hand, the increasing deployment of light detection and ranging (LiDAR) and hyperspectral sensors onboard UAVs may offer new opportunities to assess postfire carbon stocks and timber supply in forest ecosystems with the potential to meet the maximum acceptable error of 10%–15% required by forest practitioners (Fassnacht et al., 2025). This level of accuracy is well below the current standards typically achieved using airborne or spaceborne active and optical data (Kattenborn et al., 2015; Schumacher et al., 2019; Fernández-Guisuraga et al., 2022a; Goldbergs, 2023), particularly in areas affected by wildfire disturbance (Fassnacht et al., 2025). Likewise, the high spatial, spectral, and structural resolution data provided by sensors onboard UAVs, as well as terrestrial laser scanning (TLS) sensors, can serve as ground-truth reference data to augment postfire forest inventories outside sampled area (Ståhl et al., 2016; Matasci et al., 2018; Campbell et al., 2021). By integrating these products with wall-to-wall optical satellite data through upscaling procedures, large-scale maps depicting the impact of wildfires on aboveground carbon stocks and timber supply could be generated with higher accuracy and reduced uncertainty compared to approaches relying solely on field data, as this data fusion approach may broaden the ecological representativeness of ground-based reference datasets (Puliti et al., 2018).

12.3 Remote sensing–based assessment of taxonomic and functional plant diversity in postfire landscapes

Remote sensing techniques can serve as effective tools for assessing plant species diversity (Wang and Gamon, 2019), a key ecosystem property linked to ecosystem functioning, stability, and resilience (Gamfeldt et al., 2013; Spasojevic et al., 2016). Indeed, biodiversity is widely acknowledged as a key factor supporting both ecosystem services and the ecological functions and processes on which these services depend across different ecosystem types and spatiotemporal scales (Pettorelli et al., 2018; Gonzalez et al., 2020; Garland et al., 2021). As a result, there is a growing tendency to incorporate the value of management strategies aimed at preventing biodiversity loss into the assessment of ecosystem services flow (Harrison et al., 2014; Brockerhoff et al., 2017).

One established method for monitoring plant species diversity involves mapping the spatial distribution of plant species or functional types through high-resolution remote sensing data (Fernández-Guisuraga et al., 2022d), from which distinct taxonomic diversity metrics can be derived in postfire landscapes. Species-level classifications require both fine spectral and spatial detail, which is typically achieved through airborne hyperspectral imagery and LiDAR data (Sankey et al., 2017; Pervin et al., 2022) acquired from platforms such as airplanes or UAVs. These technologies, together with geographic object-based image analysis (GEOBIA) approaches (Blaschke, 2010) rather than pixel-based methods, typically enable accurate discrimination of dominant plant species or functional groups with contrasting biochemical or structural traits (Asner et al., 2008; Ustin and Gamon, 2010; Fernández-Guisuraga et al., 2022). Moreover, advanced spectral unmixing models with physical basis, such as multiple endmember spectral mixture analysis (MESMA) (Roberts et al., 1998), can be applied to narrowband optical data of low to moderate spatial resolution to decompose subpixel reflectance and retrieve fractional images that allow for the estimation of plant species abundance (Dudley et al., 2015; Kibler et al., 2019).

A second approach focuses on assessing postfire functional diversity of plant species through remotely sensed plant functional traits. An increasing number of studies emphasize the importance of functional diversity, rather than taxonomic diversity alone, in explaining the relationship between ecosystem functioning and service provisioning (Violle et al., 2014; Cimoli et al., 2024). Indeed, functional diversity captures trait variation both within and between species in the functional space of plant communities and has been shown to promote ecosystem productivity, stability, and resilience (Funk et al., 2017), including in burned landscapes (Fernández-Guisuraga et al., 2023a). Plant traits at the leaf and canopy level, such as pigment concentration, dry matter content, equivalent water thickness, and leaf area index, can be retrieved from broadband or narrowband optical data in burned plant communities through the inversion of radiative transfer models (RTMs) (Fernández-Guisuraga et al., 2021; Fernández-Guisuraga et al., 2021; Fernández-Guisuraga et al., 2024a; Lazzeri et al., 2021). Even when species-level identification is impractical due to the lack of high spatial resolution optical data, a pixel-based approach typically reflects community-level functional trait means, as each pixel may contain portions or entire individuals from one or multiple species (Zheng et al., 2022). Given that the spatial resolution of most optical satellite imagery often exceeds the size of individual shrub and tree canopies, the derived functional traits correspond to aggregated community-level attributes. As a result, measures of functional diversity are likely to capture both α and β diversity patterns, reflecting the local turnover of functional traits across the landscape (Schweiger and Laliberté, 2022). Based on the retrieved plant functional traits from optical data, several functional diversity indices can be calculated. Among the most commonly used in the remote sensing literature (Hauser et al., 2021; Pacheco-Labrador et al., 2022) are the convex hull volume (CHV; functional richness), Rao's quadratic entropy (Rao's Q), as well as functional richness evenness (FEve), divergence (FDiv), and dispersion (FDis). However, when dealing with low- to moderate-resolution optical data, factors such as within-species trait variability, background effects from soil, nonphotosynthetic vegetation, burned legacies, and seasonal dynamics can introduce uncertainty into the calculation of functional diversity metrics (Anderson, 2018; Wang and Gamon, 2019), particularly in burned areas encompassing multiple vegetation assemblages with sparse or recovering vegetation canopies (Fernández-Guisuraga et al., 2021). These uncertainties should be carefully addressed through spatially explicit analyses, such as segmentation by homogeneous vegetation patches or through the application of spectral unmixing models combined with RTM inversion procedures (Fernández-Guisuraga, Calvo, et al., 2023; Fernández-Guisuraga et al., 2024b), which can help to disentangle the reflectance contribution of different land cover components within mixed pixels.

The third approach relies on the calculation of spectral diversity metrics, which provide an indirect but increasingly adopted method for assessing plant species diversity from optical remote sensing data (Féret and Asner, 2014; Wang and Gamon, 2019; Féret and de Boissieu, 2020). Rooted in the spectral variation hypothesis, this approach quantifies spatial variability in optical data as a proxy for environmental heterogeneity and, by extension, plant species diversity (Palmer et al., 2002; Rocchini et al., 2010). However, the success of this method heavily depends on factors such as sensor characteristics, vegetation type, the scale and extent of the analysis, and the choice of spectral metrics (Rocchini et al., 2018). A wide range of metrics has been proposed to quantify spectral diversity. These include metrics derived from conventional SVIs or reflectance data of specific wavebands, information theory, and classifications into spectral species (Wang and Gamon, 2019). Beyond context-dependent SVIs, full-spectrum metrics such as the coefficient of variation, spectral angle mapper, or CHV offer alternatives for assessing plant species diversity by summarizing spectral heterogeneity by using statistical metrics of variability (Dahlin, 2016; Wang et al., 2016; Frye et al., 2021; Van Cleemput et al., 2023). An alternative method involves classifying narrowband or broadband optical data into spectral species through supervised or unsupervised classification schemes (Wang et al., 2018; Féret and de Boissieu, 2020). Rather than identifying actual species, these spectral species act as proxies for biological species and enable the direct estimation of plant species richness, as well as alpha and beta diversity metrics without field calibration data (Féret and Asner, 2014). Moreover, compared to information-based metrics, the spectral species approach tends to be less affected by the spectral signal of background components (Gholizadeh et al., 2018). However, it presents limitations when applied to coarser spatial resolution optical data in heterogeneous plant communities, such as those typical of Mediterranean burned landscapes, where multiple species often coexist within a single pixel. In such cases, resolving individual plants is unrealistic, and therefore, the underlying assumptions of the method are not fully satisfied (Somers and Asner, 2014; Fassnacht et al., 2016). In addition, feature selection in image transformation methods aimed at reducing image noise and improving computational efficiency over large scenes is a subjective process left to the user's discretion and should be carried out with caution (Féret and de Boissieu, 2020).

AI disclosure

During the preparation of this work, the author(s) used ChatGPT in order to improve the readability and linguistic clarity of the text. The final content was reviewed and validated by the authors to ensure its accuracy and scientific integrity. After using this tool/service, the author(s) reviewed and edited the content as needed and take(s) full responsibility for the content of the publication.

References

Adkins, J., Docherty, K.M., Gutknecht, J.L.M., Miesel, J.R., 2020. How do soil microbial communities respond to fire in the intermediate term? Investigating direct and indirect effects associated with fire occurrence and burn severity. Science of the Total Environment 745. Available at https://doi.org/10.1016/j.scitotenv.2020.140957, www.elsevier.com/locate/scitotenv.

Anderson, C.B., 2018. Biodiversity monitoring, earth observations and the ecology of scale. Ecology Letters 21 (10), 1572–1585. Available at https://doi.org/10.1111/ele.13106, http://onlinelibrary.wiley.com/journal/10.1111/(ISSN)1461-0248.

Asner, G.P., Jones, M.O., Martin, R.E., Knapp, D.E., Hughes, R.F., 2008. Remote sensing of native and invasive species in Hawaiian forests. Remote Sensing of Environment 112 (5), 1912–1926. Available at https://doi.org/10.1016/j.rse.2007.02.043.

Badía, D., López-García, S., Martí, C., Ortíz-Perpiñá, O., Girona-García, A., Casanova-Gascón, J., 2017. Burn effects on soil properties associated to heat transfer under contrasting moisture content. Science of the Total Environment 01-602, 1119–1128. Available at https://doi.org/10.1016/j.scitotenv.2017.05.254.

Baeza, M.J., Valdecantos, A., Alloza, J.A., Vallejo, V.R., 2007. Human disturbance and environmental factors as drivers of long-term post-fire regeneration patterns in Mediterranean forests. Journal of Vegetation Science 18 (2), 243–252. Available at https://doi.org/10.1111/j.1654-1103.2007.tb02535.x.

Beltrán-Marcos, D., Suárez-Seoane, S., Fernández-Guisuraga, J.M., Fernández-García, V., Marcos, E., Calvo, L., 2023. Relevance of UAV and sentinel-2 data fusion for estimating topsoil organic carbon after forest fire. Geoderma 430, 116290. Available at https://doi.org/10.1016/j.geoderma.2022.116290.

Blaschke, T., 2010. Object based image analysis for remote sensing. ISPRS Journal of Photogrammetry and Remote Sensing 65 (1), 2–16. Available at https://doi.org/10.1016/j.isprsjprs.2009.06.004.

Bright, B.C., Hudak, A.T., Kennedy, R.E., Braaten, J.D., Henareh Khalyani, A., 2019. Examining post-fire vegetation recovery with landsat time series analysis in three western north American forest types. Fire Ecology 15 (1). Available at https://doi.org/10.1186/s42408-018-0021-9, http://fireecology.org/journal/journal-archive/.

Brockerhoff, E.G., Barbaro, L., Castagneyrol, B., Forrester, D.I., Gardiner, B., González-Olabarria, J.R., et al., 2017. Forest biodiversity, ecosystem functioning and the provision of ecosystem services. Biodiversity and Conservation 26 (13), 3005–3035. Available at https://doi.org/10.1007/s10531-017-1453-2, http://www.springerlink.com/content/0960-3115.

Cabello, J., Fernández, N., Alcaraz-Segura, D., Oyonarte, C., Piñeiro, G., Altesor, A., et al., 2012. The ecosystem functioning dimension in conservation: insights from remote sensing. Biodiversity and Conservation 21 (13), 3287–3305. Available at https://doi.org/10.1007/s10531-012-0370-7.

Calvo, L., Santalla, S., Valbuena, L., Marcos, E., Tárrega, R., Luis-Calabuig, E., 2008. Post-fire natural regeneration of a Pinus pinaster forest in NW Spain. Plant Ecology 197 (1), 81–90. Available at https://doi.org/10.1007/s11258-007-9362-1.

Campbell, M.J., Dennison, P.E., Kerr, K.L., Brewer, S.C., Anderegg, W.R.L., 2021. Scaled biomass estimation in woodland ecosystems: testing the individual and combined capacities of satellite multispectral and lidar data. Remote Sensing of Environment 262. Available at https://doi.org/10.1016/j.rse.2021.112511, www.elsevier.com/inca/publications/store/5/0/5/7/3/3.

Chafer, C.J., Santín, C., Doerr, S.H., 2016. Modelling and quantifying the spatial distribution of post-wildfire ash loads. International Journal of Wildland Fire 25 (2), 249–255. Available at https://doi.org/10.1071/WF15074, http://www.publish.csiro.au/nid/126.htm.

Chergui, B., Fahd, S., Santos, X., Pausas, J.G., 2018. Socioeconomic factors drive fire-regime variability in the Mediterranean Basin. Ecosystems 21 (4), 619–628. Available at https://doi.org/10.1007/s10021-017-0172-6, http://www.springerlink.com/content/1432-9840.

Cimoli, E., Lucieer, A., Malenovský, Z., Woodgate, W., Janoutová, R., Turner, D., et al., 2024. Mapping functional diversity of canopy physiological traits using UAS imaging spectroscopy. Remote Sensing of Environment 302, 113958. Available at https://doi.org/10.1016/j.rse.2023.113958.

Cruz-López, M.I., Manzo-Delgado, L.D.L., Aguirre-Gómez, R., Chuvieco, E., Equihua-Benítez, J.A., 2019. Spatial distribution of forest fire emissions: a case study in three Mexican ecoregions. Remote Sensing 11 (10). Available at https://doi.org/10.3390/rs11101185, https://res.mdpi.com/remotesensing/remotesensing-11-01185/article_deploy/remotesensing-11-01185.pdf?filename=&attachment=1.

Dahlin, K.M., 2016. Spectral diversity area relationships for assessing biodiversity in a wildland-agriculture matrix. Ecological Applications 26 (8), 2756–2766. Available at https://doi.org/10.1002/eap.1390, www.esajournals.org.

Dawson, T.P., Cutler, M.E.J., Brown, C., 2016. The role of remote sensing in the development of SMART indicators for ecosystem services assessment. Biodiversity 17 (4), 136–148. Available at https://doi.org/10.1080/14888386.2016.1246384, http://www.tandfonline.com/loi/tbid20.

De Araujo Barbosa, C.C., Atkinson, P.M., Dearing, J.A., 2015. Remote sensing of ecosystem services: a systematic review. Ecological Indicators 52, 430–443. Available at https://doi.org/10.1016/j.ecolind.2015.01.007, http://www.elsevier.com/locate/ecolind.

Duane, A., Castellnou, M., Brotons, L., 2021. Towards a comprehensive look at global drivers of novel extreme wildfire events. Climatic Change 165 (3-4). Available at https://doi.org/10.1007/s10584-021-03066-4.

Dudley, K.L., Dennison, P.E., Roth, K.L., Roberts, D.A., Coates, A.R., 2015. A multi-temporal spectral library approach for mapping vegetation species across spatial and temporal phenological gradients. Remote Sensing of Environment 167, 121–134. Available at https://doi.org/10.1016/j.rse.2015.05.004, www.elsevier.com/inca/publications/store/5/0/5/7/3/3.

Enright, N.J., Fontaine, J.B., Bowman, D.M.J.S., Bradstock, R.A., Williams, R.J., 2015. Interval squeeze: altered fire regimes and demographic responses interact to threaten woody species persistence as climate changes. Frontiers in Ecology and the Environment 13 (5), 265–272. Available at https://doi.org/10.1890/140231, http://www.esajournals.org/doi/pdf/10.1890/140231.

Fairman, T.A., Bennett, L.T., Nitschke, C.R., 2019. Short-interval wildfires increase likelihood of resprouting failure in fire-tolerant trees. Journal of Environmental Management 231,

59–65. Available at https://doi.org/10.1016/j.jenvman.2018.10.021, https://www.sciencedirect.com/journal/journal-of-environmental-management.

Fassnacht, F.E., Mager, C., Waser, L.T., Kanjir, U., Schäfer, J., Buhvald, A.P., et al., 2025. Forest practitioners' requirements for remote sensing-based canopy height, wood-volume, tree species, and disturbance products. Forestry: An International Journal of Forest Research 98 (2), 233–252. Available at https://doi.org/10.1093/forestry/cpae021.

Fassnacht, F.E., Latifi, H., Stereńczak, K., Modzelewska, A., Lefsky, M., Waser, L.T., et al., 2016. Review of studies on tree species classification from remotely sensed data. Remote Sensing of Environment 186, 64–87. Available at https://doi.org/10.1016/j.rse.2016.08.013, www.elsevier.com/inca/publications/store/5/0/5/7/3/3.

Feng, T., Duncanson, L., Hancock, S., Montesano, P., Skakun, S., Wulder, M.A., et al., 2024. Characterizing fire-induced forest structure and aboveground biomass changes in boreal forests using multitemporal lidar and landsat. IEEE Journal of Selected Topics in Applied Earth Observations and Remote Sensing 17, 10108–10125. Available at https://doi.org/10.1109/JSTARS.2024.3400218, http://ieeexplore.ieee.org/xpl/RecentIssue.jsp?punumber=4609443.

Féret, J.B., Asner, G.P., 2014. Mapping tropical forest canopy diversity using high-fidelity imaging spectroscopy. Ecological Applications 24 (6), 1289–1296. Available at https://doi.org/10.1890/13-1824.1, http://www.esajournals.org/doi/pdf/10.1890/13-1824.1.

Féret, J.B., de Boissieu, F., 2020. BiodivMapR: an r package for α- and β-diversity mapping using remotely sensed images. Methods in Ecology and Evolution 11 (1), 64–70. Available at https://doi.org/10.1111/2041-210X.13310, http://onlinelibrary.wiley.com/journal/10.1111/(ISSN)2041-210X.

Fernández-García, V., Fulé, P.Z., Marcos, E., Calvo, L., 2019. The role of fire frequency and severity on the regeneration of Mediterranean serotinous pines under different environmental conditions. Forest Ecology and Management 444, 59–68. Available at https://doi.org/10.1016/j.foreco.2019.04.040, www.elsevier.com/inca/publications/store/5/0/3/3/1/0.

Fernández-García, V., Marcos, E., Reyes, O., Calvo, L., 2020. Do fire regime attributes affect soil biochemical properties in the same way under different environmental conditions? Forests 11 (3), 274. Available at https://doi.org/10.3390/f11030274.

Fernández-Guisuraga, J.M., Calvo, L., Quintano, C., Fernández-Manso, A., Fernandes, P.M., 2023. Fractional vegetation cover ratio estimated from radiative transfer modeling outperforms spectral indices to assess fire severity in several Mediterranean plant communities. Remote Sensing of Environment 290. Available at https://doi.org/10.1016/j.rse.2023.113542, www.elsevier.com/inca/publications/store/5/0/5/7/3/3.

Fernández-Guisuraga, J.M., Calvo, L., Pérez-Rodríguez, L.A., Suárez-Seoane, S., 2024a. Integrating physical-based models and structure-from-motion photogrammetry to retrieve fire severity by ecosystem strata from very high resolution UAV imagery. Fire 7 (9). Available at https://doi.org/10.3390/fire7090304, www.mdpi.com/journal/fire.

Fernández-Guisuraga, J.M., Suárez-Seoane, S., Fernandes, P.M., Fernández-García, V., Fernández-Manso, A., Quintano, C., et al., 2022a. Pre-fire aboveground biomass, estimated from LiDAR, spectral and field inventory data, as a major driver of burn severity in maritime pine (Pinus pinaster) ecosystems. Forest Ecosystems 9. Available at https://doi.org/10.1016/j.fecs.2022.100022, https://www.journals.elsevier.com/forest-ecology-and-management.

Fernández-Guisuraga, J.M., Calvo, L., Fernandes, P.M., Suárez-Seoane, S., 2022b. Short-term recovery of the aboveground carbon stock in iberian shrublands at the extremes of an environmental gradient and as a function of burn severity. Forests 13 (2). Available at https://doi.org/10.3390/f13020145, https://www.mdpi.com/1999-4907/13/2/145/pdf.

Fernández-Guisuraga, J.M., Marcos, E., Sáenz de Miera, L.E., Ansola, G., Pinto, R., Calvo, L., 2023a. Short-term responses of ecosystem multifunctionality to fire severity are modulated by fire-induced impacts on plant and soil microbial communities. Science of the Total Environment 898. Available at https://doi.org/10.1016/j.scitotenv.2023.165477, www.elsevier.com/locate/scitotenv.

Fernández-Guisuraga, J.M., Marcos, E., Calvo, L., 2023b. The footprint of large wildfires on the multifunctionality of fire-prone pine ecosystems is driven by the interaction of fire regime attributes. Fire Ecology 19 (1). Available at https://doi.org/10.1186/s42408-023-00193-4, https://www.springer.com/journal/42408.

Fernández-Guisuraga, J.M., Suárez-Seoane, S., Calvo, L., 2021. Radiative transfer modeling to measure fire impact and forest engineering resilience at short-term. ISPRS Journal of Photogrammetry and Remote Sensing 176, 30–41. Available at https://doi.org/10.1016/j.isprsjprs.2021.04.002, www.elsevier.com/inca/publications/store/5/0/3/3/4/0.

Fernández-Guisuraga, J.M., Marcos, E., Suárez-Seoane, S., Calvo, L., 2022c. ALOS-2 L-band SAR backscatter data improves the estimation and temporal transferability of wildfire effects on soil properties under different post-fire vegetation responses. Science of the Total Environment 842, 156852. Available at https://doi.org/10.1016/j.scitotenv.2022.156852.

Fernández-Guisuraga, J.M., Calvo, L., Huerta, S., Marcos, E., 2023b. Burnt wood management enhances soil multifunctionality at the medium term after a large wildfire in north-west Spain. International Journal of Wildland Fire 32 (1), 99–110. Available at https://doi.org/10.1071/wf22091.

Fernández-Guisuraga, J.M., González-Pérez, I., Reguero-Vaquero, A., Marcos, E., 2024b. Estimating grassland biophysical parameters in the cantabrian mountains using radiative transfer models in combination with multiple endmember spectral mixture analysis. Remote Sensing 16 (23), 4547. Available at https://doi.org/10.3390/rs16234547.

Fernández-Guisuraga, J.M., Verrelst, J., Calvo, L., Suárez-Seoane, S., 2021. Hybrid inversion of radiative transfer models based on high spatial resolution satellite reflectance data improves fractional vegetation cover retrieval in heterogeneous ecological systems after fire. Remote Sensing of Environment 255, 112304. Available at https://doi.org/10.1016/j.rse.2021.112304.

Fernández-Guisuraga, J.M., Calvo, L., 2023. Fuel build-up promotes an increase in fire severity of reburned areas in fire-prone ecosystems of the western Mediterranean basin. Fire Ecology 19 (1). Available at https://doi.org/10.1186/s42408-023-00232-0.

Fernández-Guisuraga, J.M., Calvo, L., Suárez-Seoane, S., 2022d. Monitoring post-fire neighborhood competition effects on pine saplings under different environmental conditions by means of UAV multispectral data and structure-from-motion photogrammetry. Journal of Environmental Management 305, 114373. Available at https://doi.org/10.1016/j.jenvman.2021.114373.

Franco-Manchón, I., Salo, K., Oria-de-Rueda, J.A., Bonet, J.A., Martín-Pinto, P., 2019. Are wildfires a threat to fungi in European Pinus forests? A case study of boreal and Mediterranean forests. Forests 10 (4). Available at https://doi.org/10.3390/f10040309, https://res.mdpi.com/forests/forests-10-00309/article_deploy/forests-10-00309.pdf?filename=&attachment=1.

Frye, H.A., Aiello-Lammens, M.E., Euston-Brown, D., Jones, C.S., Kilroy Mollmann, H., Merow, C., et al., 2021. Plant spectral diversity as a surrogate for species, functional and phylogenetic diversity across a hyper-diverse biogeographic region. Global Ecology and Biogeography 30 (7), 1403–1417. Available at https://doi.org/10.1111/geb.13306, http://onlinelibrary.wiley.com/journal/10.1111/(ISSN)1466-8238.

Funk, J.L., Larson, J.E., Ames, G.M., Butterfield, B.J., Cavender-Bares, J., Firn, J., et al., 2017. Revisiting the holy grail: using plant functional traits to understand ecological processes. Biological Reviews 92 (2), 1156–1173. Available at https://doi.org/10.1111/brv.12275, http://onlinelibrary.wiley.com/journal/10.1111/(ISSN)1469-185X.

Gamfeldt, L., Snäll, T., Bagchi, R., Jonsson, M., Gustafsson, L., Kjellander, P., et al., 2013. Higher levels of multiple ecosystem services are found in forests with more tree species. Nature Communications 4. Available at https://doi.org/10.1038/ncomms2328.

García-Llamas, P., Geijzendorffer, I.R., García-Nieto, A.P., Calvo, L., Suárez-Seoane, S., Cramer, W., 2019. Impact of land cover change on ecosystem service supply in mountain systems: a case study in the Cantabrian Mountains (NW of Spain). Regional Environmental Change 19 (2), 529–542. Available at https://doi.org/10.1007/s10113-018-1419-2, http://springerlink.metapress.com/app/home/journal.asp?wasp=64cr5a4mwldrxj984xaw&referrer=parent&backto=browsepublicationsresults,451,542;.

García-Llamas, P., Taboada, Á., Fernández-Guisuraga, J.M., Suárez-Seoane, S., Baeza, M.J., Reyes, O., et al., 2024. Plant fire-adaptive traits mediate long-term fire recurrence impact on the potential supply capacity of ecosystem services and their resilience. Journal of Environmental Management 365. Available at https://doi.org/10.1016/j.jenvman.2024.121662, https://www.sciencedirect.com/science/journal/03014797.

Garland, G., Banerjee, S., Edlinger, A., Miranda Oliveira, E., Herzog, C., Wittwer, R., et al., 2021. A closer look at the functions behind ecosystem multifunctionality: a review. Journal of Ecology 109 (2), 600–613. Available at https://doi.org/10.1111/1365-2745.13511, http://onlinelibrary.wiley.com/journal/10.1111/(ISSN)1365-2745.

Gholizadeh, H., Gamon, J.A., Zygielbaum, A.I., Wang, R., Schweiger, A.K., Cavender-Bares, J., 2018. Remote sensing of biodiversity: soil correction and data dimension reduction methods improve assessment of α-diversity (species richness) in prairie ecosystems. Remote Sensing of Environment 206, 240–253. Available at https://doi.org/10.1016/j.rse.2017.12.014, www.elsevier.com/inca/publications/store/5/0/5/7/3/3.

Goldbergs, G., 2023. Comparison of canopy height metrics from airborne laser scanner and aerial/satellite stereo imagery to assess the growing stock of hemiboreal forests. Remote Sensing 15 (6), 1688. Available at https://doi.org/10.3390/rs15061688.

Gonzalez, A., Germain, R.M., Srivastava, D.S., Filotas, E., Dee, L.E., Gravel, D., et al., 2020. Scaling-up biodiversity-ecosystem functioning research. Ecology Letters 23 (4), 757–776. Available at https://doi.org/10.1111/ele.13456, http://onlinelibrary.wiley.com/journal/10.1111/(ISSN)1461-0248.

Griffin, D., Chen, J., Anderson, K., Makar, P., McLinden, C.A., Dammers, E., et al., 2024. Biomass burning CO emissions: exploring insights through TROPOMI-derived emissions and emission coefficients. Atmospheric Chemistry and Physics 24 (17), 10159–10186. Available at https://doi.org/10.5194/acp-24-10159-2024, http://www.atmos-chem-phys.net/volumes_and_issues.html.

Harrison, P.A., Berry, P.M., Simpson, G., Haslett, J.R., Blicharska, M., Bucur, M., et al., 2014. Linkages between biodiversity attributes and ecosystem services: a systematic review. Ecosystem Services 9, 191–203. Available at https://doi.org/10.1016/j.ecoser.2014.05.006.

Hart, S.C., DeLuca, T.H., Newman, G.S., MacKenzie, M.D., Boyle, S.I., 2005. Post-fire vegetative dynamics as drivers of microbial community structure and function in forest soils. Forest Ecology and Management 220 (1-3), 166–184. Available at https://doi.org/10.1016/j.foreco.2005.08.012, www.elsevier.com/inca/publications/store/5/0/3/3/1/0.

Hauser, L.T., Féret, J.B., An Binh, N., van der Windt, N., Sil, Â.F., Timmermans, J., et al., 2021. Towards scalable estimation of plant functional diversity from Sentinel-2: in-situ validation in a heterogeneous (semi-)natural landscape. Remote Sensing of Environment 262. Available at https://doi.org/10.1016/j.rse.2021.112505, www.elsevier.com/inca/publications/store/5/0/5/7/3/3.

Hemes, K.S., Norlen, C.A., Wang, J.A., Goulden, M.L., Field, C.B., 2023. The magnitude and pace of photosynthetic recovery after wildfire in California ecosystems. Proceedings of the National Academy of Sciences of the United States of America 120 (15). Available at https://doi.org/10.1073/pnas.2201954120, https://www.pnas.org/doi/10.1073/pnas.2201954120.

Huerta, S., Marcos, E., Fernández-García, V., Calvo, L., 2022. Short-term effects of burn severity on ecosystem multifunctionality in the northwest Iberian Peninsula. Science of the Total Environment 844, 157193. Available at https://doi.org/10.1016/j.scitotenv.2022.157193.

Hurteau, M.D., Brooks, M.L., 2011. Short- and long-term effects of fire on carbon in US dry temperate forest systems. Bioscience 61 (2), 139–146. Available at https://doi.org/10.1525/bio.2011.61.2.9.

Karavani, A., Boer, M.M., Baudena, M., Colinas, C., Díaz-Sierra, R., Pemán, J., et al., 2018. Fire-induced deforestation in drought-prone Mediterranean forests: drivers and unknowns from leaves to communities. Ecological Monographs 88 (2), 141–169. Available at https://doi.org/10.1002/ecm.1285, http://esajournals.onlinelibrary.wiley.com/hub/journal/10.1002/%28ISSN%291557-7015/issues/.

Kattenborn, T., Maack, J., Faßnacht, F., Enßle, F., Ermert, J., Koch, B., 2015. Mapping forest biomass from space – fusion of hyperspectral EO1-hyperion data and tandem-X and worldView-2 canopy height models. International Journal of Applied Earth Observation and Geoinformation 35, 359–367. Available at https://doi.org/10.1016/j.jag.2014.10.008.

Keeley, J.E., Brennan, T., Pfaff, A.H., 2008. Fire severity and ecosytem responses following crown fires in California shrublands. Ecological Applications 18 (6), 1530–1546. Available at https://doi.org/10.1890/07-0836.1, http://www.esajournals.org/doi/pdf/10.1890/07-0836.1.

Kibler, C.L., Parkinson, A.M.L., Peterson, S.H., Roberts, D.A., D'Antonio, C.M., Meerdink, S.K., et al., 2019. Monitoring post-fire recovery of chaparral and conifer species using field surveys and landsat time series. Remote Sensing 11 (24). Available at https://doi.org/10.3390/rs11242963, https://res.mdpi.com/d_attachment/remotesensing/remotesensing-11-02963/article_deploy/remotesensing-11-02963.pdf.

Lazzeri, G., Frodella, W., Rossi, G., Moretti, S., 2021. Multitemporal mapping of post-fire land cover using multiplatform PRISMA hyperspectral and sentinel-UAV multispectral data: insights from case studies in portugal and Italy. Sensors 21 (12), 3982. Available at https://doi.org/10.3390/s21123982.

Lecina-Diaz, J., Chas-Amil, M.-L., Aquilué, N., Sil, Â., Brotons, L., Regos, A., et al., 2023. Incorporating fire-smartness into agricultural policies reduces suppression costs and ecosystem services damages from wildfires. Journal of Environmental Management 337, 117707. Available at https://doi.org/10.1016/j.jenvman.2023.117707.

Lee, C., Schlemme, C., Murray, J., Unsworth, R., 2015. The cost of climate change: ecosystem services and wildland fires. Ecological Economics 116, 261–269. Available at https://doi.org/10.1016/j.ecolecon.2015.04.020.

Liu, M., Popescu, S., 2022. Estimation of biomass burning emissions by integrating ICESat-2, landsat 8, and sentinel-1 data. Remote Sensing of Environment 280, 113172. Available at https://doi.org/10.1016/j.rse.2022.113172.

Lucas-Borja, M.E., González-Romero, J., Plaza-Álvarez, P.A., Sagra, J., Gómez, M.E., Moya, D., et al., 2019. The impact of straw mulching and salvage logging on post-fire runoff and soil erosion generation under Mediterranean climate conditions. Science of the Total Environment 654, 441–451. Available at https://doi.org/10.1016/j.scitotenv.2018.11.161.

Marcos, B., Gonçalves, J., Alcaraz-Segura, D., Cunha, M., Honrado, J.P., 2023. Assessing the resilience of ecosystem functioning to wildfires using satellite-derived metrics of post-fire trajectories. Remote Sensing of Environment 286, 113441. Available at https://doi.org/10.1016/j.rse.2022.113441.

Masenyama, A., Mutanga, O., Dube, T., Bangira, T., Sibanda, M., Mabhaudhi, T., 2022. A systematic review on the use of remote sensing technologies in quantifying grasslands ecosystem services. GIScience & Remote Sensing 59 (1), 1000–1025. Available at https://doi.org/10.1080/15481603.2022.2088652.

Mataix-Solera, J., Cerdà, A., Arcenegui, V., Jordán, A., Zavala, L.M., 2011. Fire effects on soil aggregation: a review. Earth-Science Reviews 109 (1-2), 44–60. Available at https://doi.org/10.1016/j.earscirev.2011.08.002.

Matasci, G., Hermosilla, T., Wulder, M.A., White, J.C., Coops, N.C., Hobart, G.W., et al., 2018. Large-area mapping of Canadian boreal forest cover, height, biomass and other structural attributes using Landsat composites and lidar plots. Remote Sensing of Environment 209, 90–106. Available at https://doi.org/10.1016/j.rse.2017.12.020, www.elsevier.com/inca/publications/store/5/0/5/7/3/3.

Meyer, S.T., Koch, C., Weisser, W.W., 2015. Towards a standardized rapid ecosystem function assessment (REFA). Trends in Ecology and Evolution 30 (7), 390–397. Available at https://doi.org/10.1016/j.tree.2015.04.006, https://www.sciencedirect.com/journal/trends-in-ecology-and-evolution.

Moghli, A., Santana, V.M., Baeza, M.J., Pastor, E., Soliveres, S., 2022. Fire recurrence and time since last fire interact to determine the supply of multiple ecosystem services by mediterranean forests. Ecosystems 25 (6), 1358–1370. Available at https://doi.org/10.1007/s10021-021-00720-x, http://www.springerlink.com/content/1432-9840.

Morán-Ordóñez, A., Bugter, R., Suárez-Seoane, S., de Luis, E., Calvo, L., 2013. Temporal changes in socio-ecological systems and their impact on ecosystem services at different governance scales: a case study of heathlands. Ecosystems 16 (5), 765–782. Available at https://doi.org/10.1007/s10021-013-9649-0.

Moreira, F., Viedma, O., Arianoutsou, M., Curt, T., Koutsias, N., Rigolot, E., et al., 2011. Landscape – wildfire interactions in southern Europe: Implications for landscape management. Journal of Environmental Management 92 (10), 2389–2402. Available at https://doi.org/10.1016/j.jenvman.2011.06.028.

Nolè, A., Rita, A., Spatola, M.F., Borghetti, M., 2022. Biogeographic variability in wildfire severity and post-fire vegetation recovery across the European forests via remote sensing-derived spectral metrics. Science of The Total Environment 823, 153807. Available at https://doi.org/10.1016/j.scitotenv.2022.153807.

Pacheco-Labrador, J., Migliavacca, M., Ma, X., Mahecha, M., Carvalhais, N., Weber, U., et al., 2022. Challenging the link between functional and spectral diversity with radiative transfer modeling and data. Remote Sensing of Environment 280. Available at https://doi.org/10.1016/j.rse.2022.113170, www.elsevier.com/inca/publications/store/5/0/5/7/3/3.

Palmer, M.W., Earls, P.G., Hoagland, B.W., White, P.S., Wohlgemuth, T., 2002. Quantitative tools for perfecting species lists. Environmetrics (London, Ont.) 13 (2), 121–137. Available at https://doi.org/10.1002/env.516.

Pascual, A., Guerra-Hernández, J., 2023. An integrated assessment of carbon emissions from forest fires beyond impacts on aboveground biomass. A showcase using airborne lidar and GEDI data over a megafire in Spain. Journal of Environmental Management 345, 118709. Available at https://doi.org/10.1016/j.jenvman.2023.118709.

Paula, A., Figueiredo, A., Pulido, F., Castro, P., 2023. Could spaceborne multispectral data be used as SMART proxies' indicators for agroforestry ecosystem services? – A systematic review. Ecological Indicators 156, 111162. Available at https://doi.org/10.1016/j.ecolind.2023.111162.

Paula, S., Ojeda, F., 2009. Belowground starch consumption after recurrent severe disturbance in three resprouter species of the genus Erica. Botany 87 (3), 253–259. Available at https://doi.org/10.1139/b08-134.

Pausas, J.G., Pratt, R.B., Keeley, J.E., Jacobsen, A.L., Ramirez, A.R., Vilagrosa, A., et al., 2016. Towards understanding resprouting at the global scale. New Phytologist 209 (3), 945–954. Available at https://doi.org/10.1111/nph.13644, http://www.wiley.com/bw/editors.asp?ref=0028-646X&site=1.

Pausas, J.G., Keeley, J.E., 2014. Evolutionary ecology of resprouting and seeding in fire-prone ecosystems. New Phytologist 204 (1), 55–65. Available at https://doi.org/10.1111/nph.12921, http://www.wiley.com/bw/editors.asp?ref=0028-646X&site=1.

Pausas, J.G., Keeley, J.E., 2021. Wildfires and global change. Frontiers in Ecology and the Environment 19 (7), 387–395. Available at https://doi.org/10.1002/fee.2359, http://onlinelibrary.wiley.com/journal/10.1002/(ISSN)1540-9309.

Pausas, J.G., Keeley, J.E., 2019. Wildfires as an ecosystem service. Frontiers in Ecology and the Environment 17 (5), 289–295. Available at https://doi.org/10.1002/fee.2044. Available from: http://onlinelibrary.wiley.com/journal/10.1002/(ISSN)1540-9309.

Pereira, P., Bogunovic, I., Zhao, W., Barcelo, D., 2021. Short-term effect of wildfires and prescribed fires on ecosystem services. Current Opinion in Environmental Science & Health 22, 100266. Available at https://doi.org/10.1016/j.coesh.2021.100266.

Pervin, R., Robeson, S.M., MacBean, N., 2022. Fusion of airborne hyperspectral and LiDAR canopy-height data for estimating fractional cover of tall woody plants, herbaceous vegetation, and other soil cover types in a semi-arid savanna ecosystem. International Journal of Remote Sensing 43 (10), 3890–3926. Available at https://doi.org/10.1080/01431161.2022.2105176, https://www.tandfonline.com/loi/tres20.

Pettorelli, N., Laurance, W.F., O'Brien, T.G., Wegmann, M., Nagendra, H., Turner, W., 2014. Satellite remote sensing for applied ecologists: opportunities and challenges. Journal of Applied Ecology 51 (4), 839–848. Available at https://doi.org/10.1111/1365-2664.12261, http://www.blackwellpublishing.com/journals/JPE.

Pettorelli, N., Schulte to Bühne, H., Tulloch, A., Dubois, G., Macinnis-Ng, C., Queirós, A.M., et al., 2018. Satellite remote sensing of ecosystem functions: opportunities, challenges and way forward. Remote Sensing in Ecology and Conservation 4 (2), 71–93. Available at https://doi.org/10.1002/rse2.59, http://onlinelibrary.wiley.com/journal/10.1002/(ISSN)2056-3485.

Pinardi, M., Stroppiana, D., Caroni, R., Parigi, L., Tellina, G., Free, G., et al., 2023. Assessing the impact of wildfires on water quality using satellite remote sensing: the lake Baikal case study. Frontiers in Remote Sensing 4. Available at https://doi.org/10.3389/frsen.2023.1107275.

Puliti, S., Saarela, S., Gobakken, T., Ståhl, G., Næsset, E., 2018. Combining UAV and sentinel-2 auxiliary data for forest growing stock volume estimation through hierarchical model-based inference. Remote Sensing of Environment 204, 485–497. Available at https://doi.org/10.1016/j.rse.2017.10.007.

Quintano, C., Calvo, L., Fernández-Manso, A., Suárez-Seoane, S., Fernandes, P.M., Fernández-Guisuraga, J.M., 2023. First evaluation of fire severity retrieval from PRISMA hyperspectral data. Remote Sensing of Environment 295. Available at https://doi.org/10.1016/j.rse.2023.113670, www.elsevier.com/inca/publications/store/5/0/5/7/3/3.

Raftoyannis, Y., Nocentini, S., Marchi, E., Sainz, R.C., Guemes, C.G., Pilas, I., et al., 2014. Perceptions of forest experts on climate change and fire management in European Mediterranean forests. IForest 7 (1), 33–41. Available at https://doi.org/10.3832/ifor0817-006, http://www.sisef.it/iforest/pdf/?id=ifor0817-006.

Reddy, A.D., Hawbaker, T.J., Wurster, F., Zhu, Z., Ward, S., Newcomb, D., et al., 2015. Quantifying soil carbon loss and uncertainty from a peatland wildfire using multi-temporal LiDAR. Remote Sensing of Environment 170, 306–316. Available at https://doi.org/10.1016/j.rse.2015.09.017.

Rincón, A., Pueyo, J.J., 2010. Effect of fire severity and site slope on diversity and structure of the ectomycorrhizal fungal community associated with post-fire regenerated Pinus pinaster Ait. seedlings. Forest Ecology and Management 260 (3), 361–369. Available at https://doi.org/10.1016/j.foreco.2010.04.028.

Roberts, D.A., Gardner, M., Church, R., Ustin, S., Scheer, G., Green, R.O., 1998. Mapping Chaparral in the Santa Monica mountains using multiple endmember spectral mixture models. Remote Sensing of Environment 65 (3), 267–279. Available at https://doi.org/10.1016/s0034-4257(98)00037-6.

Rocchini, D., Luque, S., Pettorelli, N., Bastin, L., Doktor, D., Faedi, N., et al., 2018. Measuring β-diversity by remote sensing: a challenge for biodiversity monitoring. Methods in Ecology and Evolution 9 (8), 1787–1798. Available at https://doi.org/10.1111/2041-210X.12941, http://onlinelibrary.wiley.com/journal/10.1111/(ISSN)2041-210X.

Rocchini, D., Balkenhol, N., Carter, G.A., Foody, G.M., Gillespie, T.W., He, K.S., et al., 2010. Remotely sensed spectral heterogeneity as a proxy of species diversity: recent advances and open challenges. Ecological Informatics 5 (5), 318–329. Available at https://doi.org/10.1016/j.ecoinf.2010.06.001.

Rodrigues, M., Cunill Camprubí, À., Balaguer-Romano, R., Coco Megía, C.J., Castañares, F., Ruffault, J., et al., 2023. Drivers and implications of the extreme 2022 wildfire season in southwest Europe. Science of the Total Environment 859. Available at https://doi.org/10.1016/j.scitotenv.2022.160320, www.elsevier.com/locate/scitotenv.

Romeo, F., Marziliano, P.A., Turrión, M.B., Muscolo, A., 2020. Short-term effects of different fire severities on soil properties and Pinus halepensis regeneration. Journal of Forestry Research 31 (4), 1271–1282. Available at https://doi.org/10.1007/s11676-019-00884-2, https://www.springer.com/journal/11676.

Rosero-Vlasova, O.A., Vlassova, L., Pérez-Cabello, F., Montorio, R., Nadal-Romero, E., 2018. Modeling soil organic matter and texture from satellite data in areas affected by wildfires and cropland abandonment in Aragón, Northern Spain. Journal of Applied Remote Sensing 12 (4). Available at https://doi.org/10.1117/1.JRS.12.042803, http://www.spie.org/x3636.xml.

Sáenz de Miera, L.E., Pinto, R., Gutierrez-Gonzalez, J.J., Calvo, L., Ansola, G., 2020. Wildfire effects on diversity and composition in soil bacterial communities. Science of the Total Environment 726. Available at https://doi.org/10.1016/j.scitotenv.2020.138636, www.elsevier.com/locate/scitotenv.

Salo, K., Kouki, J., 2018. Severity of forest wildfire had a major influence on early successional ectomycorrhizal macrofungi assemblages, including edible mushrooms. Forest Ecology and Management 415-416, 70–84. Available at https://doi.org/10.1016/j.foreco.2017.12.044.

Sankey, T., Donager, J., McVay, J., Sankey, J.B., 2017. UAV lidar and hyperspectral fusion for forest monitoring in the southwestern USA. Remote Sensing of Environment 195, 30–43.

Available at https://doi.org/10.1016/j.rse.2017.04.007, www.elsevier.com/inca/publications/store/5/0/5/7/3/3.

Schirpke, U., Ghermandi, A., Sinclair, M., Van Berkel, D., Fox, N., Vargas, L., et al., 2023. Emerging technologies for assessing ecosystem services: a synthesis of opportunities and challenges. Ecosystem Services 63, 101558. Available at https://doi.org/10.1016/j.ecoser.2023.101558.

Schmerbeck, J., Kohli, A., Seeland, K., 2015. Ecosystem services and forest fires in India—context and policy implications from a case study in Andhra Pradesh. Forest Policy and Economics 50, 337–346. Available at https://doi.org/10.1016/j.forpol.2014.09.012.

Schumacher, J., Rattay, M., Kirchhöfer, M., Adler, P., Kändler, G., 2019. Combination of multi-temporal sentinel 2 images and aerial image based canopy height models for timber volume modelling. Forests 10 (9), 746. Available at https://doi.org/10.3390/f10090746.

Schweiger, A.K., Laliberté, E., 2022. Plant beta-diversity across biomes captured by imaging spectroscopy. Nature Communications 13 (1). Available at https://doi.org/10.1038/s41467-022-30369-6, http://www.nature.com/ncomms/index.html.

Seidl, R., Rammer, W., Spies, T.A., 2014. Disturbance legacies increase the resilience of forest ecosystem structure, composition, and functioning. Ecological Applications 24 (8), 2063–2077. Available at https://doi.org/10.1890/14-0255.1, http://www.esajournals.org/doi/pdf/10.1890/14-0255.1.

Shi, C., Li, Y., Zhang, T., Wang, H., Wu, L., Suriguga, et al., 2023. Light grazing intensity enhances ecosystem services in semi-arid grasslands through plant trait associations. Journal of Environmental Management 348, 119375. Available at https://doi.org/10.1016/j.jenvman.2023.119375.

Smith-Tripp, S.M., Coops, N.C., Mulverhill, C., White, J.C., Axelson, J., 2024. Landsat assessment of variable spectral recovery linked to post-fire forest structure in dry sub-boreal forests. ISPRS Journal of Photogrammetry and Remote Sensing 208, 121–135. Available at https://doi.org/10.1016/j.isprsjprs.2024.01.008, https://www.sciencedirect.com/science/journal/09242716.

Somers, B., Asner, G.P., 2014. Tree species mapping in tropical forests using multi-temporal imaging spectroscopy: wavelength adaptive spectral mixture analysis. International Journal of Applied Earth Observation and Geoinformation 31 (1), 57–66. Available at https://doi.org/10.1016/j.jag.2014.02.006.

Spasojevic, M.J., Bahlai, C.A., Bradley, B.A., Butterfield, B.J., Tuanmu, M.N., Sistla, S., et al., 2016. Scaling up the diversity-resilience relationship with trait databases and remote sensing data: the recovery of productivity after wildfire. Global Change Biology 22 (4), 1421–1432. Available at https://doi.org/10.1111/gcb.13174, www.blacksci.co.uk/~cgilib/jnlpage.bin?Journal=gcb&File=gcb&Page=aims.

Ståhl, G., Saarela, S., Schnell, S., Holm, S., Breidenbach, J., Healey, S.P., et al., 2016. Use of models in large-area forest surveys: comparing model-assisted, model-based and hybrid estimation. Forest Ecosystems 3 (1). Available at https://doi.org/10.1186/s40663-016-0064-9, https://forestecosyst.springeropen.com/.

Starbuck, C.M., Berrens, R.P., McKee, M., 2006. Simulating changes in forest recreation demand and associated economic impacts due to fire and fuels management activities. Forest Policy and Economics 8 (1), 52–66. Available at https://doi.org/10.1016/j.forpol.2004.05.004.

Supriya, L., 2017. Ecosystems could once bounce back from wildfires. Now, they're being wiped out for good. Science (New York, N.Y.). Available at https://doi.org/10.1126/science.aar8089.

Taboada, A., García-Llamas, P., Fernández-Guisuraga, J.M., Calvo, L., 2021. Wildfires impact on ecosystem service delivery in fire-prone maritime pine-dominated forests. Ecosystem Services 50, 101334. Available at https://doi.org/10.1016/j.ecoser.2021.101334.

Tejedor, E., Benito, G., Serrano-Notivoli, R., González-Rouco, F., Esper, J., et al., 2024. Recent heatwaves as a prelude to climate extremes in the western Mediterranean region. Npj Climate and Atmospheric Science 7 (1). Available at https://doi.org/10.1038/s41612-024-00771-6.

Thom, D., Seidl, R., 2016. Natural disturbance impacts on ecosystem services and biodiversity in temperate and boreal forests. Biological Reviews 91 (3), 760–781. Available at https://doi.org/10.1111/brv.12193.

Ustin, S.L., Gamon, J.A., 2010. Remote sensing of plant functional types. New Phytologist 186 (4), 795–816. Available at https://doi.org/10.1111/j.1469-8137.2010.03284.x.

Van Cleemput, E., Adler, P., Suding, K.N., 2023. Making remote sense of biodiversity: what grassland characteristics make spectral diversity a good proxy for taxonomic diversity? Global Ecology and Biogeography 32 (12), 2177–2188. Available at https://doi.org/10.1111/geb.13759, http://onlinelibrary.wiley.com/journal/10.1111/(ISSN)1466-8238.

Violle, C., Reich, P.B., Pacala, S.W., Enquist, B.J., Kattge, J., 2014. The emergence and promise of functional biogeography. Proceedings of the National Academy of Sciences of the United States of America 111 (38), 13690–13696. Available at https://doi.org/10.1073/pnas.1415442111, http://www.pnas.org/content/111/38/13690.full.pdf+html.

Wang, R., Gamon, J.A., Schweiger, A.K., Cavender-Bares, J., Townsend, P.A., Zygielbaum, A.I., et al., 2018. Influence of species richness, evenness, and composition on optical diversity: a simulation study. Remote Sensing of Environment 211, 218–228. Available at https://doi.org/10.1016/j.rse.2018.04.010, www.elsevier.com/inca/publications/store/5/0/5/7/3/3.

Wang, R., Gamon, J.A., Emmerton, C.A., Li, H., Nestola, E., Pastorello, G.Z., et al., 2016. Integrated analysis of productivity and biodiversity in a southern Alberta prairie. Remote Sensing 8 (3). Available at https://doi.org/10.3390/rs8030214, http://www.mdpi.com/2072-4292/8/3/214/pdf.

Wang, R., Gamon, J.A., 2019. Remote sensing of terrestrial plant biodiversity. Remote Sensing of Environment 231, 111218. Available at https://doi.org/10.1016/j.rse.2019.111218.

Yapp, G., Walker, J., Thackway, R., 2010. Linking vegetation type and condition to ecosystem goods and services. Ecological Complexity 7 (3), 292–301. Available at https://doi.org/10.1016/j.ecocom.2010.04.008.

Zheng, Z., Zeng, Y., Schuman, M.C., Jiang, H., Schmid, B., Schaepman, M.E., et al., 2022. Individual tree-based vs pixel-based approaches to mapping forest functional traits and diversity by remote sensing. International Journal of Applied Earth Observation and Geoinformation 114. Available at https://doi.org/10.1016/j.jag.2022.103074, http://www.elsevier.com/locate/jag.

Quantification of biomass burning emissions based on remote sensing

13

Roberto Fraile, Ana Calvo, and Carlos Blanco-Alegre
Department of Physics, Universidad de León, León, Spain

13.1 Introduction

Biomass burning represents a significant source of emissions into the atmosphere, directly contributing to global processes such as climate change, air quality degradation, and disruption of biogeochemical cycles (Andreae and Merlet, 2001). This practice includes both natural forest fires and controlled or accidental burning for agricultural, forestry, and domestic purposes, especially in tropical, subtropical, and boreal regions. When burned, biomass emits a complex mixture of gases and aerosols (Andreae, 2019; Kaiser et al., 2012; Zhang et al., 2022), including greenhouse gases (such as CO_2, CH_4, and N_2O), air pollutants (such as CO, NO_x, and volatile organic compounds), as well as suspended particulate matter, particularly particles smaller than 10 or 2.5 μm (PM10 and PM2.5, respectively), which are the most harmful to human health (Alonso-Blanco et al., 2018; Tang et al., 2017).

Emissions from biomass burning account for approximately 10%–15% of total global anthropogenic greenhouse gas emissions, according to recent estimates (Crippa et al., 2023; Granier et al., 2011). Additionally, the aerosols emitted affect incoming solar radiation and modify cloud formation, thereby altering the Earth's radiative balance (Harrison et al., 2018; Rosenfeld et al., 2014). At the regional level, these emissions have direct consequences for human health (Adam et al., 2021; Zou et al., 2019), especially in populations that are frequently exposed to smoke episodes.

Against this backdrop, there is a critical need to accurately quantify emissions from biomass burning with high spatial and temporal resolution. However, traditional methods based on national inventories, statistical estimates, or in situ spot measurements face significant limitations (Junghenn Noyes et al., 2020), such as poor spatial coverage, low temporal resolution, and a lack of representativeness for scattered or large-scale events. In this scenario, remote-sensing techniques emerge as fundamental tools for overcoming these limitations, offering an unprecedented perspective on the spatial and temporal dynamics of fires and their associated emissions.

Remote sensing, understood as the remote acquisition of information about the Earth's surface using sensors mounted on satellites, aerial, or terrestrial platforms, enables the monitoring of key parameters for estimating emissions: detection of active burning sites, identification and delimitation of the burned area, estimation of biomass before the fire, and indirect measurement of gas and particle concentrations in the atmosphere. Through the processing and analysis of multispectral, thermal, and

Advances in Remote Sensing for Forest Fire Management. DOI: https://doi.org/10.1016/B978-0-443-41692-7.00017-2
© 2027 Elsevier Ltd. All rights are reserved, including those for text and data mining, AI training, and similar technologies.

hyperspectral images, it is possible to feed emission models based on specific emission factors (EFs) for each type of vegetation and combustion conditions. Likewise, the availability of satellite missions with high spatial and temporal resolution, such as Moderate-resolution imaging spectroradiometer (MODIS), Visible infrared imaging radiometer suite (VIIRS), Sentinel-2, or Tropospheric monitoring instrument (TROPOMI), has significantly improved the accuracy and coverage of estimates.

The integration of remote-sensing data with atmospheric and transport models has significantly expanded the ability to estimate not only the volume of gases and particles emitted, but also their transport, chemical transformation, and radiative effects on both regional and global scales. Initiatives such as the Global fire emissions database (GFED), the Fire Inventory from NCAR (FINN), the Fire radiative power emissions database (FRP-ED), and the FireCCI project of ESA's Climate Change Initiative program are examples of how satellite observation data are integrated to generate global inventories of fire emissions.

Despite these advances, substantial uncertainties remain. The accuracy of estimates still depends heavily on the quality of input data (Van Der Werf et al., 2017), including biomass maps, fuel type characterization, and associated EFs. In addition, atmospheric conditions, such as cloud cover or the presence of dense smoke, can limit the observational capabilities of optical sensors. The spatial heterogeneity of fires, their rapid and nonlinear evolution, and the scarcity of field-based validation measurements continue to hinder the generation of fully reliable results.

This chapter presents a detailed overview of the main remote-sensing techniques employed to quantify atmospheric emissions from biomass burning. It includes current methodologies, integrated models that translate remote observations into quantitative emissions estimates, emissions databases, and case studies focused on validating the use of satellite observations. In addition, the latest technological advances, remaining limitations, and future prospects for improving the accuracy, coverage, and applicability of these tools are discussed.

The aim is to provide a comprehensive and up-to-date overview of the state of the art in this field, highlighting its relevance to climate change science, environmental risk management, public health, and policy-making for mitigation and adaptation.

13.2 Emission quantification via integrated remote-sensing models

13.2.1 Conceptual framework and key variables

Quantifying emissions from biomass burning events poses a significant challenge to atmospheric science and environmental monitoring. Reliable estimates are critical for understanding regional air quality, climate forcing, and biogeochemical cycling. Two main approaches are used to estimate fire emissions: (1) bottom-up models estimate emissions from activity data such as burned area, fuel load, combustion completeness (CC), and EFs, making them highly dependent on input datasets, and (2) top-down models infer emissions from atmospheric observations of trace gases and aerosols combined with inverse transport modeling. While bottom-up approaches are process-based and widely applied in inventories, top-down methods provide independent constraints and are essential for evaluating and adjusting bottom-up estimates (Wiggins et al., 2021).

Within this framework, satellite remote sensing has become a key tool for determining global biomass burning emissions, primarily by deriving essential parameters such as the area burned, the amount of fuel consumed, and the completeness of combustion. These parameters are then used in bottom-up models to calculate the amount of carbon and associated trace gases released during fire events. The most widely used bottom-up approach for quantifying biomass burning emissions related to the carbon fraction is described by the following equation:

$$C = A \cdot B \cdot CC \cdot fc \tag{13.1}$$

where C is the total carbon emitted (g C), A is the burned area (ha), B is the available fuel load (kg dry matter per ha), CC is the combustion completeness (dimensionless, 0–1), and fc is the fraction of carbon in dry biomass (typically 0.45–0.55 g C kg^{-1} dry matter) (Liu and Popescu, 2022).

In this approach, the estimation of biomass burning emissions is related to the carbon fraction, and each variable in the equation is derived from a combination of satellite remote-sensing products, empirical models, and field observations. The accuracy of the final carbon estimate critically depends on the robustness of each component. Below, the main variables and their current methods of estimation are summarized:

- Burned area (A) refers to the total surface affected by fire, generally expressed in hectares, and is most often estimated from multispectral satellite data. This estimation relies on change detection methods that evaluate reflectance differences before and after a fire event. Nevertheless, several sources of uncertainty can affect accuracy, such as the presence of clouds, smoke plumes, or the challenge of identifying fires that are small in extent or low in intensity (Giglio et al., 2018). Among the most frequently applied datasets are: (1) the MODIS Burned Area Product (MCD64A1), which delivers global daily coverage at 500 m resolution through the combination of reflectance and thermal anomaly information; (2) Sentinel-2 multispectral instrument (MSI) and Landsat-8 9 OLI, both providing finer spatial resolutions between 10 and 30 m, which are particularly useful for delineating burned patches in heterogeneous or agricultural regions; and (3) VIIRS, with an intermediate resolution of 375 m, which complements MODIS in detecting active fire perimeters.
- Fuel load (B) corresponds to the quantity of dry biomass that can potentially burn per unit of surface area, usually expressed in kilograms of dry matter per hectare. Its estimation relies on several data sources and approaches: (1) land cover classifications (e.g., MODIS Land Cover Type, Copernicus Global Land Service) assign vegetation types to characteristic fuel load values obtained from field observations; (2) biomass mapping products, generated through radar (Sentinel-1), light detection and ranging (LiDAR) (ICESat-2), or integrated datasets such as globbiomass—a global project that maps above-ground biomass (AGB) at ~100 m resolution to support research on climate, carbon balance, and forest management—or the Global Ecosystem Dynamics Investigation (GEDI) mission, which delivers high-resolution LiDAR profiles of forest structure to improve AGB estimates and carbon accounting; (3) empirical approaches, which rely on vegetation indices such as normalized

difference vegetation index (NDVI) and other productivity metrics to model spatial variations in fuel availability; and (4) alternative methods rely on dynamic global vegetation models to represent AGB and its temporal variability (Santoro et al., 2021).

- *CC* is defined as the proportion of available fuel consumed during a fire. Values range between 0 (no burning) and 1 (complete burning). *CC* is influenced by factors such as fuel moisture, fire intensity, and whether the fire occurs at ground level or in the canopy. To constrain *CC* values across ecosystems, data from experimental burns, postfire field assessments, and laboratory combustion studies are often employed (van Leeuwen et al., 2014). Numerous investigations have confirmed a relationship between fire severity, typically estimated using remote-sensing metrics such as the differenced normalized burn ratio (dNBR), and combustion efficiency, that is, the proportion of living biomass consumed in fire events. For instance, Balde et al. (2023) identified strong correlations between Landsat-derived dNBR and ground-based combustion efficiency in Mediterranean forests. Their results showed values ranging from 0.44 to 0.81 in coniferous forests and from 0.64 to 0.86 in broadleaf forests, depending on the severity category. These outcomes emphasize the direct effect of fire severity on *CC* and demonstrate the usefulness of satellite-based severity indicators for refining bottom-up emission models. There are several methodologies for obtaining *CC* values, the most notable of which are: (1) biome-specific parameters: derived from experimental burning trials and reported literature values, with typical ranges of 0.85–0.9 for savanna grasslands and 0.3–0.5 for peatlands; (2) remote-sensing indicators: thermal sensor products (e.g., MODIS, VIIRS, and Geostationary Operational Environmental Satellite [GOES]) provide fire radiative power (FRP), which can be linked to combustion strength. This enables near-real-time estimates of CO_2 and CO (Freeborn et al., 2022; Wooster et al., 2005); and (3) model-based methods integrate structural fuel data from LiDAR and fire severity estimates from NIR/SWIR reflectance and emission ratios to generate dynamic *CC* values.
- Carbon fraction (*fc*) is the proportion of dry biomass mass that is carbon, generally assumed to be within the range of 0.45–0.55. This value varies slightly depending on the type of fuel: around 0.50–0.52 for woody biomass; closer to 0.45–0.48 for grasses and crop residues; and up to 0.55 in some cases for peat and organic soils. These values are obtained from laboratory analysis of biomass samples, and although they are often treated as fixed, more advanced emission models may vary *fc* according to land cover or ecosystem type (Liu and Popescu, 2022).

Although quantifying the total amount of carbon released during biomass burning is fundamental, the composition of the emitted trace gases and aerosols is what primarily drives atmospheric and climate impacts. To estimate the release of specific chemical species, bottom-up models utilize EFs, which represent the amount of a given compound emitted per unit of dry biomass consumed (typically expressed in grams of the compound per kilogram of dry matter). Therefore, gaseous and particulate emissions (E_i) can be calculated as follows:

$$E_i = C \cdot \frac{EF_i}{M_C} \tag{13.2}$$

where E_i is the emission of species i (g), C is the total carbon emitted (g C, derived from Eq. 13.1), EF_i is the emission factor of species i (g species kg^{-1} dry matter burned), and M_C is the mass of carbon in g kg^{-1} of dry biomass consumed (Akagi et al., 2011).

EFs can vary substantially depending on the type of ecosystem, the condition of the fuel, and the combustion phase (flaming versus smoldering). For instance, fires resulting from tropical deforestation typically exhibit high CO_2 and NO_x emissions due to intense flaming, whereas peatland and boreal fires produce high levels of CO and CH_4 under smoldering conditions (Van Der Werf et al., 2017). Additionally, Vicente et al. (2013) documented a wide variety of smoke particle constituents, further illustrating the complexity of EF characterization across fire types. Updated compilations, such as the work of Andreae (2019), provide biome-specific EF datasets covering a wide range of gases (CO_2, CO, CH_4, nonmethane volatile organic compounds, NO_x, SO_2), as well as aerosol components (organic carbon [OC], black carbon [BC], $PM_{2.5}$).

Although bottom-up methods remain the standard approach for estimating biomass burning emissions, top-down approaches are gaining importance as an independent means of validation. The use of satellite observations has increased to estimate EF with instruments such as TROPOMI, Ozone Monitoring Instrument (OMI), and Infrared Atmospheric Sounding Interferometer (IASI) being used to measure trace gas concentrations in smoke plumes. These instruments enable the direct measurement of enhancements in CO, NO_2, SO_2, and formaldehyde over regions affected by fires (Fu et al., 2022; Gonzalez Abad et al., 2019). These datasets, combined with plume dispersion modeling, can refine regional emission estimates and provide independent inventory validation. Despite these advances, there remains high uncertainty, which makes it essential to integrate satellite retrievals, field-based EF databases, and dynamic fire models to improve future fire emission inventories.

13.2.2 Global inventories and intercomparison of emission datasets

As can be seen, most calculations are derived from datasets, with numerous global emission datasets being produced operationally using satellite inputs, notably. The main databases are the following: GFED, FINN, Global Fire Assimilation System (GFAS), Fire Energetics and Emissions Research (FEER), and Quick Fire Emissions Dataset (QFED). Despite the extensive use of these inventories, substantial discrepancies persist in their estimates of key atmospheric pollutants, such as OC and BC. For instance, emissions reported for the same region and period can vary by a factor of up to four (Van Der Werf et al., 2017; Wiedinmyer et al., 2011).

Recent studies have continued to highlight and quantify these divergences. Pan et al. (2020) conducted a comprehensive intercomparison of six global inventories—GFED3.1/4s, FINN1.5, GFAS1.2, FEER1.0, and QFED2.4—and reported substantial variability in both OC and BC estimates, driven mainly by differences in burned area detection methods, satellite inputs, and EFs. As demonstrated by Wiedinmyer et al. (2023), the FINN2.5 model typically indicates high levels of CO and CO_2 emissions, accompanied by lower levels of BC, in contrast to the QFED2.5 model. This discrepancy can be attributed to variations in the assumptions concerning combustion processes and the allocation of fuel types. Hua et al. (2024) conducted a further examination of the impact of these discrepancies on modeled atmospheric pollutants.

The study demonstrated that emissions from FINN2.5, GFED4s, and QFED2.5 produce considerable variability in surface pollutant concentrations simulated by CESM2-CAM6 (a widely used climate model of NCAR), with regional biases of up to 60% for OC and BC identified. Moreover, Anderson et al. (2024) conducted a comparative analysis of the Global Forest Fire Emissions Prediction System (GFFEPS) with multiple inventories, revealing substantial variations in carbon emission estimates. Notably, FINN2.5 reported estimates reaching up to 4231 Tg C yr^{-1}, which exceeds the upper bound estimated by GFFEPS by more than double. These discrepancies not only compromise cross-model comparability but also propagate uncertainties in regional air quality and climate impact assessments, emphasizing the urgent need for harmonized inventory methodologies and improved observational constraints.

Furthermore, several studies have evaluated the performance of fire emission inventories using atmospheric observations and model-based diagnostics. Independent performance evaluations employing aerosol optical depth (AOD) observations from Multiangle imaging spectro radiometer (MISR) and Aerosol robotic network (AERONET) have demonstrated that QFED2.4 generally exhibits a stronger correlation with observed AOD, followed by FEER1.0. However, it has been observed that all inventories tend to underestimate aerosol loading during periods of intense boreal and tropical fire seasons (Petrenko et al., 2017; Tosca et al., 2013).

Finally, a critical limitation of current inventories lies in the propagation of uncertainties across variables. Errors in burned area detection, fuel load estimation, CC, and EFs are multiplicative rather than additive, often resulting in order-of-magnitude differences between inventories. Explicit quantification of error propagation is therefore essential to evaluate the robustness of emission datasets. In this context, the Aerosol Comparisons between Observations and Models (AeroCom) initiative (https://aerocom.met.no/) has conducted multimodel intercomparisons that include fire emissions from various inventories and assess how they propagate into aerosol loading simulations. These comparisons reveal that model divergence is often driven not only by emissions but also by assumptions about: (1) OA/OC (organic aerosol/OC) conversion ratios; (2) aerosol optical properties and mass extinction coefficients; and (3) dry and wet deposition rates. Such parameters exert a substantial influence on the concentrations of simulated AOD and surface particulate matter. Experiments conducted by AeroCom indicate that enhanced constraints on the microphysical properties of aerosols, in conjunction with improved fire emissions, have the potential to reduce the substantial variability observed in model outputs (Kipling et al., 2016). In a recent study, Petrenko et al. (2025) utilized updated MODIS AOD observations and AeroCom Phase III simulations to reevaluate inventory performance, revealing a persistent underestimation of AOD in tropical regions and substantial variability in modeled aerosol distributions, contingent on the selected emission dataset. These biases have substantial implications for air quality modeling and radiative forcing assessments, reinforcing the need for better calibration using observed aerosol distributions.

13.3 Verification with on-site measurements: case studies

Wildfire emission estimates are essential inputs for atmospheric models, but their validation is hindered by the scarcity of high-quality in situ data. In response, recent years have seen a growing number of studies directly comparing satellite-derived emissions with field measurements, either from ground-based stations or airborne platforms. These efforts are crucial for improving the reliability of satellite products and refining emission inventories.

For instance, Stockwell et al. (2022) developed a novel method for estimating carbon emissions in wildfire smoke plumes. This method integrates LiDAR-based aerosol extinction profiles with in situ carbon measurements from wildfires in the western United States. This approach enabled direct comparison with satellite-based inventories and demonstrated strong agreement with the Fuel2Fire inventory (r^2=0.93). While other commonly used inventories also correlated well, they showed significant over and underestimations. Airborne carbon monoxide measurements closely matched TROPOMI satellite estimates. EFs for primary gases and aerosols were derived from satellite-based FRP, including several compounds that had not previously been reported for temperate forest fuels.

Building on previous validation efforts, Wiggins et al. (2021) compared methodologies for estimating wildfire smoke emissions. Traditionally, emissions are calculated using either bottom-up or top-down models. Using high-resolution data from the NASA/NOAA fire influence on regional to global environments and air quality (FIREX-AQ) campaign, which combined in situ and remote airborne measurements from wildfires in the western United States and prescribed/agricultural fires in the southeast, they evaluated these traditional methods alongside a new in situ-based approach. The study revealed that global, low-resolution inventories, such as GFED and FEER, exhibited systematic biases and weaker correlations compared to higher-resolution datasets, including in situ data, Fuel2Fire, and high spectral resolution lidar-geostationary operational environmental satellites (HSRL-GOES). Key discrepancies were linked to uncertainties in fuel load estimates, combustion efficiency, EFs, and the temporal resolution of satellite data.

In Europe, Castagna et al. (2023) compared wildfire emission estimates derived from satellite data with ground-based measurements in Calabria, Italy. They found that the GFED4s satellite data overestimated the burned area by around 18%, most likely because cropland fires were detected remotely but not reported locally (Turco et al., 2019). Although temporal trends aligned well, remote sensing significantly underestimated dry matter and emissions of CO, PM2.5, and BC by between 66% and 97%. Key limitations included low spatial resolution and difficulty in detecting fires in coastal areas. The study also found that outdated fuel load data or land use changes had a minimal impact on estimates. It emphasized the importance of carefully validating satellite data before incorporating it into atmospheric models.

Similarly, Galizia et al. (2021) evaluated the accuracy of wildfire datasets obtained through remote sensing in the southwestern Mediterranean region. They compared three satellite-derived global fire-event datasets—Fire Atlas, which reconstructs individual fire events and their spatiotemporal progression; FRY, a database of fire-patch functional traits describing burned-patch morphology; and GlobFire, a global wildfire event database generated by clustering satellite burned-area detections—with a harmonized ground-based dataset compiled from fire agency records. Although remote sensing effectively captured the variability of wildfires over time, it significantly underestimated the area burned and the number of fires—by 38% and 96%, respectively—particularly for small fires. Agreement with ground data improved notably for fires exceeding 100 ha, suggesting that these datasets are more reliable for analyzing large-scale events or aggregated time periods than for smaller fires at regional scales.

In tropical regions, Reddington et al. (2019) analyzed particulate emissions from Amazonian biomass burning by combining surface, aircraft, and satellite data collected during the South American biomass burning analysis (SAMBBA) campaign. They compared these observations with those from the aerosol model Global Model of Aerosol Processes (GLOMAP), incorporating three satellite-derived fire emission datasets (FINN, GFED, and GFAS) to explore emission uncertainties. Despite accurately reproducing vertical aerosol mass profiles, GLOMAP generally underestimated AOD relative to satellite and ground data. This suggests that AOD underestimation primarily stems from calculation uncertainties, particularly those related to water uptake and relative humidity, with errors reaching ~90%. The authors caution against relying too heavily on AOD comparisons to constrain fire-related particulate emissions.

Turco et al. (2019) compared satellite-derived burned area products (FireCCI51, MCD64A1, and MCD45A1) with ground-based gridded records from the European Forest Fire Information System (EFFIS) across the Mediterranean region. Their analysis assessed statistical discrepancies at various spatial and temporal scales. Temporal correlations were moderate to high (0.5–0.6), particularly at coarser spatial resolutions (≥1.0 degree). However, absolute burned area estimates differed significantly between products. Filtering out fires detected in urban and agricultural areas significantly improved the consistency between the satellite and ground data. Despite quantitative differences, spatial fire patterns were broadly consistent, with correlation improving as resolution decreased. FireCCI51 showed the best performance, supporting its use in fire monitoring. The study recommends accounting for uncertainties in satellite-based estimates, particularly in regional or global assessments, and stresses the importance of filtering out human-influenced land covers to improve data reliability.

These cases demonstrate the necessity of enhancing the accuracy and reliability of forest fire emission estimates. This objective requires the integration of satellite remote sensing, which provides global coverage, with field campaigns that supply detailed validation data. The combined use of these elements serves to reduce uncertainties and enhance the representation of fire emissions in atmospheric models.

13.4 Sources of uncertainty and recent methodological advances

The formulations and inventories previously discussed are used in many satellite-derived global fire products, which often exhibit varying levels of uncertainty and resolution, depending on the capabilities of the sensors and the postprocessing algorithms used. Despite the conceptual simplicity of the models, multiple sources of uncertainty significantly affect emission estimates.

- Burned area misclassification: identifying burned areas via satellite is prone to errors of commission and omission, particularly in regions with cloud cover or frequent agricultural fires (Giglio et al., 2018; Roy et al., 2019).
- Fuel load represents the next key source of uncertainty, as it depends on indirect biomass data. Biomass availability varies greatly depending on vegetation type, season, and prior disturbance history. Static, land-cover-based models often fail to capture variability in fuel loads from one year to the next, particularly in savannas or peatlands. Dynamic vegetation models and LiDAR-informed fuel mapping offer improvements (D'Este et al., 2021).
- CC variability: the proportion of biomass that is combusted depends on fire intensity, moisture content, and vegetation structure. For instance, surface fires in boreal forests often leave significant amounts of unburned organic material, whereas crown fires can consume nearly all the biomass available (van Leeuwen et al., 2014).
- Sensor detection limitations: satellite detection of active fires and burn scars may be hindered by canopy cover, smoke, and the observation angle. This can result in small or understory fires escaping detection (Johnston et al., 2018).
- The application of global-average EFs can lead to significant biases when extrapolated to heterogeneous landscapes. Regional field campaigns, such as those conducted in the Amazon and Central Africa, have demonstrated that variations in combustion efficiency and fuel composition can alter EF values by a factor of up to two (Akagi et al., 2011; Andreae, 2019).

In recent years, LiDAR-based approaches have become increasingly relevant in reducing uncertainties in burned area and biomass estimation, offering significant improvements over traditional optical-only methods. Instruments such as NASA's ICESat-2 utilize photon-counting LiDAR technology to provide fine-resolution measurements of canopy height, vertical structure, and ground elevation. These parameters enable the quantification of AGB before and after fire events, providing a more direct, spatially explicit assessment of fuel consumption. Integrating LiDAR-derived structural metrics with optical sensors, such as Landsat and Sentinel-2, and synthetic aperture radar (SAR) systems, like Sentinel-1, enhances the detection of burn severity and postfire vegetation loss, even under cloudy or smoky conditions (Gale et al., 2022). Moreover, recent advances in artificial intelligence and machine learning algorithms are increasingly applied to remote-sensing datasets to improve burned area mapping, fuel load prediction, and emission estimates, while unmanned aerial systems (UAS) provide high-resolution, flexible monitoring capabilities that complement satellite observations, particularly at local and regional scales (Boroujeni et al., 2024; Melitha et al., 2025).

The combination of active (LiDAR and SAR) and passive (optical) remote sensing has been shown to reduce the uncertainty in biomass loss estimates to below 5%–10%. This represents a significant improvement over MODIS-based burned area products, which typically exhibit higher omission errors and coarser resolution (Chen et al., 2022; Chuvieco et al., 2020). These integrated approaches enhance the spatial resolution of emission estimates, enabling a more accurate representation of fire impacts across heterogeneous landscapes. Furthermore, they refine CC parameters in bottom-up models by providing detailed insights into the amount and structure of biomass available for combustion. As these technologies become more widely available and computationally accessible, their role in global fire emission inventories is expected to grow significantly.

In Mediterranean ecosystems, prefire LiDAR metrics have also been shown to be effective in predicting fire severity. As demonstrated by Fernández-Guisuraga et al. (2021), canopy height and vertical density are strongly associated with severity levels, achieving up to 82% classification accuracy in pine forests. More recently, Fernández-Guisuraga and Calvo (2023) found that increased understory fuel density, measured via LiDAR, correlates with higher fire severity in maritime pine and holm oak stands during reburns. These findings lend support to the use of LiDAR as a pivotal instrument in modeling fire behavior and enhancing emission estimates at fine spatial scales.

Despite the advances brought about by satellite-based models and LiDAR in terms of emission estimates from biomass burning, several limitations persist: (1) the underdetection of small or low-severity fires; (2) uncertainties in fuel load linked to the reliance on static vegetation maps; (3) difficulties in constraining *CC* remotely, and; (4) discrepancies among global inventories due to input data and modeling assumptions. As outlined in the preceding sections, these factors continue to constrain the accuracy and comparability of fire emission inventories.

13.5 Future prospects

The limitations discussed in the previous section present challenges and opportunities for improving fire emission estimation. Advances in satellite technologies, modeling frameworks, and ground validation could enhance accuracy, spatial and temporal coverage, and applicability across different ecological contexts.

A central priority is the integration of multiple satellite data sources. Combining optical, thermal, SAR, LiDAR, and hyperspectral observations can improve the characterization of preburn biomass, burned area delineation, and CC. Constellations of small satellites (e.g., CubeSats) and UAS are expanding monitoring capacity, and new missions such as NASA's NISAR and ICESat-2 will provide critical insights into vegetation structure and fire severity. These advances will help to address the ongoing issue of small fires being underdetected in fragmented and agricultural landscapes.

Estimates of fuel load and composition also require refinement. The current reliance on generalized datasets and average EFs introduces significant uncertainty. Improving model realism will require the development of detailed geospatial biomass databases and the incorporation of phenological and fuel moisture data from hyperspectral and microwave sensors. Field campaigns will remain essential, particularly in ecosystems such as peatlands, tropical dry forests, and boreal understories, where EFs are poorly understood. Coordinated, standardized observations using portable spectrometers, airborne LiDAR, and flux towers are critical for calibration and validation.

Another promising direction is the assimilation of multisensor observations into dynamic vegetation and fire models. Hybrid approaches combining bottom-up estimates with top-down constraints can reconcile empirical and process-based methods. Machine learning can enhance the detection, segmentation, and predictive modeling of these factors, enabling the development of adaptive, near-real-time applications for air quality forecasting, early warning systems, and climate monitoring. At the inventory level, harmonization remains a pressing task. Global databases such as GFED, FINN, QFED, and GFAS still differ by up to a factor of four in carbonaceous emissions, undermining the reliability of climate and air quality simulations. Future intercomparison projects, particularly within AeroCom, should incorporate uncertainties from both emissions and aerosol microphysics. Transitioning toward ensemble-based approaches—where multiple inventories are combined and weighted according to their observational performance—offers a pathway to more robust near-real-time products.

In summary, the future of fire emission monitoring is contingent on the convergence of technological innovation, AI-driven analytics, hybrid modeling, and coordinated international collaboration. Advancements in sensing technologies, ground validation processes, and harmonized inventories are poised to mitigate uncertainties and enhance the reliability of global datasets. These developments will directly benefit air quality monitoring, climate change mitigation, and the sustainable management of fire-prone ecosystems.

References

Adam, M.G., Tran, P.T.M., Bolan, N., Balasubramanian, R., 2021. Biomass burning-derived airborne particulate matter in Southeast Asia: a critical review. Journal of Hazardous Materials 407. Available at https://doi.org/10.1016/j.jhazmat.2020.124760, www.elsevier.com/locate/jhazmat.

Akagi, S.K., Yokelson, R.J., Wiedinmyer, C., Alvarado, M.J., Reid, J.S., Karl, T., et al., 2011. Emission factors for open and domestic biomass burning for use in atmospheric models. Atmospheric Chemistry and Physics 11 (9), 4039–4072. Available at https://doi.org/10.5194/acp-11-4039-2011.

Alonso-Blanco, E., Castro, A., Calvo, A.I., Pont, V., Mallet, M., Fraile, R., 2018. Wildfire smoke plumes transport under a subsidence inversion: climate and health implications in a distant urban area. Science of the Total Environment 619-620, 988–1002. Available at https://doi.org/10.1016/j.scitotenv.2017.11.142, www.elsevier.com/locate/scitotenv.

Anderson, K., Chen, J., Englefield, P., Griffin, D., Makar, P.A., Thompson, D., 2024. The global forest fire emissions prediction system version 1.0. Geoscientific Model Development 17 (21), 7713–7749. Available at https://doi.org/10.5194/gmd-17-7713-2024, http://www.geosci-model-dev.net/volumes_and_issues.html.

Andreae, M.O., Merlet, P., 2001. Emission of trace gases and aerosols from biomass burning. Global Biogeochemical Cycles 15 (4), 955–966. Available at https://doi.org/10.1029/2000gb001382.

Andreae, M.O., 2019. Emission of trace gases and aerosols from biomass burning - an updated assessment. Atmospheric Chemistry and Physics 19 (13), 8523–8546. Available at https://

doi.org/10.5194/acp-19-8523-2019, http://www.atmos-chem-phys.net/volumes_and_issues.html.

Balde, B., Vega-Garcia, C., Gelabert, P.J., Ameztegui, A., Rodrigues, M., 2023. The relationship between fire severity and burning efficiency for estimating wildfire emissions in Mediterranean forests. Journal of Forestry Research 34 (5), 1195–1206. Available at https://doi.org/10.1007/s11676-023-01599-1, https://www.springer.com/journal/11676.

Boroujeni, S.P.H., Razi, A., Khoshdel, S., Afghah, F., Coen, J.L., O'Neill, L., et al., 2024. A comprehensive survey of research towards AI-enabled unmanned aerial systems in pre-, active-, and post-wildfire management. Information Fusion 108. Available at https://doi.org/10.1016/j.inffus.2024.102369, https://www.sciencedirect.com/science/journal/15662535.

Castagna, J., Senatore, A., Pellis, G., Vitullo, M., Bencardino, M., Mendicino, G., 2023. Uncertainty assessment of remote sensing- and ground-based methods to estimate wildfire emissions: a case study in Calabria region (Italy). Air Quality, Atmosphere & Health 16 (4), 705–717. Available at https://doi.org/10.1007/s11869-022-01300-1.

Chen, L., Ren, C., Bao, G., Zhang, B., Wang, Z., Liu, M., et al., 2022. Improved object-based estimation of forest aboveground biomass by integrating LiDAR data from GEDI and ICESat-2 with multi-sensor images in a heterogeneous mountainous region. Remote Sensing 14 (12), 2743. Available at https://doi.org/10.3390/rs14122743.

Chuvieco, E., Aguado, I., Salas, J., García, M., Yebra, M., Oliva, P., 2020. Satellite remote sensing contributions to wildland fire science and management. Current Forestry Reports 6 (2), 81–96. Available at https://doi.org/10.1007/s40725-020-00116-5.

Crippa, M., Guizzardi, D., Schaaf, E., Monforti-Ferrario, F., Quadrelli, R., Martin, A.R., et al., 2023. GHG Emissions of all World Countries – 2023. Available at https://doi.org/10.2760/953322.

D'Este, M., Elia, M., Giannico, V., Spano, G., Lafortezza, R., Sanesi, G., 2021. Machine learning techniques for fine dead fuel load estimation using multi-source remote sensing data. Remote Sensing 13 (9), 1658. Available at https://doi.org/10.3390/rs13091658.

Fernández-Guisuraga, J.M., Suárez-Seoane, S., García-Llamas, P., Calvo, L., 2021. Vegetation structure parameters determine high burn severity likelihood in different ecosystem types: a case study in a burned Mediterranean landscape. Journal of Environmental Management 288. Available at https://doi.org/10.1016/j.jenvman.2021.112462, https://www.sciencedirect.com/science/journal/03014797.

Fernández-Guisuraga, J.M., Calvo, L., 2023. Fuel build-up promotes an increase in fire severity of reburned areas in fire-prone ecosystems of the western Mediterranean Basin. Fire Ecology 19 (1). Available at https://doi.org/10.1186/s42408-023-00232-0, https://www.springer.com/journal/42408.

Freeborn, P.H., Jolly, W.M., Cochrane, M.A., Roberts, G., 2022. Large wildfire driven increases in nighttime fire activity observed across CONUS from 2003–2020. Remote Sensing of Environment 268. Available at https://doi.org/10.1016/j.rse.2021.112777.

Fu, Y., Li, R., Hu, J., Wang, Y., Duan, J., 2022. Investigating the impacts of satellite fire observation accuracy on the top-down nitrogen oxides emission estimation in northeastern Asia. Environment International 169, 107498. Available at https://doi.org/10.1016/j.envint.2022.107498.

Gale, M.G., Cary, G.J., Yebra, M., Leavesley, A.J., Van Dijk, A.I.J.M., 2022. Comparison of contrasting optical and LiDAR fire severity remote sensing methods in a heterogeneous forested landscape in south-eastern Australia. International Journal of Remote Sensing 43 (7), 2559–2580. Available at https://doi.org/10.1080/01431161.2022.2064197, https://www.tandfonline.com/loi/tres20.

Galizia, L.F., Curt, T., Barbero, R., Rodrigues, M., 2021. Assessing the accuracy of remotely sensed fire datasets across the southwestern Mediterranean Basin. Natural Hazards and Earth System Sciences 21 (1), 73–86. Available at https://doi.org/10.5194/nhess-21-73-2021.

Giglio, L., Boschetti, L., Roy, D.P., Humber, M.L., Justice, C.O., 2018. The collection 6 MODIS burned area mapping algorithm and product. Remote Sensing of Environment 217, 72–85. Available at https://doi.org/10.1016/j.rse.2018.08.005, www.elsevier.com/inca/publications/store/5/0/5/7/3/3.

Gonzalez Abad, G., Souri, A.H., Bak, J., Chance, K., Flynn, L.E., Krotkov, N.A., et al., 2019. Five decades observing Earth's atmospheric trace gases using ultraviolet and visible backscatter solar radiation from space. Journal of Quantitative Spectroscopy and Radiative Transfer 238. Available at https://doi.org/10.1016/j.jqsrt.2019.04.030, www.elsevier.com/inca/publications/store/2/7/2/.

Granier, C., Bessagnet, B., Bond, T., D'Angiola, A., van der Gon, H.D., Frost, G.J., et al., 2011. Evolution of anthropogenic and biomass burning emissions of air pollutants at global and regional scales during the 1980-2010 period. Climatic Change 109 (1), 163–190. Available at https://doi.org/10.1007/s10584-011-0154-1.

Harrison, S.P., Bartlein, P.J., Brovkin, V., Houweling, S., Kloster, S., Prentice, I.C., 2018. The biomass burning contribution to climate–carbon-cycle feedback. Earth System Dynamics 9 (2), 663–677. Available at https://doi.org/10.5194/esd-9-663-2018.

Hua, W., Lou, S., Huang, X., Xue, L., Ding, K., Wang, Z., et al., 2024. Diagnosing uncertainties in global biomass burning emission inventories and their impact on modeled air pollutants. Atmospheric Chemistry and Physics 24 (11), 6787–6807. Available at https://doi.org/10.5194/acp-24-6787-2024.

Johnston, J.M., Johnston, L.M., Wooster, M.J., Brookes, A., McFayden, C., Cantin, A.S., 2018. Satellite detection limitations of sub-canopy smouldering wildfires in the north american boreal forest. Fire 1 (2), 1–16. Available at https://doi.org/10.3390/fire1020028, https://www.mdpi.com/2571-6255/1/2/28/pdf.

Junghenn Noyes, K., Kahn, R., Sedlacek, A., Kleinman, L., Limbacher, J., Li, Z., 2020. Wildfire smoke particle properties and evolution, from space-based multi-angle imaging. Remote Sensing 12 (5). Available at https://doi.org/10.3390/rs12050769 769–769.

Kaiser, J.W., Heil, A., Andreae, M.O., Benedetti, A., Chubarova, N., Jones, L., et al., 2012. Biomass burning emissions estimated with a global fire assimilation system based on observed fire radiative power. Biogeosciences 9 (1), 527–554. Available at https://doi.org/10.5194/bg-9-527-2012.

Kipling, Z., Stier, P., Johnson, C.E., Mann, G.W., Bellouin, N., Bauer, S.E., et al., 2016. What controls the vertical distribution of aerosol? Relationships between process sensitivity in HadGEM3-UKCA and inter-model variation from AeroCom Phase II. Atmospheric Chemistry and Physics 16 (4), 2221–2241. Available at https://doi.org/10.5194/acp-16-2221-2016, http://www.atmos-chem-phys.net/volumes_and_issues.html.

van Leeuwen, T.T., van der Werf, G.R., Hoffmann, A.A., Detmers, R.G., Rücker, G., French, N.H.F., et al., 2014. Biomass burning fuel consumption rates: a field measurement database. Biogeosciences 11 (24), 7305–7329. Available at https://doi.org/10.5194/bg-11-7305-2014.

Liu, M., Popescu, S., 2022. Estimation of biomass burning emissions by integrating ICESat-2, Landsat 8, and Sentinel-1 data. Remote Sensing of Environment 280, 113172. Available at https://doi.org/10.1016/j.rse.2022.113172.

Melitha, G.S., Kashaigili, J.J., Mugasha, W.A., 2025. Integrating UAV, Sentinel-2, and ALOS PALSAR-2 data for improving above-ground biomass estimation in Miombo woodlands using machine learning algorithms. International Journal of Remote Sensing 46 (13), 4796–4831. Available at https://doi.org/10.1080/01431161.2025.2505247, https://www.tandfonline.com/loi/tres20.

Pan, X., Ichoku, C., Chin, M., Bian, H., Darmenov, A., Colarco, P., et al., 2020. Six global biomass burning emission datasets: intercomparison and application in one global aerosol model. Atmospheric Chemistry and Physics 20 (2), 969–994. Available at https://doi.org/10.5194/acp-20-969-2020.

Petrenko, M., Kahn, R., Chin, M., Bauer, S.E., Bergman, T., Bian, H., et al., 2025. Biomass burning emission analysis based on MODIS aerosol optical depth and AeroCom multi-model simulations: implications for model constraints and emission inventories. Atmospheric Chemistry and Physics 25 (3), 1545–1567. Available at https://doi.org/10.5194/acp-25-1545-2025, http://www.atmos-chem-phys.net/volumes_and_issues.html.

Petrenko, M., Kahn, R., Chin, M., Limbacher, J., 2017. Refined use of satellite aerosol optical depth snapshots to constrain biomass burning emissions in the GOCART model. Journal of Geophysical Research: Atmospheres 122 (20). Available at https://doi.org/10.1002/2017jd026693.

Reddington, C.L., Morgan, W.T., Darbyshire, E., Brito, J., Coe, H., Artaxo, P., et al., 2019. Biomass burning aerosol over the Amazon: analysis of aircraft, surface and satellite observations using a global aerosol model. Atmospheric Chemistry and Physics 19 (14), 9125–9152. Available at https://doi.org/10.5194/acp-19-9125-2019, http://www.atmos-chem-phys.net/volumes_and_issues.html.

Rosenfeld, D., Andreae, M.O., Asmi, A., Chin, M., De Leeuw, G., Donovan, D.P., et al., 2014. Global observations of aerosol-cloud-precipitation-climate interactions. Reviews of Geophysics 52 (4), 750–808. Available at https://doi.org/10.1002/2013RG000441, http://agupubs.onlinelibrary.wiley.com/agu/journal/10.1002/(ISSN)1944-9208/.

Roy, D.P., Huang, H., Boschetti, L., Giglio, L., Yan, L., Zhang, H.H., et al., 2019. Landsat-8 and Sentinel-2 burned area mapping - a combined sensor multi-temporal change detection approach. Remote Sensing of Environment 231. Available at https://doi.org/10.1016/j.rse.2019.111254, www.elsevier.com/inca/publications/store/5/0/5/7/3/3.

Santoro, M., Cartus, O., Carvalhais, N., Rozendaal, D.M.A., Avitabile, V., Araza, A., et al., 2021. The global forest above-ground biomass pool for 2010 estimated from high-resolution satellite observations. Earth System Science Data 13 (8), 3927–3950. Available at https://doi.org/10.5194/essd-13-3927-2021, www.earth-system-science-data.net.

Stockwell, C.E., Bela, M.M., Coggon, M.M., Wiggins, E., Gargulinski, E.M., Shingler, T., et al., 2022. Airborne emission rate measurements validate remote sensing observations and emission inventories of Western U.S. wildfires. Environmental Science and Technology 56 (12), 7564–7577. Available at https://doi.org/10.1021/acs.est.1c07121, http://pubs.acs.org/journal/esthag.

Tang, G., Zhao, P., Wang, Y., Gao, W., Cheng, M., Xin, J., et al., 2017. Mortality and air pollution in beijing: the long-term relationship. Atmospheric Environment 150, 238–243. Available at https://doi.org/10.1016/j.atmosenv.2016.11.045.

Tosca, M.G., Randerson, J.T., Zender, C.S., 2013. Global impact of smoke aerosols from landscape fires on climate and the Hadley circulation. Atmospheric Chemistry and Physics 13 (10), 5227–5241. Available at https://doi.org/10.5194/acp-13-5227-2013.

Turco, M., Herrera, S., Tourigny, E., Chuvieco, E., Provenzale, A., 2019. A comparison of remotely-sensed and inventory datasets for burned area in Mediterranean Europe. International Journal of Applied Earth Observation and Geoinformation 82, 101887. Available at https://doi.org/10.1016/j.jag.2019.05.020.

Van Der Werf, G.R., Randerson, J.T., Giglio, L., Van Leeuwen, T.T., Chen, Y., Rogers, B.M., et al., 2017. Global fire emissions estimates during 1997-2016. Earth System Science Data 9 (2), 697–720. Available at https://doi.org/10.5194/essd-9-697-2017, www.earth-system-science-data.net.

Vicente, A., Alves, C., Calvo, A.I., Fernandes, A.P., Nunes, T., Monteiro, C., et al., 2013. Emission factors and detailed chemical composition of smoke particles from the 2010 wildfire season. Atmospheric Environment 71, 295–303. Available at https://doi.org/10.1016/j.atmosenv.2013.01.062.

Wiedinmyer, C., Akagi, S.K., Yokelson, R.J., Emmons, L.K., Al-Saadi, J.A., Orlando, J.J., et al., 2011. The fire INventory from NCAR (FINN): a high resolution global model to estimate the emissions from open burning. Geoscientific Model Development 4 (3), 625–641. Available at https://doi.org/10.5194/gmd-4-625-2011.

Wiedinmyer, C., Kimura, Y., Mcdonald-Buller, E.C., Emmons, L.K., Buchholz, R.R., Tang, W., et al., 2023. The fire inventory from NCAR version 2.5: an updated global fire emissions model for climate and chemistry applications. Geoscientific Model Development 16 (13), 3873–3891. Available at https://doi.org/10.5194/gmd-16-3873-2023, http://www.geosci-model-dev.net/volumes_and_issues.html.

Wiggins, E.B., Anderson, B.E., Brown, M.D., Campuzano-Jost, P., Chen, G., Crawford, J., et al., 2021. Reconciling assumptions in bottom-up and top-down approaches for estimating aerosol emission rates from wildland fires using observations From FIREX-AQ. Journal of Geophysical Research: Atmospheres 126 (24). Available at https://doi.org/10.1029/2021jd035692.

Wooster, M.J., Roberts, G., Perry, G.L.W., Kaufman, Y.J., 2005. Retrieval of biomass combustion rates and totals from fire radiative power observations: FRP derivation and calibration relationships between biomass consumption and fire radiative energy release. Journal of Geophysical Research Atmospheres 110 (24), 1–24. Available at https://doi.org/10.1029/2005JD006318.

Zhang, A., Liu, Y., Goodrick, S., Williams, M.D., 2022. Duff burning from wildfires in a moist region: different impacts on PM2.5 and ozone. Atmospheric Chemistry and Physics 22 (1), 597–624. Available at https://doi.org/10.5194/acp-22-597-2022, http://www.atmos-chem-phys.net/volumes_and_issues.html.

Zou, Y., O'Neill, S.M., Larkin, N.K., Alvarado, E.C., Solomon, R., Mass, C., et al., 2019. Machine learning-based integration of high-resolution wildfire smoke simulations and observations for regional health impact assessment. International Journal of Environmental Research and Public Health 16 (12). Available at https://doi.org/10.3390/ijerph16122137, https://www.mdpi.com/1660-4601/16/12/2137/pdf.

Part III

Case studies on forest fire management in fire-prone regions worldwide

Near-real-time monitoring and prediction of fire intensity, burned area, and burn severity in the forecast danger system of Mexico

14

Daniel José Vega-Nieva[1], Jaime Briseño Reyes[1], Carlos Briones-Herrera[1], Adrián Silva-Cardoza[1], Diana Aimé Tinoco-Orozco[1], Mesías Edwin Domínguez[1], Norma Angélica Monjarás-Vega[1], Marcela Rosas-Chavoya[1], José Javier Corral-Rivas[1], Marín Pompa-García[1], Pablito Marcelo López-Serrano[2], César Alberto Robles Gutiérrez[3], Yair Ricardez[3], Fabiola Esquerra[3], Juan Miguel Campos Muñoz[3], Alicia Verónica Salas[3], Ursula Berenice García Herrera[3], Enrique J. Jardel-Pelaez[4], Oscar Enrique Balcázar Medina[4], Diego Rafael Perez-Salicrup[5], María Isabel Cruz López[6], Martín Cuahutle Cuahutle[6], Rainer Ressl[6], Lilia de Lourdes Manzo-Delgado[7], Colvert Gomez[7], Martin Ibarra Ochoa[8], Juan Gabriel Álvarez-González[9], Ana Daría Ruiz-González[9], William Matthew Jolly[10], Robert E. Burgan[10], Ernesto Alvarado-Celestino[11], and Sean A. Parks[12]

[1]Faculty of Forestry, University of Juarez del Estado de Durango, Durango, Mexico, [2]Faculty of Forestry and Timber Industry, University of Juarez del Estado de Durango, Durango, Mexico, [3]National Forestry Commission, Zapopan, Jalisco, Mexico, [4]South Coast University Centre, University of Guadalajara Atlas de Navarro, Autlán de Navarro, Mexico, [5]Institute for Research into Ecosystems and Sustainability. Autonomous University of Mexico, Morelia, Mexico, [6]National Commission for the Understanding and Use of Biodiversity, Mexico City, Mexico, [7]National Earth Observation Laboratory, Autonomous University of Mexico, Mexico City, Mexico, [8]National Meteorological Service, National Water Commission, Mexico City, Mexico, [9]Department of Agroforestry Engineering, Higher Polytechnic School of Engineering, University of Santiago de Compostela, Lugo, Spain, [10]Rocky Mountain Research Station, USDA Forest Service, Missoula, MT, United States, [11]School of Environmental and Forest Sciences, University of Washington, Seattle, WA, United States, [12]Aldo Leopold Wilderness Research Institute, Rocky Mountain Research Station, USDA Forest Service, Missoula, MT, United States

14.1 The Forest Fire Danger Prediction System of Mexico

The Forest Fire Danger Prediction System of Mexico (SPPIF) is an operational decision-support tool for forest fire management in Mexico (Vega-Nieva et al., 2021). In particular, it offers information about fuel dryness, expected number and location of fire events, burned area, and expected fire intensity. The maps, updated daily from

Advances in Remote Sensing for Forest Fire Management. DOI: https://doi.org/10.1016/B978-0-443-41692-7.00003-2
© 2027 Elsevier Ltd. All rights are reserved, including those for text and data mining, AI training, and similar technologies.

weather, active fire, and fire suppression information, support decision-making for fire detection planning, suppression prioritization, orient prescribed and agricultural burn timing, and aid in planning fuel treatment and long-term strategic fire prevention, suppression, and restoration resources allocation. It is publicly available online at: https://forestales.ujed.mx/incendios/

14.2 Near-real-time monitoring and forecasting of burned area and fire intensity from MODIS and VIIRS active fires in the forest forecast danger system of Mexico

14.2.1 Near-real-time monitoring of active fire perimeters

SPPIF shows active fires, filtered by land use—forest, agricultural, human interface—from Moderate Resolution Imaging Spectroradiometer (MODIS), Visible Infrared Imaging Radiometer Suite (VIIRS) (from the CONABIO antenna), and Geostationary Operational Environmental Satellite (GOES) (received at 10 minutes intervals from the antenna of Laboratorio Nacional de Observación de la Tierra [LANOT], University of the Middle Atlantic [UNAM]). Active fires are colored by date, with the last day shown in red, allowing to identify the position of the fire front every day (Fig. 14.1). For each satellite pass, near-real-time active fire perimeters are calculated using convex hull algorithms calibrated for the fuels of Mexico against MODIS and Sentinel 2 burned area by (Briones-Herrera et al., 2020; Briones-Herrera et al., 2022). They are used for preliminary burned area assessment and to monitor daily fire spread (Fig. 14.1). The number of fire perimeters is also shown in SPPIF, with the information updated in near-real time for every pass of MODIS and VIIRS (Fig. 14.2).

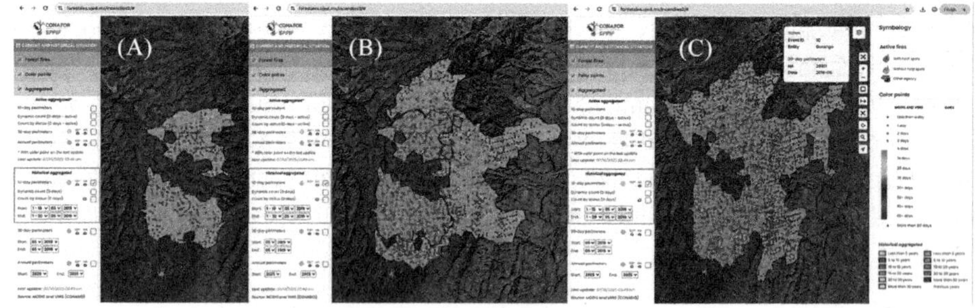

Figure 14.1 Near-real-time (NRT) Visible Infrared Imaging Radiometer Suite active fire perimeters for the wildfire of Bayacora, Durango (Mexico) in the periods 01–10 (A), 11–20 (B), and 21–30 (C) of May 2019. NRT burned area is shown by clicking on the perimeter (Fig. 14.1C). Map lines delineate study areas and do not necessarily depict accepted national boundaries.

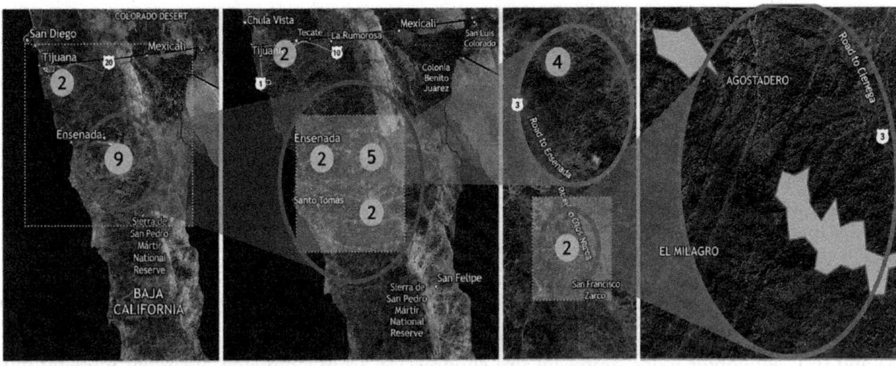

Figure 14.2 Dynamic count of active fire perimeters at different zoom scales in SPPIF for the period of August 1st–10th, 2020, in Baja California (Mexico).

14.2.2 Near-real-time monitoring of fire intensity

Beyond monitoring fire occurrence and burned area, it is important to evaluate fire intensity, given its importance in influencing flame length (Johnston et al., 2017), fire suppression difficulty (Jolly and Freeborn, 2017), impacts (Sparks et al., 2017), and emissions (Wooster et al., 2005). SPPIF shows the kernel density (KD) of VIIRS active fire intensity (FRP_N, in MW km^{-2}). This allows us to visualize in near-real-time, before Sentinel 2 imagery is available, which fires are burning more intensely, enabling daily decisions in fire suppression, prioritizing and assigning proportional resources for fire management (Fig. 14.3).

14.2.3 Where do fires burn more intensely? Predicting maximum fire intensity from fuel type and aboveground biomass in Mexico

SPPIF includes the map of maximum expected fire intensity (Fig. 14.4A) from fuel type and aboveground biomass (Tinoco-Orozco et al., 2025). Most intense fires occur at intermediate total biomass levels, in areas with higher available fine fuel load, and higher wind penetration (Fig. 14.4B). Interestingly, VIIRS KD FRP_N (MW km^{-2}) has been found to be highly correlated to the area burned at high severity from Sentinel 2 (S2) (Domínguez, 2023) (Fig. 14.4C).

14.2.4 Daily forecasting of number of fires, burned area, and fire intensity

14.2.4.1 Forecasts of number of fire perimeters and burned area by state

Based on observed MODIS fuel dryness (Vega-Nieva et al., 2018) and autoregressive models considering the observed number of VIIRS fire perimeters and burned area

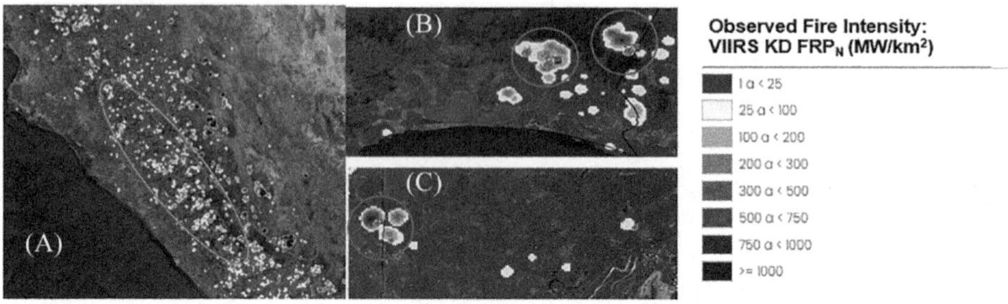

Figure 14.3 Maximum observed fire intensity (kernel density [KD] FRP_N, MW km^{-2}) for NW Mexico (A) and KD FRP_N (MW km^{-2}) for wildfires in Oaxaca (B) and Yucatan (C). In figures (B) and (C), helicopter icons represent fire suppression from Comisión Nacional Forestal with aerial equipment.

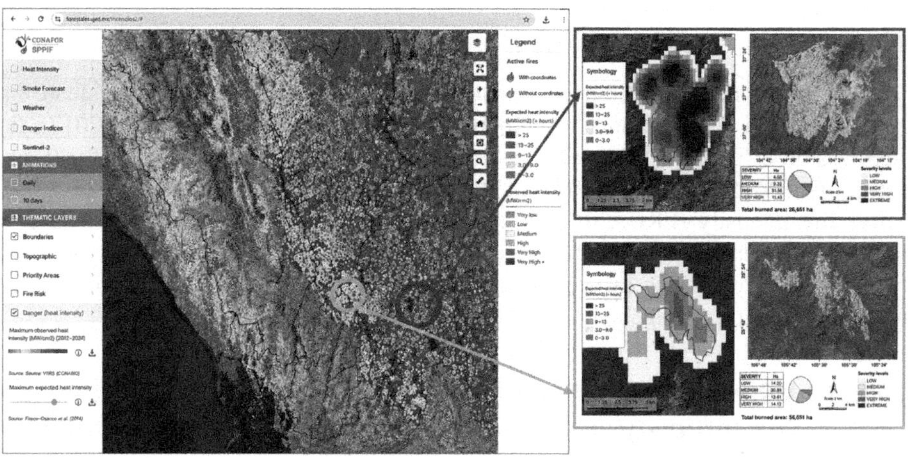

Figure 14.4 Maximum expected fire intensity from fuel type and aboveground biomass (A) and (D) against maximum observed Visible Infrared Imaging Radiometer Suite (VIIRS) kernel density (KD) FRP_N (MW km^{-2}) (B). VIIRS KD FRP_N and S2 burn severity maps for selected fires, marked in (B), are shown in (C).

from the previous days (e.g., Vega-Nieva et al., 2024), SPPIF generates daily forecasts of the expected number of fires and burned area for the next 10 days for each state of Mexico (Fig. 14.5).

14.2.4.2 Daily forecasts of fire intensity (MW km^{-2})

SPPIF provides daily forecasts of VIIRS FRP_N KD (MW km^{-2}), based on observed MODIS fuel dryness (Vega-Nieva et al., 2018), wind (National Meteorological Service [SMN] Weather Research and Forecasting [WRF] forecast), fuel type, and biomass (Vega-Nieva et al., 2023; Tinoco-Orozco et al., 2025) (Fig. 14.6).

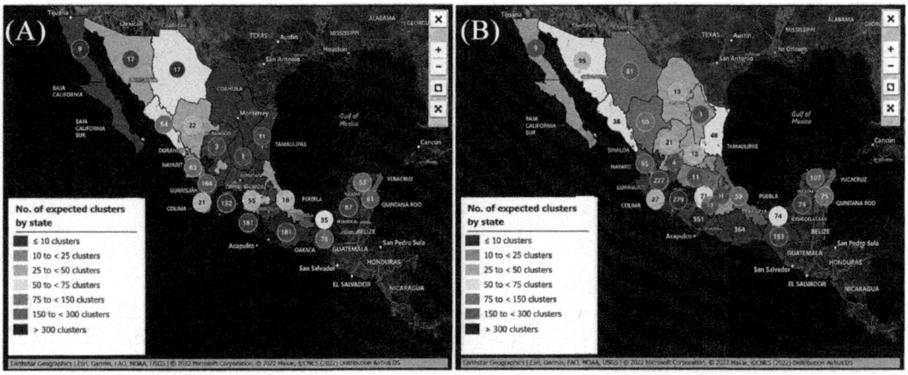

Figure 14.5 Expected and observed number of active fire perimeters by state of Mexico for the period of May 10th–20th for the years 2015 (A) and 2022 (B). Numbers in circles show the observed number of Visible Infrared Imaging Radiometer Suite active fire perimeters 10 days after the forecast. Map lines delineate study areas and do not necessarily depict accepted national boundaries.

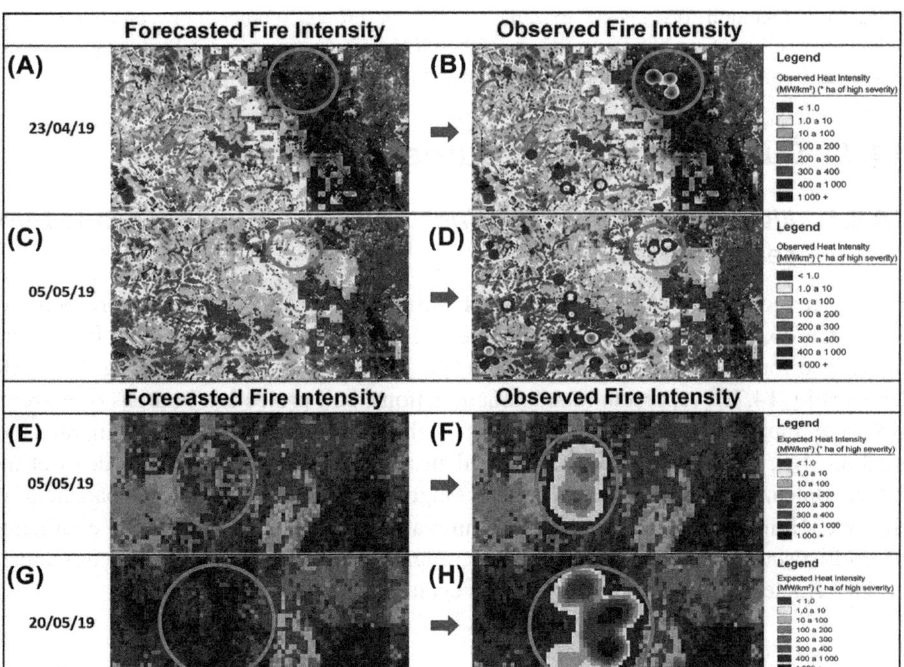

Figure 14.6 Forecasted (A, C, E, G) and observed (B, D, F, H) daily fire intensity for wildfires in Tepehuanes, Durango (A–D), and Bayacora, Durango (Mexico) (E, H) for selected dates.

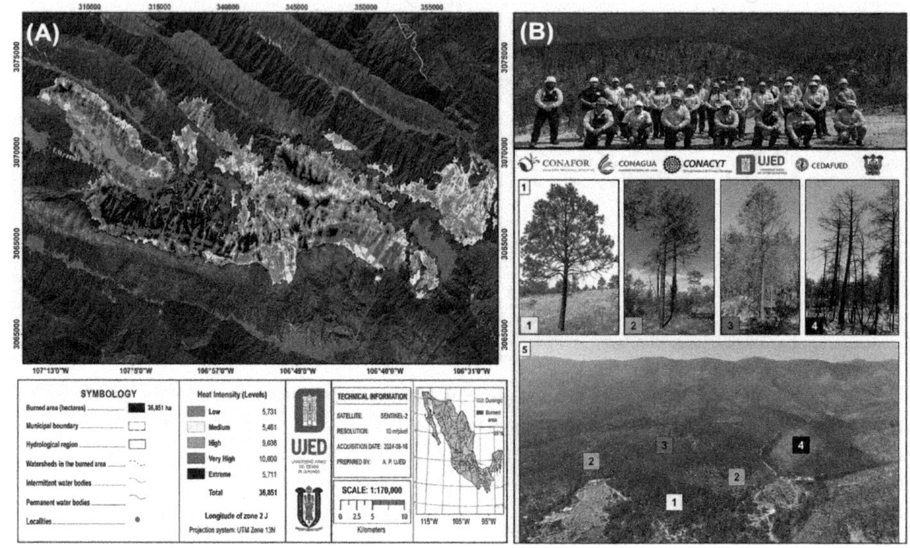

Figure 14.7 Sentinel 2 burn severity (BS) map (A) and field survey of BS (B) in the 2021 fire of Pinalosa, Nuevo León (Mexico). Black, red, orange, and yellow colors in (A) and (B) represent vegetation BS levels: extreme (>90% of crown scorched or consumed), high (61%–90% crown scorched), medium (30%–60% crown scorched), low (<30% crown scorched, burned). Green (A) indicates unburned patches.

14.3 Burn severity and burned area mapping

14.3.1 Field validation of Sentinel-2 and multispectral drone burn severity maps

Based on more than 1000 field plots of burn severity (BS) using the protocols of (Silva-Cardoza et al., 2021), biome-specific thresholds for Sentinel 2 have been developed for the arid to temperate and tropical ecosystems in Mexico (Fig. 14.7). BS levels (Fig. 14.7B) represent management actions: soil protection in areas of extreme BS, tree mortality monitoring in areas of high BS, pest monitoring in areas of medium BS, and low mortality expected in areas of low BS (Silva-Cardoza et al., 2022; Balcázar Medina et al., 2025). Ongoing work, funded by Department of Science, Humanities, Technology and Innovation (SECIHTI), aims at the evaluation of multispectral and light detection and ranging (LIDAR) imagery from unmanned aerial vehicles (UAV) to map soil and vegetation BS (Fig. 14.8).

14.3.2 Automatic mapping of Sentinel-2 burn severity and burn area in Forest Fire Risk Prediction System (SPIFF)

Using the Google Earth Engine code and the semiautomated burned area tools developed by (Briones-Herrera et al., 2023), Sentinel-2 RGB, BS, and initial burned

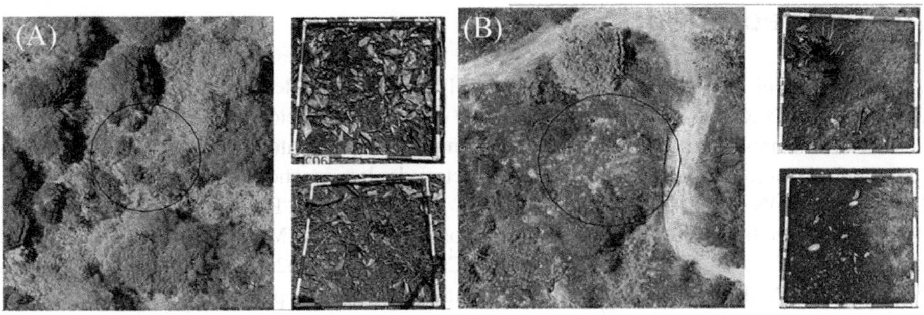

Figure 14.8 Unmanned aerial vehicles RGB images for plots of low (A) and high (B) vegetation and soil burn severity (BS) in a wildfire in Tlaxcala, Mexico. Squares show the field sampling of soil BS.

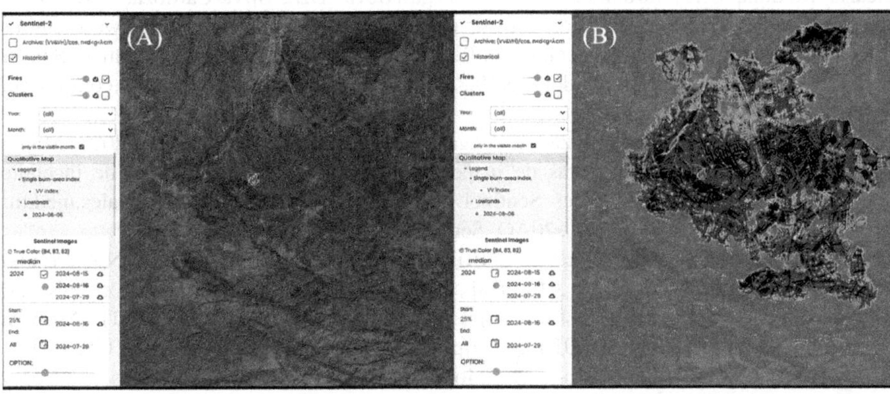

Figure 14.9 Sentinel 2 RGB (A) and burn severity (B) maps for the 2024 fire in El Carrizo, Baja California (Mexico), available for visualization and download in SPPIF. Map lines delineate study areas and do not necessarily depict accepted national boundaries.

area are automatically downloaded to SPPIF for the suppressed fires by Comisión Nacional Forestal (CONAFOR) (Fig. 14.1). Initial automatic S2 burn perimeters from SPPIF are downloaded and supervised by CONAFOR state, regional, and national fire management centers. The burned area tools of SPPIF (SPPIF-S2-BA) are widely used by CONAFOR: only in 2024, more than 4000 fire perimeters were mapped by CONAFOR technicians from all the states of Mexico, using the supervised SPPIF-S2-BA semiautomatic tools of (Briones-Herrera et al., 2023) (Fig. 14.9).

Furthermore, BS maps are used to guide postfire restoration prioritization in large fires. For example, field-validated Sentinel 2 BS maps shown in Fig. 14.7A were used to guide the planning of soil emergency rehabilitation actions in more than 2000 hectares that were severely burned in Pinalosa fire in 2021 (Fig. 14.7B).

Acknowledgments

The development and enhancement of the SPPIF has been funded by the projects: SECIHTI-MADTEC-2025-R-36 (2025–26): "Maduración del Sistema Nacional de Predicción de Peligro de Incendios Forestales de México, empleando nuevas fuentes satelitales y drones para la alerta temprana, manejo y monitoreo de los impactos de los incendios forestales en México"; CONAFOR/CONACYT-2018-C02-B-S131553 (2019–23): "Reforzamiento al Sistema Nacional de Predicción de Peligro de Incendios Forestales de México para el pronóstico de conglomerados y área quemada"; CONAFOR/CONACYT 252620: "Desarrollo de un Sistema Nacional de Predicción de Peligro de Incendios Forestales para México"; US Northern Command, CONAFOR, and United States Forest Service (USFS) (2024–25): "Reforzamiento y actualización del Sistema de Predicción de Peligro de Incendios Forestales de México."

References

Balcázar Medina, O.E., Jardel Peláez, E.J., Vega-Nieva, D.J., Silva-Cardoza, A.I., Cuevas Guzmán, R., 2025. Mapping soil burn severity and crown scorch percentage with sentinel-2 in seasonally dry deciduous Oak and Pine Forests in Western Mexico. Remote Sensing 17 (13). Available at https://doi.org/10.3390/rs17132307, http://www.mdpi.com/journal/remotesensing.

Briones-Herrera, C.I., Silva-Cardoza, A., Vega-Nieva, D.J., Briseño-Reyes, J., 2023. Manual de usuario de las herramientas de mapeo de área quemada y severidad de incendios forestales a partir de imágenes Sentinel. 45 pg. Available at http://forestales.ujed.mx/incendios2/descargas/Manual%20AQ_Sentinel_v1_1_071021.pdf.

Briones-Herrera, C.I., Vega-Nieva, D.J., Briseño-Reyes, J., Monjarás-Vega, N.A., López-Serrano, P.M., Corral-Rivas, J.J., et al., 2022. Fuel-specific aggregation of active fire detections for rapid mapping of forest fire perimeters in Mexico. Forests 13 (1). Available at https://doi.org/10.3390/f13010124, https://www.mdpi.com/1999-4907/13/1/124/pdf.

Briones-Herrera, C.I., Vega-Nieva, D.J., Monjarás-Vega, N.A., Briseño-Reyes, J., López-Serrano, P.M., Corral-Rivas, J.J., et al., 2020. Near real-time automated early mapping of the perimeter of large forest fires from the aggregation of VIIRS and MODIS active fires in Mexico. Remote Sensing 12 (12). Available at https://doi.org/10.3390/RS12122061, https://www.mdpi.com/2072-4292/12/12/2061.

Domínguez, M.E., 2023. Análisis de la intensidad calórica y severidad de los incendiso forestales en la República Mexicana. PhD Thesis. 77 pg. Maestría en Geomática Aplicada a los Recursos Forestales y Ambientales. Facultad de Ciencias Forestales, Universidad Juárez del Estado de Durango.

Johnston, J.M., Wooster, M.J., Paugam, R., Wang, X., Lynham, T.J., Johnston, L.M., 2017. Direct estimation of Byram's fire intensity from infrared remote sensing imagery. International Journal of Wildland Fire 26 (8), 668–684. Available at https://doi.org/10.1071/WF16178, http://www.publish.csiro.au/nid/126.htm.

Jolly, W.M., Freeborn, P.H., 2017. Towards improving wildland firefighter situational awareness through daily fire behaviour risk assessments in the US Northern Rockies and Northern Great Basin. International Journal of Wildland Fire 26 (7), 574–586. Available at https://doi.org/10.1071/wf16153.

Silva-Cardoza, A.I., Vega-Nieva, D.J., Briseño-Reyes, J., Briones-Herrera, C.I., López-Serrano, P.M., Corral-Rivas, J.J., et al., 2022. Evaluating a new relative phenological correction and the effect of sentinel-based earth engine compositing approaches to map fire severity and burned area. Remote Sensing 14 (13). Available at https://doi.org/10.3390/rs14133122, https://www.mdpi.com/2072-4292/14/13/3122/pdf?version=1656495196.

Silva-Cardoza, A.I., Vega-Nieva, D.J., López-Serrano, P.M., Corral-Rivas, J.J., Briseño Reyes, J., Briones-Herrera, C.I., et al., 2021. Metodología para la evaluación de la severidad de incendios forestales en campo, en ecosistemas de bosque templado de México. CONAFOR-CONACYT-2018-C02-B-S-131553. Available at http://forestales.ujed.mx/incendios2/php/publicaciones_documentos/7_3_Silva%20et%20al%2020211201_Metodologia_severidad_v1.pdf.

Sparks, A.M., Smith, A.M.S., Talhelm, A.F., Kolden, C.A., Yedinak, K.M., Johnson, D.M., 2017. Impacts of fire radiative flux on mature Pinus ponderosa growth and vulnerability to secondary mortality agents. International Journal of Wildland Fire 26 (1), 95–106. Available at https://doi.org/10.1071/WF16139, http://www.publish.csiro.au/nid/126.htm.

Tinoco-Orozco, D.A., Vega-Nieva, D.J., Briseño-Reyes, J., Dominguez-Amaya, M.E., Silva-Cardoza, A.I., Briones-Herrera, C.I., et al., 2025. Where do fires burn more intensely? Modeling and mapping maximum MODIS fire radiative power from aboveground biomass by fuel type in Mexico. Fire 8 (2). Available at https://doi.org/10.3390/fire8020054, www.mdpi.com/journal/fire.

Vega-Nieva, D.J., Briseño-Reyes, J., Briones-Herrera, C.I., Monjarás, N., Silva-Cardoza, A., et al., 2021. Manual de Usuario del Sistema de Predicción de Peligro de Incendios Forestales de México. Available at http://forestales.ujed.mx/incendios2/php/publicaciones_documentos/1_1-%20MANUAL_DE_USUARIO_SPPIF_v15_DV290820.pdf.

Vega-Nieva, D.J., Briseño-Reyes, J., López-Serrano, P.M., Corral-Rivas, J.J., Pompa-García, M., Cruz-López, M.I., et al., 2024. Autoregressive forecasting of the number of forest fires using an accumulated MODIS-based fuel dryness index. Forests 15 (1). Available at https://doi.org/10.3390/f15010042, http://www.mdpi.com/journal/forests/.

Vega-Nieva, D., Briseño-Reyes, J., Nava-Miranda, M., Calleros-Flores, E., López-Serrano, P., Corral-Rivas, J., et al., 2018. Developing models to predict the number of fire hotspots from an accumulated fuel dryness index by vegetation type and region in Mexico. Forests 9 (4), 190. Available at https://doi.org/10.3390/f9040190.

Vega-Nieva, D.J., Briseño-Reyes, Vega-Nieva, D.J., Briseño-Reyes, et al., 2023. Documentación y código del índice de pronóstico de la intensidad calórica esperada. Informe del Proyecto CONAHCYT "Reforzamiento al Sistema Nacional de Predicción de Peligro de Incendios Forestales de México para el pronóstico de conglomerados y área quemada", con clave B-S-131553, convocatoria CONAFOR-2018-C02. 14pg. Informe del Proyecto CONAHCYT "Reforzamiento al Sistema Nacional de Predicción de Peligro de Incendios Forestales de México para el pronóstico de conglomerados y área quemada".

Wooster, M.J., Roberts, G., Perry, G.L., Kaufman, Y.J., 2005. Retrieval of biomass combustion rates and totals from fire radiative power observations: FRP derivation and calibration relationships between biomass consumption and fire radiative energy release. Journal of Geophysical Research: Atmospheres 110.

Wildfires in Alaska: insights into fire occurrence, postfire impacts, and air quality implications

15

Tatiana Loboda[1], *Dong Chen*[1], *Jordan Caraballo-Vega*[1,2], *Allison Bredder*[1], *and Elijah Orland*[1,3]
[1]Department of Geographical Sciences, University of Maryland, College Park, MD, United States, [2]NASA Goddard Space Flight Center, Greenbelt, MD, United States, [3]University of Maryland - Baltimore County, Baltimore, MD, United States

15.1 Introduction

Global wildfire patterns have exhibited a complex and evolving trajectory in recent decades. While some regions have experienced a decline in burned areas, primarily due to changes in land management and agricultural practices in savannas and grasslands, other areas have witnessed a significant increase in wildfire frequency and intensity. Boreal and temperate zones have experienced a substantial spatial and temporal expansion of fire-prone conditions, attributed to rising temperatures and prolonged fire seasons (Abatzoglou et al., 2019). Alaska has witnessed an increase in both the frequency and size of wildfires, largely driven by warmer temperatures and drier conditions. Data from the Alaska Interagency Coordination Center (AICC) indicates a stark increase in fire activity: from 2001 to 2020, Alaska experienced an average of 502 fires per year, burning an average of 1.5 million acres annually. This represents a substantial increase compared to the 1981–2000 period, which averaged 566 fires and 0.7 million acres burned per year. Notably, three out of the four largest fire seasons in terms of area burned have occurred since 2000. Moreover, wildfire occurrence has increased within Alaskan tundra—an ecoregion where fire occurrence was historically rare (Hu et al., 2015)—with a corresponding increase in the potential to release vast amounts of stored carbon in deep organic soils into the atmosphere, further exacerbating climate change (Moubarak et al., 2023).

Wildfire management in Alaska presents unique challenges due to the distinct characteristics of its ecosystems and vast, remote landscapes. The predominance of boreal forests and tundra, with their carbon-rich soils and susceptibility to intense burning, differentiates Alaskan fire regimes from those in the lower 48 states. Furthermore, the remoteness and limited accessibility of much of Alaska pose logistical hurdles for fire detection, suppression, and management efforts. The vastness of the state, coupled with sparse infrastructure and limited resources, necessitates innovative approaches and collaborative strategies to effectively address the growing wildfire threat in Alaska (Alaska Statewide Operating Plan, 2023).

Advances in Remote Sensing for Forest Fire Management. DOI: https://doi.org/10.1016/B978-0-443-41692-7.00015-9
© 2027 Elsevier Ltd. All rights are reserved, including those for text and data mining, AI training, and similar technologies.

Earth observations derived from satellite remote sensing have become instrumental in Alaska's fire management, providing critical data for fire monitoring, fuel mapping, and postfire impact assessment. Satellite-derived active-fire products have been incorporated into operational fire management systems like the AICC Fire Information Map, aiding decision-making processes for fire suppression and resource allocation. The Moderate Resolution Imaging Spectroradiometer (MODIS) aboard the NASA's Terra and Aqua satellites has been extensively utilized for near-real-time fire detection and monitoring. MODIS active-fire products (MCD14A1; Giglio et al., 2016), along with data from the Visible Infrared Imaging Radiometer Suite (VIIRS) (VNP14IMG; Schroeder et al., 2014), enable rapid identification of fire hotspots and the assessment of fire spread and intensity. The enhanced orbital overlap of polar orbiters in the high northern latitudes (HNL) allows for subdaily observations of fire activity in Alaska at a spatial resolution far exceeding that of geostationary systems available at mid-latitudes, but it requires different approaches to monitoring ongoing fire activity. Furthermore, Landsat observations, with the multidecadal historical archive and moderate spatial resolution, have been instrumental in mapping vegetation types and fuels (e.g., National Land Cover Dataset—Dewitz, 2021; LANDFIRE Existing Vegetation Type, Cover, and Height—landfire.gov), which are crucial for understanding fire behavior and developing fire management strategies. NASA's Arctic Boreal Vulnerability Experiment (ABoVE) campaign has enabled the collection of critical field and airborne observations in the HNL ecosystems and allowed for the development of new enhanced algorithms for fuel mapping from Landsat (e.g., He et al., 2019; Macander et al., 2017) and hyperspectral Airborne Visible/Infrared Imaging Spectrometer (AVIRIS) Next Generation (NG) AVIRIS-NG is the name of an instrument (AVIRIS-NG); Smith et al., 2021) observations. Finally, recent computational advancements and rapid growth of machine learning (ML) and artificial intelligence (AI) algorithms, combined with extensive multi-decadal archives of satellite-based fire activity observations, have established the basis for developing innovative spatially explicit data-driven models for fire ignition forecasting (e.g., Earnest et al., 2024; Chen et al., 2024; Alkhatib et al., 2023).

In this chapter, we describe the context of the current fire monitoring capabilities and highlight the novel developments in the use of satellite observations for monitoring and forecasting fire ignition patterns, characteristics of ongoing fire activity, postfire impacts on natural ecosystems, and the impact of wildfire emissions on air quality.

15.2 Historical context of contemporary fire monitoring capabilities

Although never intended specifically for fire monitoring, the Advanced Very High Resolution Radiometer (AVHRR) sensors by the National oceanic and atmospheric administration (NOAA), launched in 1978, provided the first satellite-based platform for active-fire detection (Kalluri et al., 2021). AVHRR sensors detected light in multiple channels ranging from 0.58 to 11.5 or 12.5 micrometers (μm), depending on the generation, capturing reflectance and emissivity in visible and long-wave infrared (IR) light, respectively. The utility of the AVHRR observations for fire detection was proposed by Dozier (1981), who demonstrated that flaming fires sharply increase emittance in the mid-wave IR channel while providing a more muted response in the long-wave IR channel. The resulting equations proposed by Dozier provided a framework for using the 3.7 μm (mid-wave IR) and 11 μm (long-wave IR) channels on the AVHRR sensor to derive both the temperature and subpixel area of a thermal body detected from space. While providing an extremely beneficial demonstration of active-fire detection at an approximate pixel resolution of 1 km, the AVHRR's 3.7 μm channel saturated at comparatively low temperatures (Kaufman et al., 1998), limiting its applicability to areas with low background temperatures such as nighttime burning, and it would otherwise risk frequent commission errors (i.e., false positive detections) during the day from bright and/or hot surfaces (Malingreau, 1990; Langaas, 1992; Giglio et al., 1999).

The subsequent launch of the MODIS represented the beginning of the contemporary era for spaceborne active-fire detection. Positioned on board two satellites—Terra and Aqua, the MODIS sensor includes two channels specifically designated for active-fire detection at 4 and 11 μm with higher saturation temperatures of 450K and 400K, respectively (Kaufman et al., 1998; Justice et al., 2002; Townshend and Justice, 2002). With high orbital overlap within HNL regions, MODIS delivers a minimum of eight daily overpasses in Alaska and its active-fire detection data, with an approximate 1 km pixel resolution, has become the cornerstone of fire monitoring and fire record. The observations are available within 2–4 hours after initial detection—an effort that first began in the early years following launch (Descloitres et al., 2002; Justice et al., 2002).

Recognizing the immense value and success of satellite fire monitoring, these capabilities were transferred from the scientific (NASA) to the operational (NOAA) domain with the launch of the VIIRS sensor on the NOAA Suomi NPP (SNPP) satellite in 2012. VIIRS offers an improved spatial resolution of active-fire detections (~375 m) and enhanced geolocation accuracy (Schroeder et al., 2014). Now onboard three satellites—SNPP, NOAA-20, and NOAA-21—VIIRS sensors, together with MODIS sensors, provide observations over some parts of Alaska 16–20 times per day from a total of five individual platforms, enabling an unprecedented continuity of wildfire monitoring. However, with the anticipated loss of MODIS observations due to the decommissioning of both Terra and Aqua satellites, this high temporal record will be interrupted until two additional NOAA satellites carrying VIIRS instruments launch.

The use of geostationary satellites in HNL regions is very limited. Although the Geostationary Operational Environmental Satellite (GOES)-West captures some of Alaska's area within its field of view, the extremely oblique viewing angle at the edge stretches the already coarse nominal 2 km pixel to nearly 8 km in size, making fire detection extremely challenging.

15.3 Forecasting fire ignition patterns

Wildfire ignitions across Alaska's boreal forests and tundra are increasing in frequency and spatial extent, reflecting rapid changes in climate, vegetation, and atmospheric dynamics. Lightning is the dominant ignition source in Alaska, although a small fraction of ignitions is associated with human activity, particularly near transport corridors and settlements (State of Alaska Department of Natural Resources, 2022). Warming temperatures, altered precipitation regimes, and expanding shrub and tree cover have amplified ignition potential in regions that were historically nonflammable (Baker and Halsey, 2022). In recent decades, lightning strikes have expanded poleward and now occur across tundra and coastal zones (Veraverbeke et al., 2017; Song et al., 2024). Forecasting when and where ignitions are likely to occur, whether triggered by lightning or human activity, has become a central challenge for understanding and managing northern fire regimes (Veraverbeke et al., 2017; Peterson et al., 2010). Recent computational advancements and rapid growth of ML and AI algorithms, combined with extensive multidecadal archives of satellite-based fire activity observations, have enabled the development of innovative spatially explicit data-driven models needed to tackle fire ignition forecasting.

Approaches to modeling cloud-to-ground (CG) lightning are most frequently based on relationships with parameters related to atmospheric convective properties and cloud microphysical properties. Considering a highly limited network of meteorological stations in HNL regions, most successful approaches rely on integrating the observed CG strikes with dynamic Numerical Weather Prediction models in a statistical or ML framework (He and Loboda, 2020). The Alaska Lightning Detection Network (ALDN; Fronterhouse, 2012) provides a multidecadal record of CG lightning strikes with a detection efficiency better than 5 km and positional accuracy higher than 70% (Reap, 1991; Dissing and Verbyla, 2003). These observations are complemented by high-resolution simulations from the Weather Research and Forecasting (WRF) model, with parameterizations that best describe Alaskan conditions (He and Loboda, 2020; He et al., 2021). Using random forest models, we established a strong potential for modeling CG lightning strikes in Alaskan tundra (He and Loboda, 2020), an approach we subsequently expanded to boreal forests using an ensemble of ML models, including random forest, XGBoost, and temporally aware neural network (NN) algorithms. Our initial validation results for the previously unseen 2023 fire season showed a very strong performance for the approach, with the Probability of Detection (POD) exceeding 97% and a False Alarm rate of ~5%.

While CG lightning is a source, the probability of ignition is governed by the complex interactions among fuels, weather, and topography. Our modeling framework unites multiple complementary datasets within a cohesive 1 km analysis grid. Our response variable—the ignition point—is extracted from satellite observations of fire activity as the first detection within the contiguous in space and time cluster of observations. Active-fire detections are derived from the VIIRS VNP14IMG product (Schroeder et al., 2014, Giglio et al., 2016). The Fire Events Data Suite (FEDS) algorithm groups contiguous VIIRS thermal anomalies into events and designates the earliest detection as the ignition point, producing a temporally consistent record of ignition locations (Csiszar et al., 2014). The suite of predictor variables includes the CG lightning forecasts described above, WRF-simulated meteorological predictors, surface heterogeneity and fuel load, and topography. The meteorological inputs include temperature, humidity, wind, and precipitation, derived indices from the

Canadian Forest Fire Danger Rating System, and convective available potential energy (CAPE) and mid-tropospheric temperature (T850) (Bieniek et al., 2020). Surface heterogeneity is represented through LANDFIRE Fuel Vegetation Type (FuelVT) data (Ryan and Opperman, 2013), providing fractional fuel composition aggregated to 1 km, and by terrain metrics derived from the 5 m interferometric synthetic aperture radar (IFSAR) digital elevation model (DEM) (US Geological Survey, 2022). Slope, aspect, and drainage indices follow the (Kasischke and Hoy, 2012) classification, distinguishing well-drained versus poorly drained surfaces that strongly influence fuel moisture. Human accessibility is incorporated via impervious-surface fraction from the National Land Cover Dataset (US Geological Survey, 2023) and the global Human Footprint Index (Gassert et al., 2023), capturing the limited but consequential anthropogenic ignition potential in populated areas.

The expanded fire ignition model fuses atmospheric, environmental, and anthropogenic drivers using a multimodal deep-learning architecture (Fig. 15.1). The system integrates convolutional and transformer modules with an Adaptive Fourier Neural Operator (AFNO) attention mechanism to capture complex spatiotemporal dependencies (Guibas et al., 2021). A physics-informed loss function enforces temporal and spatial continuity, while a Bayesian AFNO branch quantifies pixel-wise uncertainty, producing calibrated probabilistic outputs. Training data span 2012–22,

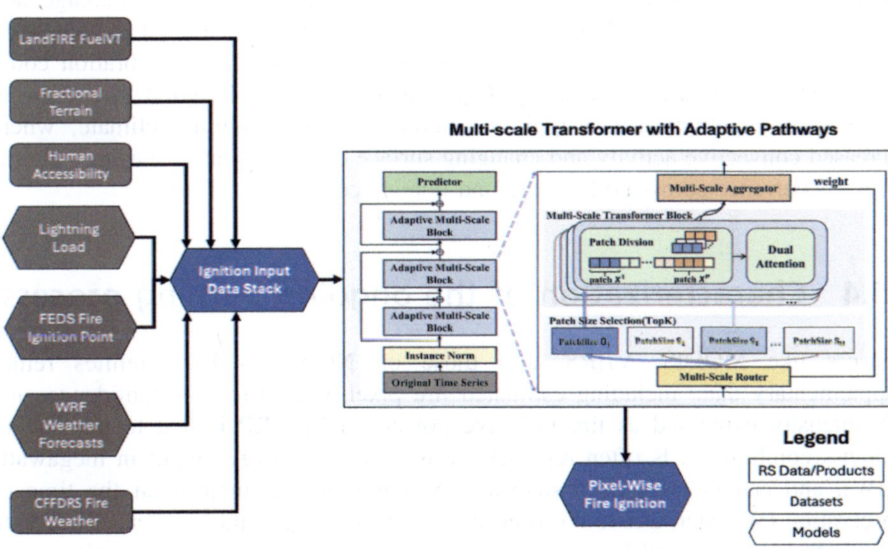

Figure 15.1 Fire ignition conceptual workflow. Data from fuels, terrain, lightning load, ignition, weather, and fire indices are used as input. Example of a spatially and temporally aware transformer model used for data fusion and ignition forecasting.
Source: Chen, Y., Hantson, S., Andela, N., Coffield, S.R., Graff, C.A., Morton, D.C., et al., 2022. California wildfire spread derived using VIIRS satellite observations and an object-based tracking system. Scientific Data. 9, (1), https://doi.org/10.1038/s41597-022-01343-0, www.nature.com/sdata/.

validated against 2023–24 fire-season observations, generating daily ignition-probability maps at 1 km resolution for both biomes.

Feature-importance analyses across both the lightning and ignition models reveal a connected sequence of atmospheric and surface processes influencing fire potential in Alaska. For lightning occurrence, mid-tropospheric temperature (T850), near-surface temperature (T2), wind shear, and moisture anomalies were dominant predictors, underscoring the role of convective instability and dryness in generating CG lightning. When these variables were extended into the ignition framework, lightning probability, elevation, precipitation, and fuel composition emerged as the strongest predictors, highlighting the interaction between convective activity and surface fuel conditions. Together, these findings demonstrate that ignition probability in Alaska depends on the combined influence of large-scale atmospheric energy, local topography, and the spatial distribution of burnable fuels.

Both the lightning and ignition models achieved high predictive skill, with area under the curve (AUC) values exceeding 0.90 for 24–48-hour forecasts and remaining above 0.80 at 10-day lead times. POD values were consistently above 0.85 during active-fire periods, indicating reliable forecast performance across short and medium ranges. Biome-specific analyses revealed distinct ignition regimes: tundra fires are primarily controlled by lightning activity and atmospheric dryness, while boreal regions show greater sensitivity to fuel heterogeneity, soil drainage, and antecedent precipitation. Spatial commission errors were concentrated in areas with complex terrain and mixed vegetation, suggesting that localized calibration could further improve predictive accuracy. Collectively, these results point to a growing influence of lightning-driven ignitions under Alaska's warming climate, where increased convective activity and changing surface conditions are reshaping patterns of fire occurrence across both tundra and boreal ecosystems.

15.4 Characterization of the ongoing burning process

Sensors like MODIS, VIIRS, and those on NOAA GOES satellites return supplementary data, including estimated fire pixel area, detection confidence, and fire intensity expressed as fire radiative power (FRP). FRP is tied to the rate of biomass combustion, is often represented as a rate of energy output in megawatts (MW), and can be used as a snapshot estimator of fire intensity at the time of observation (Wooster, 2003; Wooster et al., 2005). For geostationary satellites with extremely high temporal frequency imaging, FRP can be integrated to estimate the total energy output of the fire, referred to as fire radiative energy (FRE), but their relatively coarse ~2 km pixel resolution limits the ability to pinpoint this energy to a discrete subpixel source. On the other hand, MODIS and VIIRS sensors can provide information on FRP at comparatively higher spatial resolutions (375 m–1 km), but their limited overpass times, even at high latitudes, restrict FRP measurements to only a few times each day. Additionally, these overpasses occur during times often not amenable to capturing a fire's diurnal cycle, limiting the ability of these sensors to observe burning at peak intensities (Giglio, 2007; Andela et al., 2015; Wiggins et al., 2020).

Notwithstanding, FRP can be used to monitor a wide array of natural processes, including emissions from burning (Houghton, 2012; Ichoku and Ellison, 2014; Li et al., 2022), total biomass combustion (Li et al., 2018), and burn severity (Heward et al., 2013; Hantson et al., 2022; Ponomarev et al., 2023; Chatzopoulos-Vouzoglanis et al., 2024; Orland et al., 2025). Examples of this work in the boreal include investigating the prevalence of overwintering fires (Scholten et al., 2021), monitoring trends in fire intensity (Barrett and Kasischke, 2013; Li et al., 2024), quantifying emissions from burning (Vermote et al., 2009; Tanimoto et al., 2015; Wiggins et al., 2016; Ponomarev et al., 2020), investigating atmospheric smoke injection rates (Peterson et al., 2014), and even studying the occurrence of rare pyrocumulonimbus clouds (Gatebe et al., 2012; Peterson et al., 2017; Kablick et al., 2018).

The extensive record provided by these sensors has inspired many approaches to reconstruct historic fire progression and map fire evolution in near real-time (NRT). Historic examples include the use of MODIS active-fire data to map fire spread in the boreal region and beyond (Loboda and Csiszar, 2007; Giglio et al., 2009), as well as NRT approaches to provide up-to-date information on a fire's perimeter as it continues to evolve (Chen et al., 2022) (Fig. 15.2). This approach was recently

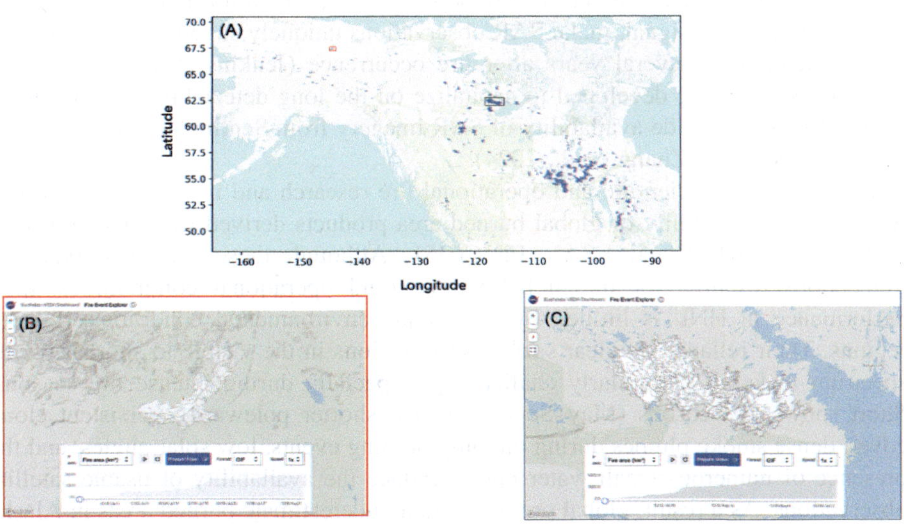

Figure 15.2 Fire perimeter mapping from active-fire observations within the NASA's Fire Event Explorer tool: https://earthdata.nasa.gov/dashboard/tools/fire-event-explorer. (A) Fires in the boreal region from January to October 2025. (B)–(C) Two examples demonstrating the detail provided by object-based fire tracking methods, where each line represents 12-hour perimeter growth. The Christian Fire in Northeast Alaska (red) shows multiple ignition points, while an unnamed, long-duration fire in Canada (black) illustrates prolonged burning patterns typical of the region. Map lines delineate study areas and do not necessarily depict accepted national boundaries.

adapted by Liu et al. (2024) to utilize GOES satellites to provide hourly spread information at a coarser resolution. Recently, a landmark study by Balch et al. (2024) additionally demonstrated the outsized role of fast-moving wildfires in inducing structural damage within the wildland-urban interface, highlighting the increasing importance of an object-based tracking approach that can quantify fire spread rates, as opposed to discrete pixel detection.

15.5 Assessing postfire impacts

Similarly to globally accepted practices, satellite observations serve as the primary source of information about the postfire effect. Imagery within visible, near-IR, and short-wave IR serves as the most commonly used data for mapping burn area extent, assessing the impact of fire on ecosystems, and monitoring postfire ecosystem recovery and change. However, because of the long history of availability of radar observations from the Canadian RADARSAT missions (the first satellite launched in 1995) and the European Remote Sensing Satellite (ERS-1 and 2) in the boreal and tundra domains of North America, in Alaska, synthetic aperture radar (SAR) imagery has been frequently used to identify burned areas (e.g., Bourgeau-Chavez et al., 1997) and postfire impacts specific to boreal and tundra biomes (e.g., Kasischke et al., 2007). The distinct changes in postfire soil moisture, which are not masked by the vegetation recovery signal, make SAR observations uniquely suited to map historical burns and monitor several years after fire occurrence (Jenkins et al., 2014). New approaches are being developed to capitalize on the long detectability of the signal from SAR and the wide availability of SAR imagery from Sentinel-1 in concert with other SAR sensors (Zhang et al., 2024).

The majority of scientific and operational fire research and monitoring, however, relies on the availability of global burned area products derived from MODIS data, such as MCD64A1 and ESA's FireCCI51. Although these are undoubtedly a tremendous resource for the global wildfire and operational communities, their performance in HNL is limited by the unique environmental conditions of these regions. Their reliance on clear surface observations in the visible to short-wave IR spectrum makes it particularly challenging, especially during intense fire seasons. Short snow-free seasons (May–September and shorter poleward), persistent cloud cover, dense smoke plumes during ongoing burning events, low solar angles, and the presence of numerous small water bodies reduce the availability of usable satellite observations. As a result, global products often underestimate burned areas in Alaska and across the Arctic and boreal zones. In a 2021 study (Chen et al., 2021), we evaluated MCD64A1 and FireCCI51 against independent fire records in Alaska and Canada between 2001 and 2015. Our results showed that MCD64A1 missed a substantial portion of the total burned area across boreal forests and tundra, amounting to 74,768 km², an area larger than the Republic of Ireland. Although FireCCI51 performed better, it omitted nearly 40% of burned areas north of 60°N. The omission of such extensive areas of fire impact substantially affects subsequent assessments of carbon and energy budgets, not only at the regional but also at the global scale. Fire extent is a key parameter in assessing the fire-induced carbon loss, postfire modification of ecosystem functioning, and surface albedo change. To improve the burned area estimates, we developed a regionally adapted Arctic Boreal Burned Area (ABBA; Fig. 15.3) (Loboda et al., 2024) dataset using an algorithm that is tailored to the specifics of wildfire occurrence in the HNL regions, including the extremely limited clear surface views during the fire season. By incorporating MODIS surface reflectance data from both the fire year and the following spring, ABBA is less constrained by seasonal imaging limitations and provides substantially lower omission error. A recent study intercomparing the satellite estimates of burned area from 12 publicly available datasets for Arctic Boreal North America and Russia recommends regionally focused burned area assessments, including ABBA and ABoVE-fire emission database (FED), as the most appropriate datasets to use for studies in this region (Clelland et al., 2024). Although not designed for operational fire suppression due to latency, ABBA offers an essential tool for research and postfire management in Alaska, including quantifying carbon emissions and

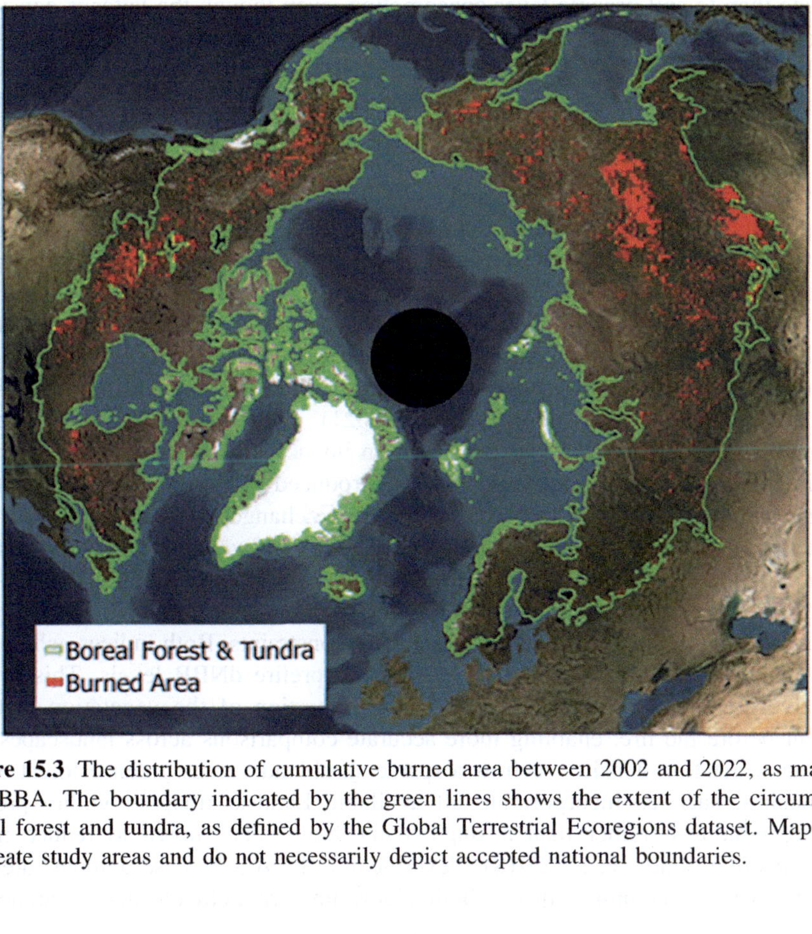

Figure 15.3 The distribution of cumulative burned area between 2002 and 2022, as mapped by ABBA. The boundary indicated by the green lines shows the extent of the circumpolar boreal forest and tundra, as defined by the Global Terrestrial Ecoregions dataset. Map lines delineate study areas and do not necessarily depict accepted national boundaries.

projecting future fire regimes. New approaches are being developed to capitalize on the extensive archive of the optical imagery from Landsat and the increased frequency of observations afforded by Sentinel-2 through advanced ML approaches to develop novel datasets of historical and contemporary burned area mapping at higher resolution (e.g., Potter et al., 2024).

While mapping fire perimeters and burned areas provides a measure of fire occurrence, assessing burn severity is critical for understanding ecological impacts and recovery trajectories. Burn severity describes the degree of change to vegetation and soils, influencing carbon release, hydrology, and succession. The most widely used remote sensing metric for severity is the differenced Normalized Burn Ratio (dNBR), based on the differenced signal in the near IR and short-wave IR at ~2 μm derived from Landsat imagery, which has become an operationally applied assessment across all ecosystems within the United States despite the highly varying degree of correlation with field-based measurements (French et al., 2008). As with burned area mapping, dNBR and its derivatives face notable challenges in HNL, including Alaska. As can be expected from an optical index, dNBR is more sensitive to canopy damage than to soil organic layer consumption, even though the latter is considered the primary determinant of severity in black spruce-dominated boreal forests of Alaska (Kasischke et al., 2008). Moreover, dNBR saturates at high severity levels, underrepresenting the severity of extreme fires, which are common in Alaska. However, the particularly limiting factor in the HNL areas is the high sensitivity of the index to phenological differences between pre and postfire images. In Alaska, where the growing season is short and cloud-free imagery is scarce, selecting optimal image pairs is particularly difficult. Even "optimal" pairs, defined by anniversary timing and minimal cloud cover, can produce inconsistent severity estimates for the same fire (Chen et al., 2020).

A systematic analysis of the availability of optimal Landsat image pairs across North America showed that Alaska is among the most challenging regions, especially in the pre-2000 period, with many fires lacking usable image pairs that meet timing and quality requirements (Chen et al., 2021). The resultant incorporation of suboptimal images in the assessment of burn impacts introduces biases as large as the fire signal itself. This means that errors introduced by using poor-quality satellite images can create a false impression of landscape changes that are equal to or even greater than the actual changes caused by the fire. Improvements to severity indices, including relative dNBR (RdNBR; Miller and Thode, 2007) and the Relativized Burn Ratio (RBR; Parks et al., 2014), partially address these limitations by adjusting for prefire vegetation conditions and mathematical constraints. Both indices achieve this by dividing dNBR by a value representing the prefire dNBR levels. This process effectively contextualizes the severity as a proportion of the vegetation that was present before the fire, enabling more accurate comparisons across landscapes with varying vegetation density, such as a sparse shrubland versus a dense forest. Other approaches, such as incorporating all available Landsat imagery within pre and postfire windows (rather than a single pair), have been shown to reduce inconsistencies (e.g., Parks et al., 2018). Nevertheless, the Alaska case demonstrates that robust severity mapping requires both algorithmic refinements and awareness of image availability constraints. The research highlights the need for innovative approaches, possibly combining Landsat with higher-frequency sensors (e.g., Sentinel-2), to better capture postfire severity in Arctic and boreal environments.

15.6 Quantifying smoke pollution

Wildfire-driven emissions are a widely acknowledged component of air quality degradation and adverse human health impacts (Loboda et al., 2023). Biomass burning produces a range of emissions, including carbon monoxide (CO), carbon dioxide (CO_2), methane (CH_4), nonmethane hydrocarbons (NMHC), nitrogen oxide (NOx), and volatile organic compounds (VOCs), alongside large volumes of particulate matter (PM) and water vapor (National Research Council, 2004). Fine $PM_{2.5}$ is considered the primary air pollutant of concern from wildfires (Alaska Department of Environmental Conservation, 2017). The small size of the particles is of particular importance both in terms of the capacity for long-range transport in the atmosphere and penetration into the human lungs, causing substantial health concerns. Smoke pollution monitoring is routinely carried out by the Environmental Protection Agency (EPA) with near-real-time assessments delivered to the public through the AirNow Fire and Smoke web platform (fire.airnow.gov). The EPA is mandated to operate an air monitoring network that tracks six "criteria" pollutants, which include CO, lead, nitrogen dioxide, ozone, PM, and sulfur dioxide (US Environmental Protection Agency, 2015). These assessments incorporate measurements at the EPA's regulatory stations, satellite observations of Aerosol Optical Depth (AOD), derived assessments of concentrations of various pollutants in the air, and the diverse network of Low-Cost Air Quality Monitors (LCAQM) deployed by various public agencies or private citizens.

Monitoring air pollution in Alaska again presents a unique set of challenges. Federal Reference Methods (FRMs) and Federal Equivalent Methods (FEMs) are used by the EPA to accurately assess compliance with air quality standards associated with the six air pollutants. These methods specify the instruments, standards, and procedures that typically require highly trained experts to install and maintain the equipment and perform sampling. In Alaska, where there is a sparse population over a large geographic area, there are a limited number of FRM and FEM sites due to siting criteria that generally prioritize placement in close proximity to population centers (Alaska Department of Environmental Conservation, 2024). This makes the observations from the LCAQM not only a beneficial source of information but also a critical one, increasing the observation network density from seven regulatory stations to over 100 point-source ground sampling locations. However, even with over a hundred stations distributed opportunistically across the state, the ground-based assessments are entirely inadequate for air quality monitoring across the state as a whole, emphasizing the importance of satellite observations.

Previous research by Chen et al. (2023) addressed this challenge by developing a modeling framework integrating the Hybrid Single-Particle Lagrangian Integrated Trajectory (HYSPLIT) model (Draxler and Hess, 1998; Stein et al., 2015), the

Wildland Fire Emissions Inventory System (WFEIS) (French et al., 2014), and the ABoVE Wildfire Date of Burning dataset (Loboda et al., 2017). Their study produced daily, gridded $PM_{2.5}$ concentrations for Alaska (2001–15), explicitly attributing emissions to wildfire sources. This 15-year product demonstrated that wildfire smoke is the dominant driver of $PM_{2.5}$ concentrations in Alaska during the fire season and provided the first spatially continuous dataset quantifying fire-driven air quality impacts across the state's rural and urban regions.

Numerous studies have shown a potent ability of satellite AOD measurements to reflect surface $PM_{2.5}$ concentrations. However, there are key challenges to estimating smoke using a satellite-data approach in the HNL region: incomplete coverage by geostationary satellites and high-quality AOD retrievals are substantially limited by low sun angles and persistent cloud cover. This necessitates the development of modeling approaches that can provide daily temporal resolution, required for air quality assessments, and fill observational gaps, particularly for smoke-related $PM_{2.5}$ concentrations. Several satellite-derived datasets that estimate $PM_{2.5}$ over Alaska are publicly available (Table 15.1). These datasets demonstrate the trade-off between spatial and temporal resolution, required to overcome the challenges posed by Alaska's environmental settings: daily data is available at coarse resolutions (greater than 10 km resolution), while finer-scale data (1 km resolution) is available in monthly or annual averages. The FINNv2.5 provides daily 1 km data, but it reports emissions at individual fire point sources as opposed to $PM_{2.5}$ concentrations across the landscape, which are needed to evaluate potential population exposure.

Building upon previous work (Chen et al., 2023; Bredder et al., 2025) advanced wildfire-driven air quality modeling by focusing on higher-resolution, near-real-time prediction. Specifically, Bredder et al. developed a framework that incorporates MODIS active-fire observations, which are less impacted by cloud cover and low observational angles compared to AOD, to backcast wildfire-driven daily $PM_{2.5}$ concentrations at 1 km resolution. In this approach, measurements at the regulatory monitors (FEM/FRM) are supplemented by the observations from LCAQM—PurpleAir sensors. PurpleAir measurements are processed to bring their estimates closer to those of the regulatory stations (Barkjohn et al., 2021). In contrast to the retrospective, physics-based simulations in Chen et al. (2023), Bredder et al.'s approach used a machine-learning-based random forest framework to estimate surface $PM_{2.5}$ directly from fire activity and meteorological predictors. This allowed for rapid and computationally efficient estimation of daily smoke impacts, with comparable accuracy to models using AOD data but without the same dependence on satellite retrievals.

MODIS active-fire product (MOD/MYD14) was used to derive daily counts, mean FRP, and a compound "Blaze" metric (count × mean FRP) around each air quality monitoring station. To capture distance decay and smoke transport, fires were partitioned into short-, mid-, and long-range buffer rings of varying distances (0–100/200/300; 100–500/600/700; 500–1000/600–1000/700–1000 km across three tests) and split into upwind versus downwind sectors. The result is 18 fire variables per day (three metrics × three distances × two wind sectors). This design isolates the

Table 15.1 Datasets related to $PM_{2.5}$ over Alaska.

Dataset	Spatial resolution	Temporal resolution	Units	Sources
Simulated Fine Particulate Matter ($PM_{2.5}$) estimates over Alaska, 2001–15	0.10 degrees	Daily, May–September in 2015–19	$\mu g\ m^{-3}$	Chen et al. (2024)
Global Estimates $PM_{2.5}$ (V6.GL.02.02)	0.01 degrees, 0.1 degrees	Annual and monthly mean, 1998–2022	$\mu g\ m^{-3}$	Shen et al. (2024)
Global/Regional Estimates (V5.GL.05.02)	0.01 degrees, 0.1 degrees	Annual and monthly mean and uncertainty, 1998–2023	$\mu g\ m^{-3}$	
North American Regional Estimates with Composition (V5.NA.04.02)	0.01 degrees	Annual, monthly, and biweekly mean total $PM_{2.5}$, compositional components, and biomass burning contributions, 2000–22	$\mu g\ m^{-3}$	van Donkelaar et al. (2024)
MERRA-2, Single-Level, Country-Level Surface $PM_{2.5}$ Monthly Mean Products V1	0.5 × 0.625 degrees	Monthly, 1980–2025	$\mu g\ m^{-3}$	Global Modeling and Assimilation Office (2021)
Copernicus Atmosphere Monitoring Service Global Reanalysis, European centre for medium-range weather forecasts (ECMWF) Atmospheric Composition Reanalysis 4)	0.75 degrees	Daily, 2003–24	$kg\ m^{-3}$	Copernicus (2025)
Global Fire Emissions Database, Version 4.1	0.25 degrees	Annual (1997–2016), monthly (1995–2016), saily/3-hourly (2000–15)	$g\ m^{-2}\ month^{-1}$	Randerson et al. (2018)
FINNv2.5 Gridded Emissions (Moderate Resolution Imaging Spectroradiometer and Visible Infrared Imaging Radiometer Suite)	0.1 degrees	Daily, 2002–23	Molecules $cm^{-2}\ s^{-1}$	Wiedinmyer and Emmons (2022)
FINNv2.5 Near-Real-Time Emissions (each fire)	1 km (each fire point)	Daily, 2018–present	$kg\ day^{-1}$	Wiedinmyer and Emmons (2022)

Source: Table reproduced from Bredder, A., Loboda, T.V., Chen, D., 2025. Hindcasting PM2.5 in Alaska During Wildfire Seasons. Geophysical Research Letters.

contribution of nearby fire activity (which is hypothesized to dominate surface concentrations) from more distant fire activity.

We trained 72 random forest regression models that included different combinations of fire metrics (Count, FRP, Blaze, or all combined), AOD treatment (excluded, high-quality only, or imputed), distance-ring definitions, and wind on/off. Fivefold cross-validation and held-out prediction sets were used to assess performance. Across all runs, cross-validated R^2 values ranged from approximately 0.74 to 0.83 on training and 0.73 to 0.87 on prediction, with root mean square error (RMSE) typically around 7 $\mu g\ m^{-3}$. When present, high-quality AOD was the single strongest predictor (Gini importance >0.65) and yielded the highest R^2, but with a sharply reduced sample size (~30% of daily observations available) due to high levels of missingness in AOD observations. Imputed AOD improved coverage yet consistently ranked below at least one fire predictor and reduced explained variance by roughly between 0.04 and 0.10 R^2 relative to high-quality AOD models.

In contrast, fire variables were robust and consistently influential, even within models excluding AOD. Simple fire count repeatedly outperformed FRP and Blaze when all were included, indicating that the number of active detections within transport-relevant space is a strong, stable proxy for ambient smoke. Accounting for wind provided modest gains in explained variance but reliably lowered RMSE and mean absolute error (MAE), suggesting fewer large residuals when transport direction is acknowledged. Importance profiles concentrated on the nearest fire ring; variables beyond ~700 km rarely achieved high importance, implying that most day-to-day Alaskan smoke burden at a location is governed by fires within several hundred kilometers.

The results show the relative magnitude of smoke episodes when compared with colocated EPA time series, particularly in high-fire years, though extremes (>150 $\mu g\ m^{-3}$) are underpredicted, which is likely a function of limited training samples in this range. Land cover, roads, and population contributed little once fire and meteorology were included; Normalized Difference Vegetation Index (NDVI) occasionally registered weak, nonzero importance. Taken together, these patterns favor a computationally lean, AOD-independent configuration (fire count in distance rings with basic meteorology) to backcast $PM_{2.5}$ across Alaska.

15.7 Conclusions

Satellite observations are the primary source of information about wildfire occurrence, its characteristics, impact on ecosystems, and air quality in Alaska. The vastness and remoteness of Alaska's tundra and boreal biomes necessitate the reliance on satellite fire monitoring for fire and resource management needs as well as scientific research. The geographic and environmental settings of this area present substantial challenges and opportunities for using satellite observations. Short snow-free summers, long periods of limited solar illumination and low sun elevation angles, and extensive, persistent cloud cover during snow-free periods make mapping fire impact on the land surface extremely challenging. Arctic boreal regions are poorly served by observations from existing geostationary platforms, which could improve clear surface imaging due to their exceptional temporal frequency of observations. However, the large orbital overlap from coarse-resolution polar-orbiting platforms supports subdaily detection and monitoring of fire activity. Moreover, active remote sensing instruments, such as radar and lidar, make a particularly notable contribution to improved understanding of wildfire occurrence in these biomes.

The unique settings of Alaska's boreal and tundra biomes require the development of methods and datasets that counterbalance the regional observational limitations and capitalize on local ecosystem responses to build more accurate assessments. While global algorithms have revolutionized fire science across the world, it is evident that regional studies benefit from regionally adapted burned area mapping datasets in Alaska. The specifics of wildfire occurrence and the lack of other widely distributed and influential sources of air pollution in Alaska also allow for developing highly effective approaches to fire-driven air quality degradation monitoring, which are not necessarily transferable to other geographic areas outside HNL. The available long-term archive of satellite observations in these ecosystems continues to provide the basis for developing novel approaches to understanding and modeling fire ignition and propagation using ML and AI methods. It is possible that the impressive success of our predictive models for fire ignition rests largely on the dominance of CG lightning as the ignition source, which can be successfully modeled from the state of the atmosphere and other landscape variables in Alaska. A broader contribution from the anthropogenic sources of ignition, which is commonly observed in other regions of the world, would likely reduce the predictive accuracy of fire ignition forecasting.

As the recognition of the importance of the rapidly changing fire regimes in HNL to the global system grows, more effort is devoted to developing novel algorithms and conceptualizing and implementing new satellite constellations to benefit wildfire monitoring. High temporal resolution of wildfire monitoring can be achieved either through launching High Elliptical Orbit (HEOs) geostationary satellites (either in the Molnyia or Tundra orbits) or launching a very large constellation of low-cost polar-orbiting satellites, similar to the FireSat constellation currently envisioned by a nongovernmental organization—the Earth Fire Alliance (www.earthfirealliance.org). As with the opening of the Landsat data archive in 2008 or the launch of the European Space Agency's Sentinel satellites, increased availability of satellite observations will likely lead to an exponential growth in wildfire monitoring and forecasting capabilities, making tremendous contributions to fire management and fire science, and by extension to the Earth System as a whole.

References

Abatzoglou, J.T., Williams, A.P., Barbero, R., 2019. Global emergence of anthropogenic climate change in fire weather indices. Geophysical Research Letters 46 (1), 326–336. Available at https://doi.org/10.1029/2018GL080959, http://agupubs.onlinelibrary.wiley.com/hub/journal/10.1002/(ISSN)1944-8007/.

Alaska Department of Environmental Conservation, 2017. 2016 Alaska Wildfire Emissions Inventory. https://dec.alaska.gov/media/6383/2016-wildfire-inventory-report.pdf Last accessed 6/27/2026.

Alaska Department of Environmental Conservation, 2024. 2024 Annual Air Quality Monitoring Network Plan. https://dec.alaska.gov/media/kmrfthes/adec_2024-annual-network-plan_final_20240627.pdf, accessed on July 13, 2026.

Alaska Statewide Operating Plan 2023, 2023.

Alkhatib, R., Sahwan, W., Alkhatieb, A., Schütt, B., 2023. A brief review of machine learning algorithms in forest fires science. Applied Sciences 13 (14), 8275. Available at https://doi.org/10.3390/app13148275.

Andela, N., Kaiser, J.W., van der Werf, G.R., Wooster, M.J., 2015. New fire diurnal cycle characterizations to improve fire radiative energy assessments made from MODIS observations. Atmospheric Chemistry and Physics 15 (15), 8831–8846. Available at https://doi.org/10.5194/acp-15-8831-2015.

Baker, B., Halsey, R.W., 2022. California chaparral and woodlands. Imperiled: The Encyclopedia of Conservation 1-3. Elsevier, United States, pp. 196–207. Available at https://www.sciencedirect.com/book/9780128211397, https://doi.org/10.1016/B978-0-12-821139-7.00013-1.

Balch, J.K., Iglesias, V., Mahood, A.L., Cook, M.C., Amaral, C., DeCastro, A., et al., 2024. The fastest-growing and most destructive fires in the US (2001 to 2020). Science (New York, N.Y.) 386 (6720), 425–431. Available at https://doi.org/10.1126/science.adk5737, https://www.science.org/doi/10.1126/science.adk5737.

Barkjohn, K.K., Gantt, B., Clements, A.L., 2021. Development and application of a United States-wide correction for PM2.5 data collected with the PurpleAir sensor. Atmospheric Measurement Techniques 14 (6), 4617–4637. Available at https://doi.org/10.5194/amt-14-4617-2021, http://www.atmos-meas-tech.net/volumes_and_issues.html.

Barrett, K., Kasischke, E.S., 2013. Controls on variations in MODIS fire radiative power in Alaskan boreal forests: Implications for fire severity conditions. Remote Sensing of Environment 130, 171–181. Available at https://doi.org/10.1016/j.rse.2012.11.017.

Bieniek, P.A., Bhatt, U.S., York, A., Walsh, J.E., Lader, R., Strader, H., et al., 2020. Lightning variability in dynamically downscaled simulations of alaska's present and future summer climate. Journal of Applied Meteorology and Climatology 59 (6), 1139–1152. Available at https://doi.org/10.1175/JAMC-D-19-0209.1, https://journals.ametsoc.org//jamc/article-pdf/59/6/1139/4961820/jamcd190209.pdf.

Bourgeau-Chavez, L.L., Harrell, P.A., Kasischke, E.S., French, N.H.F., 1997. The detection and mapping of alaskan wildfires using a spaceborne imaging radar system. International Journal of Remote Sensing 18 (2), 355–373. Available at https://doi.org/10.1080/014311697219114.

Bredder, A., Loboda, T.V., Chen, D., 2025. Hindcasting PM2.5 in alaska during wildfire seasons. Geophysical Research Letters.

Chatzopoulos-Vouzoglanis, K., Reinke, K.J., Soto-Berelov, M., Jones, S.D., 2024. Are fire intensity and burn severity associated? Advancing our understanding of FRP and NBR metrics from Himawari-8/9 and Sentinel-2. International Journal of Applied Earth Observation and Geoinformation 127. Available at https://doi.org/10.1016/j.jag.2024.103673, https://www.sciencedirect.com/science/journal/15698432.

Chen, D., Loboda, T.V., Hall, J.V., 2020. A systematic evaluation of influence of image selection process on remote sensing-based burn severity indices in North American boreal forest and tundra ecosystems. ISPRS Journal of Photogrammetry and Remote Sensing

159, 63–77. Available at https://doi.org/10.1016/j.isprsjprs.2019.11.011, www.elsevier.com/inca/publications/store/5/0/3/3/4/0.

Chen, D., Shevade, V., Baer, A., Loboda, T.V., 2021. Missing Burns in the High Northern Latitudes: The Case for Regionally Focused Burned Area Products. Remote Sensing 13, 4145. Available at https://doi.org/10.3390/rs13204145.

Chen, Y., Hantson, S., Andela, N., Coffield, S.R., Graff, C.A., Morton, D.C., et al., 2022. California wildfire spread derived using VIIRS satellite observations and an object-based tracking system. Scientific Data 9 (1). Available at https://doi.org/10.1038/s41597-022-01343-0, www.nature.com/sdata/.

Chen, D., Billmire, M., Loughner, C.P., Bredder, A., French, N.H.F., Kim, H.C., et al., 2023. Simulating spatio-temporal dynamics of surface PM2.5 emitted from Alaskan wildfires. Science of the Total Environment 898. Available at https://doi.org/10.1016/j.scitotenv.2023.165594, www.elsevier.com/locate/scitotenv.

Chen, R., He, B., Li, Y., Fan, C., Yin, J., Zhang, H., et al., 2024. Estimation of potential wildfire behavior characteristics to assess wildfire danger in southwest China using deep learning schemes. Journal of Environmental Management 351, 120005. Available at https://doi.org/10.1016/j.jenvman.2023.120005.

Clelland, A.A., Marshall, G.J., Baxter, R., Potter, S., Talucci, A.C., Rady, J.M., et al., 2024. Annual and seasonal patterns of burned area products in Arctic-Boreal North America and Russia for 2001–2020. Remote Sensing 16 (17). Available at https://doi.org/10.3390/rs16173306, http://www.mdpi.com/journal/remotesensing.

Copernicus, 2025 - Copernicus Atmosphere Monitoring Service (2020): CAMS global reanalysis (EAC4). Copernicus Atmosphere Monitoring Service (CAMS) Atmosphere Data Store. Available at https://doi.org/10.24381/d58bbf47.

Csiszar, I., Schroeder, W., Giglio, L., Ellicott, E., Vadrevu, K.P., Justice, C.O., et al., 2014. Active fires from the Suomi NPP visible infrared imaging radiometer suite: product status and first evaluation results. Journal of Geophysical Research: Atmospheres 119 (2), 803–816. Available at https://doi.org/10.1002/2013jd020453.

Descloitres, J., Sohlberg, R., Owens, J., Giglio, L., Justice, C., Carroll, M., et al., 2002. The MODIS rapid response project. International Geoscience and Remote Sensing Symposium (IGARSS) 2, 1191–1192.

Dewitz, J., 2021. National land cover database (NLCD) 2019 products (ver. 2.0, June 2021). US Geol. Surv. https://www.usgs.gov/data/national-land-cover-database-nlcd-2019-products. https://doi.org/10.5066/P9KZCM54.

Dissing, D., Verbyla, D.L., 2003. Spatial patterns of lightning strikes in interior Alaska and their relations to elevation and vegetation. Canadian Journal of Forest Research 33 (5), 770–782. Available at https://doi.org/10.1139/x02-214.

Dozier, J., 1981. A method for satellite identification of surface temperature fields of subpixel resolution. Remote Sensing of Environment 11 (C), 221–229. Available at https://doi.org/10.1016/0034-4257(81)90021-3.

Draxler, R.R., Hess, G.D., 1998. An overview of the HYSPLIT_4 modelling system for trajectories, dispersion and deposition. Australian Meteorological Magazine 47 (4), 295–308.

Earnest, B.L., McGovern, A., Karstens, C., Jirak, I., 2024. Part 1: Improving wildfire occurrence prediction for CONUS using deep learning and fire weather variables. Artificial Intelligence for the Earth Systems. Available at https://doi.org/10.1175/AIES-D-23-0057.1.

French, N.H.F., Kasischke, E.S., Hall, R.J., Murphy, K.A., Verbyla, D.L., Hoy, E.E., et al., 2008. Using Landsat data to assess fire and burn severity in the North American boreal forest region: an overview and summary of results. International Journal of Wildland Fire 17 (4), 443–462. Available at https://doi.org/10.1071/WF08007.

French, N.H.F., McKenzie, D., Erickson, T., Koziol, B., Billmire, M., Endsley, K.A., et al., 2014. Modeling regional-scale wildland fire emissions with the Wildland Fire Emissions Information System. Earth Interactions 18 (16), 1–26. Available at https://doi.org/10.1175/EI-D-14-0002.1, http://journals.ametsoc.org/doi/pdf/10.1175/EI-D-14-0002.1.

Fronterhouse, B.A., 2012. Alaska Lightning Detection Network (ALDN) briefing document.

Gassert, F., Venter, O., Watson, J.E.M., Brumby, S.P., Mazzariello, J.C., Atkinson, S.C., et al., 2023. An operational approach to near real time global high resolution mapping of the terrestrial Human Footprint. Frontiers in Remote Sensing 4. Available at https://doi.org/10.3389/frsen.2023.1130896, https://www.frontiersin.org/journals/remote-sensing.

Gatebe, C.K., Varnai, T., Poudyal, R., Ichoku, C., King, M.D., 2012. Taking the pulse of pyrocumulus clouds. Atmospheric Environment 52, 121–130. Available at https://doi.org/10.1016/j.atmosenv.2012.01.045.

Giglio, L., 2007. Characterization of the tropical diurnal fire cycle using VIRS and MODIS observations. Remote Sensing of Environment 108 (4), 407–421. Available at https://doi.org/10.1016/j.rse.2006.11.018.

Giglio, L., Kendall, J.D., Justice, C.O., 1999. Evaluation of global fire detection algorithms using simulated avhrr infrared data. International Journal of Remote Sensing 20 (10), 1947–1985. Available at https://doi.org/10.1080/014311699212290.

Giglio, L., Loboda, T., Roy, D.P., Quayle, B., Justice, C.O., 2009. An active-fire based burned area mapping algorithm for the MODIS sensor. Remote Sensing of Environment 113 (2), 408–420. Available at https://doi.org/10.1016/j.rse.2008.10.006.

Giglio, L., Schroeder, W., Justice, C.O., 2016. The collection 6 MODIS active fire detection algorithm and fire products. Remote Sensing of Environment 178, 31–41. Available at https://doi.org/10.1016/j.rse.2016.02.054, www.elsevier.com/inca/publications/store/5/0/5/7/3/3.

Global Modeling and Assimilation Office, 2021; MERRA-2avgM_2d_pm25_admin0,2d, Single-Level, Country-Level Surface PM2.5 Monthly Mean Products V1 (M2_TMAX_PM25) at GES DISC. Available at https://doi.org/10.5067/FY616726UXSR.

Guibas, J., Mardani, M.,Li, Z., Tao, A., Aanandkumar, A., Catanzaro, B., 2021. Adaptive fourier neural operators: efficient token mixers for transformers. arXiv, United States. Available at https://arxiv.org. https://doi.org/10.48550/arxiv.2111.13587.

Hantson, S., Andela, N., Goulden, M.L., Randerson, J.T., 2022. Human-ignited fires result in more extreme fire behavior and ecosystem impacts. Nature Communications 13 (1). Available at https://doi.org/10.1038/s41467-022-30030-2, http://www.nature.com/ncomms/index.html.

He, J., Loboda, T.V., 2020. Modeling cloud-to-ground lightning probability in Alaskan tundra through the integration of Weather Research and Forecast (WRF) model and machine learning method. Environmental Research Letters 15 (11), 115009. Available at https://doi.org/10.1088/1748-9326/abbc3b.

He, J., Loboda, T.V., Jenkins, L., Chen, D., 2019. Mapping fractional cover of major fuel type components across Alaskan tundra. Remote Sensing of Environment 232, 111324. Available at https://doi.org/10.1016/j.rse.2019.111324.

He, J., Loboda, T.V., Chen, D., French, N.H.F., 2021. Cloud-to-ground lightning and near-surface fire weather control wildfire occurrence in Arctic tundra. Geophysical Research Letters 48, e2021GL096814. Available at https://doi.org/10.1029/2021GL096814.

Heward, H., Smith, A.M.S., Roy, D.P., Tinkham, W.T., Hoffman, C.M., Morgan, P., et al., 2013. Is burn severity related to fire intensity? Observations from landscape scale remote sensing. International Journal of Wildland Fire 22 (7), 910–918. Available at https://doi.org/10.1071/WF12087.

Houghton, R.A., 2012. Carbon emissions and the drivers of deforestation and forest degradation in the tropics. Current Opinion in Environmental Sustainability 4 (6), 597–603. Available at https://doi.org/10.1016/j.cosust.2012.06.006.

Hu, F.S., Higuera, P.E., Duffy, P., Chipman, M.L., Rocha, A.V., Young, A.M., et al., 2015. Arctic tundra fires: natural variability and responses to climate change. Frontiers in Ecology and the Environment 13 (7), 369–377. Available at https://doi.org/10.1890/150063, http://www.esajournals.org/doi/pdf/10.1890/150063.

Ichoku, C., Ellison, L., 2014. Global top-down smoke-aerosol emissions estimation using satellite fire radiative power measurements. Atmospheric Chemistry and Physics 14 (13), 6643–6667. Available at https://doi.org/10.5194/acp-14-6643-2014.

Jenkins, L., Bourgeau-Chavez, L., French, N., Loboda, T., Thelen, B., 2014. Development of methods for detection and monitoring of fire disturbance in the alaskan tundra using a two-decade long record of synthetic aperture radar satellite images. Remote Sensing 6 (7), 6347–6364. Available at https://doi.org/10.3390/rs6076347.

Justice, C.O., Giglio, L., Korontzi, S., Owens, J., Morisette, J.T., Roy, D., et al., 2002. The MODIS fire products. Remote Sensing of Environment 83 (1-2), 244–262. Available at https://doi.org/10.1016/s0034-4257(02)00076-7.

Kablick, G., Fromm, M., Miller, S., Partain, P., Peterson, D., Lee, S., et al., 2018. The great slave lake PyroCb of 5 August 2014: observations, simulations, comparisons with regular convection, and impact on UTLS water vapor. Journal of Geophysical Research: Atmospheres 123 (21), 12–352. Available at https://doi.org/10.1029/2018JD028965, http://agupubs.onlinelibrary.wiley.com/hub/jgr/journal/10.1002/(ISSN)2169-8996/.

Kalluri, S., Cao, C., Heidinger, A., Ignatov, A., Key, J., Smith, T., 2021. The advanced very high resolution radiometer: contributing to earth observations for over 40 years. Bulletin of the American Meteorological Society 102 (2), E351. Available at https://doi.org/10.1175/bams-d-20-0088.1.

Kasischke, E.S., Hoy, E.E., 2012. Controls on carbon consumption during Alaskan wildland fires. Global Change Biology 18 (2), 685–699. Available at https://doi.org/10.1111/j.1365-2486.2011.02573.x.

Kasischke, E.S., Bourgeau-Chavez, L.L., Johnstone, J.F., 2007. Assessing spatial and temporal variations in surface soil moisture in fire-disturbed black spruce forests in Interior Alaska using spaceborne synthetic aperture radar imagery - Implications for post-fire tree recruitment. Remote Sensing of Environment 108 (1), 42–58. Available at https://doi.org/10.1016/j.rse.2006.10.020.

Kasischke, E.S., Turetsky, M.R., Ottmar, R.D., French, N.H.F., Hoy, E.E., Kane, E.S., 2008. Evaluation of the composite burn index for assessing fire severity in Alaskan black spruce forests. International Journal of Wildland Fire 17 (4), 515–526. Available at https://doi.org/10.1071/WF08002.

Kaufman, Y.J., Justice, C.O., Flynn, L.P., Kendall, J.D., Prins, E.M., Giglio, L., et al., 1998. Potential global fire monitoring from EOS-MODIS. Journal of Geophysical Research Atmospheres 103 (24), 32215–32238. Available at https://doi.org/10.1029/98JD01644, http://onlinelibrary.wiley.com/journal/10.1002/(ISSN)2169-8996.

Langaas, S., 1992. Temporal and spatial distribution of savanna fires in senegal and the Gambia, West Africa, 1989-90, derived from multi-temporal AVHRR night images. International Journal of Wildland Fire 2 (1), 21. Available at https://doi.org/10.1071/wf9920021.

Li, F., Zhang, X., Kondragunta, S., Roy, D.P., 2018. Investigation of the fire radiative energy biomass combustion coefficient: a comparison of polar and geostationary satellite retrievals over the conterminous United States. Journal of Geophysical Research: Biogeosciences 123 (2), 722–739. Available at https://doi.org/10.1002/2017JG004279, http://onlinelibrary.wiley.com/journal/10.1002/(ISSN)2169-8961.

Li, F., Zhang, X., Kondragunta, S., Lu, X., Csiszar, I., Schmidt, C.C., 2022. Hourly biomass burning emissions product from blended geostationary and polar-orbiting satellites for air quality forecasting applications. Remote Sensing of Environment 281, 113237. Available at https://doi.org/10.1016/j.rse.2022.113237.

Li, Y., Janssen, T.A.J., Chen, R., He, B., Veraverbeke, S., 2024. Trends and drivers of Arctic-boreal fire intensity between 2003 and 2022. Science of the Total Environment 926, 172020. Available at https://doi.org/10.1016/j.scitotenv.2024.172020.

Liu, T., Randerson, J.T., Chen, Y., Morton, D.C., Wiggins, E.B., Smyth, P., et al., 2024. Systematically tracking the hourly progression of large wildfires using GOES satellite observations. Earth System Science Data 16 (3), 1395–1424. Available at https://doi.org/10.5194/essd-16-1395-2024, www.earth-system-science-data.net.

Loboda, T.V., Csiszar, I.A., 2007. Reconstruction of fire spread within wildland fire events in Northern Eurasia from the MODIS active fire product. Global and Planetary Change 56 (3-4), 258–273. Available at https://doi.org/10.1016/j.gloplacha.2006.07.015.

Loboda, T.V., Hall, J.V., Baer, A., 2017. Arctic-boreal vulnerability experiment (ABoVE)ABoVE: wildfire date of burning within fire scars across Alaska and Canada, 2001-2019. Available from: https://doi.org/10.3334/ORNLDAAC/1559.

Loboda, T.V., French, N.H.F., Puett, R.C., 2023. Bridging Geophysical and Health Sciences to Study the Impacts of Biomass Burning on Human Well-Being Landscape Fire, Smoke, and Health: Linking Biomass Burning Emissions to Human Well-Being. Wiley, pp. 1–6. Available at https://doi.org/10.1002/9781119757030.ch1.

Loboda, T., Hall, J., Chen, D., Hoffman-Hall, Shevade, V., Argueta, F., et al., 2024. Arctic boreal annual burned area, circumpolar boreal forest and tundra, V2, 2002-2022. Available from: https://doi.org/10.3334/ORNLDAAC/2328.

Macander, M.J., Frost, G.V., Nelson, P.R., Swingley, C.S., 2017. Regional quantitative cover mapping of tundra plant functional types in Arctic Alaska. Remote Sensing 9 (10). Available at https://doi.org/10.3390/rs9101024, http://www.mdpi.com/2072-4292/9/10/1024/pdf.

Malingreau, J.-P., 1990. The Contribution of Remote Sensing to the Global Monitoring of Fires in Tropical and Subtropical Ecosystems. Springer Science and Business Media LLC, pp. 337–370. Available at https://doi.org/10.1007/978-3-642-75395-4_15.

Miller, J.D., Thode, A.E., 2007. Quantifying burn severity in a heterogeneous landscape with a relative version of the delta Normalized Burn Ratio (dNBR). Remote Sensing of Environment 109 (1), 66–80. Available at https://doi.org/10.1016/j.rse.2006.12.006.

Moubarak, M., Sistla, S., Potter, S., Natali, S.M., Rogers, B.M., 2023. Carbon emissions and radiative forcings from tundra wildfires in the Yukon-Kuskokwim River Delta, Alaska. Biogeosciences 20 (8), 1537–1557. Available at https://doi.org/10.5194/bg-20-1537-2023, http://www.biogeosciences.net/volumes_and_issues.html.

National Research Council, 2004. Air Quality Management in the United States. National Academies Press, Washington, D.C. Available at https://doi.org/10.17226/10728.

Orland, E., McCabe, T.D., Chen, Y., Scholten, R.C., Becker, Z., Loehman, R.A., et al., 2025. Near real-time indicators of burn severity in the western U.S. from active fire tracking. Fire Ecology 21 (1). Available at https://doi.org/10.1186/s42408-025-00407-x, https://www.springer.com/journal/42408.

Parks, S., Dillon, G., Miller, C., 2014. A new metric for quantifying burn severity: the relativized burn ratio. Remote Sensing 6 (3), 1827–1844. Available at https://doi.org/10.3390/rs6031827.

Parks, S.A., Holsinger, L.M., Voss, M.A., Loehman, R.A., Robinson, N.P., 2018. Mean composite fire severity metrics computed with google earth engine offer improved accuracy and expanded mapping potential. Remote Sensing 10 (6). Available at https://doi.org/10.3390/rs10060879, https://www.mdpi.com/2072-4292/10/6/879/pdf.

Peterson, D., Wang, J., Ichoku, C., Remer, L.A., 2010. Effects of lightning and other meteorological factors on fire activity in the North American boreal forest: implications for fire weather forecasting. Atmospheric Chemistry and Physics 10 (14), 6873–6888. Available at https://doi.org/10.5194/acp-10-6873-2010.

Peterson, D., Hyer, E., Wang, J., 2014. Quantifying the potential for high-altitude smoke injection in the North American boreal forest using the standard MODIS fire products and subpixel-based methods. Journal of Geophysical Research: Atmospheres 119 (6), 3401–3419. Available at https://doi.org/10.1002/2013jd021067.

Peterson, D.A., Hyer, E.J., Campbell, J.R., Solbrig, J.E., Fromm, M.D., 2017. A conceptual model for development of intense pyrocumulonimbus in western North America. Monthly Weather Review 145 (6), 2235–2255. Available at https://doi.org/10.1175/MWR-D-16-0232.1, http://journals.ametsoc.org/doi/pdf/10.1175/MWR-D-16-0232.1.

Ponomarev, E.I., Litvintsev, K.Y., Ponomareva, T.V., Shvetsov, E.G., Yakimov, N.D., 2020. Satellite monitoring of the wildfire in Siberia and fire emissions estimation. Sovremennye problemy distantsionnogo zondirovaniya Zemli iz kosmosa 17 (6), 45–50. Available at https://doi.org/10.21046/2070-7401-2020-17-6-45-50.

Ponomarev, E.I., Zabrodin, A.N., Shvetsov, E.G., Ponomareva, T.V., 2023. Wildfire Intensity and Fire Emissions in Siberia. Fire 6 (7). Available at https://doi.org/10.3390/fire6070246, www.mdpi.com/journal/fire.

Potter, S., Yang, Y., Burrell, A., Talucci, A., Veraverbeke, S., Randerson, J.T., et al., 2024. Mapping Arctic-Boreal Burned Area in North America Using a Convolutional Neural Network with Landsat and Sentinel-2 Imagery. SSRN, United States SSRN. Available at https://www.ssrn.com/index.cfm/en/, https://doi.org/10.2139/ssrn.4803815.

Randerson, J.T., van der Werf, G.R., Giglio, L., Collatz, G.J., Kasibhatla., P.S., 2018. Global Fire Emissions Database, Version 4, (GFEDv4). ORNL DAAC, Oak Ridge, Tennessee, USA. Available at https://doi.org/10.3334/ORNLDAAC/1293.

Reap, R.M., 1991. Climatological characteristics and objective prediction of thunderstorms over Alaska. Weather and Forecasting 6 (3), 309–319. Available at https://doi.org/10.1175/1520-0434(1991)006<0309:ccaopo>2.0.co;2.

Ryan, K.C., Opperman, T.S., 2013. LANDFIRE - A national vegetation/fuels data base for use in fuels treatment, restoration, and suppression planning. Forest Ecology and Management 294, 208–216. Available at https://doi.org/10.1016/j.foreco.2012.11.003.

Scholten, R.C., Jandt, R., Miller, E.A., Rogers, B.M., Veraverbeke, S., 2021. Overwintering fires in boreal forests. Nature 593 (7859), 399–404. Available at https://doi.org/10.1038/s41586-021-03437-y, http://www.nature.com/nature/index.html.

Schroeder, W., Oliva, P., Giglio, L., Csiszar, I.A., 2014. The New VIIRS 375m active fire detection data product: algorithm description and initial assessment. Remote Sensing of Environment 143, 85–96. Available at https://doi.org/10.1016/j.rse.2013.12.008.

Shen, S., Li, C., van Donkelaar, A., Jacobs, N., Wang, C., Martin, R., 2024. Enhancing Global Estimation of Fire Particulate Matter Concentrations by Including Geophysical a Priori Information in Deep Learning. ACS ES&T Air 1 (5), 332–345. Available at https://doi.org/10.1021/acsestair.3c00054.

Smith, C.W., Panda, S.K., Bhatt, U.S., Meyer, F.J., 2021. Improved boreal forest wildfire fuel type mapping in interior alaska using aviris-ng hyperspectral data. Remote Sensing 13 (5), 1–15. Available at https://doi.org/10.3390/rs13050897, https://www.mdpi.com/2072-4292/13/5/897/pdf.

Song, Y., Xu, C., Li, X., Oppong, F., 2024. Lightning-induced wildfires: an overview. Fire 7 (3), 79. Available at https://doi.org/10.3390/fire7030079.

State of Alaska Department of Natural Resources, 2022. Division of Forestry and Fire Protection: Fire Assignments. https://forestry.alaska.gov/fire/assignments, accessed October 15, 2025.

Stein, A.F., Draxler, R.R., Rolph, G.D., Stunder, B.J.B., Cohen, M.D., Ngan, F., 2015. NOAA's HYSPLIT atmospheric transport and dispersion modeling system. Bulletin of the American Meteorological Society 96 (12), 2059–2077. Available at https://doi.org/10.1175/bams-d-14-00110.1.

Tanimoto, H., Ikeda, K., Folkert Boersma, K., van der A, R.J., Garivait, S., 2015. Interannual variability of nitrogen oxides emissions from boreal fires in Siberia and Alaska during 1996–2011 as observed from space. Environmental Research Letters 10 (6), 065004. Available at https://doi.org/10.1088/1748-9326/10/6/065004.

Townshend, J.R.G., Justice, C.O., 2002. Towards operational monitoring of terrestrial systems by moderate-resolution remote sensing. Remote Sensing of Environment 83 (1-2), 351–359. Available at https://doi.org/10.1016/S0034-4257(02)00082-2.

U.S. Environmental Protection Agency 2015. Preamble to the Integrated Science Assessments (No. EPA/600/R-15/067). Office of Research and Development, Washington, DC.

U.S. Geological Survey, 2022. 5 Meter Alaska Digital Elevation Models (DEMs) - USGS National Map 3DEP Downloadable Data Collection.

U.S. Geological Survey, 2023. Annual National Land Cover Database (NLCD) Collection 1 Fractional Impervious Surface. https://doi.org/10.5066/P94UXNTS.

van Donkelaar, A., Martin, R., Ford, B., Li, C., Pappin, A., Shen, S., et al., 2024. North American fine particulate matter chemical composition for 2000–2022 from satellites, models, and monitors: The changing contribution of wildfires. ACS Es&t Air 1 (12), 1589–1600. Available at https://doi.org/10.1021/acsestair.4c00151.

Veraverbeke, S., Rogers, B.M., Goulden, M.L., Jandt, R.R., Miller, C.E., Wiggins, E.B., et al., 2017. Lightning as a major driver of recent large fire years in North American boreal forests. Nature Climate Change 7 (7), 529–534. Available at https://doi.org/10.1038/nclimate3329, http://www.nature.com/nclimate/index.html.

Vermote, E., Ellicott, E., Dubovik, O., Lapyonok, T., Chin, M., Giglio, L., et al., 2009. An approach to estimate global biomass burning emissions of organic and black carbon from MODIS fire radiative power. Journal of Geophysical Research Atmospheres 114 (18) Available at https://doi.org/10.1029/2008JD011188, http://onlinelibrary.wiley.com/journal/10.1002/(ISSN)2169-8996.

Wiedinmyer, C., Emmons, L., 2022: Fire INventory from NCAR (FINN) v2.5(MODIS), MOZART VOC speciation (v2.5), Zenodo [code], https://doi.org/10.5281/zenodo.7868652 (last access: 7 July 2023), 2023.

Wiggins, E.B., Veraverbeke, S., Henderson, J.M., Karion, A., Miller, J.B., Lindaas, J., et al., 2016. The influence of daily meteorology on boreal fire emissions and regional trace gas variability. Journal of Geophysical Research: Biogeosciences 121 (11), 2793–2810. Available at https://doi.org/10.1002/2016jg003434.

Wiggins, E.B., Soja, A.J., Gargulinski, E., Halliday, H.S., Pierce, R.B., Schmidt, C.C., et al., 2020. High temporal resolution satellite observations of fire radiative power reveal link between fire behavior and aerosol and gas emissions. Geophysical Research Letters 47

(23). Available at https://doi.org/10.1029/2020GL090707, http://agupubs.onlinelibrary.wiley.com/hub/journal/10.1002/(ISSN)1944-8007/.

Wooster, M., 2003. Fire radiative energy for quantitative study of biomass burning: derivation from the BIRD experimental satellite and comparison to MODIS fire products. Remote Sensing of Environment 86 (1), 83–107. Available at https://doi.org/10.1016/s0034-4257(03)00070-1.

Wooster, M.J., Roberts, G., Perry, G.L.W., Kaufman, Y.J., 2005. Retrieval of biomass combustion rates and totals from fire radiative power observations: FRP derivation and calibration relationships between biomass consumption and fire radiative energy release. Journal of Geophysical Research Atmospheres 110 (24), 1–24. Available at https://doi.org/10.1029/2005JD006318, http://onlinelibrary.wiley.com/journal/10.1002/(ISSN)2169-8996.

Zhang, P., Hu, X., Ban, Y., Nascetti, A., Gong, M., 2024. Assessing sentinel-2, sentinel-1, and ALOS-2 PALSAR-2 data for large-scale wildfire-burned area mapping: insights from the 2017–2019 Canada wildfires. Remote Sensing 16 (3), 556. Available at https://doi.org/10.3390/rs16030556.

Burned area mapping in Italy using multitemporal composite optical and synthetic aperture radar (SAR) data and machine learning

16

Giandomenico De Luca[1] *and Giuseppe Modica*[2]
[1]Institute of BioEconomy (IBE), National Research Council of Italy (CNR), Sesto Fiorentino, Italy, [2]Department of Veterinary Sciences, University of Messina, Messina, Italy

16.1 Introduction

16.1.1 Wildland fire in Italy: an overview

The spatial distribution of wildland fires in Italy is highly complex, as the country encompasses one of the most heterogeneous mosaics of vegetation types, bioclimates, and land-use systems in Europe (Blasi et al., 2014; Elia et al., 2022; Pesaresi et al., 2017). Extending between 35° and 47° North in the center of the Mediterranean basin and shaped by millennia of human activity, Italy hosts a diverse range of rural landscapes (Agnoletti, 2013), spanning climatic conditions from subarid thermo-Mediterranean zones to the alpine treeline ecotone (Pavari, 1916; De Philippis, 1937). This biogeographic and climatic heterogeneity is directly reflected in the variability of fire regime parameters—such as frequency, intensity, and seasonality—(Michetti and Pinar, 2019; Romano and Ursino, 2020), further modulated by anthropogenic drivers, including land-use change and rural depopulation (Salis et al., 2022; Spadoni et al., 2023; Ascoli et al., 2021).

The Mediterranean climate dominates most of the territory, sustaining vegetation formations such as evergreen sclerophyllous shrublands and forests, which represent some of the most fire-prone ecosystems in the region. Consequently, the majority of wildfire events are concentrated within the thermo- and meso-Mediterranean belts of the peninsula (Dadkhah et al., 2025; Elia et al., 2022; San-Miguel-Ayanz et al., 2025). However, fire occurrence extends beyond lowland Mediterranean ecosystems, also impacting higher-elevation environments (De Luca and Modica, 2023), even volcanic ecosystems (Saulino et al., 2020), where supra- and oro-Mediterranean habitats are commonly situated within protected areas such as national parks. At the extreme, wildfires are also recorded in alpine regions, where a secondary winter fire season may arise as a result of frozen soil moisture, vegetation desiccation during dormancy, prolonged dryness, and local climatic phenomena such as föhn winds and temperature inversions (Conedera et al., 2018; Valese et al., 2014; Wastl et al., 2013).

According to the most recent European fire statistics (EFFIS - Annual Statistics for Italy, 2025), the long-term trend in Italy indicates an overall increase in both total

Advances in Remote Sensing for Forest Fire Management. DOI: https://doi.org/10.1016/B978-0-443-41692-7.00012-3
© 2027 Elsevier Ltd. All rights are reserved, including those for text and data mining, AI training, and similar technologies.

burned area and number of wildland fires over the past two decades, resulting in substantial ecological damage and significant landscape transformations, including in sensitive ecosystems. During the 2024 fire season (San-Miguel-Ayanz et al., 2025), almost 508 km^2 of wildlands were affected by fires, distributed across land-cover types as follows: agriculture (32.76%), broadleaf forest (12.32%), sclerophyllous vegetation (10.23%), transitional woodlands (9.01%), coniferous forest (2.87%), mixed forest (2.00%), other natural land (30.46%), and artificial and other land covers (0.34%). Sicily, Calabria, Lazio, Sardinia, and Campania accounted for the majority of the burned surface (>66%; up to 85% when considering only the June–September period), while approximately 30% of the total affected area involved protected forest ecosystems. The province of Reggio Calabria was particularly impacted, with 10.3 km^2 of forest burned—representing 41% of the total forest area burned in Calabria and 10% of the national forest area affected. In addition, nearly one-quarter of the total burned surface (13,507 ha) occurred within Natura 2000 sites, corresponding to 0.19% of Italy's terrestrial Natura 2000 network (San-Miguel-Ayanz et al., 2025).

16.1.2 The role of optical and synthetic aperture radar remote sensing

Mapping all wildland burned areas over the course of an entire fire season is a fundamental first step for thoroughly assessing and understanding their ecological and socio-economic impacts, as well as the contributing factors influencing fire propagation (ignition and spread) and fire regime (Crowley et al., 2019).

Remote sensing, particularly that employing optical satellite imagery, has been extensively applied for burned area detection at multiple spatial scales across Italian territory, ranging from local (Alcaras et al., 2022; Lanorte et al., 2013; Smiraglia et al., 2020) to regional (Lasaponara, 2006) or national scale (Filipponi, 2019; Pulvirenti et al., 2020). Its adoption has been facilitated by the growing availability of satellite platforms offering diverse spatial, spectral, and temporal resolutions, coupled with the development of increasingly sophisticated algorithms and processing software (Chuvieco et al., 2019; Chen et al., 2024). The effectiveness of optical sensors in postfire mapping is primarily due to the high sensitivity of the visible (VIS), near-infrared (NIR), and shortwave infrared (SWIR) spectral regions to fire-induced surface changes (Pereira, Sá, et al., 1999). In the immediate aftermath of a fire, burned vegetation exhibits a distinctive spectral signature resulting from the combined effects of reduced green biomass, exposure of bare soil, the presence of ash and charcoal, and changes in surface temperature and moisture content (Badía et al., 2017; Hudak et al., 2007; Smith et al., 2005; Pereira, Sá, et al., 1999; Quintano et al., 2020; De Luca et al., 2023).

Complementary to optical data, synthetic aperture radar (SAR) active sensors have demonstrated strong performance in burned area mapping (De Luca et al., 2021; Lasaponara and Tucci, 2019), in some cases being integrated with optical datasets to enhance detection accuracy (Stroppiana et al., 2015). SAR backscatter is sensitive to variations in surface structure and dielectric properties, making it well suited for detecting fire-induced changes (Imperatore et al., 2017; Tanase et al., 2011; Gimeno et al., 2004). In Mediterranean environments with dry summers, SAR-based approaches leverage cross-polarized backscatter (e.g., vertical-horizontal, VH; horizontal-vertical, HV) that typically decreases after fires due to reduced volumetric scattering and moisture content (Carreiras et al., 2020; De Luca et al., 2021; Ruiz-Ramos et al., 2018; Tanase et al., 2010b; Tanase et al., 2010a). Conversely, co-polarized channels (vertical-vertical, VV; horizontal-horizontal, HH) can be affected by increased soil exposure following vegetation loss (Imperatore et al., 2017). Some studies have leveraged dual-polarimetric indices to exploit the complementary information provided by both polarization modes (De Luca et al., 2020; De Luca et al., 2021; De Luca et al., 2022; Martins et al., 2016).

16.1.3 The multitemporal compositing criteria: principles and literature review

Multitemporal image compositing enables the detection of fires occurring on different days throughout a fire season while maintaining spectral separability between burned and unburned areas over extended periods (Chuvieco et al., 2005; Barbosa et al., 1998; Ql and Kerr, 1997; Sousa et al., 2003). Originally developed for land-cover and land-use classification to enhance the spectral contrast of healthy vegetation (Cabral et al., 2003; Holben, 1986; Ql and Kerr, 1997), these methods have been widely adapted and progressively refined for burned area detection (Chuvieco et al., 2005; Miettinen and Liew, 2008; Pereira et al., 2017; Silva et al., 2004; Barbosa et al., 1998; Fernández et al., 1997; Pereira et al., 1999; Tansey et al., 2004a, b; Sousa et al., 2003). Refinements have targeted common artifacts, particularly cloud and cloud shadow contamination. Examples include coupling the lowest reflectance in Red or SWIR wavelengths with the multi-temporal Normalized Difference Vegetation Index (NDVI) criterion (Cabral et al., 2003; Chuvieco et al., 2005; Ql and Kerr, 1997) or selecting the minimum NIR reflectance (minNIR) from the time series (Chuvieco et al., 2005; Miettinen and Liew, 2008; Sousa et al., 2003), exploiting the low VIS–NIR reflectance of burned and charred biomass compared to high-reflective clouds. However, minNIR selection often accentuates cloud shadows (Chuvieco et al., 2005; Pereira et al., 2017; Pereira et al., 1999; Pereira et al., 2016; Sousa et al., 2003) due to spectral similarity. Alternative strategies include selecting the highest NIR reflectance among the three lowest red reflectances (Cabral et al., 2003; Chuvieco et al., 2005; Stroppiana et al., 2002). Comparative analyses (Pereira et al., 2016; Pereira et al., 2017) identified the second-lowest NIR (2ndMinNIR) as the most effective compromise between burned/unburned separability and reduced cloud shadow confusion, followed by the minNIRmaxSWIR criterion. A recent development by De Luca et al. (2022) applied the 2ndMinNIR criterion integrated with SAR backscatter data (2ndMinNIR+SAR), demonstrating the contribution of SAR information in improving burned area detection.

16.1.4 Aim of the study

This study aims to apply, optimize, and evaluate the advanced multitemporal compositing approach proposed by De Luca et al. (2022), which integrates the 2ndMinNIR criterion with SAR data to enhance burned area detection in heterogeneous Mediterranean landscapes of southern Italy. The study area encompasses multiple bioclimatic zones and forest habitat types, partly within the homonymous National Park. The robustness of the approach is assessed across contrasting vegetation structures, fire regimes, and topographic settings, while final burned area classification is performed through a geographic object-based image analysis (GEOBIA) framework.

16.2 Study area

The analysis focused on wildfire events that occurred in the Aspromonte Mountain region (Calabria, southernmost continental Italy; 38°16′ N, 15°84′ E) during the peak fire season of July–August 2021. According to European Forest Fire Information System (EFFIS) statistics, the fires affected a total area exceeding 140 km², of which

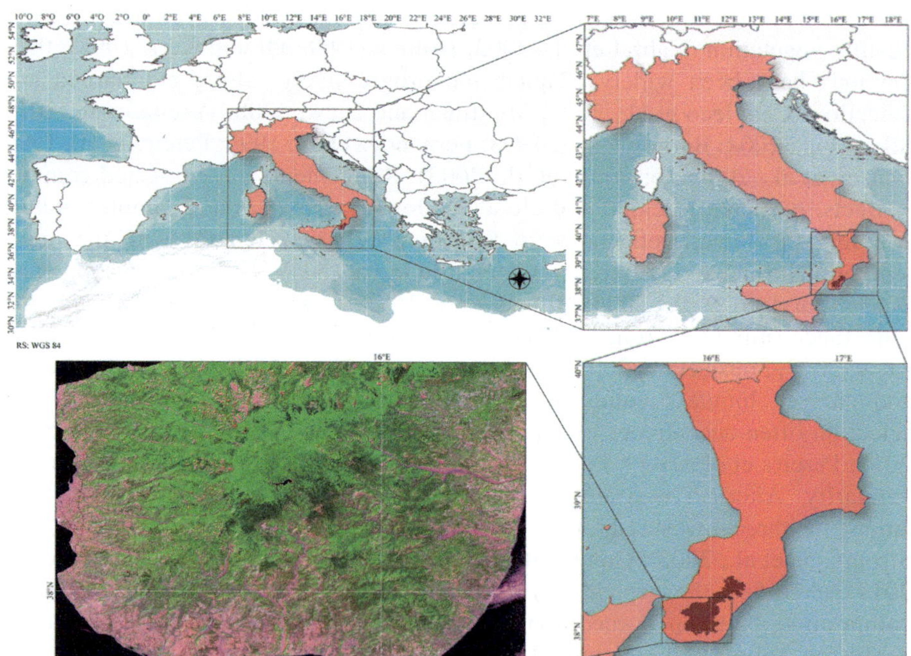

Figure 16.1 Study area. The false-color RGB image in the bottom-left panel shows the footprint of the satellite acquisitions. Map lines delineate study areas and do not necessarily depict accepted national boundaries.

approximately 73 km^2 were within the perimeter of the Aspromonte National Park (Fig. 16.1). The Aspromonte Mountains form the last continental extension of the Apennine range, with the National Park extending over 641.53 km^2 along a NE–SW axis. The bioclimate varies with altitude, with additional local variability driven by orographic factors such as proximity to the sea and slope exposition, resulting in high floristic heterogeneity. The park includes 21 Special Areas of Conservation belonging to the Natura 2000 network (Council Directive 92/43/EEC on the Conservation of Natural Habitats). Below approximately 1100 m a.s.l., conditions range from thermo-Mediterranean to supra-Mediterranean, with a humidity gradient from dry to humid. Here, sclerophyllous and Mediterranean deciduous forests prevail (*Lauretum–Castanetum*), including mesophilous and deciduous formations dominated by oak stands (*Quercus ilex, Q. suber, Q. virgiliana, Q. frainetto, Q. pubescens*), chestnut groves (*Castanea sativa*), and riparian woodlands in wetter areas (*Salix* spp., *Populus* spp.). At higher elevations, the landscape transitions to temperate bioclimates, dominated by temperate deciduous forests (beech, *Fagus sylvatica*; oak, *Q. petraea*) and oro-Mediterranean conifers (*Pinus nigra* subsp. *laricio*, *Abies alba*) (*Fagetum*), which extend up to the highest peaks (1956 m) (Spampinato, 2014).

16.3 Materials and methods

16.3.1 Satellite datasets and preprocessing

The Sentinel-2 Level-2A (Bottom of Atmosphere) dataset was preprocessed and downloaded using Google Earth Engine (GEE). A total of 20 images acquired between July 15 and September 15, 2021, were available. The search period was extended by a few weeks before and after the fire event to ensure, on one hand, sufficient coverage of the final part of the fire season and to prevent the composite criterion from discarding the latest acquisitions, on the other hand, to acquire the prefire scene. Only the Red (B4), NIR (B8), and SWIR (B12) bands were selected; the latter was resampled from 20 to 10 m spatial resolution using bilinear interpolation. Images were grouped by acquisition date.

The Sentinel-1 Ground Range Detected (GRDH) products, acquired in Interferometric Wide (IW) mode, were also downloaded from GEE for the same sensing period as the Sentinel-2 dataset. Dual-polarization data (VV and VH) were used. The images were already radiometrically calibrated (β^0), terrain-corrected, and coregistered to Sentinel-2 images (Google Earth Engine GEE, 2025).

16.3.2 Multitemporal compositing

In this study, the 2ndMinNIR criterion was applied to the optical dataset. The method operates on a time series of coregistered images, each corresponding to a different acquisition date. First, for each pixel location (same x–y position), the algorithm inspects the NIR values and selects the date with the second-lowest NIR value. The NIR value from this date is then assigned to the composite image, and the corresponding Red (B4) and SWIR (B12) values from the same date are also retained. This approach preserves spectral consistency across bands while selecting dates close to the minimum NIR response—indicative of burned biomass—while avoiding extreme low values caused by noise, residual clouds, or other anomalies. A further threshold, concerning exclusion of Red band reflectance above 6000 and below 300 to reduce cloud and cloud shadow contamination during the execution of the algorithm, has been introduced in this study.

For each pixel in the final composite, the acquisition date on which the 2ndMinNIR criterion was met was also recorded. Based on this information, for each pixel in the Sentinel-1 dataset, the value from the image with the first acquisition date following the corresponding Sentinel-2 compositing date was selected. This ensured that both datasets, optical and SAR, provided information acquired at approximately the same time.

The compositing procedure was implemented in a Python-based workflow, designed to be easily adaptable to various temporal scales and satellite datasets.

16.3.3 Object-based burned area mapping

The burned area mapping was carried out using a GEOBIA approach. Image segmentation was performed with the large-scale mean shift (LSMS) algorithm (Michel et al., 2015), while classification was conducted using the random forests (RF) machine learning algorithm (Breiman, 2001). These were implemented through the Orfeo ToolBox (OTB) v.7.0.0 (OTB, 2025) and Scikit-learn Python library v.0.2423.1 (Pedregosa et al., 2011), respectively.

16.3.3.1 Input layer creation

The normalized burn ratio (NBR; Equation 20.1) was calculated for both the prefire and the postfire composite images, and their difference (ΔNBR; Equation 20.2) was derived to enhance burn-related change signals. Similarly, delta layers were generated for the SAR composites (Equation 20.3), computed separately by polarization (VH, VV) and orbit (ascending, descending). All difference layers, together with the postfire NBR, were subsequently integrated and used as input for the segmentation and classification phases.

$$\mathrm{NBR} = \frac{\mathrm{NIR} - \mathrm{SWIR}}{\mathrm{NIR} + \mathrm{SWIR}} \tag{16.1}$$

$$\Delta\mathrm{NBR} = \mathrm{NBR}_{\mathrm{post}} - \mathrm{NBR}_{\mathrm{pre}} \tag{16.2}$$

$$\Delta\sigma^{0}{}_{\mathrm{orbit,polarization}} = \sigma^{0}{}_{\mathrm{orbit,polarization,\ post}} - \sigma^{0}{}_{\mathrm{orbit,polarization,\ pre}} \tag{16.3}$$

16.3.3.2 Object-based image segmentation and classification

The LSMS is a nonparametric segmentation algorithm (Fukunaga and Hostetler, 1975; Michel et al., 2015), consisting of four steps: (1) smoothing; (2) segmentation; (3) merging small-size segments; (4) vectorization (conversion to shapefile) and spectral feature extraction (mean and variance). Parameter settings were defined according to De Luca et al. (2022): the range radius, the spectral threshold to regroup two neighboring pixels—computed as Euclidean distance; the spatial radius, the number of neighboring pixels to search; and the minimum segment size, set to 10 pixels.

Using spectral statistics (mean—m_s—and variance—v_s) extracted for each segment during the fourth step of the LSMS procedure, the RF classifier was trained on a set of 1000 reference segments, equally divided between burned and unburned classes. Hyperparameter optimization was performed by testing a predefined set of parameter combinations (De Luca et al., 2022), with the configuration yielding the highest out-of-bag (OOB) accuracy during training selected for the final model.

16.3.4 Burned area classification assessment

Segmentation performance was examined through visual inspection, considering a burned area to be correctly delineated when it was entirely enclosed within one or more polygons without inclusion of unburned patches.

To evaluate the classification results, the derived burned area map was compared with the official EFFIS burned area dataset for the 2021 fire season, which represents the only available reference source at national scale. From this comparison, agreement categories were defined as true positives (TP), false negatives (FN), and false positives (FP). On this basis, *precision* (Equation 20.4), *recall* (Equation 20.5), and F-score (Equation 20.6) metrics were computed to provide a quantitative measure of consistency with the official data (Goutte and Gaussier, 2005; Shufelt, 1999; Sokolova et al., 2006), while visual inspection complemented the assessment by supporting the interpretation of segmentation and classification outcomes.

$$\text{precision} = \frac{|\text{TP}|}{|\text{TP} + \text{FP}|} \tag{16.4}$$

$$\text{recall} = \frac{|\text{TP}|}{|\text{TP} + \text{FN}|} \tag{16.5}$$

$$F - \text{score} = 2\cdot\frac{\text{precision}\cdot\text{recall}}{\text{precision} + \text{recall}} \tag{16.6}$$

16.4 Results

16.4.1 Multitemporal composite

The top panel of Fig. 16.2 shows the composite image generated with the 2ndMinNIR criterion (left) and the ΔNBR obtained by subtracting the prefire from the postfire index (right). Areas affected by fire events are distinctly recognizable in both outputs: in the false-color RGB composite (R=B12, G=B8, B=B8), they appear as enhanced red patches, resulting from a decline in NIR reflectance and a concomitant increase in SWIR reflectance, whereas in the grayscale ΔNBR image, they are highlighted by darker values, indicating stronger spectral changes relative to the surrounding unburned pixels.

Figure 16.2 Outcomes of the multitemporal composite derived from the June–September fire season. Top row: optical domain (Sentinel-2), with the multitemporal composite (left) and the difference between post (multitemporal optical composite) and prefire normalized burn ratio (ΔNBR; right). Middle row: synthetic aperture radar (SAR) domain (Sentinel-1), with the multitemporal composite (left) and the difference between post (multitemporal SAR composite) and prefire backscatter in ascending VV polarization ($\Delta\sigma^0$Asc,VV; right). Bottom row: two illustrative areas are shown at finer scale, highlighting the segmentation performed with the large-scale mean shift algorithm. Their corresponding locations are outlined in yellow in the upper panels.

The middle panel presents the SAR composite in VV polarization and ascending orbit, derived using the criterion based on the acquisition date closest to that of the 2ndMinNIR (left), along with the corresponding delta image ($\Delta\sigma_{Asc,VV}$) between post and prefire periods (right). Accounting that it provides the clearest SAR-based outcome, white patches corresponding to burned areas are clearly visible. As expected for VV polarization, fire events resulted in an increase in backscatter, whereas a decrease was observed in VH polarization. This behavior is further reflected in the averaged spectral features reported in Table 16.1.

16.4.2 Geographic object-based image analysis results

The segmentation process resulted in 631,079 objects, including extraneous areas such as the sea and image borders. Two overviews focusing on the burned area segments are shown in the bottom panels. Visually, it is evident that the segments accurately enclose burned pixels, including unburned islands within the main fire perimeters, as well as the smallest patches, with a minimum object size set to 10 pixels. Consistent with the findings of De Luca et al. (2022), the LSMS approach was also able to isolate low-severity areas, unburned-burned transition areas, and regions where the fire only partially affected forest tree crowns (surface fire).

The RF parameters were kept at their default values in scikit-learn (min_samples_split=2; min_samples_leaf=1; max_features="auto"), as initial tests yielded an OOB estimated error of 0.98, which was used as a fast indicator of predictive performance.

The resulting classification is presented in Fig. 16.3, along with the map of the spatial distribution of the assessment metrics (TP, FN, FP). Four exemplifying focus scenes are also reported. Quantitatively, the classification achieved a precision of 0.82, a recall (omission) of 0.70, and an F-score—representing the overall accuracy of the map—of 0.76.

The resulting burned area was estimated at 120.74 km^2, of which 57.45 km^2 was located within the Aspromonte National Park.

The main errors, both commission (FP) and omission (FN), were primarily concentrated along the fringes of the mapped burned areas—which exhibited a high degree of overlap with the EFFIS maps (>80%)—corresponding to the propagation margins of the fire perimeters. Visual inspection of the classified map suggests that a portion of FN errors may involve ambiguous zones, such as transitional areas or low-severity surface burns, whereas several patches were recorded as omissions despite being clearly unburned. A closer inspection of the map (Fig. 16.3, detailed overviews), in fact, reveals that the areas labeled as FN do not actually show signs of fire, suggesting that these errors are more likely attributable to inaccuracies in the reference perimeters rather than to the classification outcome. Outside the large burned patches, only commission errors were observed, represented by small, isolated false-positive patches distributed along the scene edges. However, in this case, detailed visual inspection suggests that some of these areas actually show evidence of fire, indicating possible omissions in the reference database rather than misclassifications by the proposed method.

Table 16.1 Averaged spectral features (mean, m_s; variance, v_s) for burned and unburned classes, calculated for each image layer employed as input for the classification process. The last column reports the feature importance values obtained from the random forest classifier.

Spectral feature	Image layer	Burned	Unburned	Feature importance (Gini)
Mean (m_s)	ΔNBR	μ −0.60 σ ±0.20	μ −0.12 σ ±0.10	0.449
	NBR	μ −0.16 σ ±0.17	μ 0.26 σ ±0.16	0.333
	$\Delta\sigma_{Desc,VH}$	μ −0.62 σ ±1.50	μ −0.52 σ ±1.50	0.012
	$\Delta\sigma_{Desc,VV}$	μ −0.42 σ ±1.39	μ −0.51 σ ±1.29	0.015
	$\Delta\sigma_{Asc,VH}$	μ −0.46 σ ±1.52	μ −0.42 σ ±1.40	0.015
	$\Delta\sigma_{Asc,VV}$	μ 0.51 σ ±1.33	μ −0.30 σ ±1.24	0.030
Variance (v_s)	ΔNBR	μ 0.013 σ ±0.009	μ 0.005 σ ±0.008	0.050
	NBR	μ 0.01 σ ±0.009	μ 0.006 σ ±0.007	0.032
	$\Delta\sigma_{Desc,VH}$	μ 10.285 σ ±4.528	μ 8.504 σ ±4.806	0.017
	$\Delta\sigma_{Desc,VV}$	μ 7.332 σ ±3.897	μ 6.377 σ ±3.393	0.016
	$\Delta\sigma_{Asc,VH}$	μ 8.125 σ ±3.557	μ 7.429 σ ±3.897	0.014
	$\Delta\sigma_{Asc,VV}$	μ 6.298 σ ±2.799	μ 5.911 σ ±3.041	0.015

The averaged spectral features (m_s and v_s), calculated separately for burned and unburned classes and for each image layer, are reported in Table 16.1, along with the corresponding feature importance values obtained from the RFclassification. Notably, the spectral separation between the averaged m_s values of burned and unburned classes is consistently greater for optical layers than for SAR-derived layers, with the only exception being the $\Delta\sigma°$ copolarized ascending layer, which shows a difference of 0.70 and is the only SAR layer approaching the optical layers in terms of feature importance (0.030). The ΔNBR, which exhibited the highest feature importance (0.449), showed a decrease of −0.48 for the burned class (−0.60) compared to the unburned class (−0.12), followed by the NBR with a reduction of 0.42, highlighting their strong separability relative to the SAR layers. The variance feature (v_s) follows the same pattern, confirming that the two optical layers are the most important contributors for both spectral metrics.

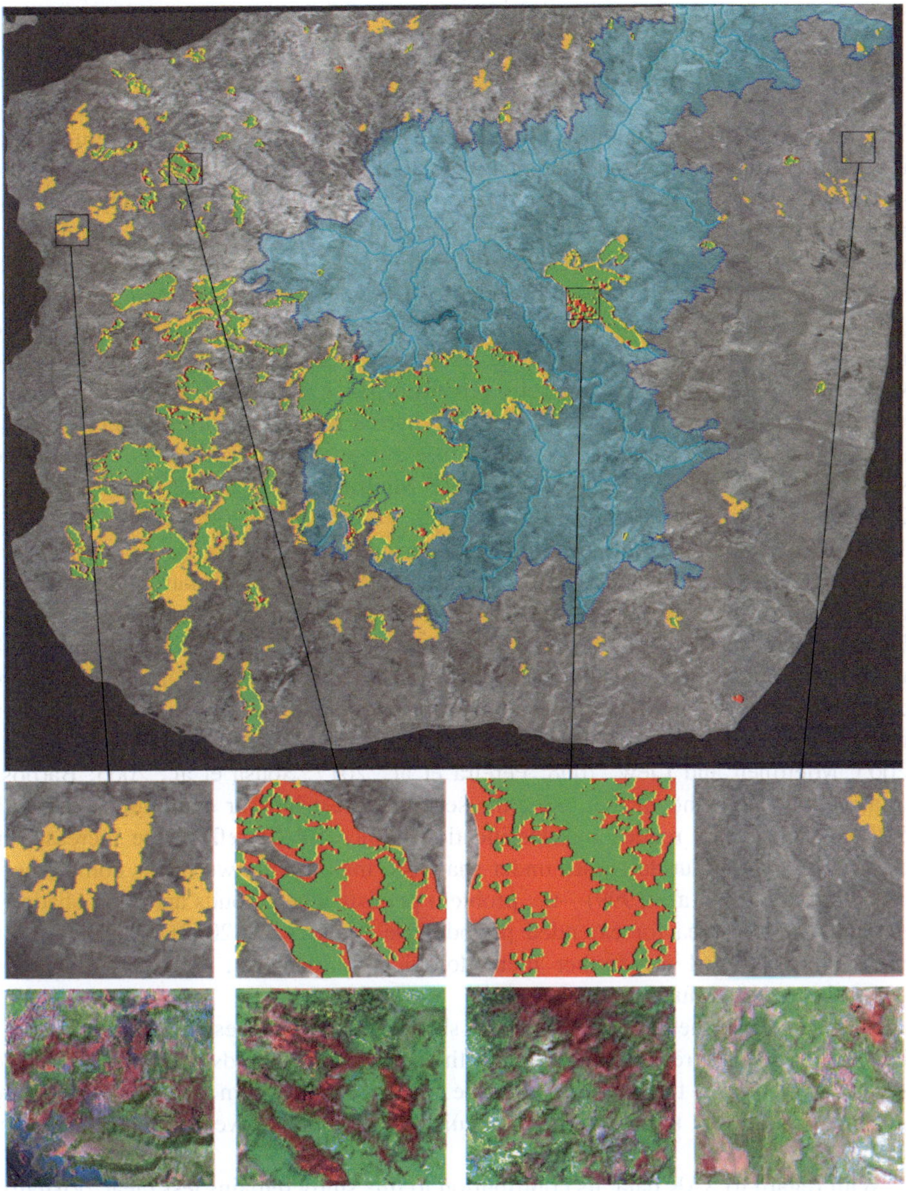

Figure 16.3 Spatial distribution of classification assessment metrics (true positives, in green; false negatives, in red; false positives, in yellow) for the burned area map. The blue outline indicates the perimeter of the Aspromonte National Park. Detailed overviews of selected focus areas are shown at the bottom.

16.5 Discussions

16.5.1 Image compositing

The final composites accounted for all pixels showing low NIR values during the investigated timeframe. This approach tends to augment separability of burned from unburned surfaces and to represent every fire event within the study window, including those that occurred at the beginning of the season, without temporal bias. In fact, NIR reflectance typically reaches its minimum in the days immediately after fire extinction, concomitantly with Red and SWIR wavelengths increasing reflectance (Miettinen et al., 2013; Pereira et al., 1999). The use of a single postfire acquisition to map long fire sequences reduces the detectability of areas burned earlier in the season. This reduction stems from postdisturbance natural processes, including vegetation recovery, chemical alterations of charred material, delayed plant mortality, and seasonal senescence, all of which present spectral response comparable to stressed vegetation (Gallagher et al., 2020; Miettinen et al., 2013; Rodman et al., 2021; van Dijk et al., 2021; Fraser et al., 2000; Verbyla et al., 2008; Pereira, Sá, et al., 1999). On the other hand, selecting the lowest NIR values improves burned area detection across extended periods but simultaneously reduces background vegetation reflectance (Cabral et al., 2003). Nevertheless, this uniform reduction generally becomes problematic only when further analyses, such as vegetation health monitoring or postfire regrowth assessment, are required. Another issue arises from cloud shadows, which also suppress NIR reflectance and can cause misclassification if the lowest NIR values are used directly. Although several studies (Chuvieco et al., 2005; Miettinen and Liew, 2008; Pereira et al., 2017; Sousa et al., 2003; Barbosa et al., 1998) have shown that using the second-lowest NIR or combining NIR with SWIR values helps reduce misclassification—since SWIR reflectance increases in burned vegetation but decreases under shadows, making the two bands complementary (Pereira, Sá, et al., 1999)—the problem persists when clouds and their shadows remain over the area for extended periods (De Luca et al., 2022). This limitation highlights the need for further methodological refinements, some of which are currently under evaluation.

In this study, the cloud and cloud shadows were successfully addressed by integrating a Red threshold that ensured the exclusion of clouds and their associated shadows. However, this condition can be achieved only when long time series are available, as at least three cloud-free acquisitions per pixel (even in unburned areas) are required.

With regard to SAR data, its contribution to this study remains less clear. Visually, the results did not match those of De Luca et al. (2022), although the dual-polarized Radar Vegetation Index (RVI) was used in that case. However, given the expected limitations of SAR performance in areas with complex topography (Richards, 2009), the use of single polarizations was considered a more cautious approach in this case. Among the tested layers, copolarization (VV) appeared to provide the strongest signal, consistent with its sensitivity to backscatter from exposed bare soil, which highlights canopy removal after burning (Carreiras et al., 2020; Saatchi, 2019).

Nonetheless, SAR layers retained a high level of speckle noise (Fig. 16.2). Unlike the relatively flat Portuguese landscapes analyzed in De Luca et al. (2022), the Aspromonte Mountains are characterized by steep relief and varying aspect, which likely exacerbated the topographic effects and reduced the clarity of SAR-based burn detection. This point deserves further investigation in future work. It is also worth noting that the SAR images used were downloaded from GEE already preprocessed, a choice that reduces processing time but may involve trade-offs in terms of flexibility and optimal parameter settings. Although separate evaluations of optical versus SAR sources were not performed in this study, visual inspection suggests that optical data already provided a clear delineation of fire perimeters. The optical sensors are, in fact, more effective for burned area mapping because of the more robust relationship between the optical wavelengths and the effects of fire on vegetation (De Luca et al., 2022; Pereira, Sá, et al., 1999; Stroppiana et al., 2015). This advantage, however, holds true only as long as sufficient cloud-free acquisitions are available. The 2021 fire season over the Aspromonte was relatively cloudy, further complicated by the area's exposure to diverse Mediterranean climatic influences (humid airflows, strong seasonal winds, and frequent atmospheric instability). Even so, the Mediterranean climate generally guarantees a sufficient number of clear-sky summer days. In contrast, applying this approach in temperate continental regions (e.g., Central and Northern Europe) would require additional refinements, particularly in exploiting SAR data more effectively. It should be noted, nevertheless, that the influence of environmental factors on the spectral contrast between burned and unburned areas over time may vary across climate bio-regions (Avetisyan et al., 2023), potentially affecting the number of cloud-free images required compared to Mediterranean standards.

16.5.2 Geographic object-based image analysis assessment

The main strength of the GEOBIA approach lies in aggregating information at the object level, using the average spectral values of pixels within each segment instead of single-pixel observations. This reduces the influence of outliers and mitigates typical biases of pixel-based classifications, thereby increasing the robustness of the results (Modica et al., 2021; Radoux and Defourny, 2008). This property is particularly relevant when dealing with SAR imagery, where speckle noise remains a persistent limitation (Polychronaki et al., 2013).

In this study, both mean and variance of the spectral layers were tested as input features for burned/unburned discrimination. In agreement with De Luca et al. (2022), the mean consistently proved to be more informative than variance, confirming its role as the key driver of classification accuracy.

It is important to emphasize that this study did not perform a formal accuracy assessment in the rigorous technical sense. Rather, the comparison with EFFIS data should be interpreted as a consistency check against the only official reference available at the European scale. To complement this evaluation, visual inspection of the classified maps was conducted, providing additional evidence of the results' reliability. Although there was a high level of overlap (>80%), visual interpretation diverged from the official EFFIS perimeters in areas where commission or omission errors were detected. Previous studies have already highlighted the potential overestimation of EFFIS datasets (Llorens et al., 2021). Most detected errors—both commission and omission—were spatially concentrated along the margins of burned areas, suggesting that these discrepancies may reflect minor perimeter misalignments rather than failures to detect fire. Outside the main fire polygons, only isolated false-positive errors were observed; however, visual inspection of satellite images (Fig. 16.3) indicated that many of these patches corresponded to actual burned areas not recorded in the EFFIS database for the June–September period. A similar issue was reported in Portugal by De Luca et al. (2022), where FP were often associated with small agricultural clearings.

16.6 Conclusions

This study optimized and validated a methodology for detecting burned areas over an entire fire season, combining multitemporal compositing with GEOBIA, implemented using open-source tools and freely available Sentinel-2 optical and Sentinel-1 SAR imagery. The approach was applied in the heterogeneous Mediterranean landscapes of the Aspromonte Mountains, Southern Italy, characterized by complex topography, diverse vegetation, and fires occurring over multiple days. The multitemporal compositing, integrating the second-minimum NIR criterion with temporally aligned SAR layers, effectively reduced cloud interference and enhanced burned/unburned separability. The GEOBIA framework accurately delineated burned perimeters, unburned islands, and low-severity patches, achieving high consistency with EFFIS data (precision=0.82, recall=0.70, F-score=0.76). Visual inspection revealed additional burned areas not captured in the official dataset, highlighting the value of high-resolution imagery and careful validation. Although SAR contributed limited improvements in mountainous terrain, optical bands (NIR and SWIR), especially in the NBR index, were sufficient for reliable burned area mapping.

Future work should explore improved SAR integration in complex terrains, cloud-computing implementations for faster processing, and high-resolution reference datasets to enhance validation. Overall, this study demonstrates the robustness of the method for systematic, high-resolution burned area detection in Mediterranean fire-prone landscapes, while also highlighting its scalability and transferability for regional- to national-scale applications, capable of capturing small fires and fine structural details.

References

Agnoletti, M., 2013. Italian historical rural landscapes: dynamics, data analysis and research findings. Environmental History (Netherlands) 1, 3–87. Available at https://doi.org/10.1007/978-94-007-5354-9_1, https://www.springer.com/series/10168.

Alcaras, E., Costantino, D., Guastaferro, F., Parente, C., Pepe, M., 2022. Normalized burn ratio plus (NBR+): a new index for sentinel-2 imagery. Remote Sensing 14 (7), 1727. Available at https://doi.org/10.3390/rs14071727.

Ascoli, D., Moris, J.V., Marchetti, M., Sallustio, L., 2021. Land use change towards forests and wooded land correlates with large and frequent wildfires in Italy. Annals of Silvicultural Research 46 (2), 177–188. Available at https://doi.org/10.12899/asr-2264, https://journals-crea.4science.it/index.php/asr/article/view/2264/pdf.

Avetisyan, D., Stankova, N., Dimitrov, Z., 2023. Assessment of spectral vegetation indices performance for post-fire monitoring of different forest environments. Fire 6 (8), 290. Available at https://doi.org/10.3390/fire6080290.

Badía, D., López-García, S., Martí, C., Ortíz-Perpiñá, O., Girona-García, A., Casanova-Gascón, J., 2017. Burn effects on soil properties associated to heat transfer under contrasting moisture content. Science of the Total Environment 601-602, 1119–1128. Available at https://doi.org/10.1016/j.scitotenv.2017.05.254.

Barbosa, P.M., Pereira, J.M.C., Grégoire, J.M., 1998. Compositing criteria for burned area assessment using multitemporal low resolution satellite data. Remote Sensing of Environment 65 (1), 38–49. Available at https://doi.org/10.1016/S0034-4257(98)00016-9.

Blasi, C., Capotorti, G., Copiz, R., Guida, D., Mollo, B., Smiraglia, D., et al., 2014. Classification and mapping of the ecoregions of Italy. Plant Biosystems - An International Journal Dealing with all Aspects of Plant Biology 148 (6), 1255–1345. Available at https://doi.org/10.1080/11263504.2014.985756.

Breiman, L., 2001. Random forests. Machine Learning 45 (1), 5–32. Available at https://doi.org/10.1023/A:1010933404324, https://doi.org/10.1023/A:1010933404324.

Cabral, A., De Vasconcelos, M.J.P., Pereira, J.M.C., Bartholomé, E., Mayaux, P., 2003. Multitemporal compositing approaches for SPOT-4 vegetation. International Journal of Remote Sensing 24 (16), 3343–3350. Available at https://doi.org/10.1080/0143116031000075936, https://www.tandfonline.com/loi/tres20.

Carreiras, J.M.B., Quegan, S., Tansey, K., Page, S., 2020. Sentinel-1 observation frequency significantly increases burnt area detectability in tropical SE Asia. Environmental Research Letters 15 (5). Available at https://doi.org/10.1088/1748-9326/ab7765, https://iopscience.iop.org/article/10.1088/1748-9326/ab7765.

Chen, Y., Morton, D.C., Randerson, J.T., 2024. Remote sensing for wildfire monitoring: insights into burned area, emissions, and fire dynamics. One Earth 7 (6), 1022–1028. Available at https://doi.org/10.1016/j.oneear.2024.05.014, https://www.sciencedirect.com/science/journal/25903322.

Chuvieco, E., Ventura, G., Martín, M.P., Gómez, I., 2005. Assessment of multitemporal compositing techniques of MODIS and AVHRR images for burned land mapping. Remote Sensing of Environment 94 (4), 450–462. Available at https://doi.org/10.1016/j.rse.2004.11.006.

Chuvieco, E., Mouillot, F., van der Werf, G.R., San Miguel, J., Tanasse, M., Koutsias, N., et al., 2019. Historical background and current developments for mapping burned area from satellite Earth observation. Remote Sensing of Environment 225, 45–64. Available at https://doi.org/10.1016/j.rse.2019.02.013, www.elsevier.com/inca/publications/store/5/0/5/7/3/3.

Conedera, M., Krebs, P., Valese, E., Cocca, G., Schunk, C., Menzel, A., et al., 2018. Characterizing Alpine pyrogeography from fire statistics. Applied Geography 98, 87–99. Available at https://doi.org/10.1016/j.apgeog.2018.07.011.

Crowley, M.A., Cardille, J.A., White, J.C., Wulder, M.A., 2019. Generating intra-year metrics of wildfire progression using multiple open-access satellite data streams. Remote Sensing of Environment 232. Available at https://doi.org/10.1016/j.rse.2019.111295, www.elsevier.com/inca/publications/store/5/0/5/7/3/3.

Dadkhah, H., Rana, D., Ghaderpour, E., Mazzanti, P., 2025. Analyzing wildfire patterns and climate interactions in Campania, Italy: a multi-sensor remote sensing study. Ecological Informatics 90, 103249. Available at https://doi.org/10.1016/j.ecoinf.2025.103249.

De Luca, G., Modica, G., 2023. Canopy fire effects estimation using sentinel-2 imagery and deep learning approach. A case study on the aspromonte National Park. Studies in Computational Intelligence 1088. Springer Science and Business Media Deutschland GmbH, Italy, pp. 403–417. Available at https://www.springer.com/series/7092, 10.1007/978-3-031-25755-1_27.

De Luca, G., Modica, G., Fattore, C., Lasaponara, R., 2020. Unsupervised burned area mapping in a protected natural site. An approach using SAR sentinel-1 data and K-mean algorithm. Lecture Notes in Computer Science (including subseries Lecture Notes in Artificial Intelligence and Lecture Notes in Bioinformatics) 12253. Springer Science and Business Media Deutschland GmbH, Italy, pp. 63–77. Available at https://www.springer.com/series/558, 10.1007/978-3-030-58814-4_5.

De Luca, G., Silva, J.M.N., Modica, G., 2021. A workflow based on Sentinel-1 SAR data and open-source algorithms for unsupervised burned area detection in Mediterranean ecosystems. GIScience and Remote Sensing 58 (4), 516–541. Available at https://doi.org/10.1080/15481603.2021.1907896, http://www.tandfonline.com/loi/tgrs20?open=50&repitition=0#vol_50.

De Luca, G., Silva, J.M.N., Modica, G., 2022. Regional-scale burned area mapping in Mediterranean regions based on the multitemporal composite integration of Sentinel-1 and Sentinel-2 data. GIScience and Remote Sensing 59 (1), 1678–1705. Available at https://doi.org/10.1080/15481603.2022.2128251, http://www.tandfonline.com/loi/tgrs20?open=50&repitition=0#vol_50.

De Luca, G., Modica, G., Silva, J.M.N., Praticò, S., Pereira, J.M.C., 2023. Assessing tree crown fire damage integrating linear spectral mixture analysis and supervised machine learning on Sentinel-2 imagery. International Journal of Digital Earth 16 (1), 3162–3198. Available at https://doi.org/10.1080/17538947.2023.2243900, http://www.tandfonline.com/toc/tjde20/current.

De Philippis, A., 1937. Classificazioni ed Indici del Clima, in Rapporto Alla Vegetazione Forestale Italiana. Giornale botanico italiano 44 (1), 1–169. Available at https://doi.org/10.1080/11263503709438750.

EFFIS - Annual Statistics for Italy, 2025. 9 2025. Available from: http://URLhttps://forest-fire.emergency.copernicus.eu/apps/effis.statistics/estimates.

Elia, M., Giannico, V., Ascoli, D., Argañaraz, J.P., D'Este, M., Spano, G., et al., 2022. Uncovering current pyroregions in Italy using wildfire metrics. Ecological Processes 11 (1). Available at https://doi.org/10.1186/s13717-022-00360-6, http://www.ecologicalprocesses.com/.

Fernández, A., Illera, P., Casanova, J.L., 1997. Automatic mapping of surfaces affected by forest fires in Spain using AVHRR NDVI composite image data. Remote Sensing of Environment 60 (2), 153–162. Available at https://doi.org/10.1016/S0034-4257(96)00178-2.

Filipponi, F., 2019. Exploitation of sentinel-2 time series to map burned areas at the national level: a case study on the 2017 Italy wildfires. Remote Sensing 11 (6), 622. Available at https://doi.org/10.3390/rs11060622.

Fraser, R.H., Li, Z., Cihlar, J., 2000. Hotspot and NDVI differencing synergy (HANDS): a new technique for burned area mapping over boreal forest. Remote Sensing of Environment 74 (3), 362–376. Available at https://doi.org/10.1016/S0034-4257(00)00078-X.

Fukunaga, K., Hostetler, L., 1975. The estimation of the gradient of a density function, with applications in pattern recognition. IEEE Transactions on Information Theory 21 (1), 32–40. Available at https://doi.org/10.1109/TIT.1975.1055330.

Gallagher, M.R., Skowronski, N.S., Lathrop, R.G., McWilliams, T., Green, E.J., 2020. An improved approach for selecting and validating burn severity indices in forested landscapes. Canadian Journal of Remote Sensing 46 (1), 100–111. Available at https://doi.org/10.1080/07038992.2020.1735931, http://www.tandfonline.com/loi/ujrs20.

Gimeno, M., San-Miguel-Ayanz, J., Schmuck, G., 2004. Identification of burnt areas in Mediterranean forest environments from ERS-2 SAR time series. International Journal of Remote Sensing 25 (22), 4873–4888. Available at https://doi.org/10.1080/01431160412331269715, https://www.tandfonline.com/loi/tres20.

Google Earth Engine (GEE), Sentinel-1 Algorithms, 2025. 2025 9 2025 9. Available from: https://developers.google.com/earth-engine/guides/sentinel1.

Goutte, C., Gaussier, E., 2005. A probabilistic interpretation of precision, recall and F-score, with implication for evaluation. Lecture Notes in Computer Science 3408, 345–359. Available at https://doi.org/10.1007/978-3-540-31865-1_25, http://www.springeronline.com/sgw/cda/frontpage/0,11855,4-40356-69-1180994-0,00.html.

Holben, B.N., 1986. Characteristics of maximum-value composite images from temporal AVHRR data. International Journal of Remote Sensing 7 (11), 1417–1434. Available at https://doi.org/10.1080/01431168608948945.

Hudak, A.T., Morgan, P., Bobbitt, M.J., Smith, A.M.S., Lewis, S.A., Lentile, L.B., et al., 2007. The relationship of multispectral satellite imagery to immediate fire effects. Fire Ecology 3 (1), 64–90. Available at https://doi.org/10.4996/fireecology.0301064.

Imperatore, P., Azar, R., Calo, F., Stroppiana, D., Brivio, P.A., Lanari, R., et al., 2017. Effect of the vegetation fire on backscattering: an investigation based on sentinel-1 observations. IEEE Journal of Selected Topics in Applied Earth Observations and Remote Sensing 10 (10), 4478–4492. Available at https://doi.org/10.1109/JSTARS.2017.2717039, http://ieeexplore.ieee.org/xpl/RecentIssue.jsp?punumber=4609443.

Lanorte, A., Danese, M., Lasaponara, R., Murgante, B., 2013. Multiscale mapping of burn area and severity using multisensor satellite data and spatial autocorrelation analysis. International Journal of Applied Earth Observation and Geoinformation 20 (1), 42–51. Available at https://doi.org/10.1016/j.jag.2011.09.005.

Lasaponara, R., 2006. Estimating spectral separability of satellite derived parameters for burned areas mapping in the Calabria region by using SPOT-Vegetation data. Ecological Modelling 196 (1-2), 265–270. Available at https://doi.org/10.1016/j.ecolmodel.2006.02.025.

Lasaponara, R., Tucci, B., 2019. Identification of burned areas and severity using SAR sentinel-1. IEEE Geoscience and Remote Sensing Letters 16 (6), 917–921. Available at https://doi.org/10.1109/lgrs.2018.2888641.

Llorens, R., Sobrino, J.A., Fernández, C., Fernández-Alonso, J.M., Vega, J.A., 2021. A methodology to estimate forest fires burned areas and burn severity degrees using sentinel-2 data. Application to the October 2017 fires in the Iberian Peninsula. International Journal of Applied Earth Observation and Geoinformation 95. Available at https://doi.org/10.1016/j.jag.2020.102243, http://www.elsevier.com/locate/jag.

Martins, F.da S.R.V., dos Santos, J.R., Galvão, L.S., Xaud, H.A.M., 2016. Sensitivity of ALOS/PALSAR imagery to forest degradation by fire in northern Amazon. International Journal of Applied Earth Observation and Geoinformation 49, 163–174. Available at https://doi.org/10.1016/j.jag.2016.02.009.

Michel, J., Youssefi, D., Grizonnet, M., 2015. Stable mean-shift algorithm and its application to the segmentation of arbitrarily large remote sensing images. IEEE Transactions on Geoscience and Remote Sensing 53 (2), 952–964. Available at https://doi.org/10.1109/TGRS.2014.2330857.

Michetti, M., Pinar, M., 2019. Forest fires across Italian regions and implications for climate change: a panel data analysis. Environmental and Resource Economics 72 (1), 207–246. Available at https://doi.org/10.1007/s10640-018-0279-z.

Miettinen, J., Liew, S.C., 2008. Comparison of multitemporal compositing methods for burnt area detection in Southeast Asian conditions. International Journal of Remote Sensing 29 (4), 1075–1092. Available at https://doi.org/10.1080/01431160701281031, https://www.tandfonline.com/loi/tres20.

Miettinen, J., Hyer, E., Chia, A.S., Kwoh, L.K., Liew, S.C., 2013. Detection of vegetation fires and burnt areas by remote sensing in insular Southeast Asian conditions: current status of knowledge and future challenges. International Journal of Remote Sensing 34 (12), 4344–4366. Available at https://doi.org/10.1080/01431161.2013.777489, https://www.tandfonline.com/loi/tres20.

Modica, G., De Luca, G., Messina, G., Praticò, S., 2021. Comparison and assessment of different object-based classifications using machine learning algorithms and UAVs multispectral imagery: a case study in a citrus orchard and an onion crop. European Journal of Remote Sensing 54 (1), 431–460. Available at https://doi.org/10.1080/22797254.2021.1951623.

OTB, 2025. 2025 9. Available from: https://www.orfeo-toolbox.org/.

Pavari, A., 1916. Studio preliminare sulla coltura di specie forestali esotiche in Italia. Prima parte (generale). Annali del Regio Istituto Superiore Nazionale Forestale I (1913-15), 221.

Pedregosa, F., Varoquaux, G., Gramfort, A., Michel, V., Thirion, B., Grisel, O., et al., 2011. Scikit-learn: machine learning in Python. Journal of Machine Learning Research 12, 2825–2830. Available at http://jmlr.csail.mit.edu/papers/volume12/pedregosa11a/pedregosa11a.pdf.

Pereira, A.A., Pereira, J.M.C., Libonati, R., Oom, D., Carvalho, L.M.T., 2016. Multitemporal image compositing evaluation algorithms for PROBA-V burned area analysis. Proba-V Symposium 2015–2016.

Pereira, A.A., Libonati, R., Oom, D., Tavares, L.M.C., Pereira, J.M., 2017. Avaliação de técnicas de compósitos multitemporais em imagens PROBA-V para o mapeamento de áreas queimadas. Anais Do XVIII Simpósio Brasilero de Sesorlamento Remoto - SBSR.

Pereira, J.M.C., Sousa, A.M.O., Sá, A.C.L., Pilar Martín, M., Chuvieco, E., 1999. Regional-Scale Burnt Area Mapping in Southern Europe using NOAA-AVHRR 1 km Data. Springer Science and Business Media LLC, pp. 139–155. Available at 10.1007/978-3-642-60164-4_8.

Pereira, J.M.C., Sá, A.C.L., Sousa, A.M.O., Silva, J.M.N., Santos, T.N., Carreiras, J.M.B., 1999. Spectral Characterisation and Discrimination of Burnt Areas. Springer Science and Business Media LLC, pp. 123–138. Available at 10.1007/978-3-642-60164-4_7.

Pesaresi, S., Biondi, E., Casavecchia, S., 2017. Bioclimates of Italy. Journal of Maps 13 (2), 955–960. Available at https://doi.org/10.1080/17445647.2017.1413017.

Polychronaki, A., Gitas, I.Z., Veraverbeke, S., Debien, A., 2013. Evaluation of ALOS PALSAR imagery for burned area mapping in greece using object-based classification. Remote Sensing 5 (11), 5680–5701. Available at https://doi.org/10.3390/rs5115680, http://www.mdpi.com/2072-4292/5/11/5680/pdf.

Pulvirenti, L., Squicciarino, G., Fiori, E., Fiorucci, P., Ferraris, L., Negro, D., et al., 2020. An Automatic processing chain for near real-time mapping of burned forest areas using sentinel-2 data. Remote Sensing 12 (4), 674. Available at https://doi.org/10.3390/rs12040674.

Ql, J., Kerr, Y., 1997. On current compositing algorithms. Remote Sensing Reviews 15 (1-4), 235–256. Available at https://doi.org/10.1080/02757259709532340.

Quintano, C., Fernández-Manso, A., Roberts, D.A., 2020. Enhanced burn severity estimation using fine resolution ET and MESMA fraction images with machine learning algorithm. Remote Sensing of Environment 244, 111815. Available at https://doi.org/10.1016/j.rse.2020.111815.

Radoux, J., Defourny, P., 2008. Quality assessment of segmentation results devoted to object-based classification. Lecture Notes in Geoinformation and Cartography 257–271. Available at https://doi.org/10.1007/978-3-540-77058-9_14, http://link.springer.com/bookseries/7418.

Richards, J.A., 2009. Remote Sensing with Imaging Radar Signals and Communication Technology. Springer, Berlin Heidelberg. Available at 10.1007/978-3-642-02020-9.

Rodman, K.C., Andrus, R.A., Veblen, T.T., Hart, S.J., 2021. Disturbance detection in landsat time series is influenced by tree mortality agent and severity, not by prior disturbance. Remote Sensing of Environment 254. Available at https://doi.org/10.1016/j.rse.2020.112244, www.elsevier.com/inca/publications/store/5/0/5/7/3/3.

Romano, N., Ursino, N., 2020. Forest fire regime in a Mediterranean ecosystem: unraveling the mutual interrelations between rainfall seasonality, soil moisture, drought persistence, and biomass dynamics. Fire 3 (3), 49. Available at https://doi.org/10.3390/fire3030049.

Ruiz-Ramos, J., Marino, A., Boardman, C.P., 2018. Using sentinel 1-SAR for monitoring long term variation in burnt forest areas. International Geoscience and Remote Sensing Symposium (IGARSS) 2018, 4901–4904. Available at https://doi.org/10.1109/IGARSS.2018.8518960.

Saatchi, S., 2019. SAR methods for mapping and monitoring forest biomass. The SAR Handbook. Comprehensive Methodologies for Forest Monitoring and Biomass Estimation. Available at https://doi.org/10.25966/nr2c-s697.

Salis, M., Del Giudice, L., Jahdi, R., Alcasena-Urdiroz, F., Scarpa, C., Pellizzaro, G., et al., 2022. Spatial patterns and intensity of land abandonment drive wildfire hazard and likelihood in Mediterranean agropastoral areas. Land 11 (11), 1942. Available at https://doi.org/10.3390/land11111942.

San-Miguel-Ayanz, J., Durrant, T., Boca, R., Maianti, P., Libertà, G., Felix Oom, D.J., et al., 2025. Advance report on forest fires in Europe, Middle East and North Africa 2024. https://doi.org/10.2760/1264626.

Saulino, L., Rita, A., Migliozzi, A., Maffei, C., Allevato, E., Garonna, A.P., et al., 2020. Detecting burn severity across mediterranean forest types by coupling medium-spatial resolution satellite imagery and field data. Remote Sensing 12 (4). Available at https://doi.org/10.3390/rs12040741, https://res.mdpi.com/d_attachment/remotesensing/remotesensing-12-00741/article_deploy/remotesensing-12-00741-v2.pdf.

Shufelt, J., 1999. Performance evaluation and analysis of monocular building extraction from aerial imagery. IEEE Transactions on Pattern Analysis and Machine Intelligence 21 (4), 311–326. Available at https://doi.org/10.1109/34.761262.

Silva, J.M.N., Cadima, J.F.C.L., Pereira, J.M.C., Grégoire, J.M., 2004. Assessing the feasibility of a global model for multi-temporal burned area mapping using SPOT-VEGETATION data. International Journal of Remote Sensing 25 (22), 4889–4913. Available at https://doi.org/10.1080/01431160412331291251, https://www.tandfonline.com/loi/tres20.

Smiraglia, D., Filipponi, F., Mandrone, S., Tornato, A., Taramelli, A., 2020. Agreement index for burned area mapping: integration of multiple spectral indices using sentinel-2 satellite images. Remote Sensing 12 (11), 1862. Available at https://doi.org/10.3390/rs12111862.

Smith, A.M.S., Wooster, M.J., Drake, N.A., Dipotso, F.M., Falkowski, M.J., Hudak, A.T., 2005. Testing the potential of multi-spectral remote sensing for retrospectively estimating fire severity in African Savannahs. Remote Sensing of Environment 97 (1), 92–115. Available at https://doi.org/10.1016/j.rse.2005.04.014.

Sokolova, M., Japkowicz, N., Szpakowicz, S., 2006. Beyond Accuracy, F-Score and ROC: A Family of Discriminant Measures for Performance Evaluation. Springer Science and Business Media LLC, WS-06-06, pp. 1015–1021. Available at 10.1007/11941439_114.

Sousa, A.M.O., Pereira, J.M.C., Silva, J.M.N., 2003. Evaluating the performance of multi-temporal image compositing algorithms for burned area analysis. International Journal of Remote Sensing 24 (6), 1219–1236. Available at https://doi.org/10.1080/01431160110114466, https://www.tandfonline.com/loi/tres20.

Spadoni, G.L., Moris, J.V., Vacchiano, G., Elia, M., Garbarino, M., Sibona, E., et al., 2023. Active governance of agro-pastoral, forest and protected areas mitigates wildfire impacts in Italy. Science of the Total Environment 890. Available at https://doi.org/10.1016/j.scitotenv.2023.164281, www.elsevier.com/locate/scitotenv.

Spampinato, G., 2014. Guida alla flora dell'Aspromonte.

Stroppiana, D., Pinnock, S., Pereira, J.M.C., Grégoire, J.M., 2002. Radiometric analysis of SPOT-VEGETATION images for burnt area detection in Northern Australia. Remote Sensing of Environment 82 (1), 21–37. Available at https://doi.org/10.1016/S0034-4257(02)00021-4.

Stroppiana, D., Azar, R., Calò, F., Pepe, A., Imperatore, P., Boschetti, M., et al., 2015. Integration of optical and SAR data for burned area mapping in Mediterranean regions. Remote Sensing 7 (2), 1320–1345. Available at https://doi.org/10.3390/rs70201320, http://www.mdpi.com/2072-4292/7/2/1320/pdf.

Tanase, M.A., Santoro, M., Wegmüller, U., de la Riva, J., Pérez-Cabello, F., 2010a. Properties of X-, C- and L-band repeat-pass interferometric SAR coherence in Mediterranean pine forests affected by fires. Remote Sensing of Environment 114 (10), 2182–2194. Available at https://doi.org/10.1016/j.rse.2010.04.021.

Tanase, M.A., Santoro, M., De La Riva, J., Pérez-Cabello, F., Le Toan, T., 2010b. Sensitivity of X-, C-, and L-band SAR backscatter to burn severity in Mediterranean pine forests. IEEE Transactions on Geoscience and Remote Sensing 48 (10), 3663–3675. Available at https://doi.org/10.1109/TGRS.2010.2049653.

Tanase, M., de la Riva, J., Santoro, M., Pérez-Cabello, F., Kasischke, E., 2011. Sensitivity of SAR data to post-fire forest regrowth in Mediterranean and boreal forests. Remote Sensing of Environment 115 (8), 2075–2085. Available at https://doi.org/10.1016/j.rse.2011.04.009.

Tansey, K., Grégoire, J.M., Binaghi, E., Boschetti, L., Brivio, P.A., Ershov, D., et al., 2004a. A global inventory of burned areas at 1 km resolution for the year 2000 derived from spot vegetation data. Climatic Change 67 (2-3), 345–377. Available at https://doi.org/10.1007/s10584-004-2800-3.

Tansey, K., Grégoire, J.M., Stroppiana, D., Sousa, A., Silva, J., Pereira, J.M.C., et al., 2004b. Vegetation burning in the year 2000: global burned area estimates from SPOT VEGETATION data. Journal of Geophysical Research D: Atmospheres 109 (14), D14. Available at https://doi.org/10.1029/2003JD003598.

van Dijk, D., Shoaie, S., van Leeuwen, T., Veraverbeke, S., 2021. Spectral signature analysis of false positive burned area detection from agricultural harvests using Sentinel-2 data.

International Journal of Applied Earth Observation and Geoinformation 97, 102296. Available at https://doi.org/10.1016/j.jag.2021.102296.
Valese, E., Conedera, M., Held, A.C., Ascoli, D., 2014. Fire, humans and landscape in the European Alpine region during the Holocene. Anthropocene 6, 63–74. Available at https://doi.org/10.1016/j.ancene.2014.06.006.
Verbyla, D.L., Kasischke, E.S., Hoy, E.E., 2008. Seasonal and topographic effects on estimating fire severity from Landsat TM/ETM+ data. International Journal of Wildland Fire 17 (4), 527–534. Available at https://doi.org/10.1071/WF08038.
Wastl, C., Schunk, C., Lüpke, M., Cocca, G., Conedera, M., Valese, E., et al., 2013. Large-scale weather types, forest fire danger, and wildfire occurrence in the Alps. Agricultural and Forest Meteorology 168, 15–25. Available at https://doi.org/10.1016/j.agrformet.2012.08.011.

Fire severity across Chilean Mediterranean forest ecosystems

17

Pablo Sarricolea[1,2,4], *David Beltrán-Marcos*[3], *Massimiliano Farris*[2], *Francisca Gutiérrez-Cáceres*[4], *Miguel Aguilera*[5], *and María Fuentealba*[6,7,8]
[1]Center for Climate and Resilience Research (CR)2, University of Chile, Santiago, Chile, [2]Department of Geography, University of Chile, Santiago, Chile, [3]Area of Ecology, Department of Biodiversity and Environmental Management, Faculty of Biological and Environmental Sciences, University of León, León, Spain, [4]Centro Regional Copernicus-LAC Chile, University of Chile, Santiago, Chile, [5]Facultad de Ciencias Agrarias y Forestales, Centro de Desarrollo del Secano Interior, Universidad Católica del Maule; Talca, Chile, [6]Institute of Geography, Pontificia Universidad Católica de Chile, Santiago, Chile, [7]Instituto de Ecología & Biodiversidad, Pontificia Universidad Católica de Chile, Santiago, Chile, [8]Centro UC Desierto de Atacama, Pontificia Universidad Católica de Chile, Santiago, Chile

17.1 Introduction

Chile is one of the countries with the greatest latitudinal extent in the world (17°S–56°S), and undoubtedly the most extensive in the Southern Hemisphere. From north to south, it encompasses tropical and subtropical desert and semiarid climates (up to approximately 32°S), Mediterranean climates (32°S–40°S), and temperate rainy climates (37°S–50°S), including high-altitude climates in the north (above 3000 m a.s.l.), ice fields (46°S–56°S), and southern forests and steppes (inland between 50°S and 56°S). This latitudinal and climatic range gives rise to a remarkable diversity of landscapes, notably the Mediterranean climate hotspot characterized by sclerophyllous and temperate forest biomes. Added to this diversity are human activities associated with industrial forestry, dominated by extensive industrial plantations of *Pinus radiata* D. Don and *Eucalyptus globulus* Labill, which, promoted through public policies, have expanded throughout much of central Chile (32°S–38°S).

In Chile, the fire regime, in terms of frequency, intensity, and extent, has changed significantly since the mid-20th century, with a sustained increase in the number of events but relative overall stability in the total area burned (Úbeda and Sarricolea, 2016). This pattern is largely driven by the expansion of highly flammable *P. radiata* and *Eucalyptus* spp. forest plantations, together with the accumulation of fuel resulting from intensive silvicultural practices. Contemporary fires, therefore, reflect a more anthropogenic and severe fire regime, with growing ecological and territorial implications. Wildfires rank among the most critical environmental hazards of the 21st century, especially in Mediterranean-type regions, such as central Chile, where

Advances in Remote Sensing for Forest Fire Management. DOI: https://doi.org/10.1016/B978-0-443-41692-7.00005-6
© 2027 Elsevier Ltd. All rights are reserved, including those for text and data mining, AI training, and similar technologies.

recurrent droughts and heatwaves exacerbate both their frequency and severity. This phenomenon has intensified over recent decades, not only due to industrial plantations but also because of the increasing urbanization of Wildland–Urban Interface (WUI) areas.

The sustained increase in temperatures and the decline in precipitation, combined with the prolonged megadrought that began in 2010, have created a highly fire-prone context between the regions of Valparaíso and La Araucanía, where more than 80% of the national population and the country's main agroforestry activities are concentrated. In this setting, wildfires pose not only an ecological threat but also a social and territorial risk, affecting biodiversity, ecosystem services, and the safety of exposed communities.

The large extent, homogeneity, fuel load, and flammability of vegetation in forest monocultures and Mediterranean sclerophyllous shrublands have favored high fire severity and rapid propagation of recent megafires (e.g., in 2017). According to the Center for Climate and Resilience Research (CR2, in González et al., 2020), the 2016–17 fire season marked a turning point in the history of Chile's fire regime, with more than 500,000 hectares burned, exceeding the historical average since 1970 by nearly tenfold. The most affected vegetation types were exotic plantations (*P. radiata* and *Eucalyptus* spp.), which accounted for approximately 50%–63% of the burned area, followed by sclerophyllous shrublands (16%–20%), native forests (13%–20%), and grasslands (8%). The predominance of fast-growing and highly flammable exotic vegetation, combined with continuous fuel accumulation and extreme climatic conditions such as prolonged heat waves, megadrought, and strong coastal winds, explains the exceptionally severe character of these megafires. Together, these factors have transformed the central Mediterranean landscape into a fire-prone system of high recurrence and intensity, producing profound ecological and social impacts that challenge current suppression and recovery capacities (Sarricolea et al., 2020; González et al., 2020).

One aspect that must be clarified is that central Chile is dominated by a Mediterranean climate (type Csb according to the Köppen–Geiger classification), characterized by rainy winters and dry summers, with a pronounced summer water deficit that strongly conditions vegetation structure and functioning (Sarricolea et al., 2017). This climatic regime supports sclerophyllous woodlands and hard-leaved shrublands, dominated by species adapted to drought and recurrent fire. In the IUCN Global Ecosystem Typology (GET); (Keith et al., 2022), these formations correspond mainly to the T3.2 seasonally dry temperate heath and shrublands, characterized by evergreen shrublands of low productivity and high physiological resistance to water stress, and to the T6.4 temperate alpine grasslands and shrublands, which prevail in coastal and Andean mountain sectors where low temperatures restrict tree growth. The overlap between the Mediterranean climatic domain and the distribution of these Mediterranean-type ecosystems highlights the territory's vulnerability to large-scale wildfires, as it combines climatic conditions conducive to ignition with plant communities highly adapted to, and functionally dependent on, periodic fire regimes (Fig. 17.1).

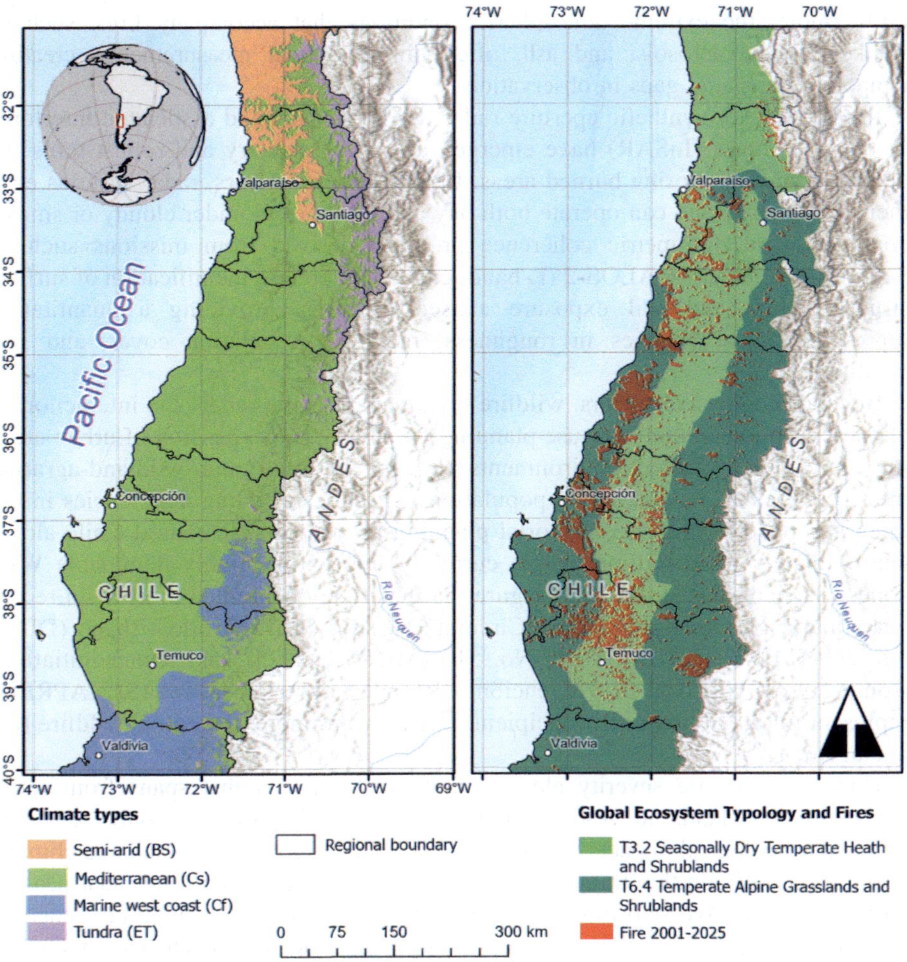

Figure 17.1 Mediterranean climate domain and ecosystem typology in central Chile, showing fire distribution from 2001 to 2025. Map lines delineate study areas and do not necessarily depict accepted national boundaries.
Source: IUCN Global Ecosystem Typology (Keith et al., 2022); MODIS Fire (2001–2025); Köppen–Geiger classification (Sarricolea et al., 2017).

Fire severity, understood as the degree of ecological damage to vegetation and soil, is a key indicator for assessing postfire effects and the resilience of forest ecosystems. Traditionally, it has been estimated through spectral indices derived from optical sensors, such as the difference normalized burn ratio (dNBR), the relative dNBR (RdNBR), or the Enhanced Vegetation Index–based dNBR (dNBR-EVI), which are commonly applied to Landsat and Sentinel-2 imagery. In Chile, however, these remotely sensed estimates have rarely been validated with field observations.

Furthermore, the extreme atmospheric conditions that accompany fires, such as smoke, clouds, aerosols, and ash, often hinder optical measurements, creating temporal and spatial gaps in observation.

In this context, synthetic aperture radar (SAR) sensors and associated interferometric techniques (InSAR) have emerged as complementary and robust tools for detecting and monitoring burned areas. Unlike optical sensors, radar systems emit their own energy and can operate both day and night, even under cloudy or smoky conditions. Interferometric coherence products derived from missions such as Sentinel-1 (C-band) or ALOS-2 (L-band) enable the precise identification of surface structural loss and soil exposure caused by fires, providing a quantitative representation of changes in roughness, moisture, vegetation cover, and soil conditions.

Beyond biophysical factors, wildfire dynamics in Chile reflect the interaction of climatic, economic, and land-use planning variables. The expansion of urban areas into rural and forested environments, known as WUI and wildland-agrarian interfaces (WAI), has increased population exposure, revealing deficiencies in the integration of fire risk into territorial planning instruments. In central Chile alone, approximately three million people currently live within highly flammable WUI zones (Sarricolea et al., 2020), a figure that has likely continued to grow in recent years alongside urban expansion. Circular División de Desarrollo Urbano (DDU) No. 269 (MINVU, 2014), DDU No. 510 (MINVU, 2025), and recent initiatives from Servicio Nacional de Prevención y Respuesta ante Desastres (SENAPRED) represent initial, although still incipient, steps toward more preventive wildfire risk management.

Understanding fire severity along a latitudinal gradient that spans from sclerophyllous shrublands to Andean temperate forests requires not only assessing immediate ecological impacts but also addressing their territorial, social, and climatic implications. Within this framework, the combined use of optical and radar sensors represents a decisive step toward a more adaptive and transformative fire management approach that is aligned with the challenges of climate change in Chile's Mediterranean landscapes.

This chapter aims to summarize current knowledge on fire severity in Chile, with particular emphasis on Mediterranean ecosystems. In this context, we seek to detect and identify the main knowledge gaps that must be addressed both in research and in strengthening institutional capacities, especially given the increasing intensity and frequency of wildfire regimes under global change. The chapter is structured as follows: first, we present a review of the current state of the art concerning the use of optical remote sensing, the effects of fire on soils, and the emerging opportunities offered by radar-based systems, including a brief radar application to the severe wildfire that occurred in February 2024 in Valparaíso. Next, we present our own case study, highlighting the potential of the Composite Burn Index (CBI) and the combined use of Sentinel-2 and PlanetScope imagery for the Coelemu area (Itata Valley), where fieldwork was integrated with advanced remote sensing techniques (Coelemu Wildfire, 2023). Finally, we provide concluding remarks that synthesize the main findings of this work.

17.2 Fire-severity research in Chile: remote-sensing advances

17.2.1 State-of-the-art in fire-severity assessment

Global scientific production on fire severity has shown steady growth, reaching more than 360 scientific articles indexed in the Web of Science (WoS) by 2024. Compared with fewer than 80 contributions in 2010, the scientific contributions have increased 4.5 times over this period. This rapid expansion is particularly evident in countries with Mediterranean-type ecosystems, where recurrent wildfires have become a large-scale socioecological problem. To characterize this trend, an exhaustive bibliographic search was conducted in the WoS Core Collection using specific topic search expressions that target studies on wildfire severity and its assessment methods (Table 17.1).

These searches were conducted without initial time restrictions, and thesaurus filters in WoS were later applied by country (author affiliation) and by keywords to identify collaboration patterns, thematic clusters, and regional differences in research.

The coauthorship and keyword cooccurrence analysis performed with the VOSviewer software reveals a highly interconnected global system led by the United States, Spain, Portugal, and Australia (Fig. 17.2) countries with a long-standing tradition in wildfire research. In contrast, Chile, despite experiencing recurrent wildfires, shows a more moderate research output, with only 42 WoS-indexed articles (around 10% of the production of Spain, the United States, or Australia), even though it is one of the five regions in the world with Mediterranean ecosystems. This positions Chile as an emerging country with still limited scientific production. Chilean publications are more recent, less numerous, and mostly integrated through

Table 17.1 Search queries used in the bibliometric analysis.

Search scope	Search string (Web of Science Core Collection, TS field)	Purpose
Global search	("wildfire" OR "forest fire" OR "vegetation fire*" OR "bushfire*") AND ("burn severity" OR "fire severity" OR Composite Burn Index [CBI] OR difference normalized burn ratio [dNBR] OR relative differenced normalized burn ratio [RdNBR] OR relativized burn ratio [RBR] OR normalized burn ratio [NBR])	Identify global studies on wildfire severity using spectral indices and field metrics.
Chilean search	("wildfire" OR "forest fire" OR "vegetation fire*" OR "bushfire*") AND ("burn severity" OR "fire severity" OR CBI OR dNBR OR RdNBR OR RBR OR NBR) AND (Chile OR Chilean)	Identify studies specifically addressing fire severity in the Chilean context.

Note: TS, Topic search field (title, abstract, author keywords, and Keywords Plus).

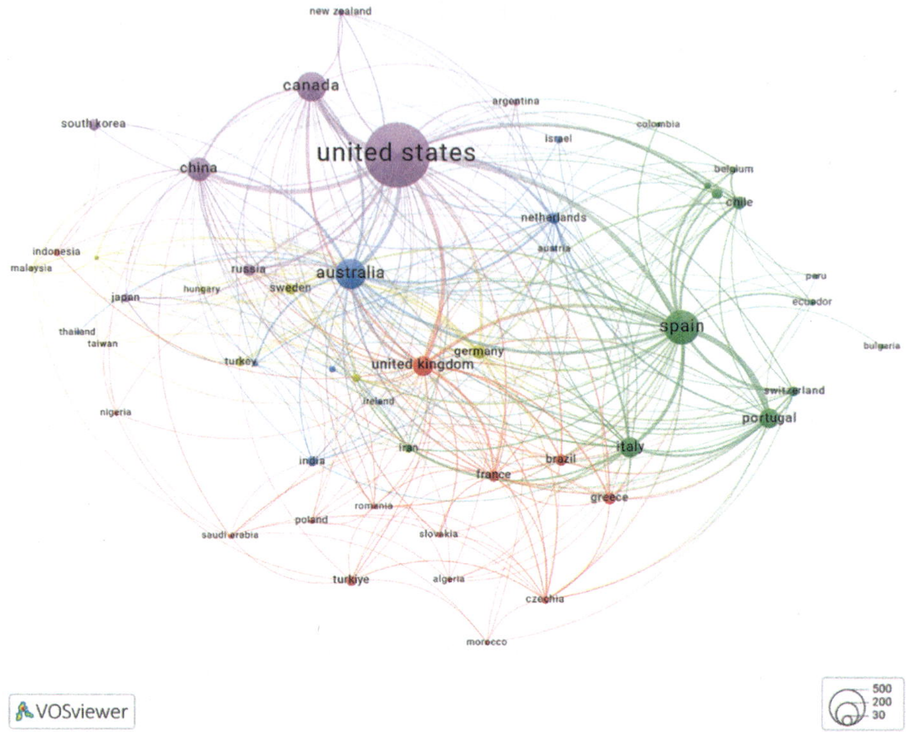

Figure 17.2 Global coauthorship network on fire-severity research derived from the Web of Science (all years–October 2025). Node size indicates publication volume, and line thickness reflects collaboration intensity. The map highlights dominant clusters led by the United States, Spain, Portugal, and Australia, and the emerging role of Chile within Mediterranean fire research.

international collaboration networks, particularly with the United States and Spain. This highlights the importance of international cooperation in the systematic study of fire severity in Mediterranean ecosystems. The gap is reflected in both the limited temporal continuity of national studies and the lack of cross-validation between different satellite data sources and field observations.

The global keyword cooccurrence analysis (Fig. 17.3) reveals three main thematic clusters that structure wildfire severity research. The red cluster, which can be described as "Ecological and biological post-fire responses," includes terms such as species, abundance, mortality, seedling, treatment, forest structure, and community composition. This axis is linked to fire ecology, natural regeneration, and changes in vegetation community structure, highlighting the interaction between fire severity, mortality, and resprouting strategies. The green cluster refers to remote sensing and fire assessment techniques (optical and radar-based), grouping terms such as remote sensing, burned area, normalized burn ratio (NBR), dNBR, RdNBR, Sentinel-2, Landsat, Normalized Difference Vegetation Index (NDVI), algorithm, map, and

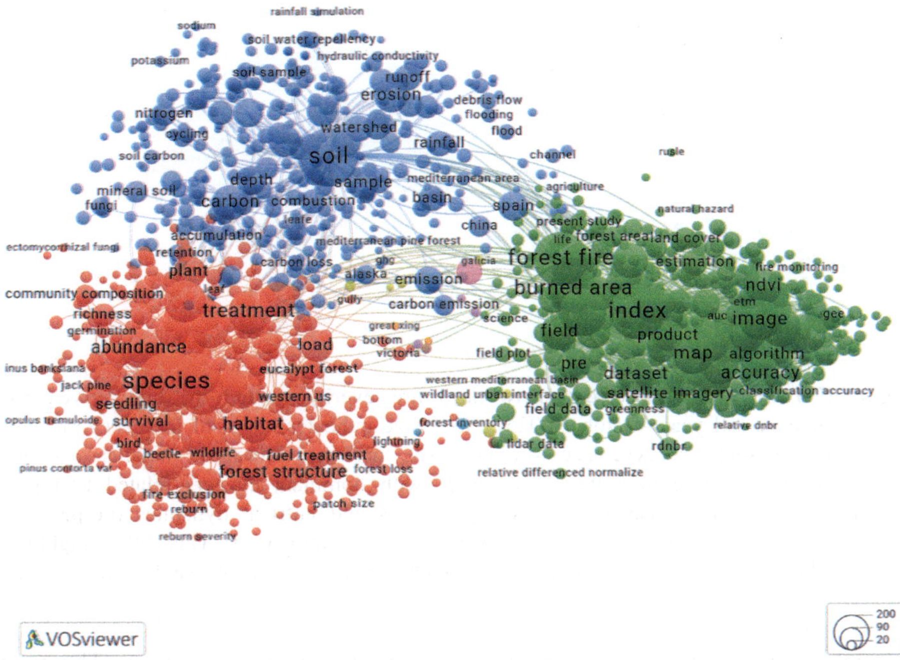

Figure 17.3 Global keyword cooccurrence network on fire-severity research derived from the Web of Science (all years–October 2025). Node size indicates keyword frequency, and line thickness represents cooccurrence strength. The visualization identifies three main thematic clusters: (1) remote sensing and mapping techniques (green), (2) soil and hydrological processes (blue), and (3) species and ecological responses to fire (red). Together, they illustrate the dominant methodological focus on spectral indices (normalized burn ratio, difference normalized burn ratio, relative differenced normalized burn ratio) and the emerging integration of soil and ecological dimensions in global fire severity studies.

accuracy. This green cluster represents the dominant methodological core, centered on the spectral quantification of severity, burned-area mapping, and the validation of satellite products. It also includes a growing number of references to SAR and Sentinel-1, reflecting the gradual incorporation of radar systems and the transition toward multisensor approaches. Finally, the blue cluster encompasses edaphic, hydrological, and biogeochemical processes, with terms such as soil, carbon, erosion, water repellency, runoff, nutrient, hydraulic conductivity, and soil property. This blue group corresponds to studies on fire impacts on soils, carbon balance, hydrophobicity, and degradation or recovery processes, and it also shows an emerging interest in integrating physical and chemical soil properties with remote sensing and hydrological modeling.

In contrast to the global panorama described above (Fig. 17.3), the thematic network corresponding to Chile (Fig. 17.4) exhibits a much smaller and more fragmented structure. Although the same three major conceptual domains can be identified—remote sensing and spectral analysis, soil processes, and ecological

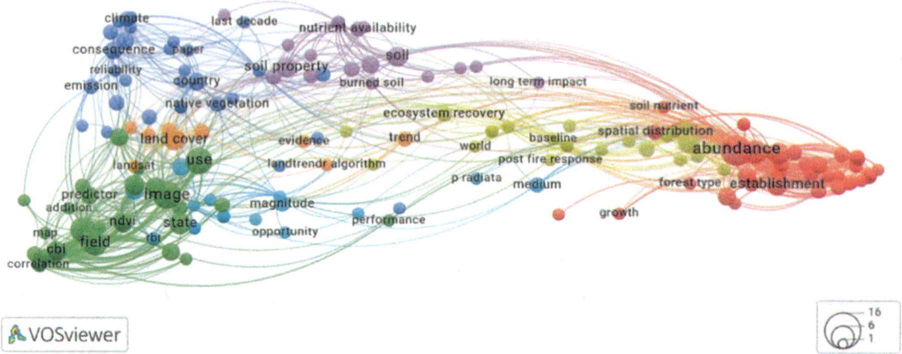

Figure 17.4 Keyword cooccurrence network on fire-severity research in Chile derived from Web of Science (all years–October 2025). Node size indicates keyword frequency, and line thickness represents cooccurrence strength. The network reveals a smaller and less interconnected structure than the global map, with dominant clusters related to remote sensing and spectral indices (green), soil properties and nutrient dynamics (purple), and ecological responses such as species abundance and regeneration (red). This pattern highlights the limited thematic integration and the emerging character of Chilean research on fire severity.

responses—they appear less interconnected and with lower link density. The green cluster, centered on terms such as Landsat, NDVI, CBI, and burned area, shows that national research has focused mainly on severity estimation using optical sensors and derived spectral indices. In contrast, topics related to soil properties, nutrients, or postfire recovery (purple cluster) are scarce, and ecological studies addressing species abundance, establishment, and regeneration (red cluster) appear isolated or have low cooccurrence.

This configuration confirms that Chilean scientific production on fire severity is still at an emerging stage, with a predominantly spectral focus and limited development in the integration of biophysical and edaphic variables. Compared to leading Mediterranean regions, such as Spain and Portugal—where studies combine optical and radar sensors with field assessments and ecological modeling—the Chilean case reveals a significant thematic and methodological gap. However, the increasing availability of active missions such as Sentinel-1 and national observation programs provides a concrete opportunity to broaden the scope of research by incorporating indicators of soil, moisture, and ecosystem recovery, enabling a more comprehensive evaluation of fire severity across the Mediterranean ecosystems of central Chile.

The most comprehensive effort published to date for the case of Chile was developed by Miranda et al. (2022), who produced an extensive national wildfire severity database. This dataset characterizes fire scars between 1985 and 2018, integrating more than 8000 burn scars larger than 10 hectares derived from Landsat imagery, supported by ignition point data provided by Corporación Nacional Forestal (CONAF). The product delivers standardized information on the extent and severity levels (RdNBR) of burned areas across the country, allowing for the characterization of fire regimes over more than three decades. However, the study acknowledges limitations associated with the exclusive use of spectral indices, which can underestimate severity beneath the canopy or confuse nonfire-related land cover changes. It therefore emphasizes the need for calibration using field observations and vegetation-type adjustments, as demonstrated through the CBI in the works of (Peña and Brenning, 2023) using Sentinel-2, (Fernández-Guisuraga et al., 2023) using Sentinel-2 and PlanetScope, and (Yáñez et al., 2025) in the Maule Region. Ultimately, there is a growing need to generate more detailed information on fire severity in Chile, ideally with context-specific values and the highest possible spatial resolution. This requires further research addressing forest types, species composition, maturity, and other structural attributes to more accurately determine fire severity across the country's climatically diverse regions.

17.2.2 Soil impacts and postfire recovery

Fire is a natural process that plays a key role in Mediterranean ecosystems (Fuentes-Ramirez et al., 2022; Ibarra-Bonilla et al., 2024; Kruger, 2014). In these environments, fire acts as a mechanism for returning nutrients to the soil and enhancing its fertility. However, in recent decades, the frequency, extent, and intensity of wildfires have increased worldwide (Cunningham et al., 2024), raising major environmental concerns.

Depending on their severity and duration, fires can disrupt essential ecosystem services and functions, such as nutrient cycling and soil fertility, which affect the soil's ability to support plant life. While low-to-moderate severity fires promote the renovation of dominant vegetation by eliminating undesired species and temporarily increasing pH and available nutrients, high-intensity fires can lead to the complete loss of organic matter, which harms soil properties (Certini, 2005; Fuentes-Ramirez et al., 2022). The situation in Chile is a major concern due to the megadrought that has been ongoing since 2010, characterized by an uninterrupted series of dry years (Garreaud et al., 2017; Garreaud et al., 2020). Furthermore, central Chile is dominated by industrial plantations that require high levels of water (White et al., 2021). This biomass is highly combustible due to water scarcity, making the region susceptible to severe fires that could potentially affect thousands of people (González et al., 2018; Sarricolea et al., 2020). Given the highly agricultural nature of this country, it is essential to improve our knowledge of fires, particularly regarding their impact on the physical, chemical, and organic properties of the soil.

Fire can alter key soil properties, affecting nutrient dynamics and its hydraulic behavior. Changes in temperature during the burning process affect the organic matter content, pH levels, and soil texture (Fuentes-Ramirez et al., 2022; Giraldo et al., 2024). This has implications for the recovery of ecosystems after fire. Table 17.2 shows the main effects on soil properties of Mediterranean ecosystems, highlighting the contributions from central Chile. Giraldo et al. (2024) investigated how fire affects the soil capacity to hold and transmit water to deeper soil horizons (Table 17.2). Soil samples were collected from matorral and Valdivian temperate forests and incinerated to recreate three levels of fire severity: low (100°C–200°C),

Table 17.2 Changes in soil properties after high intensities of fires in central Chile. Research carried out on Mediterranean ecosystems on a global scale. Contributions whose area of study is central Chile are indicated by an asterisk (*).

Soil property	Changes after fire	Methodological approach	References
Soil organic matter	Decrees, slow recovery without vegetation	Sampling of burned areas and laboratory analysis	Certini (2005); Giraldo et al. (2024); *Fuentes-Ramirez et al. (2022); DeBano (1991)
pH	Increases by ash input	Laboratory analysis	Giraldo et al. (2024); Bodí et al. (2014)
Nutrient availability (N, P, Ca, etc.)	Temporary increase, but followed by volatilization and leaching	Sampling of burned areas and laboratory analysis	Giraldo et al. (2024); *Fuentes-Ramirez et al. (2022)
Erosion and soil loss	Increase in water and wind erosion, higher surface runoff	Sampling of burned areas and laboratory analysis	Cerdà and Doerr (2005)
Microorganism communities	Increase in microbiological activity	Laboratory analysis (metagenomics)	Fuentes-Ramirez et al. (2018)
Hydrophobicity	Temporary decrease but followed by an increase in water repellency	Infiltration and water- repellency tests	Doerr et al. (2004); Doerr et al. (2007)

medium (300°C–600°C), and high (600°C–800°C). They evaluated the changes in organic matter, pH, cation exchange capacity, electrical conductivity, and texture before and after heating. The authors then examined the relationship between these results and hydraulic parameters, including infiltration, unsaturated hydraulic conductivity, and the water repellency index (R), among others. Their main finding was that the water repellency index decreased alongside a reduction in clay content, likely due to the clay's capacity to retain moisture.

Organic matter and nitrogen tend to volatilize rapidly in high-intensity fires, while other elements, such as phosphorus and potassium, are less affected. Fuentes-Ramirez et al. (2022) evaluated the effects of the 2015 fire on soil nutrient dynamics in the China Muerta National Reserve, which is in south-central Chile (38°S). Soil samples were collected between 2016 and 2018 along a transect spanning a gradient of fire severity, from low to high. Their results show that only N availability declined over time, while the Organic Matter (OM), P, and K availability increased. However, the increase in OM and nutrients can be the result of couplings in biogeochemical cycles, which in turn increase C capture and other elements by colonizing species in burned spaces (Certini, 2005). In addition, the colonizer plants can fix and immobilize nitrogen, leading to low concentrations in the soil. Although ash incorporated into the soil increases soil pH and nutrients, this may be temporary, as other factors, such as topography, wind, and precipitation, can mobilize both ash and soil particles (Giraldo et al., 2024; Bodí et al., 2014). Furthermore, fire usually reduces the stability of soil aggregates and can cause the soil to become water-repellent. This has a negative effect on infiltration, increasing overland flow and rain splash detachment (Cerdà and Doerr, 2005; Doerr et al., 2004).

Remote sensing is an essential tool for mapping and monitoring the impact of high-severity fires on soil properties (Gholinejad and Khesali, 2021; Morresi et al., 2022; Ibarra-Bonilla et al., 2024). Since satellite imagery has global coverage and periodically revisits areas, it provides time-series images that allow pre and postfire soil conditions to be identified. This information is vital for developing management strategies to aid the recovery of these areas (Gholinejad and Khesali, 2021). To improve the accuracy of mapping high-severity burn areas, Morresi et al., (2022) used satellite images to analyze ten fires that occurred in the western Italian Alps in 2017. From Sentinel-2 satellite imagery, they calculated the NBR and the RdNBR using both before and after fire images. Moderate Resolution Imaging Spectro-radiometer (MODIS) images were used to characterize the phenological conditions and help generate seasonally consistent reflectance composites. The CBI was obtained through field observations assessing vegetation and soil damage across multiple sites. The study concluded that phenology-based satellite analysis is useful for mapping burned areas under cloudy conditions where image availability is limited. Once the area affected by the fires has been defined, satellite images are also useful for monitoring the recovery of ecosystems. They allow for the evaluation of the regeneration of vegetation and soil properties. NDVI, Soil Adjusted Vegetation Index (SAVI), and even NBR can quantify the recovery of vegetation (Ibarra-Bonilla et al., 2024). NDVI is a proxy for vegetation growth and biological productivity, and the use of it together with SAVI improves the vegetation monitoring for areas with sparse vegetation, such as Mediterranean ecosystems. Ibarra-Bonilla et al. (2024) evaluated the postfire vegetation dynamics in four burned sites in 2011 in a temperate forest ecosystem in Mexico. They analyzed NDVI and SAVI data from Thematic Mapper (TM5) and the Operational Land Image 8 (OLI8) sensors for the period 2009–17 and identified the total recovery of the vegetation in 2017. Although they did not evaluate the state of soil properties, these are strongly linked to vegetation cover.

17.2.3 Emerging opportunities: radar and active sensors

Chile's Mediterranean region conditions and varied topography establish circumstances that render the nation especially vulnerable to wildfire events. The nation faced one of its most catastrophic wildfire periods in 2017, during which 114 simultaneous fires burned across Chile, rapidly expanding due to elevated temperatures, strong arid winds, and reduced vegetation moisture levels (Urrutia-Jalabert et al., 2018). Most recently, the 2017–23 fire period exhibited comparable difficulties, with regional winds, called Puelche, affecting fire conditions, emphasizing the essential need for reliable monitoring systems that can deliver prompt information for fire management operations (Hayasaka, 2024).

Conventional optical remote sensing approaches encounter substantial constraints in fire-susceptible areas due to cloud obstruction and smoke interference. With the ability to penetrate cloud cover and smoke while providing day and night imaging capabilities, SAR can serve a vital function in wildfire surveillance (Ban et al., 2020). The fusion of SAR with optical imagery through data integration methodologies provides a holistic approach to wildfire monitoring that overcomes the constraints of individual sensor approaches (Chuvieco et al., 2019).

The combination of various SAR sensors functioning at distinct frequencies and polarizations delivers complementary data regarding fire effects on vegetation and soil environments. Multiple research studies utilize different frequency bands (wavelength and frequency) to obtain various parameters based on wildfire characteristics, including C-band (Tanase et al., 2020; Alarcon-Aguirre et al., 2022) and L-band (Mutai and Chang, 2020; An et al., 2023). The primary findings and key conclusions emphasized in this research show that multisensor SAR-optical fusion methodologies can substantially enhance wildfire monitoring capabilities in Chilean Mediterranean ecosystems, with specific focus on burn-severity evaluation and postfire recovery surveillance.

The regeneration of Chile's native vegetation, as well as that of industrial plantations (such as *P. radiata* and *E. globulus*), after wildfires depends on the intensity of the fire and the characteristics of the ecosystem (whether natural or managed). There is ongoing debate over whether native vegetation, mainly sclerophyllous, can recover after fires of different severities, both in the short and long term. Thus, the first challenge is to accurately determine and assess fire severity, since this enables the development of effective postfire management measures and plans. Such assessment must consider factors such as topography, vegetation composition, and climatic conditions. In Chile's Mediterranean climate (characterized by wet winters and dry summers), vegetation moisture content strongly influences fire dynamics. Recent studies (Zhang et al., 2024) have shown that combining SAR sensors improves fire-severity assessment in comparable Mediterranean ecosystems, providing insights applicable to the Chilean context. In other words, this represents a unique opportunity for Chile to begin its systematic application.

17.2.3.1 Synthetic aperture radar principles for burn mapping and multitemporal analysis

In SAR technology, the backscattered signal represents a crucial element and depends on various parameters when the transmitted signal interacts with surface features to gather information, including surface roughness, dielectric characteristics, topography, local incidence angle, and moisture content (Awange and Kiema, 2013;

Barbera, 2014). The intensity of these variations depends on the burn intensity and vegetation type affected (Huang et al., 2016). The backscatter change detection methodology relies on determining baseline conditions prior to fire events and contrasting these with postfire measurements. The temporal period for establishing baselines is crucial, as seasonal changes in vegetation phenology and soil moisture can affect the intensity of fire-induced modifications (Imperatore et al., 2017).

The temporal development of backscatter values offers insight into burn advancement and recovery trajectories. Multitemporal examination allows differentiation between various burn-severity categories based on the intensity and duration of backscatter modifications (Siegert et al., 2001). Similarly, more suitable studies exist where uniform, high-biomass regions have experienced severe burning (Polychronaki et al., 2013). Current developments in deep learning methodologies for SAR-based wildfire monitoring, as shown by Ban et al. (2020) utilizing Sentinel-1 time series, have demonstrated encouraging outcomes for near-real-time monitoring applications. Additionally, Gorelick et al. (2017) highlighted the significance of multitemporal validation methodologies for reliable algorithm evaluation.

17.2.3.2 Polarimetric synthetic aperture radar analysis

Polarimetric SAR data offers supplementary information regarding vegetation structure and burn severity through examination of various scattering mechanisms (Cloude and Pottier, 1997). The decomposition of polarimetric characteristics into surface, volume, and double-bounce scattering elements allows detailed characterization of impacts on vegetation structure (Freeman and Durden, 1998). Chilean sclerophyllous forests display distinctive polarimetric characteristics that can be utilized for enhanced burn-severity evaluation. The combination of C-band and L-band polarimetric data offers complementary information about forest areas.

Among the most accessible radar missions are Sentinel-1, ALOS-2 PALSAR-2, and Satélite Argentino de Observación Con Microondas (SAOCOM), which provide complementary capabilities for wildfire monitoring and vegetation analysis. Their different radar bands (C and L) and revisit frequencies make them suitable for both rapid fire detection and postfire biophysical assessment. Table 17.3 summarizes the main characteristics of these radar sensors.

Optical–SAR fusion is increasingly central to burn mapping because the sensors contribute complementary information: optical data carry spectral burn signals, while SAR captures structural and moisture changes, enabling more robust fire impact mapping and near-real-time progression monitoring with time series and deep learning (Chuvieco et al., 2019; Ban et al., 2020; Hu et al., 2023). For soil moisture, coupling Sentinel-1 backscatter with Sentinel-2 vegetation indices helps account for canopy effects on surface soil moisture recovery. Recent work in Mediterranean settings shows that both L-band SAOCOM and C-band Sentinel-1 are effective for soil moisture retrieval mapping applications at regional scales (Brunelli and Mancini, 2024; Shi et al., 2025). Operational fusion requires careful preprocessing, temporal alignment, and spatial registration so that pixel-level combinations improve interpretability and reliability, a workflow increasingly supported by cloud platforms such

Table 17.3 Comparison of key synthetic aperture radar (SAR) missions and their main specifications for wildfire monitoring and vegetation assessment in Chile.

Mission	Radar band/ frequency	Wavelength	Spatial resolution	Temporal resolution	Swath width	Polarization	Incidence angle	Main advantages
Sentinel-1 (A and B)	C-band (5.405 GHz)	5.6 cm	5 × 20 m (IW), 40 × 40 m (Extra Wide or EW)	Six days (constellation)	250 km (IW), 400 km (EW)	Dual-pol (VV/VH or HH/HV)	29.1–46 degrees	High revisit rate; effective for near-real-time fire detection.
ALOS-2 PALSAR-2	L-band (1.27 GHz)	24 cm	1 × 3 m (Spotlight), 3 × 3 m (Stripmap), 100 × 100 m (ScanSAR)	14 days	25 km (Stripmap), 350 km (ScanSAR)	Full polarimetric (HH, HV, VH, VV)	8–70 degrees	Deeper vegetation penetration; detailed biophysical analysis.
SAOCOM	L-band	24 cm	10 × 10 m (Stripmap), 100 × 100 m (ScanSAR)	16 days (single sat.)	80 km (Stripmap), 320 km (ScanSAR)	Full polarimetric	17–47 degrees	Regional focus on South America; useful for monitoring vegetation and soil moisture.

as Google Earth Engine and by established multisensor fusion frameworks that span from classical concepts to modern implementations (Pohl and van Genderen, 1995; Gorelick et al., 2017; Tamiminia et al., 2020).

Evidence from other Mediterranean ecosystems is directly transferable to Chile, given similarities in climate and vegetation structure. For instance, Belenguer-Plomer et al. (2019) demonstrated in Mediterranean Europe that integrating thermal anomalies with Sentinel-1 backscatter improves burned-area detection. In boreal and temperate forests, including Canadian case studies, C-band SAR has shown strong penetration for assessing burn severity, and comparative analyses using Sentinel-2, Sentinel-1, and ALOS-2 clarify differences between burned and unburned surfaces when using postevent or bi-temporal observations, insights that are applicable to Chilean monitoring needs (Bourgeau-Chavez et al., 2002; Zhang et al., 2024).

17.2.3.3 Interferometric synthetic aperture radar analysis

Interferometric SAR uses coherence loss to indicate fire-related surface and canopy changes, supports postfire landslide vulnerability assessment, and refines burn mapping and postrecovery tracking (Tanase et al., 2010; Handwerger et al., 2019; Engdahl and Hyyppa, 2003). However, this needs to be combined with amplitude-based change detection and temporal decorrelation analysis. For example, using both C- and L-band provides complementary scale-dependent sensitivity in Mediterranean pine forests (Tanase et al., 2010). Polarimetric decompositions, such as those by Freeman and Durden (1998), distinguish surface, volume, and double-bounce scattering affected by fire, and pairing polarimetric features with machine learning has shown promise for automated burn-severity classification, adding to evidence that polarimetric responses at C- and L-band are informative over burned forests (Cloude and Pottier, 1997; Freeman and Durden, 1998; Yang et al., 2023; Tanase et al., 2014). Overall, multisensor fusion, supported by scalable cloud infrastructures and advanced SAR interferometric and polarimetric techniques, is consolidating as a practical pathway to operational fire mapping, severity assessment, and postfire hazard evaluation in Chilean Mediterranean ecosystems (Gorelick et al., 2017; Chuvieco et al., 2019; Tamiminia et al., 2020).

17.2.3.4 Brief application in Valparaíso wildfire 2024

Major wildfires broke out in central Chile during January–February 2024, with the Valparaíso Region, notably the municipalities of Valparaíso, Villa Alemana, and Quilpué, being severely affected. The burned area is estimated at approximately 11,350 Ha (Cima Vergara, 2024). The event caused 135 fatalities and 8188 affected households, positioning it among the most severe disasters recorded in the country (Cima Vergara, 2024).

To delineate the burn scar using SAR-optical data fusion, Sentinel-1 IW Single Look Complex (SLC) imagery was processed to derive interferometric coherence (VV polarization) over the pair 2024-02-02 and 2024-02-14. For the optical component, a Sentinel-2 MultiSpectral Instrument (MSI) L2A scene from 2024-02-07 was used to compute the NBR (Key and Benson, 2006), defined as (NIR–SWIR2)/(NIR+SWIR2),

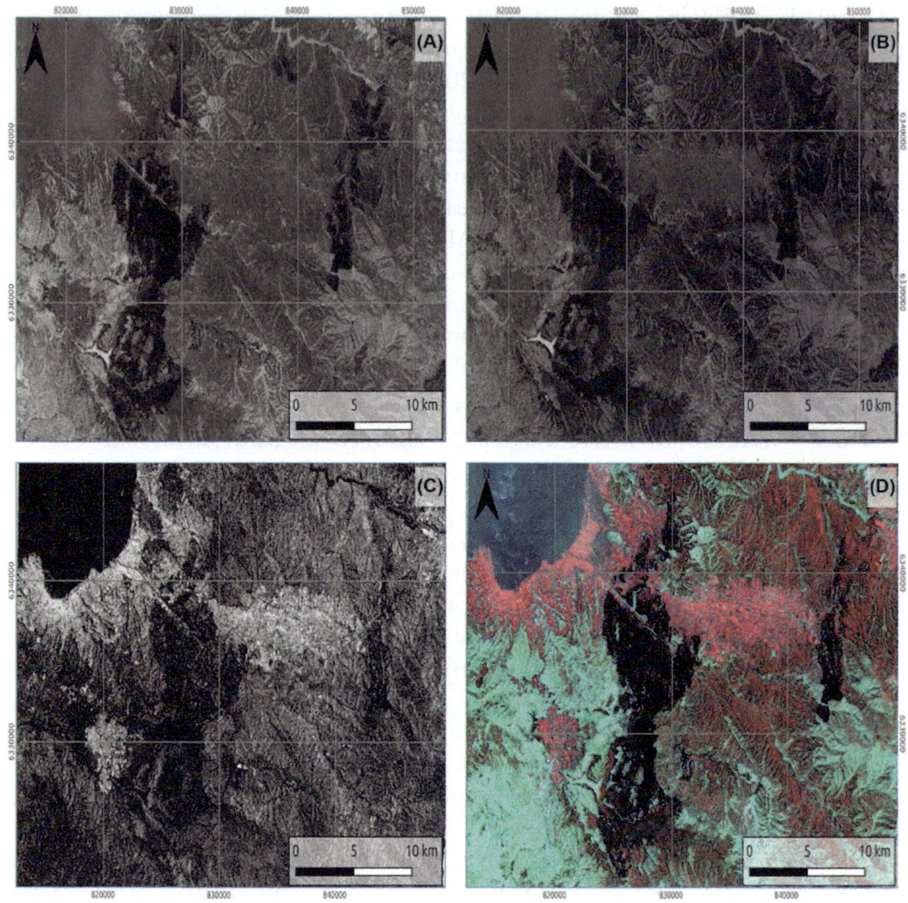

Figure 17.5 Example of synthetic aperture radar–optical data fusion for the 2024 Valparaíso wildfire. (A) normalized burn ratio (NBR); (B) Normalized Difference Moisture Index (NDMI); (C) Sentinel-1 VV coherence (pair: 2024-02-02 and 2024-02-14); and (D) RGB composite (R: coherence, G: NDMI, B: NBR).

and the Normalized Difference Moisture Index (NDMI) (Wang et al., 2008), defined as (NIR–SWIR1)/(NIR+SWIR1).

In Fig. 17.5, Section (A) shows the results of the NBR index, clearly highlighting the burned areas with values close to 0.8. Section (B) presents the NDMI index, emphasizing the moisture deficit caused by the burned surface. In this case, the focus near Villa Alemana appears with lower intensity compared to the NBR results. For the optical results in (A) and (B), a reduced cloud cover is observed, which in this case does not interfere with the identification of the burned area but underscores the importance of addressing such events with SAR sensors to avoid losing information that can be crucial for emergency management. Section (C) corresponds to the coherence estimated between both dates for the VV polarization, which shows higher reflectivity in the presence of exposed soil, as typically observed after wildfire events. Finally, Section (D) displays an RGB composite (R: Coherence, G: NDMI, B: NBR), where the burned area is represented by colors ranging from deep red to brown, facilitating the visual identification of both fire foci between the Quilpué–Valparaíso conurbation and the boundary of Villa Alemana.

17.3 Field-based severity assessment: the Itata Valley fire case study

Field-based assessments have long been considered a reliable method for evaluating the recovery of ecosystem functions and services after wildfire events, through the measurement of biophysical indicators in vegetation and soil (Key and Benson, 2006; Morgan et al., 2014). However, under the current context of global fire regime change, characterized by increasingly extensive and severe wildfires (Beltrán-Marcos et al., 2024; Duane et al., 2021), traditional field sampling campaigns face significant limitations. Field methods require a considerable amount of time and are labor-intensive, particularly for large-scale applications (Veraverbeke et al., 2011). Moreover, their localized nature hampers the capacity to capture the full spatial heterogeneity of fire impacts, especially in ecosystems composed of multiple vegetation types (Miller and Thode, 2007; Fernández-Guisuraga et al., 2023). Within this context, remote sensing approaches such as (1) mono- and bi-temporal spectral indices (Fernández-García et al., 2018; Mallinis et al., 2017), (2) spectral unmixing analysis (SMA) to decompose subpixel reflectance signals (Quintano et al., 2013), or (3) radiative transfer models (RTMs) (Fernández-Guisuraga et al., 2021) have emerged as powerful tools for monitoring and quantifying fire-induced impacts across large burned areas due to their cost-effectiveness and synoptic nature (Veraverbeke et al., 2011). Nevertheless, to obtain reliable wall-to-wall assessments of fire severity that allow for the accurate identification of those areas requiring emergency restoration actions, it is crucial to establish strong relationships with field-based measurements of actual wildfires. These links are usually derived from empirical models calibrated using field observations and remote sensing products, such as spectral indices (Fernández-García et al., 2018; Fernández-Guisuraga et al., 2023; Mallinis et al., 2017), from which continuous index values are segmented into discrete fire-severity classes (low, moderate, and high).

Given the ecological diversity of Chilean Mediterranean ecosystems and the varying degrees of land-use management in this region, the application of integrative and multistrata field-based indices such as the CBI; (Key and Benson, 2006) and its derivatives (e.g., GeoCBI; De Santis and Chuvieco, 2009) remains limited. Following the extensive wave of wildfires that affected central Chile in 2023, field-based severity assessments have been employed to better understand how agroforestry management influences fire-severity patterns (Beltrán-Marcos et al., 2025). In particular, a study area was selected in the municipality of Coelemu (Itata, Ñuble Region; Fig. 17.6), where a large wildfire that occurred in February 2023 burned approximately 4676 hectares of Mediterranean ecosystems. The affected area

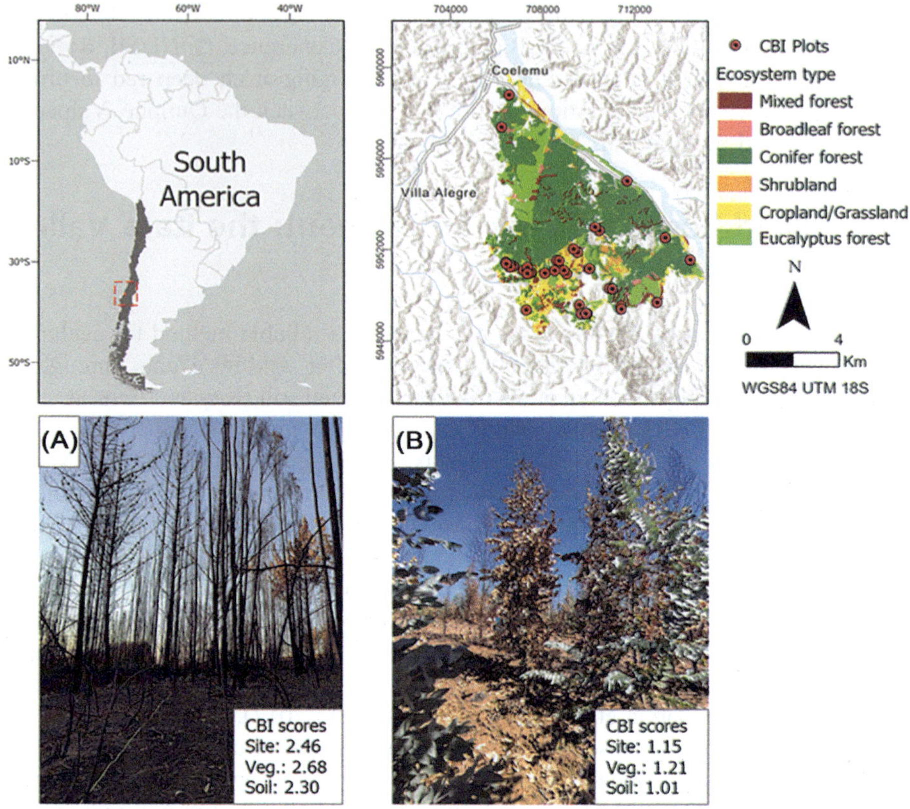

Figure 17.6 Location of Composite Burn Index (CBI) plots and types of ecosystems affected by the Coelemu wildfire that occurred in 2023 in central Chile (above); example of CBI field plots and scores at the site, vegetation, and soil levels in (A) *Pinus* industrial plantations and (B) *Eucalyptus* industrial plantations (below). Map lines delineate study areas and do not necessarily depict accepted national boundaries.

included extensive industrial plantations of *P. radiata* D. Don and *E. globulus* Labill., residual native forests dominated by *Nothofagus obliqua* (Mirb.) Oerst., shrublands primarily composed of *Genista monspessulana* L., and large areas of grasslands and croplands. This landscape had been subject to low levels of active forest management, mostly oriented toward sustaining an intensive wood production cycle (Leal-Medina et al., 2024). As a result, the affected area was characterized by homogeneous fuel structures, only interrupted by patches of agricultural land dedicated to grape cultivation. In this study site, a total of 36 field plots measuring 20 × 20 m were randomly established approximately one month after the fire in homogeneously burned patches, ensuring a minimum spacing of 100 m between plots (Fig. 17.6). These plots were assessed following the CBI protocol (Key and Benson, 2006), adapted for Mediterranean ecosystems (Fernández-García et al., 2018) to improve ecological consistency in short-term fire impact evaluations. The CBI approach is based on the visual assessment of several fire-severity attributes such as char height, ash color, or tree mortality that allow assigning semiquantitative scores (from 0—unchanged—to 3—severely burned) across the substrate (soil strata) and four vegetation strata (low vegetation <1 m, shrubs and saplings 1–5 m, intermediate trees 5–20 m, and tall trees >20 m) in the field plots. This method yields a mean plot-level CBI value that may be classified as low (CBI < 1.25), moderate (1.25 ≤ CBI ≤ 2.25), and high (CBI > 2.25), following standardized thresholds (Miller and Thode, 2007).

These field-based measurements formed the basis for calibrating satellite-derived fire-severity indices retrieved from Sentinel-2 MSI (20 m/pixel spatial resolution) and PlanetScope SuperDove (PSB.SD) (3 m/pixel spatial resolution) through ordinary least squares (OLS) models. The correlations observed with field CBI data from this study site demonstrated strong performance (Fig. 17.7), comparable to or even exceeding that reported in previous studies in European (e.g., Fernández-Guisuraga et al., 2023) and Chilean Mediterranean regions (e.g., Peña and Brenning, 2023). Among the evaluated indices, the NBR; (García and Caselles, 1991) based spectral indices such as the dNBR (Key, 2006) derived from Sentinel-2 data exhibited the highest correlation with CBI scores. PlanetScope-based indices, while generally showing slightly lower explanatory power, benefited from their finer spatial resolution, which improved performance in highly fragmented vegetation patches and field plot alignment. The integration of high-resolution remote sensing with ground-truth data provides a robust and ecologically meaningful basis for ecological impact mapping and decision-making in postfire management, particularly in complex landscapes shaped by mixed land uses and variable fuel loads (Fig. 17.8). This methodological framework offers a transferable approach for other fire-prone Mediterranean regions in Chile. In the context of increasing fire frequency, size, and

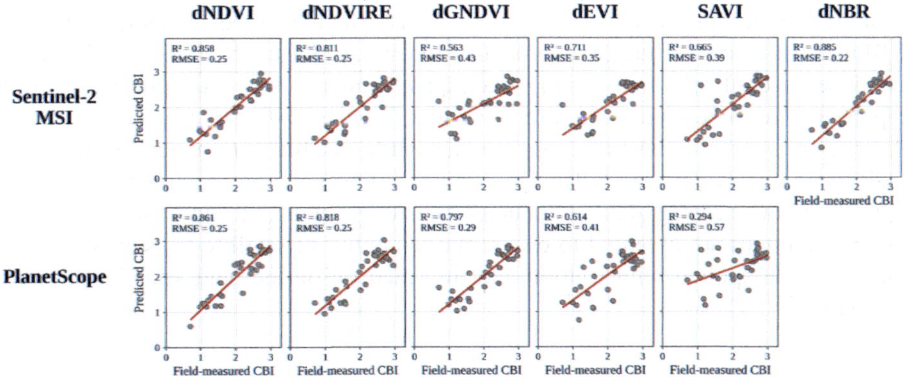

Figure 17.7 Relationship between field-measured and predicted Composite Burn Index (CBI) scores at the site level by spectral indices (difference normalized vegetation index [dNDVI], difference normalized vegetation red-edge index [dNDVIRE], difference normalized green vegetation index [dGNDVI], difference enhanced vegetation index [dEVI], Soil Adjusted Vegetation Index [SAVI], and difference normalized burn ratio [dNBR]) retrieved from Sentinel-2 MSI and PlanetScope PSB.SD satellite images.

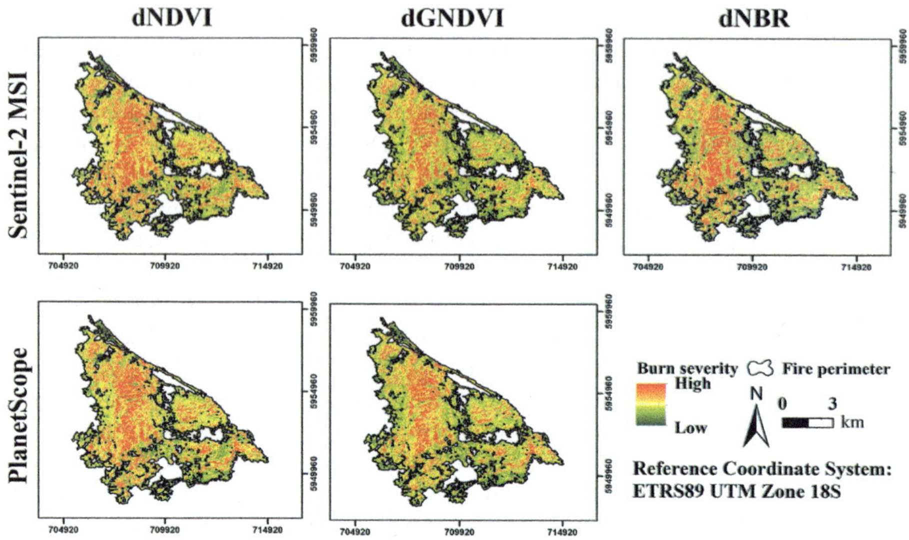

Figure 17.8 Wall-to-wall fire severity estimates for the ordinary least squares model of the difference normalized vegetation index (dNDVI), difference normalized green vegetation index (dGNDVI), and difference normalized burn ratio (dNBR) spectral indices calibrated from field data collected in Coelemu wildfire. Red areas highlight potential zones for urgent post-fire management interventions, such as erosion control. Map lines delineate study areas and do not necessarily depict accepted national boundaries.

severity, particularly under climate-driven extremes (Duane et al., 2021), this type of integrative assessment is essential to guide rapid restoration actions, evaluate ecosystem resilience, and design prefire fuel management interventions. Nonetheless, in some Mediterranean ecosystems, fire-severity effects may be decoupled across the CBI multilayered structure (e.g., broadleaf forests), so the calibration of multispectral or hyperspectral products using integrative indices may not provide a fully representative assessment of fire severity (Fernández-Guisuraga et al., 2023). Averaging severity scores across strata, while useful for standardization, can underrepresent canopy-level damage, which largely governs spectral reflectance in optical remote sensing data (Fernández-Guisuraga et al., 2021). Therefore, while these techniques are essential to ensure the reliability of satellite-based assessments, they must be applied with caution and informed by on-the-field knowledge of fire behavior.

17.4 Conclusions

Fire severity in the Chilean Mediterranean region, where native sclerophyllous forests and shrublands have been extensively replaced by industrial *P. radiata* and *Eucalyptus* plantations, represents a major scientific, political, and social challenge that demands integrated approaches combining field observations, optical and radar remote sensing, and robust geographical and ecological knowledge. Despite recent advances following the 2017 megafires, national research remains focused mainly on spectral indices derived from optical imagery, with limited incorporation of radar data and field calibration. This imbalance hinders the development of consistent severity metrics across the country's diverse ecosystems, both natural and industrial, leaving postfire severity monitoring and restoration strategies without the information needed for effective resource management.

The evidence reviewed in this chapter highlights three main priorities. First, the urgent need to locally calibrate and validate severity indices, such as the CBI, to reflect the structural and floristic diversity of Chilean vegetation. Second, the value of integrating optical and radar (SAR) data, particularly from the Sentinel-1, Sentinel-2, and PlanetScope missions, which provide an effective framework for mapping burned areas under adverse atmospheric conditions and for monitoring postfire recovery. Third, the importance of systematic integration between field and satellite data, which enables ecologically meaningful severity assessments and supports decision-making in restoration and risk-reduction planning, including the monitoring of soil impacts and their recovery over time.

The combined results from the Coelemu (2023) and Valparaíso (2024) case studies demonstrate progress toward integrating radar coherence and high-resolution spectral indices, such as those from PlanetScope. However, further laboratory analyses of soil samples are still needed to evaluate postfire conditions and damage beyond vegetation. Future efforts should consolidate these approaches within operational frameworks supported by national observation programs, standardized field protocols, and open data repositories.

Ultimately, advancing toward a more adaptive and management-oriented fire-severity monitoring system requires bridging disciplinary gaps among remote sensing, geography, soil science, and ecology. Strengthening institutional collaboration and investing in long-term monitoring will be essential to increase Chile's resilience to increasingly severe fire regimes in a warming and drying climate.

Acknowledgments

This work was supported by the Spanish Ministry of Science and Innovation in the framework of the LANDSUSFIRE project within the National Program for the Promotion of Scientific-Technical Research (2021–23) (grant numbers PID2022–139156OB-C21, PID2022–139156OB-C22) and by the Regional Government of Castile and León in the framework of the IA-FIREXTCyL project [grant number LE081P23].

This work was supported by the Chilean National Agency for Research and Development (ANID) through FONDECYT Iniciación Project 11221034 and by the Center for Climate and Resilience Research (CR2), based at the University of Chile and funded through the Actividades de Interés Nacional (AIN) program.

References

Awange, J., Kiema, J.B., 2013. Environmental Geoinformatics 10, 978–983. Available at https://link.springer.com/content/pdf/10.1007/978-3-030-03017-9.pdf.

An, K., Jones, C.E., Lou, Y., 2023. Developing a detection and monitoring framework for wildfire regimes with L-Band Polarimetric SAR. Authorea Preprints. https://doi.org/10.22541/essoar.168056839.98485943/v1.

Alarcon-Aguirre, G., Miranda Fidhel, R.F., Ramos Enciso, D., Canahuire-Robles, R., Rodriguez-Achata, L., Garate-Quispe, J., 2022. Burn Severity Assessment Using Sentinel-1 SAR in the Southeast Peruvian Amazon, a Case Study of Madre de Dios. Fire 5, 94. Available at https://doi.org/10.3390/fire5040094.

Ban, Y., Zhang, P., Nascetti, A., Bevington, A.R., Wulder, M.A., 2020. Near real-time wildfire progression monitoring with sentinel-1 SAR time series and deep learning. Scientific Reports 10 (1). Available at https://doi.org/10.1038/s41598-019-56967-x, www.nature.com/srep/index.html.

Barbera, E., 2014. COSMO-SkyMed polarimetric data for soil moisture retrieval: capability of SAR data for landslide monitoring. Available at http://www.fedoa.unina.it/9860/1/Barbera_Eliana_26.pdf.

Belenguer-Plomer, M.A., Tanase, M.A., Fernandez-Carrillo, A., Chuvieco, E., 2019. Burned area detection and mapping using sentinel-1 backscatter coefficient and thermal anomalies. Remote Sensing of Environment 233. Available at https://doi.org/10.1016/j.rse.2019.111345, www.elsevier.com/inca/publications/store/5/0/5/7/3/3.

Beltrán-Marcos, D., Suárez-Seoane, S., Fernández-Guisuraga, J.M., Azevedo, J.C., Calvo, L., 2024. Fire regime attributes shape pre-fire vegetation characteristics controlling extreme fire behavior under different bioregions in Spain. Fire Ecology 20 (1). Available at https://doi.org/10.1186/s42408-024-00276-w, https://www.springer.com/journal/42408.

Bodí, M.B., Martin, D.A., Balfour, V.N., Santín, C., Doerr, S.H., Pereira, P., Cerdà, A., Mataix-Solera, J., 2014. Wildland fire ash: Production, composition and eco-hydro-geomorphic effects. Earth-Science Reviews 130, 103–127. Available at https://doi.org/10.1016/j.earscirev.2013.12.007.

Bourgeau-Chavez, L.L., Kasischke, E.S., Brunzell, S., Mudd, J.P., Tukman, M., 2002. Mapping fire scars in global boreal forests using imaging radar data. International Journal of Remote Sensing 23 (20), 4211–4234. Available at https://doi.org/10.1080/01431160110109589, https://www.tandfonline.com/loi/tres20.

Brunelli, B., Mancini, F., 2024. Comparative analysis of SAOCOM and sentinel-1 data for surface soil moisture retrieval using a change detection method in a semiarid region (Douro River's basin, Spain). International Journal of Applied Earth Observation and Geoinformation 129, 103874. Available at https://doi.org/10.1016/j.jag.2024.103874.

Cerdà, A., Doerr, S.H., 2005. Influence of vegetation recovery on soil hydrology and erodibility following fire: an 11-year investigation. International Journal of Wildland Fire 14 (4), 423–437. Available at https://doi.org/10.1071/WF05044.

Certini, G., 2005. Effects of fire on properties of forest soils: a review. Oecologia 143 (1), 1–10. Available at https://doi.org/10.1007/s00442-004-1788-8.

Chuvieco, E., Mouillot, F., van der Werf, G.R., San Miguel, J., Tanase, M., Koutsias, N., et al., 2019. Historical background and current developments for mapping burned area from satellite earth observation. Remote Sensing of Environment 225, 45–64. Available at https://doi.org/10.1016/j.rse.2019.02.013.

Cima Vergara, M.A., 2024. Vulnerabilidad del Área Metropolitana de Valparaíso ante incendios. Revista Planeo (58). Available at https://doi.org/10.7764/plan.058.135.

Cloude, S.R., Pottier, E., 1997. An entropy based classification scheme for land applications of polarimetric SAR. IEEE Transactions on Geoscience and Remote Sensing 35 (1), 68–78. Available at https://doi.org/10.1109/36.551935.

Cunningham, C.X., Williamson, G.J., Bowman, D.M., 2024. Increasing frequency and intensity of the most extreme wildfires on Earth. Nature Ecology & Evolution 8, 1420–1425. Available at https://doi.org/10.1038/s41559-024-02452-2.

DeBano, L.F., 1991. The effect of fire on soil properties proceedings management and productivity of western-Montane. Forest Soils 151–155. Available at https://books.google.com/books?hl=es&lr=&id=PGTOhtl2380C&oi=fnd&pg=PA151&dq=The+effect+of+fire+on+soil+properties.+In+Proceedings+management+and+productivity+of+western-Montane.+Forest+Soils+&ots=8luGoaQZ5O&sig=Qx_sTsZhc7M2sN4ucYiET4CUfO8.

De Santis, A., Chuvieco, E., 2009. GeoCBI: a modified version of the composite burn index for the initial assessment of the short-term burn severity from remotely sensed data. Remote Sensing of Environment 113 (3), 554–562. Available at https://doi.org/10.1016/j.rse.2008.10.011.

Doerr, S.H., Blake, W.H., Shakesby, R.A., Stagnitti, F., Vuurens, S.H., Humphreys, G.S., et al., 2004. Heating effects on water repellency in Australian eucalypt forest soils and their value in estimating wildfire soil temperatures. International Journal of Wildland Fire 13 (2), 157–163. Available at https://doi.org/10.1071/WF03051.

Doerr, S.H., Ritsema, C.J., Dekker, L.W., Scott, D.F., Carter, D., 2007. Water repellence of soils: new insights and emerging research needs. Hydrological Processes 21 (17), 2223–2228. Available at https://doi.org/10.1002/hyp.6762.

Duane, A., Castellnou, M., Brotons, L., 2021. Towards a comprehensive look at global drivers of novel extreme wildfire events. Climatic Change 165. Available at https://doi.org/10.1007/s10584-021-03066-4.

Engdahl, M.E., Hyyppa, J.M., 2003. Land-cover classification using multitemporal ERS-1/2 insar data. IEEE Transactions on Geoscience and Remote Sensing 41, 1620–1628. Available at https://doi.org/10.1109/tgrs.2003.813271.

Fernández-García, V., Santamarta, M., Fernández-Manso, A., Quintano, C., Marcos, E., Calvo, L., 2018. Burn severity metrics in fire-prone pine ecosystems along a climatic gradient using landsat imagery. Remote Sensing of Environment 206, 205–217. Available at https://doi.org/10.1016/j.rse.2017.12.029.

Fernández-Guisuraga, J.M., Fernandes, P.M., Marcos, E., Beltrán-Marcos, D., Sarricolea, P., Farris, M., et al., 2023. Caution is needed across Mediterranean ecosystems when interpreting wall-to-wall fire severity estimates based on spectral indices. Forest Ecology and Management 546. Available at https://doi.org/10.1016/j.foreco.2023.121383, www.elsevier.com/inca/publications/store/5/0/3/3/1/0.

Fernández-Guisuraga, J.M., Verrelst, J., Calvo, L., Suárez-Seoane, S., 2021. Hybrid inversion of radiative transfer models based on high spatial resolution satellite reflectance data improves fractional vegetation cover retrieval in heterogeneous ecological systems after fire. Remote Sensing of Environment 255, 112304. Available at https://doi.org/10.1016/j.rse.2021.112304.

Freeman, A., Durden, S.L., 1998. A three-component scattering model for polarimetric SAR data. IEEE Transactions on Geoscience and Remote Sensing 36 (3), 963–973. Available at https://doi.org/10.1109/36.673687.

Fuentes-Ramirez, A., Almonacid-Muñoz, L., Muñoz-Gómez, N., Moloney, K.A., 2022. Spatio-temporal variation in soil nutrients and plant recovery across a fire-severity gradient in old-growth Araucaria-Nothofagus forests of south-central Chile. Forests 13 (3), 448. Available at https://doi.org/10.3390/f13030448.

Fuentes-Ramirez, A., Barrientos, M., Almonacid, L., Arriagada-Escamilla, C., Salas-Eljatib, C., 2018. Short-term response of soil microorganisms, nutrients and plant recovery in fire-affected Araucaria araucana forests. Applied Soil Ecology 131, 99–106. Available at https://doi.org/10.1016/j.apsoil.2018.08.010.

García, M.J.L., Caselles, V., 1991. Mapping burns and natural reforestation using thematic mapper data. Geocarto International 6 (1), 31–37. Available at https://doi.org/10.1080/10106049109354290.

Garreaud, R.D., Alvarez-Garreton, C., Barichivich, J., Boisier, J.P., Christie, D., Galleguillos, M., et al., 2017. The 2010-2015 megadrought in central Chile: impacts on regional hydroclimate and vegetation. Hydrology and Earth System Sciences 21 (12), 6307–6327. Available at https://doi.org/10.5194/hess-21-6307-2017, http://www.hydrol-earth-syst-sci.net/volumes_and_issues.html.

Garreaud, R.D., Boisier, J.P., Rondanelli, R., Montecinos, A., Sepúlveda, H.H., Veloso-Aguila, D., 2020. The Central Chile Mega Drought (2010–2018): A climate dynamics perspective. International Journal of Climatology 40, 421–439. Available at https://doi.org/10.1002/joc.6219.

Gholinejad, S., Khesali, E., 2021. An automatic procedure for generating burn severity maps from the satellite images-derived spectral indices. International Journal of Digital Earth 14 (11), 1659–1673. Available at https://doi.org/10.1080/17538947.2021.1966525.

Giraldo, C.V., Acevedo, S.E., Contreras, C.P., Santibáñez, F., Sáez, E., Calderón, F.J., et al., 2024. Effects of soil heating changes on soil hydraulic properties in central Chile. Geoderma 449. Available at https://doi.org/10.1016/j.geoderma.2024.117013, https://www.sciencedirect.com/science/journal/00167061.

González, M.E., Gómez-González, S., Lara, A., Garreaud, R., Díaz-Hormazábal, I., 2018. The 2010–2015 Megadrought and its influence on the fire regime in central and south-central Chile. Ecosphere 9 (8). Available at https://doi.org/10.1002/ecs2.2300, http://esajournals.onlinelibrary.wiley.com/hub/journal/10.1002/(ISSN)2150-8925/.

González, M.E., Sapiains, R., Gómez-González, S., Garreaud, R., Miranda, A., Galleguillos, M., Jacques, M., Pauchard, A., Hoyos, J., Cordero, L., Vásquez, F., Lara, A., Aldunce, P., Delgado, V., Arriagada, Ugarte, A.M., Sepúlveda, A., Farías, L., García, R., Rondanelli, R.,J., Ponce, R.,Vargas, F., Rojas, M., Boisier, J.P., C., Carrasco, Little, C., Osses, M., Zamorano, C., Díaz-Hormazábal, I., Ceballos, A., Guerra, E., Moncada, M., Castillo, I ., 2020. Incendios forestales en Chile: causas, impactos y resiliencia. Centro de Ciencia del Clima y la Resiliencia (CR)2, (ANID/FONDAP/15110009), 84 pp. Available at https://www.cr2.cl/incendios/.

Gorelick, N., Hancher, M., Dixon, M., Ilyushchenko, S., Thau, D., Moore, R., 2017. Google earth engine: planetary-scale geospatial analysis for everyone. Remote Sensing of Environment 202, 18–27. Available at https://doi.org/10.1016/j.rse.2017.06.031.

Handwerger, A.L., Huang, M.H., Fielding, E.J., Booth, A.M., Bürgmann, R., 2019. A shift from drought to extreme rainfall drives a stable landslide to catastrophic failure. Scientific Reports 9 (1). Available at https://doi.org/10.1038/s41598-018-38300-0, www.nature.com/srep/index.html.

Hayasaka, H., 2024. Active Wildland Fires in Central Chile and Local Winds (Puelche). Remote Sensing 16, 2605. Available at https://doi.org/10.3390/rs16142605.

Huang, H., Roy, D.P., Boschetti, L., Zhang, H.K., Yan, L., Kumar, S.S., et al., 2016. Separability analysis of sentinel-2A multi-spectral instrument (MSI) data for burned area discrimination. Remote Sensing 8 (10). Available at https://doi.org/10.3390/rs8100873, http://www.mdpi.com/journal/remotesensing.

Hu, X., Zhang, P., Ban, Y., Rahnemoonfar, M., 2023. GAN-based SAR and optical image translation for wildfire impact assessment using multi-source remote sensing data. Remote Sensing of Environment 289, 113522. Available at https://doi.org/10.1016/j.rse.2023.113522.

Ibarra-Bonilla, J.S., Pinedo-Alvarez, A., Prieto-Amparán, J.A., Siller-Clavel, P., Santellano-Estrada, E., Álvarez-Holguín, A., et al., 2024. Post-fire vegetation dynamics of a temperate mixed forest: an assessment based on the variability of Landsat spectral indices. Trees, Forests and People 17. Available at https://doi.org/10.1016/j.tfp.2024.100648, https://www.sciencedirect.com/science/journal/26667193.

Imperatore, P., Azar, R., Calo, F., Stroppiana, D., Brivio, P.A., Lanari, R., et al., 2017. Effect of the vegetation fire on backscattering: an investigation based on sentinel-1 observations. IEEE Journal of Selected Topics in Applied Earth Observations and Remote Sensing 10 (10), 4478–4492. Available at https://doi.org/10.1109/JSTARS.2017.2717039, http://ieeexplore.ieee.org/xpl/RecentIssue.jsp?punumber=4609443.

Keith, D.A., Ferrer-Paris, J.R., Nicholson, E., Bishop, M.J., Polidoro, B.A., Ramirez-Llodra, E., et al., 2022. A function-based typology for Earth's ecosystems. Nature 610 (7932), 513–518. Available at https://doi.org/10.1038/s41586-022-05318-4, https://www.nature.com/nature/.

Key, C.H., 2006. Ecological and sampling constraints on defining landscape fire severity. Fire Ecology 2 (2), 34–59. Available at https://doi.org/10.4996/fireecology.0202034.

C.H. Key, N.C. Benson, Landscape Assessment (LA) sampling and analysis methods. USDA Forest Service - General Technical Report RMRS-GTR. (164) (2006), A-1, Available at http://www.fs.fed.us/rm/pubs/rmrs_gtr164.pdf. undefined.

Kruger, F. 2014. Fire in Mediterranean Ecosystems: Ecology, Evolution and Management. International Forestry Review, 16(1), pp. 113+. Gale Academic OneFile, https://link.gale.com/apps/doc/A365981518/AONE?u=anon~3c274dde&sid=googleScholar&xid=8a1e3b04. Accessed 5 June 2026.

Leal-Medina, C., Lopatin, J., Contreras, A., González, M.E., Galleguillos, M., 2024. Post-fire *Pinus radiata* invasion in a threatened biodiversity hotspot forest: a multi-scale remote sensing assessment. Forest Ecology and Management 561, 121861. Available at https://doi.org/10.1016/j.foreco.2024.121861.

Mallinis, G., Mitsopoulos, I., Chrysafi, I., 2017. Evaluating and comparing Sentinel 2A and Landsat-8 Operational Land Imager (OLI) spectral indices for estimating fire severity in a Mediterranean pine ecosystem of Greece. GIScience & Remote Sensing 55, 1–18. Available at https://doi.org/10.1080/15481603.2017.1354803.

Ministerio de Vivienda y Urbanismo (MINVU). 2014. Circular DDU N° 269: Definición de áreas de riesgo por amenaza de incendio en los Instrumentos de Planificación Territorial. División de Desarrollo Urbano, Ministerio de Vivienda y Urbanismo de Chile. https://www.minvu.gob.cl/wp-content/uploads/2019/06/DDU-269.pdf.

Ministerio de Vivienda y Urbanismo (MINVU). 2025. Circular DDU N° 510: Guía Metodológica para la Elaboración de Estudios de Riesgo de los Planes Reguladores del Nivel Comunal e Intercomunal. División de Desarrollo Urbano, Ministerio de Vivienda y Urbanismo de Chile. https://www.aoa.cl/wp-content/uploads/2024/11/DDU-510-20-de-noviembre-del-2024.pdf.

Miller, J.D., Thode, A.E., 2007. Quantifying burn severity in a heterogeneous landscape with a relative version of the delta normalized burn ratio (dNBR). Remote Sensing of Environment 109 (1), 66–80. Available at https://doi.org/10.1016/j.rse.2006.12.006.

Miranda, A., Mentler, R., Moletto-Lobos, Í., Alfaro, G., Aliaga, L., Balbontín, D., et al., 2022. The landscape fire scars database: mapping historical burned area and fire severity in Chile. Earth System Science Data 14 (8), 3599–3613. Available at https://doi.org/10.5194/essd-14-3599-2022.

Morgan, P., Keane, R.E., Dillon, G.K., Jain, T.B., Hudak, A.T., Karau, E.C., et al., 2014. Challenges of assessing fire and burn severity using field measures, remote sensing and modelling. International Journal of Wildland Fire 23 (8), 1045–1060. Available at https://doi.org/10.1071/WF13058, http://www.publish.csiro.au/nid/126.htm.

Morresi, D., Marzano, R., Lingua, E., Motta, R., Garbarino, M., 2022. Mapping burn severity in the western Italian Alps through phenologically coherent reflectance composites derived from sentinel-2 imagery. Remote Sensing of Environment 269, 112800. Available at https://doi.org/10.1016/j.rse.2021.112800.

Mutai, S., Chang, L., 2020. Post-fire hazard detection using alos-2 radar and landsat-8 optical imagery. ISPRS Annals of the Photogrammetry, Remote Sensing and Spatial Information Sciences VI-3/W1-2020 (3), 75–82. Available at https://doi.org/10.5194/isprs-annals-vi-3-w1-2020-75-2020.

Peña, M.A., Brenning, A., 2023. Benchmarking sentinel-2-derived predictors for long-term burn severity modelling: the 2016–17 Chilean firestorm. International Journal of Remote Sensing 44 (8), 2668–2690. Available at https://doi.org/10.1080/01431161.2023.2205981, https://www.tandfonline.com/loi/tres20.

Polychronaki, A., Gitas, I., Veraverbeke, S., Debien, A., 2013. Evaluation of ALOS PALSAR Imagery for Burned Area Mapping in Greece Using Object-Based Classification. Remote Sensing 5, 5680–5701. Available at https://doi.org/10.3390/rs5115680.

Quintano, C., Fernández-Manso, A., Roberts, D.A., 2013. Multiple Endmember Spectral Mixture Analysis (MESMA) to map burn severity levels from Landsat images in Mediterranean countries. Remote Sensing of Environment 136, 76–88. Available at https://doi.org/10.1016/j.rse.2013.04.017.

Sarricolea, P., Herrera-Ossandon, M., Meseguer-Ruiz, O., 2017. Climatic regionalisation of continental Chile. Journal of Maps 13, 66–73. Available at https://doi.org/10.1080/17445647.2016.1259592.

Sarricolea, P., Serrano-Notivoli, R., Fuentealba, M., Hernández-Mora, M., de la Barrera, F., Smith, P., et al., 2020. Recent wildfires in central Chile: detecting links between burned areas and population exposure in the wildland urban interface. Science of The Total Environment 706, 135894. Available at https://doi.org/10.1016/j.scitotenv.2019.135894.

Shi, J., Yang, H., Hou, X., Zhang, H., Tang, G., Zhao, H., et al., 2025. Coupling SAR and optical remote sensing data for soil moisture retrieval over dense vegetation covered areas. PLoS One 20 (1), e0315971. Available at https://doi.org/10.1371/journal.pone.0315971.

Siegert, F., Ruecker, G., Hinrichs, A., Hoffmann, A.A., 2001. Increased damage from fires in logged forests during droughts caused by El Niño. Nature 414 (6862), 437–440. Available at https://doi.org/10.1038/35106547.

Tamiminia, H., Salehi, B., Mahdianpari, M., Quackenbush, L., Adeli, S., Brisco, B., 2020. Google earth engine for geo-big data applications: a meta-analysis and systematic review. ISPRS Journal of Photogrammetry and Remote Sensing 164, 152–170. Available at https://doi.org/10.1016/j.isprsjprs.2020.04.001.

Tanase, M.A., Belenguer-Plomer, M.A., Roteta, E., Bastarrika, A., Wheeler, J., Fernández-Carrillo, Á., et al., 2020. Burned area detection and mapping: Intercomparison of sentinel-1 and sentinel-2 based algorithms over tropical Africa. Remote Sensing 12 (2). Available at https://doi.org/10.3390/rs12020334, https://res.mdpi.com/d_attachment/remotesensing/remotesensing-12-00334/article_deploy/remotesensing-12-00334-v2.pdf.

Tanase, M.A., Santoro, M., Aponte, C., de la Riva, J., 2014. Polarimetric Properties of Burned Forest Areas at C- and L-Band. IEEE Journal of Selected Topics in Applied Earth Observations and Remote Sensing 7, 267–276. Available at https://doi.org/10.1109/jstars.2013.2261053.

Tanase, M.A., Santoro, M., Wegmüller, U., de la Riva, J., Pérez-Cabello, F., 2010. Properties of X-, C- and L-band repeat-pass interferometric SAR coherence in Mediterranean pine forests affected by fires. Remote Sensing of Environment 114 (10), 2182–2194. Available at https://doi.org/10.1016/j.rse.2010.04.021.

Úbeda, X., Sarricolea, P., 2016. Wildfires in Chile: a review. Global and Planetary Change 146, 152–161. Available at https://doi.org/10.1016/j.gloplacha.2016.10.004.

Urrutia-Jalabert, R., González, M.E., González-Reyes, Á., Lara, A., Garreaud, R., 2018. Climate variability and forest fires in central and south-central Chile. Ecosphere 9 (4). Available at https://doi.org/10.1002/ecs2.2171, http://esajournals.onlinelibrary.wiley.com/hub/journal/10.1002/(ISSN)2150-8925/.

Veraverbeke, S., Lhermitte, S., Verstraeten, W.W., Goossens, R., 2011. Evaluation of pre/post-fire differenced spectral indices for assessing burn severity in a mediterranean environment with landsat thematic mapper. International Journal of Remote Sensing 32 (12), 3521–3537. Available at https://doi.org/10.1080/01431161003752430, https://www.tandfonline.com/loi/tres20.

Wang, L., Qu, J.J., Hao, X., 2008. Forest fire detection using the normalized multi-band drought index (NMDI) with satellite measurements. Agricultural and Forest Meteorology 148, 1767–1776. Available at https://doi.org/10.1016/j.agrformet.2008.06.005.

White, D.A., Silberstein, R.P., Balocchi-Contreras, F., Quiroga, J.J., Meason, D.F., Palma, J.H.N., et al., 2021. Growth, water use, and water use efficiency of *Eucalyptus globulus* and *Pinus radiata* plantations compared with natural stands of Roble-Hualo forest in the coastal mountains of central Chile. Forest Ecology and Management 501. Available at https://doi.org/10.1016/j.foreco.2021.119676, www.elsevier.com/inca/publications/store/5/0/3/3/1/0.

Yang, X., Yang, X., Zhang, C., Wang, J., 2023. SAR image classification using markov random fields with deep learning. Remote Sensing 15 (3), 617. Available at https://doi.org/10.3390/rs15030617.

Yáñez, M.A., Garrido, P., Blanco, J., Gajardo, J., Osborne, D., Osborne, T., 2025. Effects of burn severity on the diversity and relative abundance of birds 3-years postfire in a coastal forest of Nothofagus glauca (Phil.) Krasser in central Chile. Journal for Nature Conservation 86, 126965. Available at https://doi.org/10.1016/j.jnc.2025.126965.

Zhang, P., Hu, X., Ban, Y., Nascetti, A., Gong, M., 2024. Assessing sentinel-2, sentinel-1, and ALOS-2 PALSAR-2 data for large-scale wildfire-burned area mapping: insights from the 2017–2019 Canada wildfires. Remote Sensing 16 (3), 556. Available at https://doi.org/10.3390/rs16030556.

The Australian megafires of 2019–20: impact, response, and future direction

18

Rebecca Kate Gibson[1], *Matt Wiseman*[2], *Tony Prior*[2], *Greg Summerell*[2], *Graham Moore*[2], *and Katharine Haynes*[2]
[1]NSW Department of Climate Change, Energy, the Environment and Water, Parramatta, New South Wales, Australia, [2]NSW National Parks and Wildlife Service, Parramatta, New South Wales, Australia

18.1 Introduction

The Australian megafires of 2019–20 burnt 7.2 Mha of predominantly eucalypt forests and woodlands; a globally unprecedented area of temperate forest biome (Bowman et al., 2021; Boer et al., 2020). These were the largest forest fires in recent history (Nolan et al., 2020). As a result, Australian ecosystems have been exposed to an unprecedented extent of high-severity fire, with complete consumption of aboveground biomass (Collins et al., 2021). Estimates of carbon emissions from these fires range between 612 and 830 Mt of CO_2 equivalent, which contributed to Australia's total carbon emissions for that year being 2.9 times greater than the average annual emissions from 2001 to 2018 (Bowman et al., 2021; Shiraishi and Hirata, 2021; Australian Government Department of Industry 2020, 2021). Heavy smoke blanketed densely populated areas for months, resulting in a high rate of smoke-related illness and premature deaths (Howard et al., 2020; Borchers Arriagada et al., 2020). Smoke plumes and pyrocumulonimbus clouds reached over 30 km in height, injecting large quantities of aerosols into the stratosphere, which circumnavigated the globe (Kablick et al., 2020).

The Australian megafires of 2019–20 had catastrophic impacts on biodiversity (van Eeden and Dickman, 2023; Ward et al., 2020). There was huge investment in postfire field assessments to estimate the impact of the fires on biodiversity. In one of the largest metaanalyses of postfire field survey data ever collated (Driscoll et al., 2024), analysis of over 800,000 postfire response records from over 2000 taxa revealed that the largest effects on plants and animals were in areas with frequent or recent past fires and within extensively burnt areas. Areas burnt at high severity, outside protected areas, or under extreme drought also had larger effects (Driscoll et al., 2024). The 2019–20 Australian megafires impacted the conservation status of 70–82 taxa, with a 22%–26% increase in Australian statutory lists, and uplisting for 8%–10% of threatened taxa (Legge et al., 2022).

While much has been documented on quantifying the impacts to the environment, flora, fauna, and ecosystems, including an entire book dedicated to the impacts and

Advances in Remote Sensing for Forest Fire Management. DOI: https://doi.org/10.1016/B978-0-443-41692-7.00020-2
© 2027 Elsevier Ltd. All rights are reserved, including those for text and data mining, AI training, and similar technologies.

lessons from Australia's Megafires of 2019–20 (Rumpff et al., 2023), the 2019–20 fires in Australia have also had a transformative effect on operational fire mapping and monitoring technologies within fire management agencies. This chapter presents several individual perspectives on responding to the 2019–20 fire season, offering profound, unique insights, followed by an overview of the impacts, response, and future direction related to remote sensing technologies within fire management agencies in New South Wales, where the largest impacts of the 2019–20 fires occurred.

18.1.1 Responding to the 2019–20 fires from a remote sensing scientist's perspective

Rebecca Gibson is a senior remote sensing research scientist with the NSW Department of Climate Change, Energy, the Environment and Water, who has led the research and development of the operational NSW Fire Extent and Severity Mapping (FESM) program, supported by the NSW Rural Fire Service (RFS) since 2017.

From early September 2019, there was a sudden and escalating rise in interest from across the department to access outputs from the remote sensing, machine learning-based FESM program that I had been developing over the past few years. Several very large, early-season wildfires had started to burn in northern NSW, including the Bees Nest (117,252 ha) and Raspberry Road (10,824 ha) fires with a large proportion burning across the National Park estate. By that stage, I had written up a manuscript that was in review with Remote Sensing of Environment, documenting the technical method and accuracy metrics of the FESM approach (Gibson et al., 2020). I had also built a prototype system for the semiautomated processing workflow, but it was yet to be put through its paces and was not yet operationally linked to the RFS fire management system for automated streaming of input location data and delivery of outputs. The prototype FESM system required manual input from an end user to locate the approximate area to apply the trained algorithm and map the fire extent and severity.

In October, my director arranged for me to give a briefing to the NSW National Parks and Wildlife Service (NPWS) to showcase the FESM method and offer support to map fires during the current season, to assist their rapid response assessment and prioritize postfire management actions. I left that meeting with an enormous workload to produce FESM outputs for every recent and active fire currently burning in the state. This involved manual handling of input data, which was an enormously complex undertaking with over 40 very large fires that had already started throughout August and September 2019, and new fires starting every day. Some of these fires were stopping and restarting, amalgamating with other fires, and being renamed. As such, the spatial input location data was difficult to manually unravel, but we did not have the time to establish automated systems to handle input data. Furthermore, mapping for some fires was being urgently requested before the fires were extinguished and the fire continued to spread, meaning a single fire was often being remapped multiple times at later stages, which was another significant data management challenge. All these issues made the task at hand very complex and stressful for weeks on end.

By early December, the crisis had further escalated with over 14 megafires concurrently burning across the state that would eventually reach over 100,000 ha each. In mapping the severity of these megafires, I quickly found the technical limits of the prototype processing system I had built. I was still making significant refinements in the processing efficiencies of the system, including fixing "bugs" in my code, such as a wild "infinity loop" programming error that was only realized beyond a certain, extremely large fire size, which I had never reached during my testing phase. These challenges facilitated the rapid development of a very robust operational mapping program that can withstand the most extreme fire season on record.

There was urgency from the executive level within the agency for our team to generate not only spatial outputs of the severity mapping but also frequently updated statistics on area burnt by severity class, across tenure, bioregion, and vegetation types (Fig. 18.1). There was a particularly strong drive to rapidly quantify fire impacts to endangered species such as the Koala, and to World Heritage Gondwana Rainforest areas, to inform parliamentary inquiries. We did not yet have automated systems in place to efficiently handle such requests. We were also receiving requests for fire mapping from external agencies such as the Environmental Protection Agency as well as forestry and water management agencies, adding to the data management workload. The severity mapping was actively being used to target postfire field survey locations, for example, to locate unburnt islands within a wider burnt landscape that may offer hope of threatened species populations' survival. Indeed, there are stories of remarkable success. For example, the Border Fire in southern NSW impacted one of two known sites of the Endangered Merimbula Star-Hair (Astrotricha sp. Wallagaraugh). A site visit in May 2020 confirmed that all known plants were burnt under extreme fire severity and had been killed. However, using the FESM data, small patches of previously unsurveyed habitat were identified in areas mapped as unburnt or low severity. Targeting these locations, three unrecorded plants were found, expanding the previously known population and providing options to safeguard the population with ex situ collection of material in case of recruitment failure in the areas burnt by extreme severity fire (Laura Schweickle, personal communication).

Given the wide application of the mapping in the field immediately postfire, we took the opportunity to rapidly build a FESM field validation app for staff to collect field observations of the severity mapping accuracy. Analysis of the field validation data largely aligned with the results of my testing during the development phase of FESM, which was based on high-resolution aerial photo interpretation and my own field validation campaigns. Overall, field staff found the FESM data to be highly accurate, robust, and useful for a number of field applications. However, there were notable accuracy differences between vegetation types, with low severity under dense canopy rainforests recognized as the most difficult to map accurately, reflecting inherent limitations of the satellites.

One of my most visceral memories of that time was the moment I first saw the severity map output for the Currowan fire on the south coast of NSW. The extent of extreme severity blackened my computer screen. Sitting alone in my home office late into the night, I was brought to tears knowing the ramifications of the impact of this

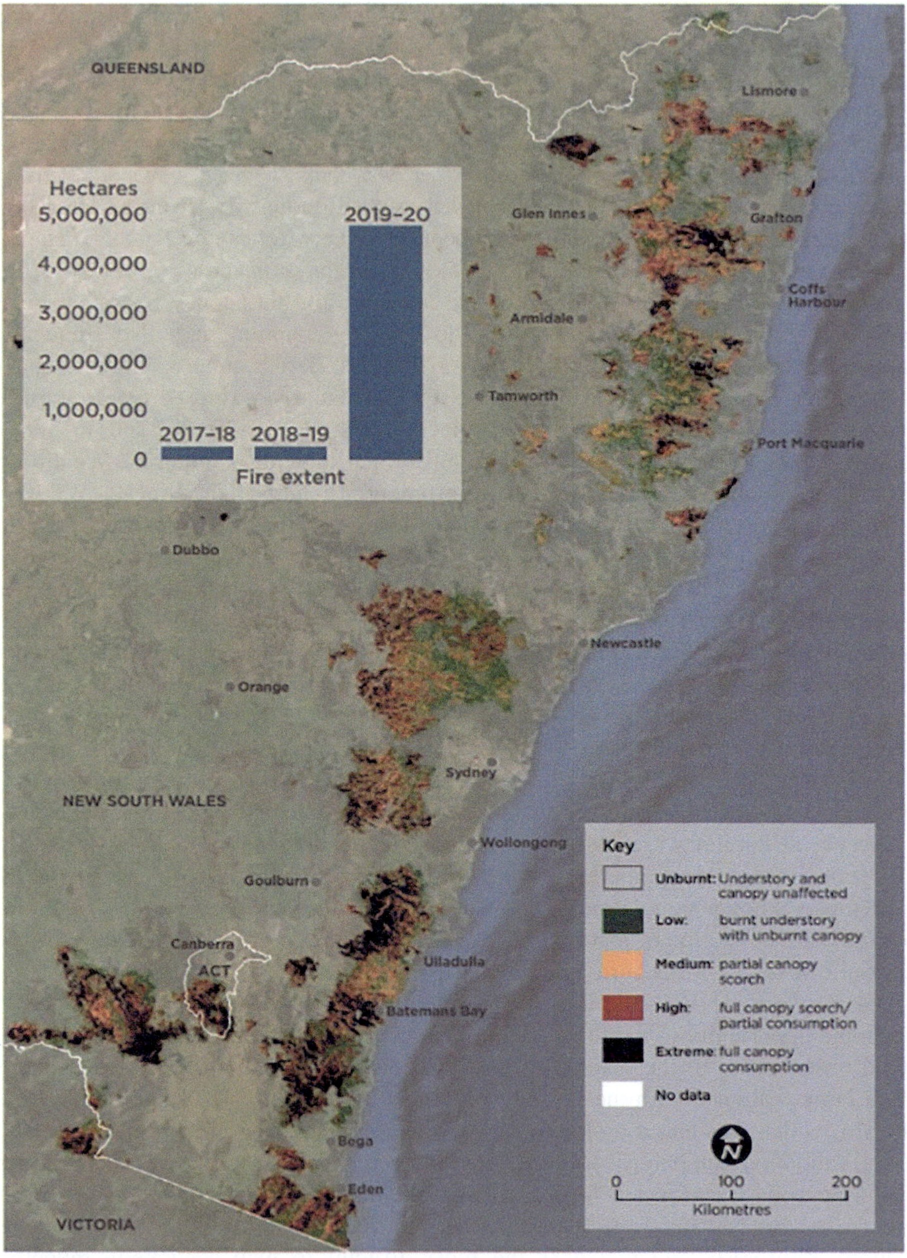

Figure 18.1 Eastern part of New South Wales, Australia, showing the extent and severity of the 2019–20 fires. The bar chart compares fire extent between the 2019–20, 2018–19, and 2017–18 fire years. Map lines delineate study areas and do not necessarily depict accepted national boundaries.

Source: From DPE, 2021. Fire extent and severity mapping: annual report for the 2019–20, 2018–19 and 2017–18 fire years." In, Department of Planning and Environment, Sydney.

fire for the forests and communities in the path of this blaze. Stories came to light over the coming days, with personal recounts of experiences of the Currowan fire from members of the public and firefighters on the ground. A fire-generated thunderstorm (Pyrocumulonimbus or PyroCb) formed over the Currowan fire, which is an extremely dangerous event with erratic fire behavior, extreme winds, lightning, tornadoes, and black hail. Over 300 houses were destroyed, and three human lives were lost. The Currowan fire burnt an incredible 315,119 ha, but unbelievably, that was the fifth largest megafire to burn in 2019–20. Proportionately, though, this fire had the largest area of high and extreme severity fire, while others further north had far greater variation among severity classes. Such variation in severity across the landscape would not have been so clearly captured without the remote sensing fire severity mapping technology.

As the fire season ended, the smoke cleared, and the drought was broken with heavy rainfall, the pace and urgency of our work did not ease. The NSW Bushfire Inquiry was soon established, and there was an enormous amount of work to contribute invited submissions as part of the NSW Bushfire Risk Management Research Hub. I also partook in interviews as requested by the Inquiry panel, as a remote sensing expert in a fire management agency. Remote sensing derived spatial data and mapping helped answer important questions such as the effectiveness of prior hazard reduction burns on reducing the severity of the megafires (Williamson et al., 2020), documented in the final report of the inquiry (Owens and O'Kane 2020). Many recommendations from the inquiry recognized that remote sensing technologies, spatial mapping, and monitoring capabilities should be significantly invested in and enhanced to improve our preparedness and response to future wildfire seasons.

Five years on, the bushfire crisis of 2019–20 continues to drive my priorities in strengthening fire mapping operating systems, monitoring and reporting systems, to better support fire management operations, preparedness, planning, and response with near real-time data and analytics (DCCEEW, NSW, 2025).

18.1.2 Responding to the 2019–20 fires from protected area managers' perspective

As fire managers with NSW National Parks and Wildlife Service (NPWS) Matthew Wiseman and Tony Prior collaborate on burn planning, planned fire implementation, the rolling 3 year burn program as well as wildfire suppression. During the 2019-20 fire season, they were involved in fire operations including Incident controller, Operations Officer, Air Attack Supervisor, Air Observer and Division Commander on the North Coast of NSW. These fires burnt 48% of the of the NSW North Coast Branch's NPWS reserves covering 616 108 ha. This experience provides a unique perspective of the fire operations, lessons learned and ongoing research.

In August 2019, after three years of drought causing deep soil dryness, litter beds in wet sclerophyll forests with rainforest communities, which are not normally available for combustion, were dry enough to carry fire. This underlying dryness, combined with days of extreme fire weather (humidity as low as 5% and strong northwesterly winds of 40–50 km hours^{-1}) drove fires to spread in some instances up to 40 km in one day (e.g., the Busby's Flat fire). On November 8, 2019, a series of lightning ignitions, combined with extreme fire weather, resulted in a complex of uncontained fires across the NSW north coast, breaking containment with updrafts combined with wind-driven spot fires up to 10 km ahead of the main fire front (see Fig. 18.2). Blackout conditions resulted from smoke blanketing the whole north coast. This grounded all firefighting aircraft and support for ground crews. Aerial line scanning aircraft and satellite-based remote sensing were also compromised, limiting access to up-to-date imagery of wildfire location, intensity, and size. As such, incident management teams (IMT) had restricted capabilities to guide wildfire suppression strategies and tactics used to prioritize the protection of built and environmental assets.

The 2019–20 bushfire event was of such a scale that it caused significant challenges in fire mapping. Once resources were stretched across multiple fire grounds, information became more difficult to obtain. Air Observers and aircraft are limited resources and could not be everywhere across such extensive fire grounds with their work prioritized by the IMT in the most critical areas. Line-scanning was also in high demand and prioritized at state level where risks were the highest. All this led to loss of situational awareness across multiple fire grounds on fire location, size, rate of spread, assets at risk, along with details on any new fires.

The 2019–20 bushfire crisis stretched IMTs to their limits. Managing all the information coming in from so many concurrent incidents made it extremely challenging for planning teams to update systems for community safety messaging, for Incident Controllers to communicate situational awareness, and for Operations teams to take appropriate action, prioritize, and respond to emerging threats.

On-ground fire suppression and containment with ground crews was most successful when in tandem with air operations, providing intel on wildfire progression, crew safety, and combined with suppression and detection of spot overs with heavy plant operations. Backburning was consistently effective when conducted with the wind on days when conditions moderated, along with the above combined air ops, ground crews, and heavy plant worked in close cooperation. There were many days when conditions did not allow for backburning, and both ground and air operations had to suspend operations to maintain crew safety or were only able to focus on the protection of life and property, not the containment of fires.

The North Coast Ranges comprise dry sclerophyll and heath communities on exposed ridges and northwest-facing slopes bisected by wet gullies/creeks, and wet sclerophyll with rainforest understorey grading to rainforest on south-facing and sheltered areas. Many experienced fire practitioners rely on these wet gullies/creeks and wet sclerophyll with rainforest to contain wildfires, where no practical trails exist to contain remote fires. In 2019, on the far north coast, wet sclerophyll with rainforest understorey communities had litter beds dry enough to carry fire for the first time since 1968 and previously in 1926. Interestingly, the 1968 fires in this region burnt a larger area than the 2019–20 fires, while the 1926 fires burnt similar areas to the 2019–20 fires. Even in these extreme conditions, with the Keetch–Byram drought index (KBDI) of 184, it was observed that fire was not burning into rainforest without emergent pyrophytic species (e.g., Brush-box, Turpentine, Corymbia, and Eucalyptus species). Upon the fires being extinguished on the far north coast, we had a unique

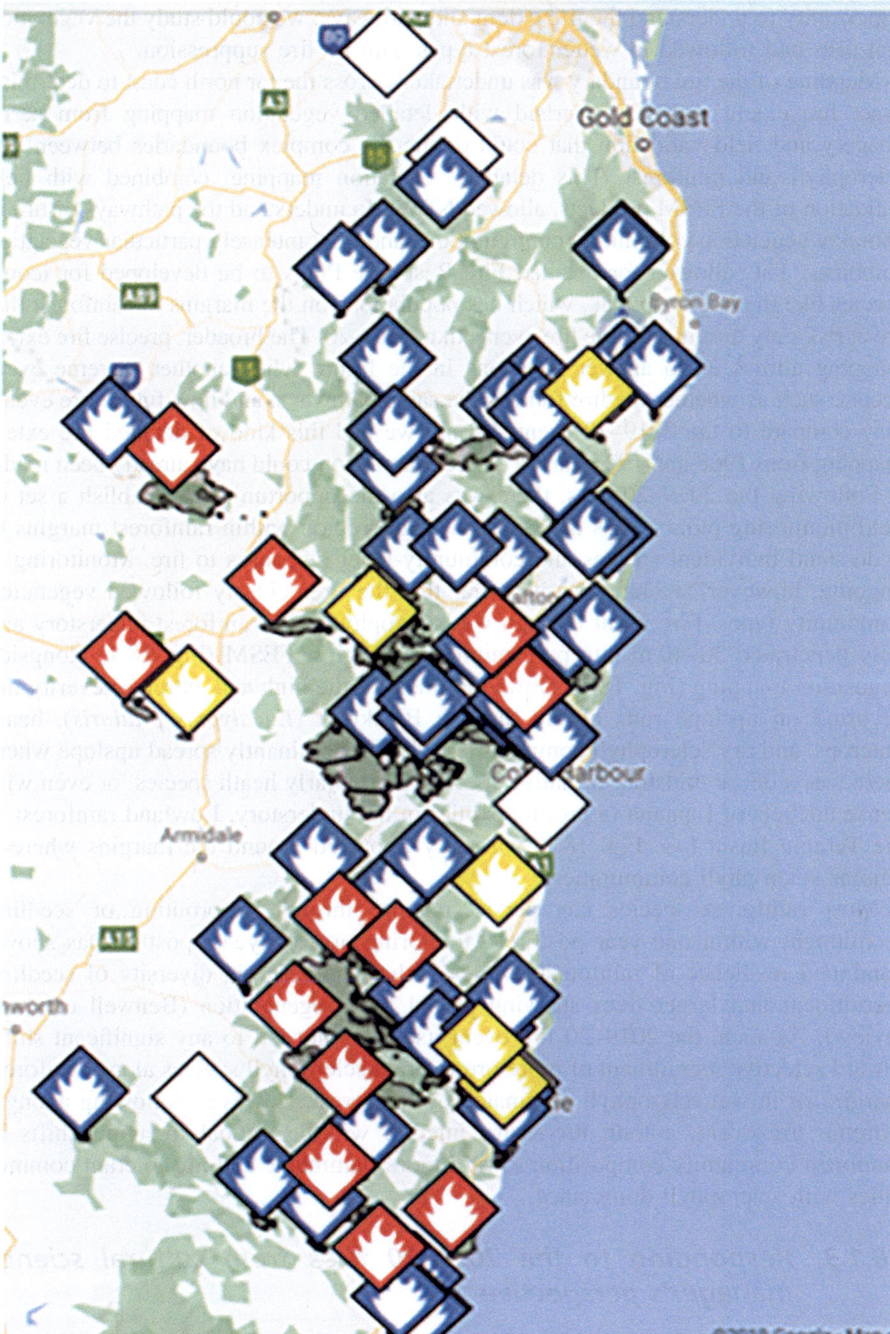

Figure 18.2 North Coast NSW on November 9, 2019, where 12 fires were under an emergency warning, 76 fires under "Watch and Act," 47 fires under "Advice" warnings, and four fires pending confirmation. Map lines delineate study areas and do not necessarily depict accepted national boundaries.

opportunity to understand the fire extent and pathways; we could study the vegetation that fires had followed in wetter forest types with no fire suppression.

Mapping of the fire boundary was undertaken across the far north coast to determine exact fire extent and was overlaid with detailed vegetation mapping from aerial imagery and field validation that could define the complex boundaries between wet sclerophyll and rainforest. This detailed vegetation mapping, combined with field validation of the FESM products, allowed NPWS to understand the pathways of fire in complex vegetation community configurations and how intensely particular vegetation combusts. This allowed for detailed Fire Response Plans to be developed for iconic species like the Nightcap Oak, which has populations on the margins of rainforest that are at risk only during extreme fire events like 2019–20. The broader, precise fire extent mapping allows us to answer questions in the future when another extreme event occurs, such as whether the fire follows the same pathways, and how future fire events may compare to the 2019–20 event. If only we had this kind of detailed fire extent mapping from 1968 and 1926, then a direct comparison could have already been made.

Following the 2019–20 fires, there was also the opportunity to establish a set of field monitoring plots across the Border Ranges region within rainforest margins to understand individual species and community-level responses to fire. Monitoring is ongoing; however, we have learnt so far that the fires closely followed vegetation community types. Fire burnt through wet sclerophyll with rainforest understory and only penetrated 30–40 m into pure rainforest margins. FESM (Fig. 18.1) alongside vegetation mapping (Fig. 18.1) shows that most of the high and extreme severity fire occurred on upslope runs of fire through Blackbutt (*Eucalyptus pilularis*), heath outcrops, and dry sclerophyll communities. Fire predominantly spread upslope where there was a dense midstory of native species, particularly heath species, or even with dense thickets of Lantana (a prolific exotic) in the understory. Lowland rainforest in the Terania Basin (see Fig. 18.1) was only impacted around the margins where it adjoins sclerophyll communities.

Most rainforest species can persist postfire through resprouting or seedling recruitment within one year postfire. Monitoring up to 5 years postfire has shown continued resilience of rainforest species with an increase in diversity of seedling recruitment and larger trees showing partial bark regeneration (Benwell et al. in review). As such, the 2019–20 fire event itself has not led to any significant shifts toward selective recruitment of more fire-tolerant sclerophyll species along rainforest margins or in wet sclerophyll dominated by rainforest. However, following a single extreme fire event, a long fire-return interval will be critical to avoid shifts in rainforest community composition toward more flammable and fire-tolerant communities with sclerophyll dominance.

18.1.3 Responding to the 2019–20 fires from cultural science manager's perspective

Graham Moore, Gurrungutti-Munji-Yuin, Djiringanj Country, Far South Coast, Yuin Elder and Senior Scientist Cultural Science, NSW Department of Climate Change, Energy, the Environment and Water (DCCEEW) and Dr Gregory Summerell, non-

Aboriginal Manager of the Fire and Culture Science team, NSW DCCEEW. Greg has partnered for many years with Aboriginal colleagues and friends who have generously shared their knowledge and understanding to enable them to see and understand a more holistic and relational way of being on, with and for Country. Graham and Greg give insight into the use of drone technology with Aboriginal Ranger crews to monitor post fire response and mitigation actions after the 2019–20 fires.

Following the 2019–20 fires, Graham and Greg undertook action research with the Merrimans Local Aboriginal Land Council (LALC) rangers on the Far South Coast of NSW on Dirringanj Country, South Eastern NSW (Djiringanj et al., 2025). The aim was to ensure the voices of Aboriginal knowledge holders and their practices were included alongside ecological and bushfire sciences and new technologies to create meaningful outcomes for people and Country; in many cases, bridging the world-views of Indigenous knowledge with these sciences (Graham, 2008). This work was established in response to the 2019–20 bushfire Inquiry (Owens and O'Kane 2020), addressing recommendations related to enabling Aboriginal communities to better engage in fire management planning, operations, and sharing of knowledge. It also aligns with the Indigenous leadership goals whereby traditional knowledge, language preservation, and cultural practices support the transmission of intergenerational knowledge and reaffirm community identity (Davies, 2025).

This work supports community-led research whereby the Aboriginal Rangers determine the best ways to contribute to regional-based planning and coordination for bushfire mitigation activities. Bush Fire Management Committees (BFMC) have representation from land management agencies and organizations that have responsibility for bushfire management, mitigation, and emergency responses, including Aboriginal Rangers. The purpose of the BFMC is to create risk mitigation plans by understanding each representative's capacity and resources to develop shared plans of action and implementation.

Maps and imagery from remote sensing technologies are commonly used in these committee meetings, yet the Aboriginal Rangers previously had little or no capacity to contribute to this conversation. Together with the rangers, we captured and processed drone imagery over culturally significant parcels of land to allow the Aboriginal Rangers to express, through imagery, their points of view for bushfire risk mitigation. This capture of imagery from a sky view was innovative, as observations could be placed within a landscape-scale context and overlain with ancillary spatial data such as the NSW FESM products (DPE, 2021; Gibson et al., 2020). A deeper understanding of the cultural aspects could also be shared. For example, repeated drone capture of imagery throughout the year was undertaken to show indicators from Country, such as flowering onset of certain species, indicating when fire should be considered to be placed within Country. The imagery also allowed for spatial awareness of what Country types exist and where the planning for mitigation works could occur. Following the 2019–20 fires, the understory of much of the Country being assessed has become impenetrable due to rapid and thick undergrowth occurring. As such, the use of drones also made the site accessible for planning purposes, such as where to put containment and control lines in preparation for cultural reset burns (Figs. 18.3 and 18.4).

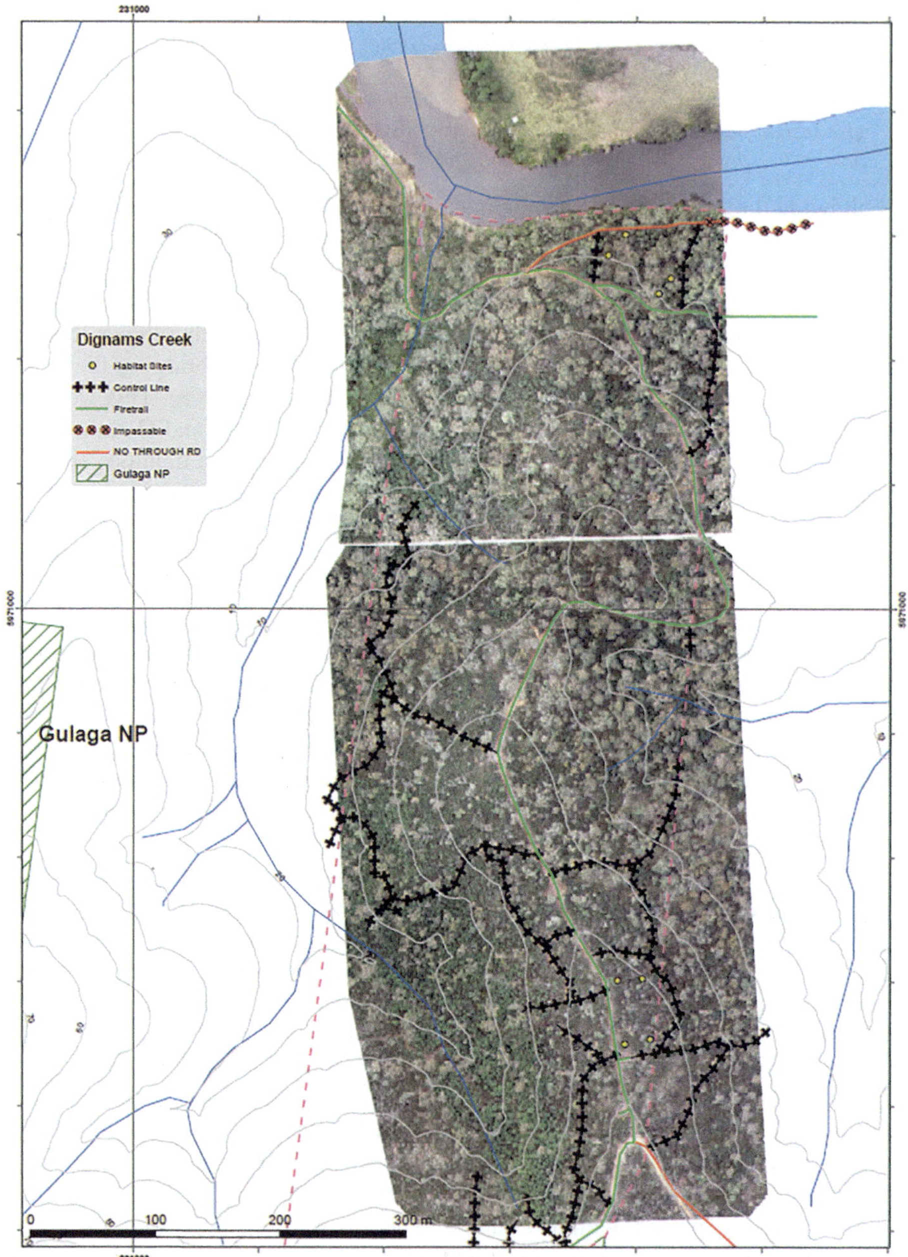

Figure 18.3 Topographic map of Dignams Creek with drone imagery overlaid, used in preparation for a cultural reset burn. Map lines delineate study areas and do not necessarily depict accepted national boundaries.

Figure 18.4 A young Aboriginal Ranger operating a drone.

There are many more reasons why drone mapping was introduced early in the project (Djiringanj et al., 2025): (1) it appeals to the young Aboriginal Rangers. (2) Many knowledge holders can no longer access remote cultural sites due to age or health reasons. The drone imagery can be provided to Community and Elders so they can immerse themselves in Country, from a sky perspective, to help them reflect on the health of Country and what needs to be done. (3) With the development of a cultural data library populated with drone data, the rangers have discussed possibilities of 3D modeling of scar trees (plus other types of cultural sites) using photogrammetry. This photogrammetry could be an insurance policy for the worst circumstance if the site were destroyed, creating a record of what was and what will be. These remote sensing technologies appeal to our younger generation of Aboriginal Rangers and have changed the way they can speak on behalf of Country from a sky perspective and share stories of Country, both within Aboriginal Communities and with government authorities planning bushfire mitigation activities.

18.2 Impacts, response, and future direction of remote sensing of fire in NSW Government

As the unprecedented 2019–20 fire season unfolded, NSW fire management agencies were unprepared to handle the impact. IMTs were under immense strain with so many fires burning simultaneously. Aerial line scan aircraft were at capacity, with many fires unable to be covered. Data management and mapping systems, such as the "fires near me" app, a major public information tool, could not properly load and display the complexity of spatial line work capturing fire extent boundaries. New fire severity mapping technology implemented robust science for highly accurate data outputs. However, the processing, data management, data analytics, and reporting systems were still under heavy development, requiring manual, time-consuming work. From a federal government perspective, information was unavailable to the same standards across state boundaries, causing great difficulties for accurate analysis and reporting in response to parliamentary inquiries. Many challenges arose for fire management agencies that had never been previously encountered due to the enormous scale of impact.

As the rain arrived and the smoke settled in early 2020, there was no pause as the NSW Bushfire Inquiry commenced. Dave Owens Australian Police Medal (APM), former Deputy Commissioner of NSW Police, and Professor Mary O'Kane AC, Independent Planning Commission Chair and former NSW Chief Scientist and Engineer, led the six-month inquiry, which reviewed the causes of, preparation for, and response to the 2019–20 bushfires (Owens and O'Kane 2020). The final report presents 76 recommendations (R1–76) for future improvements on how NSW plans, prepares for, and responds to bushfires. Numerous recommendations are directly related to improving spatial data information management and the application of remote sensing technologies, including:

- establish a national bushfire database (R3)
- establish a spatial technology acceleration program to maximize the information available from new and existing remote sensing technologies, to improve our capability to detect ignitions and monitor fire intensity and progression (R4)
- establish a whole-of-government acquisition program for imagery and light detection and ranging (LiDAR) (R18)
- develop remote sensing technologies to monitor and audit how well Asset Protection Zones and defendable spaces are being maintained (R22)
- expand NSW's Remotely Piloted Aerial Systems (RPAS) capability to enhance NSW's ability to improve situational awareness (R51)
- make improvements to the Fires Near Me app, including enlarged text, a statement about the app's limitations, and incorporating new functionality as technology improves (R68).

The NSW Bushfire Inquiry also made significant recommendations that Government establish NSW as a major world center of bush fire research, and technology development (R5), and invest in long-term ecosystem and land management monitoring, modeling, forecasting, research, and evaluation, including tracking and forecasting ecosystem change under projected changes to climate, better understanding interactions of fire with other disturbances, and better understanding the influence of land management on landscape flammability (R36). Remote sensing and fire mapping technologies can make a significant contribution to these recommendations as foundational datasets to be used in research and analysis. In response to the recommendations, significant government investment has been made in new initiatives, including:

- The NSW DCCEEW Applied Bushfire Science Program (ABSP) was established in 2022, with over $14 million in funding invested to date. The ABSP scientists are an interdisciplinary team comprised of ecologists, soil scientists, cultural scientists, social

scientists, and knowledge exchange experts working collaboratively. The aim of the program is to drive transformational change in bushfire management by embedding leading research within government. The team works collaboratively with key partners across NSW Government, particularly NPWS and RFS, to ensure research is relevant and outcomes influence decision-making and drive change. The research focuses on outcomes that will update key policies and practices for the reduction of harm to biodiversity and cultural values. Developing an in-house capacity for applied bushfire research has been a purposeful objective for DCCEEW. We believe that our close connections across agencies ensure research outcomes that meet agency needs, and our bushfire scientists are able to deliver tailored and trusted knowledge directly to fire agencies.
- In response to the recommendations of the 2020 NSW Bushfire Inquiry (R5) and the 2022 NSW Flood Inquiry, the ABSP established the NSW Bushfire and Natural Hazards Research Centre (BNHRC) in 2023. The NSW BNHRC is a state-focused collaboration between the NSW Government, Australian research institutions, NSW land and emergency management agencies, and Indigenous knowledge holders. The administration and knowledge exchange are led by the ABSP team at DCCEEW, which focuses on ensuring the translation and application of the research outcomes into new policy and practice. The center's mission is to generate high-quality, codesigned research and evidence to improve real-world practice in bushfire and natural hazards management. The BNHRC builds on the important legacy of its predecessor, the NSW Bushfire Risk Management Research Hub.

The Remote Sensing of Fire program within the Science and Insights branch of NSW DCCEEW was established prior to the 2019–20 fire season and the NSW Bushfire Inquiry, but the focus and direction of priorities in the program are now largely shaped by the inquiry recommendations and the requirements of both the ABSP and BNHRC. The FESM products are key foundational data used by both the ABSP and BNHRC, including field survey support, understanding fuel dynamics and fire behavior, and postfire ecosystem resilience. Research collaboration continues to improve the FESM program, for example, through large-scale field validation campaigns targeted at determining the limits to FESM accuracy of low severity under dense canopy. Priorities of the Remote Sensing of Fire program are focused on FESM (Gibson et al., 2020; Gibson et al., 2023) and derived products, including high-severity fire heterogeneity (Gibson et al., 2025) and postfire biomass recovery (Gibson et al., 2025; Gibson et al., 2022; Gibson and Hislop, 2022; Bendall et al., 2024; Djiringanj et al., 2023). The ultimate future goal of the Remote Sensing of Fire program is to develop a fully integrated, highly automated, complete system of remote sensing of fire, incorporating all phases from fire prediction, fire detection, fire progression mapping, fire extent and severity, high-severity fire heterogeneity, and postfire biomass recovery. The program aims to continue fostering research collaborations at the state, national, and international scale to enhance the applications of our products and services for fire research and management.

Many of the inquiry's recommendations are being addressed through new strategies, increased funding, improved coordination, and translation of research into operations. While progress is being made, more work is needed with continued investment in fire management, technology, community education, and climate adaptation. The 2019–20 Australian fires will undoubtedly serve as a benchmark for us to measure future bushfire seasons against, to see how far we have come and how much more there is still to do to improve our planning, preparedness, and response, to safeguard the community and ecosystems of NSW and Australia from the next "black summer."

Acknowledgments

Merrimans Rangers and Board members from the Merrimans Aboriginal Land Council. Chris Simpson from Science team in Science and Insights Division of DCCEEW, who processed the drone imagery and created the spatial maps.

References

Australian Government Department of Industry, Science, Energy and Resources, 2020. Estimating greenhouse gas emissions from bushfires in Australia's temperate forests; focus on 2019–20.

Australian Government Department of Industry, Science, Energy and Resources, 2021. Quarterly Update of Australia's National Greenhouse Gas Inventory: December 2020.

Bendall, E.R., Choat, B., Gibson, R.K., Nolan, R.H., 2024. Review of Forest Recovery in the Coastal IFOA Region of New South Wales Following the 2019/2020 Wildfires and Preceding Drought. Hawkesbury Institute for the Environment, Western Sydney University.

Boer, M.M., Resco de Dios, V., Bradstock, R.A., 2020. Unprecedented burn area of Australian mega forest fires. Nature Climate Change 10, 171–172.

Borchers Arriagada, N., Palmer, A.J., Bowman, D.M.J.S., Morgan, G.G., Jalaludin, B.B., Johnston, F.H., 2020. Unprecedented smoke-related health burden associated with the 2019–20 bushfires in eastern Australia. Medical Journal of Australia 213 (6), 282–283. Available at https://doi.org/10.5694/mja2.50545, www.mja.com.au.

Bowman, D.M.J.S., Williamson, G.J., Price, O.F., Ndalila, M.N., Bradstock, R.A., 2021. Australian forests, megafires and the risk of dwindling carbon stocks. Plant Cell and Environment 44 (2), 347–355. Available at https://doi.org/10.1111/pce.13916, http://onlinelibrary.wiley.com/journal/10.1111/(ISSN)1365-3040.

Collins, L., Bradstock, R.A., Clarke, H., Clarke, M.F., Nolan, R.H., Penman, T.D., 2021. The 2019/2020 mega-fires exposed Australian ecosystems to an unprecedented extent of high-severity fire. Environmental Research Letters 16 (4), 044029. Available at https://doi.org/10.1088/1748-9326/abeb9e.

Davies, B., 2025. Indigenous leadership: nurturing cultural diversity, environmental stewardship, and community empowerment. A Research Project of Merrimans Local Aboriginal Land Council. Merrimans Local Aboriginal Land Council.

DCCEEW, NSW, 2025. Remote Sensing of Fire; FESM Dashboard and Data Access. Available at https://naturalhazardsscience-hub.seed.nsw.gov.au/remote-sensing-fire.

Djiringanj, C., Foster, R., Parsons, V., Parsons, R., Campbell, Z., Moore, G., et al., 2023. NSW Post-fire Biomass Recovery Monitoring by Remote Sensing Report for 3 years following 2019–20. Department of Planning and Environment.

Djiringanj, C., Foster, R., Parsons, V., Parsons, R., Campbell, Z., Moore, G., et al., 2025. Working together to drive change: weaving caring for country practices into fire risk management on Djiringanj Country. Australian Journal of Emergency Management 40.

DPE, 2021. Fire Extent and Severity Mapping: Annual Report for the 2019–20, 2018–19 and 2017–18 Fire Years. Department of Planning and Environment, Sydney.

Driscoll, D.A., Macdonald, K.J., Gibson, R.K., Doherty, T.S., Nimmo, D.G., Nolan, R.H., et al., 2024. Biodiversity impacts of the 2019–2020 Australian megafires. Nature 635 (8040), 898–905. Available at https://doi.org/10.1038/s41586-024-08174-6, https://www.nature.com/nature/.

van Eeden, L.M., Dickman, C.R., 2023. Estimating the number of wild animals affected by Australia's 2019–20 wildfires. In: Rumpff, L., Legge, S.M., van Leeuwen, S., Wintle, B., Woinarski, J.C.Z. (Eds.), Australia's Megafires: Biodiversity Impacts and Lessons from 2019–20. CSIRO Publishing: Melbourne.

Gibson, R.K., Hislop, S., 2022. Signs of resilience in resprouting. International Journal of Wildland Fire 31, 545–557.

Gibson, R., Danaher, T., Hehir, W., Collins, L., 2020. A remote sensing approach to mapping fire severity in south-eastern Australia using sentinel 2 and random forest. Remote Sensing of Environment 240. Available at https://doi.org/10.1016/j.rse.2020.111702, www.elsevier.com/inca/publications/store/5/0/5/7/3/3.

Gibson, R.K., White, L.A., Hislop, S., Nolan, R.H., Dorrough, J., 2022. The post-fire stability index; a new approach to monitoring post-fire recovery by satellite imagery. Remote Sensing of Environment 280, 113151. Available at https://doi.org/10.1016/j.rse.2022.113151.

Gibson, Kate, R., Mitchell, A., Chang, H.-C., 2023. Image texture analysis enhances classification of fire extent and severity using sentinel 1 and 2 satellite imagery. Remote Sensing 15.

Gibson, R.K., Driscoll, D.A., Macdonald, K.J., Williamson, G.J., Nolan, R.H., Doherty, T.S., et al., 2025. Remotely sensed fire heterogeneity and biomass recovery predicts empirical biodiversity responses. Global Ecology and Biogeography 34 (4). Available at https://doi.org/10.1111/geb.70040.

Graham, M., 2008. Some thoughts about the philosophical underpinnings of aboriginal worldviews. Australian Humanities Review 45, 181–194.

Howard, Z.L., Carlson, S.J., Baldwin, Z., Johnston, F., Durrheim, D.N., Dalton, C.B., 2020. High community burden of smoke-related symptoms in the Hunter and New England regions during the 2019–2020 Australian bushfires. Public Health Research and Practice 30 (4). Available at https://doi.org/10.17061/PHRP30122007, https://www.phrp.com.au/?p=39206.

Kablick, G.P., Allen, D.R., Fromm, M.D., Nedoluha, G.E., 2020. Australian PyroCb smoke generates synoptic-scale stratospheric anticyclones. Geophysical Research Letters 47 (13). Available at https://doi.org/10.1029/2020gl088101.

Legge, S., Rumpff, L., Woinarski, J.C.Z., Whiterod, N.S., Ward, M., Southwell, D.G., et al., 2022. The conservation impacts of ecological disturbance: time-bound estimates of population loss and recovery for fauna affected by the 2019–2020 Australian megafires. Global Ecology and Biogeography 31 (10), 2085–2104. Available at https://doi.org/10.1111/geb.13473.

Nolan, R.H., Boer, M.M., Collins, L., Resco de Dios, V., Clarke, H., Jenkins, M., et al., 2020. Causes and consequences of eastern Australia's 2019–20 season of mega-fires. Global Change Biology 26 (3), 1039–1041. Available at https://doi.org/10.1111/gcb.14987, http://onlinelibrary.wiley.com/journal/10.1111/(ISSN)1365-2486.

Owens, D., O'Kane, M. 2020. Final Report of the NSW Bushfire Inquiry. NSW Government, Accessed June 2021. https://www.nsw.gov.au/nsw-government/projects-and-initiatives/nsw-bushfire-inquiry.

Rumpff, L., Legge, S.M., van Leeuwen, S., Wintle, B.A., Woinarski, J.C.Z., 2023. Australia's Megafires.

Shiraishi, T., Hirata, R., 2021. Estimation of carbon dioxide emissions from the megafires of Australia in 2019–2020. Scientific Reports 11 (1). Available at https://doi.org/10.1038/s41598-021-87721-x, www.nature.com/srep/index.html.

Ward, M., Tulloch, A.I.T., Radford, J.Q., 2020. Impact of 2019–2020 mega-fires on Australian fauna habitat' Nature Ecology and Evolution 4, 1321–1326.

Williamson, G.J., Gibson, R.K., Matala, A., Cirulis, B., 2020. Fire severity and behaviour -Fire severity and past impact. In Report prepared by the NSW Bushfire Management Research Hub for the NSW Bushfire Inquiry. Wollongong: University of Wollongong.

The Fire Atlas of Portugal: its history and uses

19

João M.N. Silva[1], *Luis Filipe Lopes*[2], *and José M.C. Pereira*[1]
[1]Forest Research Centre, Associate Laboratory TERRA, School of Agriculture, University of Lisbon, Lisbon, Portugal, [2]Forest Research Centre, School of Agriculture, University of Lisbon, Lisbon, Portugal

19.1 Introduction

In many parts of the world, fire regimes have been altered from their historical patterns and are expected to continue changing rapidly due to climate change, land-cover transformation, and shifts in land use. In Portugal, as in most Mediterranean countries, land cover is largely shaped by human activity. Currently, forests occupy about 36% of the country's area (ICNF-Instituto da Conservação da Natureza e das Florestas, 2019) compared with less than 10% at the end of the 19th century (Mather and Perreira, 2006). Eucalypt and maritime pine plantations account for 48% of the forest area (ICNF-Instituto da Conservação da Natureza e das Florestas, 2019), while cork oak and holm oak woodlands cover 34%. The remaining forest is composed of stone pine (6%), other oaks (3%), chestnut (1%), and other species (7%). Shrublands and pastures occupy an additional 31% of the land, and agriculture covers 23%. Beyond climate influences, contemporary fire regimes are strongly shaped by human activities, by suppressing natural ignitions or increasing human-caused ignitions, by increasing or reducing fuel loads (e.g., through agricultural abandonment or prescribed burning), and by modifying landscapes (e.g., urbanization, deforestation, and afforestation).

In Mediterranean regions of Europe, North America, and Australia, the frequency of extreme fire events is increasing, causing severe damage to both rural and urban areas, as well as loss of human life (e.g., California 2025, Greece 2007, Portugal 2017, Australia 2009). To address these challenges, new management strategies must be developed that focus on managing fire regimes, not individual fires (Cochrane and Bowman, 2021). This involves understanding the dynamic relationships between climate, terrain, fuels, and fire behavior, and implementing integrated management with widespread use of prescribed burning. Fire is a persistent and inherent feature of many ecosystems, requiring adaptive management rather than fire exclusion. Designing an effective fire regime management strategy involves defining the typical range of characteristics, such as fire frequency, intensity, type (e.g., surface or crown), severity, seasonality, and spatial extent. These analyses rely on data derived from burned area maps, which are typically compiled over several decades to provide robust estimates of fire regime metrics (Morgan et al., 2001).

Advances in Remote Sensing for Forest Fire Management. DOI: https://doi.org/10.1016/B978-0-443-41692-7.00021-4
© 2027 Elsevier Ltd. All rights are reserved, including those for text and data mining, AI training, and similar technologies.

Satellite sensors provide the only way to produce burned area maps across large regions and over extended time periods. Over the last few decades, remote sensing has become crucial for studying fire regimes and monitoring landscape changes at global or regional scales. Satellite observations provide objective, spatially and temporally explicit information on burn patterns and fire activity, which are essential for understanding the interaction of fire with climate and vegetation (Chuvieco et al., 2019). Although several global burned area products exist, they do not eliminate the need to develop region-specific burned area maps. In this context, the Landsat program is particularly valuable, offering long-term time series at medium spatial resolution (30 m), which is essential for analyzing spatiotemporal trends in burned areas at a scale suitable to support management interventions. Portugal is in a rather unique situation, with the availability of a very long-term (1975–present) annual, countrywide fire atlas that relies on all Landsat satellites except Landsat 1. This chapter introduces the Portuguese context in terms of fire incidence (section 19.2), then describes the history of the Portuguese fire atlas (section 19.3), its operational applications (section 19.4), supported research (section 19.5), and finally the monthly version (section 19.6) and the current version of the fire atlas (section 19.7).

19.2 The Portuguese context

Data from the European Forest Fire Information System (EFFIS) show that Portugal is the European Union country most widely affected by rural fires (Fig. 19.1). Between 1980 and 2024, the country recorded a mean annual burned area equivalent to 1.31% of its mainland territory, nearly three times higher than that of Greece, which ranks second at 0.34%. The climate of Portugal is Mediterranean-type, with

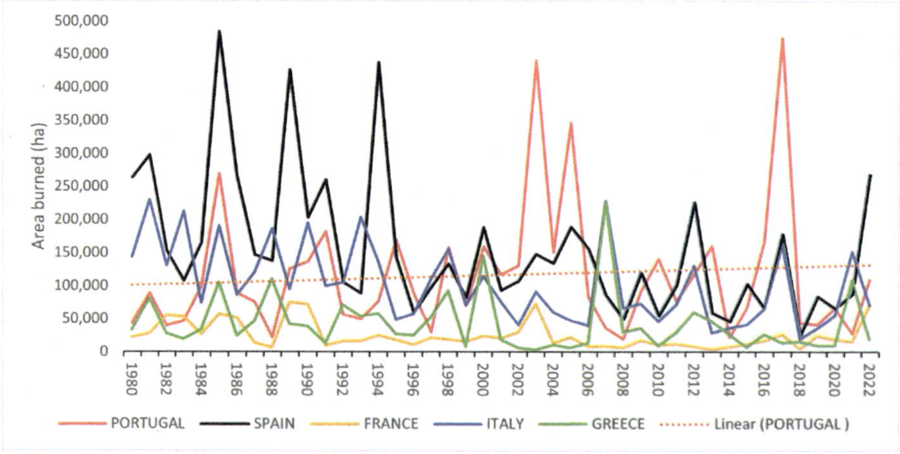

Figure 19.1 Area burned in Southern European countries over the last four decades. *Source*: Data from the European Forest Fire Information System, EFFIS.

cool, wet winters and dry, warm to hot summers. The Atlantic Ocean exerts a moderating influence on climate that allows winter temperatures to remain suitable for vegetation growth, yielding high net primary productivity. Fine fuels in forests and shrublands dry out over the summer, generating suitable conditions for the spread of large and intense wildfires (Keeley et al., 2012). This biophysical setting is aggravated by ongoing processes of rural depopulation, agricultural land abandonment, and widespread afforestation efforts initiated in the mid-20th century (Oliveira et al., 2017). Consequently, large burned areas are now increasingly concentrated in thinly populated rural regions with aging populations (Oliveira et al., 2017). The combined impact of these processes has also turned Portugal into the country experiencing the highest rate of planted forest area lost to wildfire at global level (Bousfield et al., 2023). Fig. 19.2 shows Moderate Resolution Imaging Spectroradiometer (MODIS) images for the years 2003, 2005, 2017, and 2025, the years with the largest total annual burned area in Portugal, accounting for 4.7%, 3.8%, 5.9%, and 2.8% of the country's total area, respectively.

19.3 History of the Portuguese Fire Atlas

In 1995, the Portuguese Forest Service (known at that time as *Direcção-Geral das Florestas*) requested the Forest Research Centre (*Centro de Estudos Florestais*—CEF) to develop a countrywide burned area map for the years 1993 and 1994. They were the first burned area maps developed for the entire territory of mainland Portugal derived from Landsat imagery. The Forest Service positively evaluated the quality and utility of the maps, and the Forest Research Centre was awarded annual contracts to produce the burned area map, which were renewed over the next 15 years, until 2010. The Forest Research Centre used research funding to extend the fire atlas back to 1975, beginning with Landsat 2. Landsat 1 was launched in 1972 but the images from that satellite were not available for Portugal.

During this period, the annual burned area maps were derived from late summer–autumn Landsat images, with a spatial resolution of 80 m and a minimum mapping unit of 35 ha for the period 1975–83 (MultiSpectral Scanner [MSS] era). Since 1984 (Thematic Mapper [TM]–Enhanced Thematic Mapper [ETM] era), the spatial resolution of the data has been 30 m with a minimum mapping unit of 5 ha. A single postsummer acquisition composed of eight images covering the territory of mainland Portugal was used due to the cost of the images and the limited computational capacity. In the first year, each Landsat scene was divided into four sections for processing. Later, with more powerful computers, it became possible to process the eight images individually, although it was still necessary to combine the eight maps of burned areas to create a single map of Portugal. With improved processing capabilities, the procedure began with constructing a mosaic of the eight Landsat images. This allowed burned areas to be mapped across mainland Portugal simultaneously. At that time, the focus was on mapping summer fires, as images from late summer or autumn did not adequately capture fires occurring in spring due to

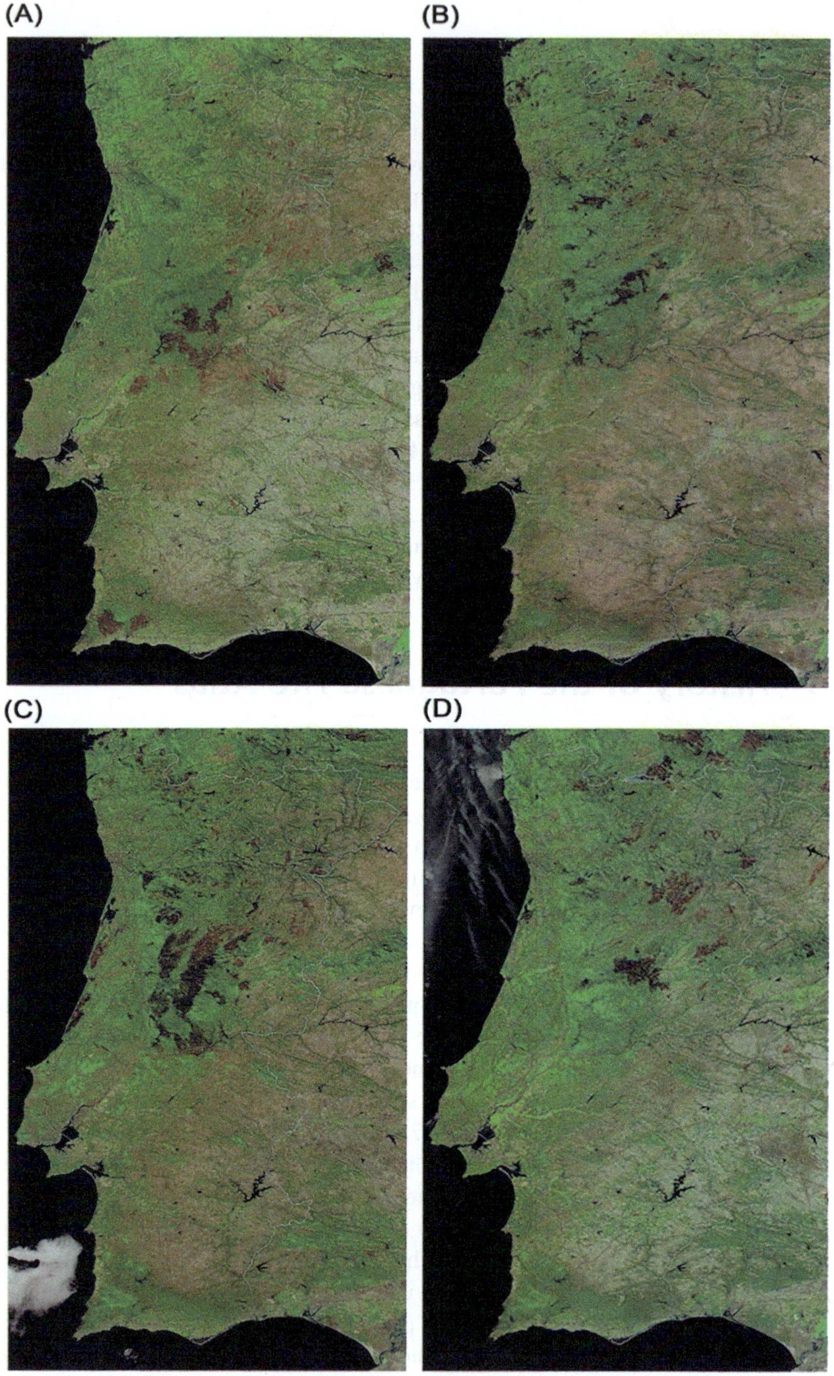

Figure 19.2 Moderate Resolution Imaging Spectroradiometer (MODIS) images over Portugal for the years with the largest area burned. (A) MODIS images over Portugal for the years 2003, (B) 2005, (C) 2017, and (D) 2025, the years with the largest area burned in Portugal, accounting for 4.7%, 3.8%, 5.9%, and 2.8% of the country's total area, respectively. The data for 2025 is provisional and refers to September 15. Map lines delineate study areas and do not necessarily depict accepted national boundaries.
Source: Data on annual burned area from the Portuguese Forest Service (Instituto da Conservação da Natureza e das Florestas, ICNF). Imagery source: NASA Worldview.

vegetation regrowth during the long interval between fire occurrence and image acquisition.

For the first years (1993–96), the annual fire perimeter maps were derived using a semiautomatic procedure, starting with supervised image classification and complemented by thorough manual editing of classification results (Sousa et al., 1997). The classifier used was very simple, the parallelepiped that builds a multidimensional box or "parallelepiped" in spectral space, based on training areas (burned and unburned areas). The classification was carried out in two steps. First, a classification was made in which the choice of training areas for burned areas was very restrictive, selecting only areas that were clearly burned. In other words, the aim was to minimize commission errors while accepting a high omission error. In the second step, a 650 m buffer zone was defined around the areas classified as burnt in the first step. A second classification of the image was then carried out, restricted to these buffer zones only. This time, the selection of training areas was less restrictive, to widen the spectral intervals and include pixels adjacent to those classified as burned in the first phase in the final map of burned areas. The first phase aimed to detect the centers of the burned areas, while the second phase allowed them to be delimited more accurately, reducing the commission errors that would occur if the entire image were used instead of just the buffer area. This technique is a precursor to others that were later used in burned area mapping (e.g., Bastarrika et al., 2011).

Later, in 1997, the simple methodology used for mapping burned areas based on a parallelepiped classifier was replaced by a more sophisticated statistical method: a classification tree. The classification and regression tree (CART) method was first introduced by Breiman et al. (1984), and at the time it was adopted for producing the Portuguese fire atlas, it was a state-of-the-art method that had been successfully applied to remotely sensed data. For example, a classification tree was used to produce one of the first global land-cover maps, derived from 1 km Advanced Very High Resolution Radiometer (AVHRR) imagery (Hansen et al., 2000). The master thesis of Sá (1999) describes the application of CART for burned area mapping in Portugal. Classification trees are a nonparametric method based on binary recursive partitioning, offering great flexibility, as they allow the same class to be reached using different sets of rules. In the case of burned areas, different rules may be that intensity was low and displayed a low spectral contrast. Additionally, different rules can be automatically developed for pixels within burned areas and pixels at the edge of burned areas, which have a mixed spectral signature due to the presence of both burned and unburned vegetation. Landsat bands and vegetation indices, such as the Normalized Difference Vegetation Index (NDVI) (Tucker, 1996) and the Normalized Difference Water Index (NDWI) (Gao, 1996), were used as input variables. In the earlier years, only images from the year being mapped were used. Later, with improved computer processing capabilities, images from both the year being mapped and the previous year were incorporated to enable change detection and improve the accuracy of the burned area maps. Accuracy of the classification algorithm was assessed with a 10-fold cross-validation procedure, allowing to report a cross-validation misclassification rate and an aggregate error matrix of the 10 trees (Breiman et al., 1984). Burned area estimates derived from image classification were compared with municipality-level field statistics provided by the Forest Service. Whenever possible, fieldwork was conducted in those municipalities exhibiting large discrepancies to investigate the causes of the differences between the two sources' estimates.

In 2011, the Forest Service began producing Portugal's annual burned area maps, combining a variety of data sources, such as field surveys, Landsat and Sentinel 2 imagery, and burned area perimeters from the EFFIS, with some loss of data and methodological consistency. Sporadically, the CEF was asked to produce the burned area maps (in 2014 and 2017).

19.4 Operational applications

Over the last 30+ years, the growing set of annual burned area maps has been used in the development of diverse operational applications, such as fire hazard mapping, assessment of fire danger at the wildland–urban interface (WUI), design and optimization of the national fuel break network, and mapping of Portugal's fire regimes for various government agencies.

19.4.1 Fire hazard mapping

After a decade of annual burned area mapping, the Forest Service requested the Forest Research Centre to develop an updated fire risk map for Portugal (Pereira and Santos, 2003). This map was officially adopted and used for fire management planning over several years. The data sources included annual burned area maps from the Portuguese Fire Atlas (1990–99), land-cover maps, topographic and climatic datasets, and demographic information. The goal was to integrate environmental, climatic, and socio-economic variables into a coherent spatial framework for evaluating wildfire risk. Fire probability across mainland Portugal was mapped by analyzing spatial patterns of fire occurrence during the 1990s. Environmental variables—such as forest and shrubland distribution, elevation, slope, temperature, precipitation, and population density—were related to the burn probability map using a regression tree CART (Breiman et al., 1984). Areas at risk were represented by large, continuous forest zones, which were overlaid on the burn probability map to produce the final fire risk map.

The analysis of the map revealed a remarkable spatial pattern: higher burn probability classes were strongly concentrated in the northern half of Portugal. The largest continuous patch of very high burn probability corresponds to a region with extensive forests and shrublands. High fire probability tends to coincide with areas that receive relatively high mean annual precipitation but also experience frequent hot and dry spells during summer. These conditions lead to the accumulation of fine fuels during wetter periods, which are then consumed during times more conducive to fire ignition and spread. Regions with intermediate population density showed the highest fire probabilities. In contrast, densely populated areas typically have smaller and more fragmented forest and shrubland patches, which limit the potential for large wildfires. From a management perspective, these findings highlight the need for region-specific prevention strategies. Key measures include targeted fuel reduction, protection of WUIs, and the integration of fire risk maps into land-use and spatial planning.

19.4.2 National fire danger assessment at wildland–urban interface

A novel definition of WUI and a methodology to map it in mainland Portugal were developed by Pereira et al. (2018). WUI mapping supports wildfire risk assessment, prevention, and land management planning. The WUI represents the boundary where urban areas meet or are in close proximity to flammable vegetation, and its characterization is essential in Portugal, a country with the highest wildfire incidence in southern Europe. The WUI was defined as a set of linear segments where urban polygons are either in direct physical contact with wildland polygons (direct interface) or are less than 100 m away from flammable vegetation (indirect interface). The analysis uses the national land-cover map, reclassified into urban and wildland (forests, shrublands, and pastures) categories, combined with degree of urbanization and burned area maps from the Portuguese Fire Atlas, complemented by data from the Forest Service. Computational geometry and spatial analysis were applied to delineate the interface along polygon vertices, resulting in detailed, scalable mapping at the national level.

Results indicate that 40,319 urban polygons in Portugal include some form of WUI, totaling 29,977 km of direct WUI and 11,192 km of indirect WUI. The distribution reveals marked regional contrasts: northwestern Portugal shows high densities of direct WUI due to intensive forest and settlement intermixing, while central and southern regions have lower densities, often with agricultural or open areas acting as buffers. Densely urbanized coastal zones have long WUI perimeters but reduced exposure, whereas sparsely populated interior areas show high WUI density and elevated fire risk. Fine-scale WUI mapping provides critical insights for fire prevention and management strategies, namely for delineating secondary fuel break networks and prioritizing protection measures for rural settlements in fire-prone landscapes.

19.4.3 Delineation of the national fuel break network

After the severe wildfire seasons of 2003 and 2005, the Forest Service proposed the creation of a national fuel break network (Rede Primária de Faixas de Gestão de Combustível, RPFGC) in Portugal (Pinho, Louro, and Paulo, 2006; Pinho, 2008). This network was designed to be implemented at the landscape scale, functioning as a structural element. The width of the strips in the primary network should not be less than 125 m, and they should preferably be located along ridgelines or valleys. The orientation of these strips should be perpendicular to the direction of winds associated with meteorological conditions favorable to large wildfires, and tree cover should not exceed 50%.

Due to the scarcity of financial resources and the limited commitment of private landowners, only 17% of the planned network (out of a total of 3539 km) had been created by 2016 (Oliveira et al., 2016). Given the high costs associated with the full implementation of the network, the CEF was requested to reassess it. This process included redimensioning the network, prioritizing its implementation in areas of greater relevance (procedure described in Almeida, [2019]). The redimensioning was based on the assignment of a priority level to each segment, allowing the selection of the most relevant ones for the new network. The priority index of each segment was calculated based on a set of variables grouped into five main themes: fire, which includes variables related to the historical incidence of wildfires in the surrounding area and the orientation of segments relative to wind direction on high-risk days; fuel, which assesses the percentage of combustible cover (shrubs and forests) in the area surrounding each segment; population hazard, which quantifies the percentage of combustible cover near settlements, serving as an indicator of the vulnerability of people and infrastructure; economic value, which considers the potential economic losses in the event of a fire; and ecological value, which accounts for the presence of classified areas of biological interest and the potential damage to native forests. The integrated priority index for each segment was obtained through a weighted evaluation of the importance of each variable and theme. Based on this index, a priority threshold was defined to determine which segments would be included in the structural network, aiming to achieve one-third of the total length of the original network. Additional segments were selected to ensure connectivity between high-priority ones, thereby establishing a coherent and functional network. As a result of this optimization of the initial primary network, the final structural network reached a total length of 3543 km.

19.4.4 Mapping of fire regimes at parish level

Pereira et al. (2022) conducted a comprehensive analysis of rural fire regimes across mainland Portugal over a 38-year period (1980–2017). The main objective was to classify and map fire regimes at the parish level to support long-term integrated fire management strategies. The study utilized georeferenced data from multiple sources, including satellite-derived annual burned area maps from the Portuguese Fire Atlas, the Portuguese Rural Fire Database (PRFD), and MODIS fire radiative power data. Seven fire-related variables were constructed at the parish level: area burned, coefficient of variation of the area burned, average size of the 10 largest burned areas, number of fires, radiative power (intensity) of the fire, duration of the fire season, and percentage of the fire season that occurs in "winter." Additionally, five antecedent variables were mapped: predominant land cover, thermal and rainfall bioclimate types, population density, and demographic change. The analysis excluded parishes with minimal fire activity, resulting in a dataset of 2222 parishes.

Principal Component Analysis and cluster analysis identified nine distinct fire regimes, which were further grouped into four macro-regimes: pastoral burning, periurban small fires, forest wildfires, and agricultural burning. Key findings include the spatial concentration of pastoral fire regimes in mountainous northern regions, periurban fire regimes in densely populated metropolitan areas, and intense forest fire regimes in depopulated central regions. These insights have significant implications for fire management, suggesting tailored strategies such as promoting winter burns in pastoral zones, reinforcing fuel breaks in forested areas, and protecting urban–rural interfaces. The study provides a foundational framework for regionalizing public policies under Portugal's National Integrated Fire Management Plan (2020–30), emphasizing the need for differentiated approaches based on local fire dynamics and socio-environmental conditions.

19.5 Atlas-supported research

Wildfires in Portugal have been extensively studied, revealing complex patterns of atmospheric emissions, fire frequency, land-cover selectivity, spatiotemporal trends of burned area, among many other topics. These investigations, based on multidecade datasets of burned areas, provide an in-depth understanding of fire dynamics and their environmental, climatic, and socio-economic impacts. Such knowledge is essential for supporting forest management policies and strategies for climate change mitigation and wildfire risk reduction.

19.5.1 Atmospheric emissions from vegetation fires in Portugal

Atmospheric emissions from wildfires in mainland Portugal were estimated over a 19-year period (1990–2008), quantifying the uncertainty associated with these estimates and identifying the variables that most strongly influence that uncertainty (Rosa et al., 2011). The study also evaluated the relationship between burned area and total emissions in order to support improved national-scale fire emission assessments. The analysis was based on burned area maps from the Portuguese Fire Atlas, national land-cover maps, national forest inventory data, biometric models, and combustion and emission factors obtained from the literature. Emissions were calculated as the product of area burnt, biomass loading per unit area, combustion factor, and emission factor, using land cover-specific values for all variables. A comprehensive uncertainty and sensitivity analysis was conducted using Monte Carlo simulations and variance decomposition techniques.

The results revealed strong interannual variability in burned area, ranging from about 9000 ha in 2008 to 440,000 ha in 2003. Corresponding greenhouse gas emissions varied between 159 and 5655 Gg CO_2 equivalents. The emission factor and the combustion factor for shrublands were identified as the key contributors to model output variance. A very strong correlation (R^2=0.95) was found between burned area and total emissions, suggesting that reliable estimates can be derived once burned area is quantified. Wildfire emissions account for 1%–9% of Portugal's total greenhouse gas output and are highly seasonal, concentrated during summer months. Improving knowledge of combustion and emission factors, enhancing fuel management, and integrating wildfire emissions into national climate mitigation and air quality strategies are essential for effective environmental management.

19.5.2 Fire frequency analysis in Portugal

A 31-year dataset (1975–2005) of annual burnt area maps, extracted from the Portuguese Fire Atlas, was used to characterize fire frequency patterns and evaluate how methodological choices, such as mapping resolution, time series length, and inclusion of censored data, affect fire interval estimates (Oliveira et al., 2012). The study area encompassed mainland Portugal, divided into 21 forest planning regions based on environmental attributes like climate, topography, land cover, and population density. Three time series were constructed to assess the impact of minimum mapping unit (5 vs 35 ha), time span (22 vs 31 years), and data completeness (complete vs censored intervals) on fire frequency modeling using the Weibull function. The Weibull model was chosen for its flexibility and ability to represent fire interval distributions and hazard functions.

Key findings include a national mean fire interval of 36 years and a median of 27.6 years. The estimated fire rotation period was 82 years, corresponding to an annual burnt area of 1.2%. Northern regions like Beira Interior Norte and Tâmega showed shorter fire intervals and higher burn rates, while southern regions such as Alentejo Central and Baixo Alentejo exhibited longer intervals and lower fire incidence. A large portion of central and northern Portugal had a fire return interval of fewer than 25 years. Including censored data significantly increased fire interval estimates, especially in low-incidence regions. Ignoring small fires has little or no effect on the estimates of Weibull function parameters because small fires, although numerous, typically represent a small fraction of the total area burnt. In terms of implications for forest management, the authors emphasize that throughout the regions with mean or median fire intervals of less than 25 years, the sustainability of maritime pine forests, managed for timber on a 35–40-year harvest cycle, is threatened. In recent decades, many landowners under pressure from fire risk have converted pine plantations to eucalypt plantations that are managed for pulpwood under a short rotation of 9–12 years.

19.5.3 Wildfire selectivity for land-cover type

Analyzing the relationship between fire and vegetation, particularly fire incidence by vegetation type, is challenging because firefighting activities often alter this relationship. Suppression efforts may prioritize protecting certain land-cover types or vegetation formations, thereby influencing observed fire patterns. The size of wildfires is a very important factor in the fire-vegetation relationship, with large fires being less selective in terms of land cover (Barros and Pereira, 2014; Nunes et al., 2005). Nunes et al. (2005), using 506 fires that occurred in Portugal, investigated whether fires select given land-cover types for burning by comparing the proportions of land-cover types present in burned areas and their respective surroundings. Results showed that fires are selective, with small fires exhibiting stronger land-cover preferences than large fires, and that there is a marked preference for shrublands followed by forests, while agriculture is avoided. Following a different methodological approach, Barros and Pereira (2014) obtained similar results. The ranking of land-cover types according to fire proneness, from less to most fire prone, was: annual crops, evergreen oak woodlands, eucalypt plantations, pine stands, and shrublands. All land-cover types exhibited reductions in fire preference as fire size increased. Other studies obtained similar rankings (e.g., Moreira et al., 2009). The lower selectivity of eucalypts compared to pine may be explained in part by their economic value and active management, and suppression effectiveness, which contributes to mitigating the effect of severe weather conditions (Barros and Pereira, 2014).

19.5.4 Spatiotemporal trends of area burnt in Portugal

Spatial and temporal dynamics of burned area across the Iberian Peninsula from 1975 to 2013 were derived to elucidate the drivers of fire regime changes and their implications for wildfire management (Silva et al., 2019). The study integrated national fire datasets, Portugal's Landsat-based Fire Atlas, and Spain's General Statistics on Forest Fires (*Estadística General de Incendios Forestales* - EGIF) fire database, harmonized within a 10 × 10 km grid framework to enable fine-scale spatial analysis. Trend detection was performed using the Theil–Sen slope estimator and the Contextual Mann–Kendall (CMK) test, which incorporates spatial autocorrelation to identify statistically significant regional trends. Supplementary datasets on fire weather (Daily Severity Rating derived from the Canadian Fire Weather Index), land cover (CORINE Land Cover map), and demographic changes (national censuses) were used to interpret climatic and anthropogenic influences on observed patterns.

The results demonstrate a pronounced spatial heterogeneity in burned area trends. In Spain, the largest significant clusters exhibited decreasing trends, located mainly in the northwestern part of the country (Galicia, Castilla y León, and Asturias autonomous communities). Valencia, Andalucía, and Extremadura also display spatial clusters of decreasing burnt area trends, reflecting the effectiveness of enhanced fire suppression policies. In contrast, Portugal displayed divergent regional patterns, with a strong increasing trend in the northwestern region and a marked decline in central Portugal. Regression analyses between burned area and the Daily Severity Rating revealed that climatic variability explains only part of the observed changes, underscoring the critical role of human factors such as land-use dynamics, population shifts, and urban expansion. Specifically, rural depopulation and agricultural abandonment were associated with declining fire activity, while the expansion of the urban–rural interface in northwestern Portugal contributed to increased fire incidence. The study underscores the need for integrated fire management strategies that account for both biophysical and socioecological dimensions. Understanding these spatiotemporal trends provides essential insights for forest management, allocation of suppression resources, and policy development aimed at mitigating future wildfire risks under changing climatic and land-use conditions.

19.6 The Monthly Atlas (1984–2021)

The Annual Fire Atlas has served in the last three decades as Portugal's primary database for burned area mapping. However, as previously noted, its operational design imposed an annual temporal resolution, typically relying on a single nationwide mosaic derived from autumn satellite imagery. This approach effectively captured large summer wildfires but systematically underestimated smaller pastoral and agricultural burns occurring in winter and early spring. In addition, methodological inconsistencies between production periods (e.g., between those years made by the Forest Research Centre—1975–2010—and the Forest Service—2011–present) and variations in spatial resolution introduced further weaknesses in the dataset. These limitations motivated the development of an improved, more consistent, and information-rich national fire record.

To overcome these issues, a new Landsat-based methodology was implemented, resulting in the Atlas of Monthly Burned Area for Portugal (Neves et al., 2023), covering 1984–2021 (i.e., starting in the TM era). Instead of delineating new perimeters, the approach focused on reprocessing and temporally refining the existing Annual Fire Atlas perimeters. The main goal was to disaggregate each annual burned area polygon into individual fire events and assign them the closest possible day and month of occurrence. This process leveraged the full Landsat archive available in Google Earth Engine (GEE). Using 7178 Landsat scenes spanning 37 years, the temporal resolution of the national fire record was enhanced to an approximately monthly scale. Fire dating was performed by analyzing the pixel-wise time series of the Normalized Burn Ratio (NBR) and selecting the date of minimum NBR value, corresponding to the moment of maximum burn impact. Temporal validation, performed using daily active fire detections from MODIS and Visible Infrared Imaging Radiometer Suite (VIIRS) sensors, demonstrated high accuracy: across the 37-year record, the Monthly Fire Atlas achieved 84% accuracy for a maximum temporal offset of 16 days, and 91% accuracy for a 32-day offset, confirming the robustness of the fire dating methodology. This refinement substantially improved temporal accuracy, enabling a clearer understanding of intra-annual fire variability and seasonal fire regimes across mainland Portugal (Fig. 19.3).

19.7 An entirely new Fire Atlas

The new Fire Atlas employs a deep learning approach to improve the detection of burned areas and address limitations identified in the previous Monthly Atlas. For each year, using GEE, an annual multiband composite was generated from Landsat 4–9 imagery, combining a quality mask, temporal smoothing, and removal of outliers. The composite was produced through a per-pixel reduction based on the maximum NBR variation, that is, the largest difference in NBR values between two consecutive values, allowing the identification of the moment of strongest spectral change typically associated with fire events. Each composite included four bands: prefire NBR, postfire NBR, postfire NBR of the previous year (to exclude previously burned areas), and Day of the Year (DOY) homogeneity (Haralick homogeneity texture describing the spatial homogeneity of the fire date). An additional DOY layer records, for each pixel, the day corresponding to the observation with maximum NBR change, providing an approximate date of fire occurrence.

Training data covered eight years, six guided by the previous fire atlas and two complemented with Portuguese Forest Service winter-fire databases to improve the

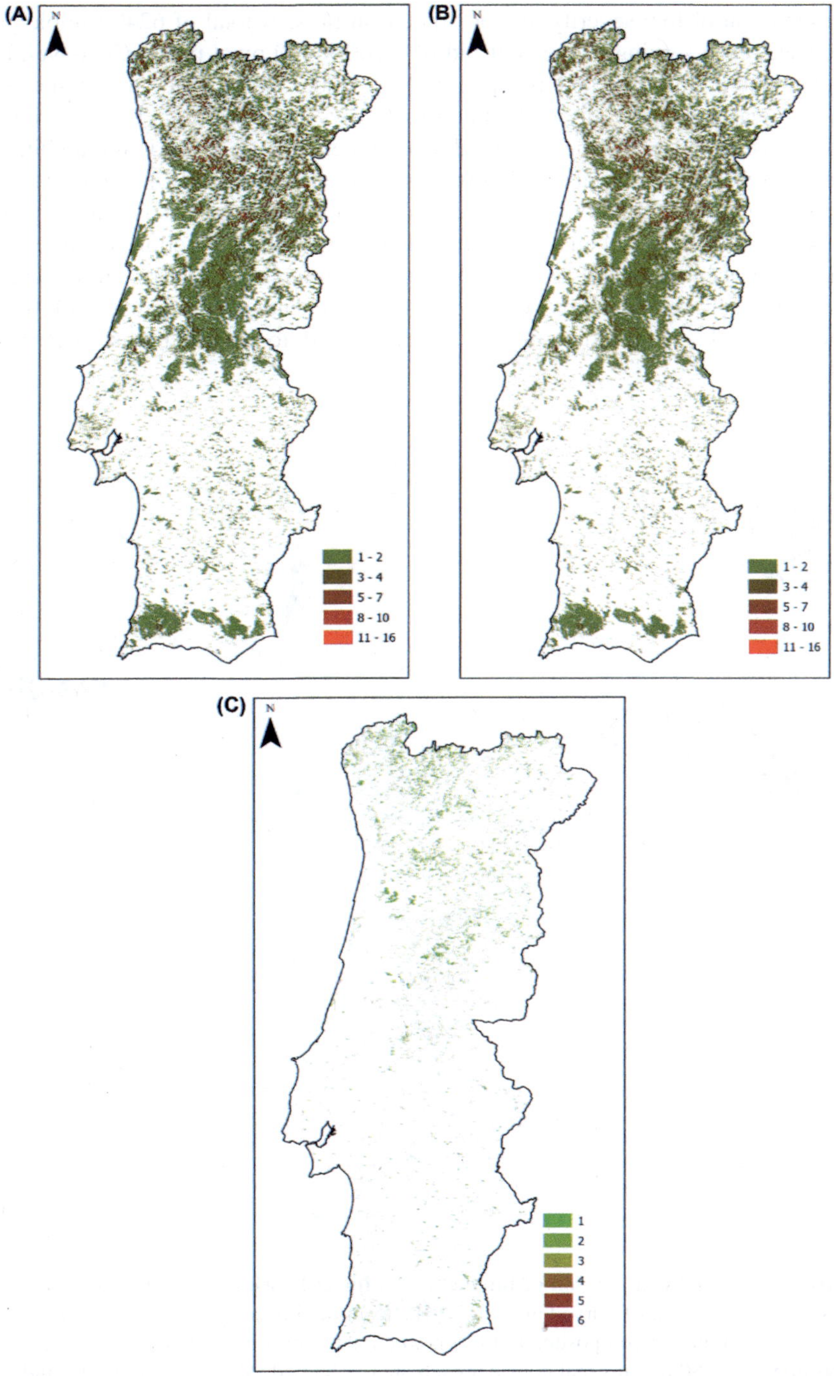

Figure 19.3 Number of times each location burned. Number of times each location burned in the period of 1984–2021, using the Monthly Fire Atlas: (A) annual estimates, (B) division into summer,and (C) winter periods. Map lines delineate study areas and do not necessarily depict accepted national boundaries.

representation of low-severity and/or off-season fires. A total of 6249 sample points were used, each serving as the centroid of a 160 × 160 pixel patch. Cross-validation was conducted through 10 bootstrapped subsets under a leave-one-out scheme, where Portugal was randomly divided into 40 equally sized regions. The classification model was an adapted convolutional neural network (CNN) U-Net architecture, generating 80 annual outputs (10 bootstraps × 8 displacements) (Fig. 19.4). As part of the ongoing model calibration, the decision threshold used to classify burned pixels is still being defined. Initial analyses reveal that the model's predictions capture burned areas more accurately and extensively than the reference datasets, influencing the precision–recall relationships. Consequently, independent validation will be conducted once this calibration process is completed to ensure robust performance evaluation.

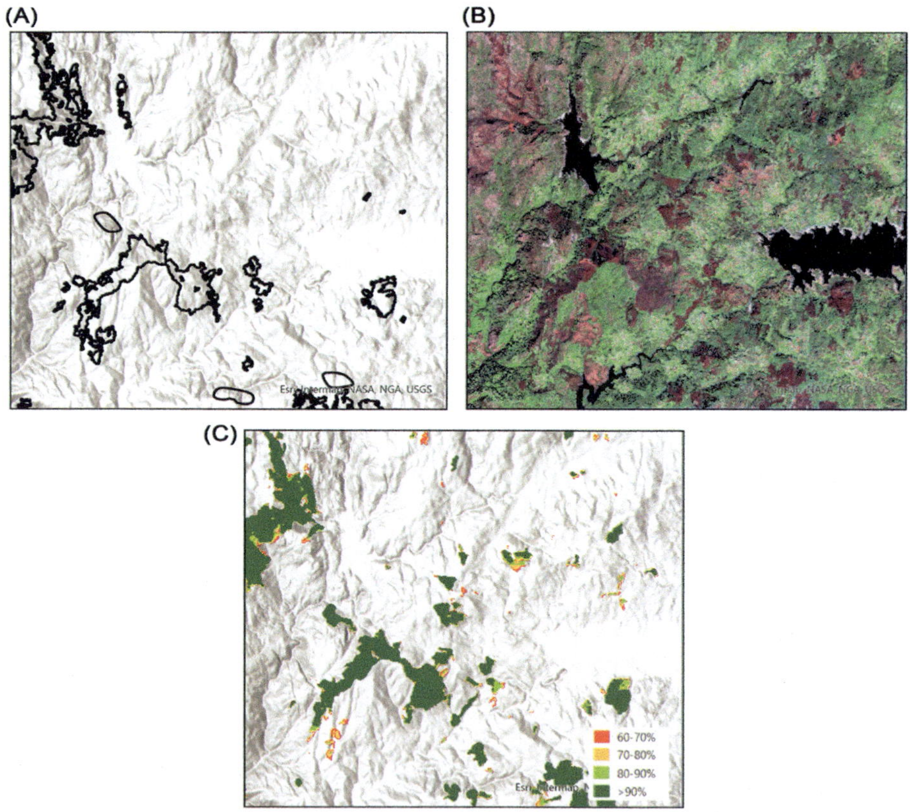

Figure 19.4 Visual comparison of burned areas for 2017 with the Forest Service dataset. (A) Visual comparison of burned areas for 2017 from the Forest Service dataset, (B) the New Atlas—Landsat annual composite, and (C) prediction. Values in prediction correspond to the proportion of the 80 outputs that consider each area as burned. Map lines delineate study areas and do not necessarily depict accepted national boundaries.

The new Atlas represents a major improvement over the previous Monthly Atlas, providing greater temporal resolution and analytical capacity (Fig. 19.5). In the original annual product, several nearby fires were often merged into large fire complexes, which obscured the true number and size of individual events. With the finer temporal detail now available, these complexes can be separated into distinct fire occurrences, improving the accuracy of fire size distribution by revealing more small fires and reducing the apparent number of extremely large ones. The enhanced temporal precision also allows for a clearer characterization of fire seasonality, distinguishing between summer wildfires, usually associated with uncontrolled burning, and winter or early spring fires related to land management practices.

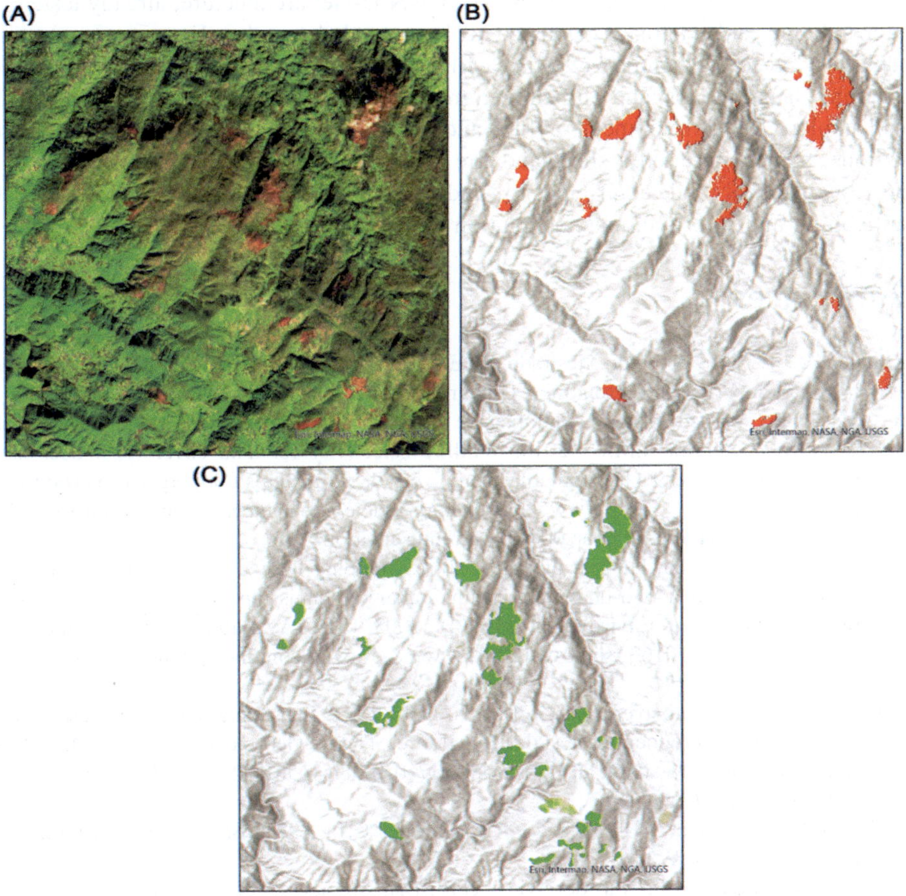

Figure 19.5 Visual comparison of a Landsat-8 image, the monthly atlas, and new atlas. (A) Visual comparison of a Landsat-8 image, (B) the monthly atlas, and (C) new atlas for 2021. Map lines delineate study areas and do not necessarily depict accepted national boundaries.

Building on these improvements, the new Atlas will provide a more accurate and detailed understanding of Portugal's fire regimes and their evolution, supporting updates to previous research based on the earlier version and fostering new lines of investigation into fire dynamics, seasonality, and management practices.

19.8 Future work

Future developments of the Portuguese Fire Atlas aim to expand its temporal and spatial coverage. Key priorities include annual updating and reprocessing of Landsat-MSS data back to 1975 using GEE, enabling the creation of a 50-year Fire Atlas. Improving training datasets and refining deep learning models will enhance the detection of low-severity winter fires. The CNN U-Net architecture, already tested in Mozambique and Angola, is planned to be extended to other European countries, such as Greece. The improved atlas will be used to generate updated fire regime mapping, hazard maps, and burned area trend analyses. Furthermore, future research will explore the development of fuel and fire management strategies in mountain regions, namely in altered shrubland fire regimes. These advancements will refine previous studies and contribute to understanding fire dynamics, improve risk mitigation, and inform climate adaptation policies across fire-prone regions.

References

Almeida, R.J.L., 2019. Tese do Mestrado em Engenharia Florestal e dos Recursos Naturais do Instituto Superior de Agronomia. Universidade de Lisboa. Lisboa Metodologia para a criação de uma rede primária de faixas de gestão de combustível em Portugal continental.

Barros, A.M.G., Pereira, J.M.C., 2014. Wildfire selectivity for land cover type: does size matter? PLoS One 9 (1) Available from. Available at https://doi.org/10.1371/journal.pone.0084760, http://www.plosone.org/article/fetchObject.action?uri=info%3Adoi%2F10.1371%2Fjournal.pone.0084760&representation=PDF.

Bastarrika, A., Chuvieco, E., Martín, M.P., 2011. Mapping burned areas from landsat TM/ETM+ data with a two-phase algorithm: balancing omission and commission errors. Remote Sensing of Environment 115 (4), 1003–1012. Available from. Available at https://doi.org/10.1016/j.rse.2010.12.005, www.elsevier.com/inca/publications/store/5/0/5/7/3/3.

Bousfield, C.G., Lindenmayer, D.B., Edwards, D.P., 2023. Substantial and increasing global losses of timber-producing forest due to wildfires. Nature Geoscience 16 (12), 1145–1150. Available from. Available at https://doi.org/10.1038/s41561-023-01323-y, https://www.nature.com/ngeo/.

Breiman, L., Friedman, J.H., Olshen, R.A., Stone, C.J., 1984. Stone, Classification and Regression Trees. Wadsworth.

Chuvieco, E., Mouillot, F., van der Werf, G.R., San Miguel, J., Tanasse, M., Koutsias, N., et al., 2019. Historical background and current developments for mapping burned area from satellite Earth observation. Remote Sensing of Environment 225, 45–64. Available from. Available at https://doi.org/10.1016/j.rse.2019.02.013, www.elsevier.com/inca/publications/store/5/0/5/7/3/3.

Cochrane, M.A., Bowman, D.M.J.S., 2021. Manage fire regimes, not fires. Nature Geoscience 14 (7), 455–457. Available from. Available at https://doi.org/10.1038/s41561-021-00791-4, http://www.nature.com/ngeo/index.html.

Gao, B.C., 1996. NDWI - A normalized difference water index for remote sensing of vegetation liquid water from space. Remote Sensing of Environment 58 (3), 257–266. Available at https://doi.org/10.1016/S0034-4257(96)00067-3.

Hansen, M.C., Sohlberg, R., Defries, R.S., Townshend, J.R.G., 2000. Global land cover classification at 1 km spatial resolution using a classification tree approach. International Journal of Remote Sensing 21 (6-7), 1331–1364. Available at https://doi.org/10.1080/014311600210209.

ICNF-Instituto da Conservação da Natureza e das Florestas, 2019. 6°Inventário Florestal Nacional-Relatório Final.

Keeley, J.E., Bond, W.J., Bradstock, R.A., Pausas, J.G., Rundel, P.W., 2012. Fire in Mediterranean Ecosystems: Ecology, Evolution and Management. Cambridge University Press. Available at https://doi.org/10.1017/CBO9781139033091.

Mather, A.S., Pereira, J.M.C., 2006. Incêndios Florestais em Portugal: Caracterização, Impactes e Prevenção,Transição florestal e fogo em Portugal. In: J.S. Pereira, J.M.C. Pereira, F.C. Rego, J.M.N. Silva, T.P. Silva, ISA Press, Lisboa.

Moreira, F., Vaz, P., Catry, F., Silva, J.S., 2009. Regional variations in wildfire susceptibility of land-cover types in Portugal: implications for landscape management to minimize fire hazard. International Journal of Wildland Fire 18 (5), 563–574. Available at https://doi.org/10.1071/WF07098.

Morgan, P., Hardy, C.C., Swetnam, T.W., Rollins, M.G., Long, D.G., 2001. Mapping fire regimes across time and space: understanding coarse and fine-scale fire patterns. International Journal of Wildland Fire 10 (3-4), 329–342. Available from. Available at https://doi.org/10.1071/wf01032, http://www.publish.csiro.au/nid/126.htm.

Neves, A.K., Campagnolo, M.L., Silva, J.M.N., Pereira, J.M.C., 2023. A Landsat-based atlas of monthly burned area for Portugal, 1984–2021. International Journal of Applied Earth Observation and Geoinformation 119 Available from. Available at https://doi.org/10.1016/j.jag.2023.103321, http://www.elsevier.com/locate/jag.

Nunes, M.C.S., Vasconcelos, M.J., Pereira, J.M.C., Dasgupta, N., Alldredge, R.J., Rego, F.C., 2005. Land cover type and fire in portugal: do fires burn land cover selectively? Landscape Ecology 20 (6), 661–673. Available at https://doi.org/10.1007/s10980-005-0070-8.

Oliveira, S.L.J., Pereira, J.M.C., Carreiras, J.M.B., 2012. Fire frequency analysis in Portugal (19752005), using Landsat-based burnt area maps. International Journal of Wildland Fire 21 (1), 48–60. Available at https://doi.org/10.1071/WF10131.

Oliveira, T.M., Barros, A.M.G., Ager, A.A., Fernandes, P.M., 2016. Assessing the effect of a fuel break network to reduce burnt area and wildfire risk transmission. International Journal of Wildland Fire 25 (6), 619–632. Available at https://doi.org/10.1071/WF15146. Available from: http://www.publish.csiro.au/nid/126.htm.

Oliveira, T.M., Guiomar, N., Baptista, F.O., Pereira, J.M.C., Claro, J., 2017. Is Portugal's forest transition going up in smoke? Land use Policy 66, 214–226. Available from. Available at https://doi.org/10.1016/j.landusepol.2017.04.046, www.elsevier.com/inca/publications/store/3/0/4/5/1/.

Pereira, J.M.C., Santos, T.N., 2003. Fire Risk and Burned Area Mapping in Portugal. Direcção-Geral das Florestas, Lisboa.

Pereira, J.M.C., Alexandre, P.M., Campagnolo, M.L., Bar-Massada, A., Radeloff, V.C., Silva, P.C., et al., 2018. Defining and mapping the wildland-urban interface in Portugal.

Advances in Forest Fire Research. Available at https://doi.org/10.14195/978-989-26-16-506_81.

Pereira, J.M.C., Silva, P.C., Melo, I., Oom, D., Baldassarre, G., Pereira, M.G. 2022 ForestWISE (Coord.) – Projetos AGIF 2021 (P32100231) 29 Cartografia de Regimes de Fogo à Escala da Freguesia (1980–2017).

Pinho, J., 2008. A Silvopastorícia na Prevenção dos Fogos RuraisOrganização do território e defesa da floresta contra incêndios. In: M.B. Moreira, J.S. Coelho, ISA Press, Lisboa ISA Press, Lisboa.

Pinho, J.R., Louro, G., Paulo, S., 2006. Incêndios florestais em Portugal: caracterização impactes e prevençãoRecuperação das Áreas Ardidas em Portugal e a Gestão do Fogo: a experiência da Equipa de Reflorestação (2004-2005). In: J.S. Pereira, J.M.C. Pereira, F.C. Rego, J.M.N. Silva, T.P. Silva, ISA Press, Lisboa ISA Press, Lisboa.

Rosa, I.M.D., Pereira, J.M.C., Tarantola, S., 2011. Atmospheric emissions from vegetation fires in Portugal (1990–2008): estimates, uncertainty analysis, and sensitivity analysis. Atmospheric Chemistry and Physics 11 (6), 2625–2640. Available at https://doi.org/10.5194/acp-11-2625-2011.

Sá, A.C. L., 1999. Tese do Mestrado em Matemática Aplicada às Ciências Biológicas, Instituto Superior de Agronomia, Universidade de Lisboa. Lisboa Comparação entre a regressão logística e as árvores de classificação na cartografia de áreas ardidas com imagens do satélite Landsat 5.

Silva, J.M.N., Moreno, M.V., Le Page, Y., Oom, D., Bistinas, I., Pereira, J.M.C., 2019. Spatiotemporal trends of area burnt in the Iberian Peninsula, 1975–2013. Regional Environmental Change 19 (2), 515–527. Available from. Available at https://doi.org/10.1007/s10113-018-1415-6, http://springerlink.metapress.com/app/home/journal.asp?wasp=64cr5a4mwldrxj984xaw&referrer=parent&backto=browsepublicationsresults,451,542;.

Sousa, A.M.O., Silva, J.M.N., Pereira, J.M.C., 1997. Cartografia de áreas ardidas em Portugal com imagens landsat 5 TM. Revista Florestal 10 (1), 19–26.

Tucker, C.J., 1996. Advances in the use of NOAA AVHRR data for land applicationsHistory of the use of AVHRR data for land applications. In: G. D'Souza, A.S. Belward, J.P. Malingreau, The Netherlands' Kluwer.

Part IV

Future directions and challenges

20 Integrating new technologies: challenges and opportunities for scalability, accessibility, and interoperability

Miguel Iglesias González and Alberto Cerrillo
R&D Department, Vexiza, León, Spain

20.1 Introduction

Global ecosystems are transitioning into a novel fire regime, distinguished by wildfires of a magnitude and intensity that are unprecedented in recent anthropogenic history. Empirical data confirm an accelerating escalation of this phenomenon. Specifically, research from the University of Maryland demonstrates that the annual extent of tree cover loss attributable to biomass combustion has increased by more than 100% since the early 2000s. 2024 was the most severe documented period for forest fires, culminating in the incineration of no less than 13.5 million hectares of forest globally (an area equivalent in size to Greece). This value exceeds the prior record, established in 2023, by a significant margin of 13% (Potapov et al., 2025).

This is not a temporary anomaly but a clear, long-term trajectory. A landmark report by the United Nations Environment Programme projects a global increase in extreme fires of up to 14% by 2030, 30% by 2050, and 50% by the end of the century (Sullivan et al., 2022). These fires are destroying ecosystems, threatening biodiversity, degrading soil and water resources, and releasing vast quantities of greenhouse gases that create a vicious feedback loop, further intensifying the very climate change that fuels them (Change, 2014).

The primary thesis of this chapter is that the potential for these technologies to mitigate the global wildfire crisis is fundamentally contingent upon surmounting three profound and interconnected challenges: scalability, accessibility, and interoperability. The promise of advancements in artificial intelligence (AI), remote sensing, and robotics is currently impeded by the considerable difficulty of deploying these solutions at the planetary scale required. Furthermore, their implementation risks establishing a bifurcated system of protection that marginalizes the most vulnerable populations. Finally, a persistent challenge is the failure of disparate technological systems and governmental agencies to achieve seamless operational synergy (Coughlan and Petty, 2012).

Advances in Remote Sensing for Forest Fire Management. DOI: https://doi.org/10.1016/B978-0-443-41692-7.00002-0
© 2027 Elsevier Ltd. All rights are reserved, including those for text and data mining, AI training, and similar technologies.

20.2 The global forest fire scenario: a system in danger

To comprehend the challenges and opportunities of technological integration, it is first necessary to understand the scale and dynamics of the problem space. The global wildfire scenario is no longer a collection of isolated regional events but a deeply interconnected system where climatic drivers, ecological impacts, and feedback loops operate on a planetary scale. The emergent paradigm is defined by an amplified risk regime, the expansion of pyrogenic landscapes, and a series of cascading effects that compromise the homeostatic integrity of foundational ecosystems (Change, 2014).

20.2.1 The intensifying fire regime: a data-driven global overview

The intensification of wildfire activity is a global phenomenon, but its characteristics depend on the region, driven by a combination of climate change, land-use patterns, and natural climate cycles. The world's northern high-latitude regions have become the epicenter of fire-related forest destruction. More than 60% of all fire-related tree cover loss between 2001 and 2024 occurred in boreal forests (Potapov et al., 2025). These regions are warming at a faster rate than the rest of the planet, a trend that directly contributes to longer fire seasons, greater fire frequency, and unprecedented severity (Post et al., 2019). Russia, for example, endured its three worst fire seasons on record between 2020 and 2024, with the 2021 season alone burning at least 5.4 million hectares, fueled by prolonged heatwaves that the Intergovernmental Panel on Climate Change (IPCC) has stated would be "practically impossible" without human-induced climate change (Potapov et al., 2025; Change, 2014). Similarly, Canada's 2023 fire season was the worst, with over 17 million hectares burned (more than triple the 10-year average), driven by higher-than-normal temperatures and persistent drought conditions (Hu, Yue and Tian, 2024; Jain et al., 2024) (Fig. 20.1).

While fire is a natural part of the boreal cycle, its increasing prevalence in tropical forests represents a dangerous new trend. These historically low-risk ecosystems are now experiencing a dramatic surge in fire activity, stoked by agricultural expansion, forest degradation, and climatic phenomena (Potapov et al., 2025; Pausas and Keeley, 2021). Over the last 24 years, fire-related tree cover loss in the tropics has increased at an average rate of 47,200 hectares per year, culminating in a peak in 2024 where over 4 million hectares were lost—more than the previous three years combined (Potapov et al., 2025). Natural climate cycles like El Niño, which cause high temperatures and below-average rainfall, heavily exacerbated the 2016 and 2024 fire seasons (Withey et al., 2018). In 2024, fires were responsible for nearly half (48%) of all tree cover loss in critical primary forests like the Amazon and Congo Basin, which are vital for global carbon storage and biodiversity. In Bolivia, agricultural expansion and drought led to over a million hectares of forest burned in 2024, a 114% increase over its previous record (Potapov et al., 2025).

This pattern demonstrates that extreme fire is no longer an issue confined to historically fire-prone regions like California or Australia. The emergence of the Arctic, tropical rainforests, and central Europe as major fire hotspots signifies a "globalization" of fire risk (Sullivan et al., 2022). This transformation has profound

Figure 20.1 Burned areas. Spanish Burned Areas 2024–25. Map lines delineate study areas and do not necessarily depict accepted national boundaries.

implications. An extreme fire season in Canada can degrade air quality thousands of miles away in Europe (Wang et al., 2024), and a drought in the Amazon can have cascading effects on global weather patterns (Marengo and Espinoza, 2016). Wildfire has thus evolved from a series of national concerns into an interconnected global security threat, demanding management frameworks that reflect this new reality.

20.2.2 Ecosystemic tipping points and cascading impacts

The damage caused by modern extreme wildfire events extends far beyond the immediate loss of trees. These events are pushing entire ecosystems toward irreversible tipping points, degrading essential natural resources, and accelerating the very climate change that ignites them.

The current fire regime is a primary driver of the global biodiversity crisis. A catastrophic 73% average decline in monitored wildlife populations since 1970, with extreme events like wildfires pushing countless species closer to extinction (Kelly et al., 2020). Extreme wildfire events cause an immediate loss of habitat, shelter, and food sources. The destruction of cavity trees and ground logs harms small mammals and nesting birds, while the loss of fruit trees leads to a decline in species that depend on them (Nasi et al., 2002). Not only in the Northern Hemisphere but also in the Southern Hemisphere, where in regions like Australia, extreme wildfire events were able to produce a 55% reduction in the 1380 species analyzed in the aftermath (Driscoll et al., 2024). These fires are not just isolated disasters; they are disrupting the fundamental structure and function of ecosystems, threatening to trigger tipping points, such as the collapse of the Amazon rainforest, with devastating consequences for global climate stability and biodiversity (Marengo and Espinoza, 2016).

A critical dynamic of the current fire regime is the self-perpetuating feedback loop between wildfires and climate change. As forests and peatlands burn, they release immense stores of carbon into the atmosphere, which exacerbates global warming and, in turn, creates hotter, drier conditions conducive to more frequent and intense fires (Lasslop et al., 2019). The scale of these emissions is staggering. The 2023 Canadian wildfires emitted over nine times the average carbon of the past two decades, contributing to almost a quarter of the entire globe's fire emissions for that period (Jones et al., 2024).

But not only during the wildfire, its impacts cascade through the environment long after the flames are extinguished. The intense heat from severe fires can fundamentally alter soil properties, combust organic matter, destroy the seed bank, and increase water repellency (Santín and Doerr, 2016). With vegetation cover removed, the charred landscape is highly susceptible to erosion. Postfire rainstorms can trigger flash floods and debris flows, carrying ash, sediment, nutrients, and pollutants into streams, rivers, and reservoirs (Vieira et al., 2015). This contamination degrades water quality for both ecosystems and human consumption, and the increased flood risk can persist for up to a decade or more until vegetation is sufficiently restored (Versini et al., 2013).

20.3 Technological advancements in wildfire management and resilience

In response to the new world context, a new generation of technologies is being developed and deployed across all phases of wildfire management (from prevention and prediction to detection, response, and recovery). These technological advancements are moving beyond traditional suppression tools to create an integrated ecosystem of sensors, data platforms, and intelligent systems designed to provide earlier warnings, greater situational awareness, and more precise interventions.

20.3.1 From detection to decision: the role of remote sensing and artificial intelligence

The fight against devastating wildfires has entered a new technological era, driven by significant advancements in satellite detection and monitoring. In response, a new generation of satellites, combined with AI, is providing firefighters and emergency managers with unprecedented tools to detect, track, and even predict wildfire behavior, marking a pivotal shift from reaction to proactive management.

A key evolution is the shift toward dedicated satellite constellations with higher temporal and spatial resolution. For years, we have relied on instruments like the Moderate Resolution Imaging Spectroradiometer (MODIS) and the Visible Infrared Imaging Radiometer Suite (VIIRS) on polar-orbiting satellites. While crucial, their revisit times over a specific area can be several hours to a day, a critical delay when a fire can explode in size. Similarly, geostationary satellites like the Geostationary

Operational Environmental Satellites (GOES) offer constant monitoring but at a lower resolution, making it difficult to spot smaller, incipient fires.

Constellations like the National Oceanic and Atmospheric Administration's (NOAA) GOES and Europe's Copernicus program offer continuous monitoring of entire continents (Wells et al., 2025; Dosiou et al., 2024). These satellites use instruments like the Advanced Baseline Imager to detect thermal anomalies, or hotspots, and measure their intensity via Fire Radiative Power (FRP), which helps estimate emissions and fire behavior. The experimental Next-Generation Fire System (NGFS) from NOAA represents a major leap forward, using AI algorithms to comb through satellite data and send geolocated alerts for new fires in as little as 1 minute from the time the heat signature reaches the satellite, detecting fires as small as a quarter of an acre (Wells et al., 2025; Delamere et al., 2025).

Another European example is related to the Meteosat Third Generation (MTG) system. Managed by EUMETSAT, this advanced series of geostationary meteorological satellites represents a quantum leap over its predecessors, offering far more powerful and precise tools for the early detection and monitoring of fires across Europe, Africa, and the Middle East (Holmlund et al., 2021).

The centerpiece of this new capability is the Flexible Combined Imager (FCI) instrument aboard the MTG-I (Imager) satellites. The FCI marks a massive improvement in three critical areas: temporal, spatial, and spectral resolution. While the previous Meteosat Second Generation (MSG) scanned the full Earth disc every 15 minutes, the FCI does so every 10 minutes. More importantly, it can perform a rapid scan of Europe (a quarter of the disc) every 2.5 minutes. This frequency drastically reduces the time it takes to detect a new fire outbreak, a crucial factor for a swift and effective response that can determine whether a fire is contained or becomes a disaster (Filizzola et al., 2025) (Fig. 20.2).

The second breakthrough lies in the integration of AI and machine learning with satellite data. Raw satellite images, particularly thermal data, can generate false positives from sources like industrial heat or reflective surfaces. Modern systems now use sophisticated AI algorithms to analyze imagery in near real time, cross-referencing it with historical data, weather patterns, and vegetation maps to confirm a fire's existence with high accuracy (Diaconu, 2023; Kukuk and Kilimci, 2021). NASA's Fire Information for Resource Management System (FIRMS) is now capable of delivering ultra-real-time data within 60 seconds of a satellite observation (Davies et al., 2019).

Furthermore, AI is not just detecting fires; it helps to model their future. By feeding real-time satellite data on fire location, intensity, FRP, and spread into predictive models, AI can forecast a fire's likely path (Preeti et al., 2021; Lattimer, Hodges and Lattimer, 2020). This allows authorities to make more strategic decisions about resource allocation, evacuation orders, and containment lines.

But AI is not only involved in the analysis and interpretation of satellite data; it is also applied in other disciplines, such as the early detection of forest fire outbreaks. With the growing network of surveillance cameras, from fixed towers in remote forests to urban-wildland interfaces, a vast network of digital eyes has been created (Heyns et al., 2019). By integrating advanced AI, particularly computer vision

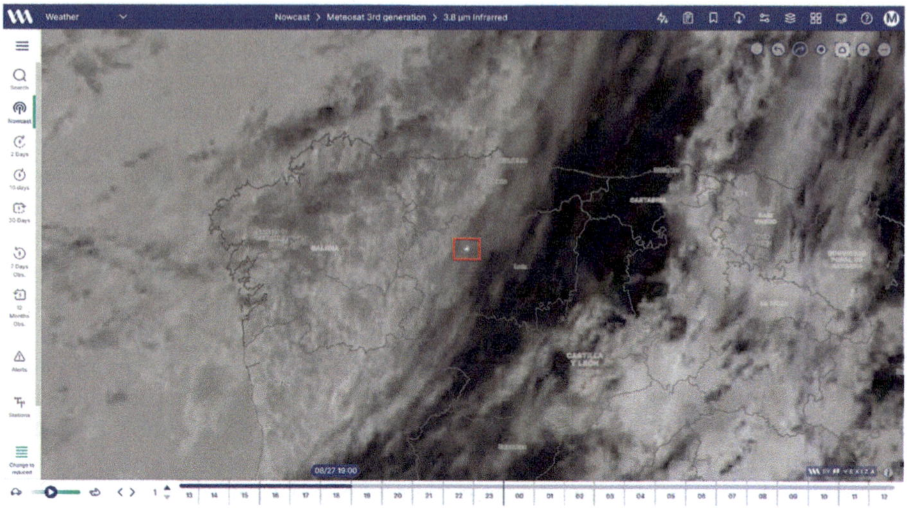

Figure 20.2 Meteosat satellite's 3.8 µm infrared channel. Presence of an active forest fire (indicated by the red box) in the northwest of the Iberian Peninsula.

algorithms, this existing infrastructure is being transformed into a proactive, round-the-clock fire watch system, enabling detection far earlier than traditional methods (Wei and Xu, 2022; Govil et al., 2020).

At the core of this technology are deep learning models, primarily convolutional neural networks (CNNs). These algorithms are trained on immense datasets containing thousands of images and video clips of various fire and smoke manifestations (Bahhar et al., 2023). This training allows the AI to learn and recognize the subtle, often dynamic characteristics of an incipient fire. It can distinguish the specific textural patterns, colors, and motion of smoke plumes as they rise above the canopy, or identify the faint flicker of a flame, even at a great distance or in challenging lighting conditions, like dawn or dusk (Almeida et al., 2022; Barmpoutis et al., 2019).

Several specialized AI architectures are applied for this task. Image classification models can analyze a camera's field of view and determine the probability that smoke or fire is present. More advanced object detection models, such as You Only Look Once (YOLO) (Casas et al., 2023) or Faster R-CNN (Barmpoutis et al., 2019), go a step further. They can pinpoint the exact location of the fire or smoke within the video frame by drawing a bounding box around it. This is critically important, as it provides immediate information on the fire's origin, reduces the rate of false alarms by ignoring irrelevant phenomena (like fog, dust, or moving vehicles), and allows automated camera systems to zoom in for more detailed verification.

To enhance accuracy, these AI systems are designed to analyze both spatial and temporal features. They do not just look at a single frame but analyze the sequence of frames to understand how a potential smoke plume is behaving (its growth, direction, and dispersion pattern), which is a key indicator of a real fire (Gunay et al., 2012; Li et al., 2018).

By automating the monitoring process, AI eliminates the limitations of human-led surveillance, which can suffer from fatigue and oversight. When the AI detects a high-probability event, it instantly sends an alert to fire response centers, complete with image captures, GPS coordinates, and other relevant data. This fusion of AI and surveillance cameras provides a powerful, scalable, and cost-effective solution, drastically reducing detection times and giving firefighters a possibility of containing a potential disaster before it escalates.

This is the role that AI plays in the field of forest fire detection, but what about prediction? It should be noted that anthropogenic factors play a decisive role in forest fire prediction, as most forest fires are caused by human activity (Menezes et al., 2022; Sari, 2023). However, despite this human influence being extremely complex to model, the inclusion of AI as a predictive agent has led to significant advances in this field. By analyzing a complex web of environmental variables, AI models can generate highly accurate forecasts on fire risk and spread (Fig. 20.3).

At its core, AI-driven prediction is an exercise in complex data fusion. Machine learning algorithms are fed vast amounts of diverse data, including:

- Meteorological data: Real-time and forecasted information on temperature, humidity, wind speed and direction, and lightning strike probability.
- Topographical data: Information about the landscape, such as slope, elevation, and aspect (the direction a slope faces), which significantly influences fire spread.
- Fuel data: Satellite-derived information on vegetation type, density, and, crucially, fuel moisture content. Metrics like the Normalized Difference Vegetation Index (NDVI) help AI gauge how dry and flammable an area is.

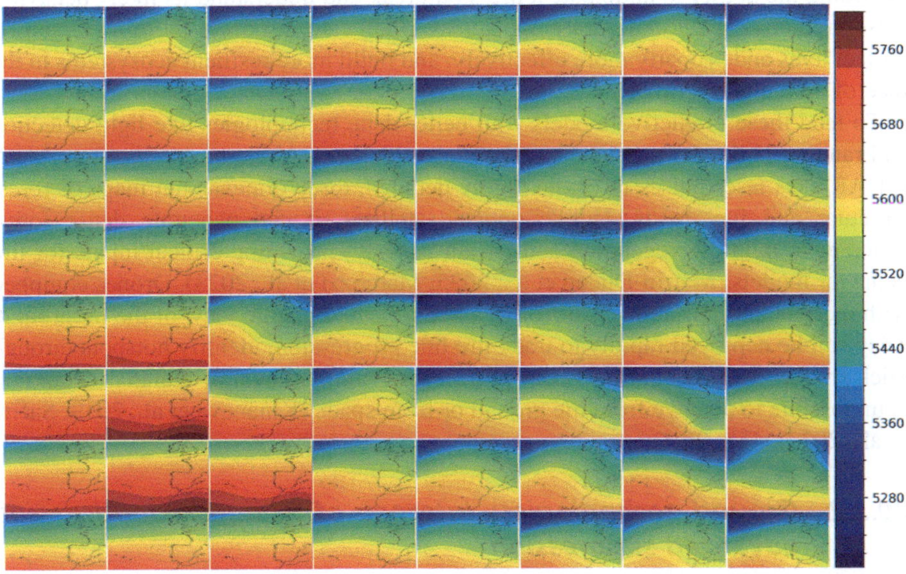

Figure 20.3 Artificial intelligence prediction. Unsupervised neural network. Map lines delineate study areas and do not necessarily depict accepted national boundaries.

- Historical fire data: Decades of records on where fires started in the past, their size, and the conditions under which they occurred.
- Human activity data: Proximity to roads, power lines, campsites, and populated areas, which are common ignition sources.

This data is used to train predictive models, primarily machine learning algorithms like random forests, gradient boosting machines, and deep neural networks (Andrianarivony and Akhloufi, 2024; Jain et al., 2020). These models are trained to recognize complex, nonlinear relationships among all these variables. They learn the subtle combination of conditions that historically preceded a major fire event, allowing them to identify high-risk patterns long before ignition occurs.

This predictive power manifests in two critical applications: fire risk assessment, where AI generates dynamic, forward-looking risk maps, often on a daily or even hourly basis. These maps highlight "hotspots" where the probability of ignition is highest. This allows fire management agencies to strategically preposition firefighters and equipment, issue targeted public warnings, and implement preventative measures like temporarily shutting down power lines in high-wind events.

And fire spread modeling, where in an active fire, AI-powered simulators predict its potential path and rate of spread over the coming hours and days. By continuously ingesting new weather data, these models can update their forecasts in near real time. This provides invaluable intelligence to incident commanders, helping them allocate resources effectively, plan safer firefighting strategies, and determine optimal evacuation routes for threatened communities.

In essence, AI is transforming wildfire management into a strategic science. By providing the ability to anticipate a fire's outbreak and understand its likely behavior, these predictive methodologies empower us to allocate resources more intelligently, protect lives and property more effectively, and build more resilient ecosystems in the face of a changing climate.

To feed these predictive models with better data, technologies like light detection and ranging (LiDAR) and unmanned aerial vehicles (UAVs), or drones, are crucial. LiDAR uses laser pulses to create highly detailed, three-dimensional maps of forest structure and fuel loads. Unlike optical satellites, it can penetrate dense forest canopies to map the vegetation and terrain beneath, providing critical inputs for fire behavior models (Martin-Ducup et al., 2025; Marcozzi et al., 2023). Drones equipped with thermal and infrared sensors offer a hyperlocal, dynamic view of the fireground. They can be deployed rapidly to track fire line progression, identify hotspots through thick smoke, and map burn severity postfire, providing a level of granular detail that is impossible to achieve from the ground or from space (Kinaneva et al., 2019; Saffre et al., 2022) (Fig. 20.4).

20.3.2 Ground technologies: Internet of Things, robotics, and advanced suppression

The Internet of Things (IoT) is revolutionizing wildfire management by providing real-time data and enabling faster, more effective responses to fire threats. This network of interconnected sensors, drones, and communication systems offers a

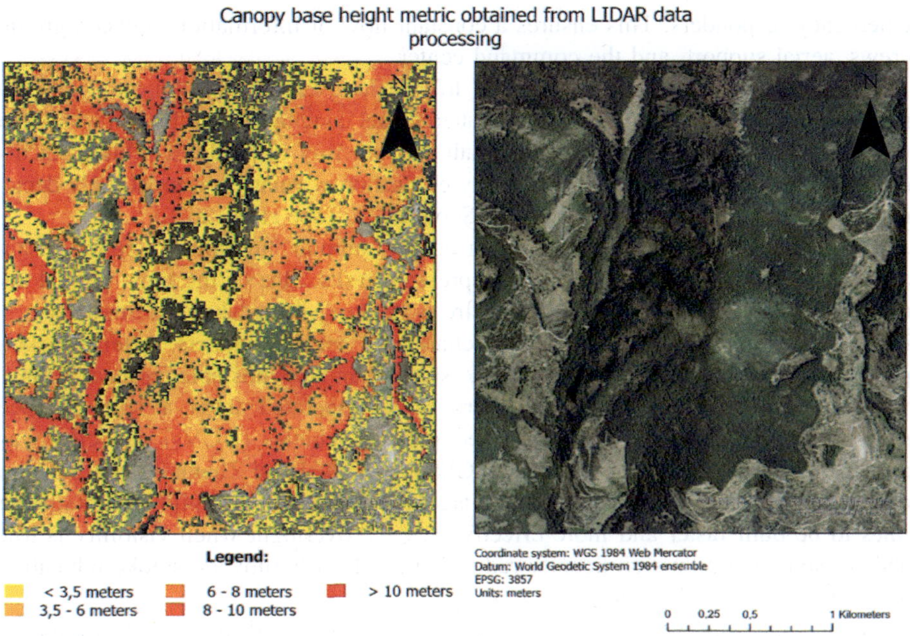

Figure 20.4 Light detection and ranging technology. Canopy base height.

significant advantage in detecting, monitoring, and ultimately mitigating the devastating impact of wildfires.

At the forefront of this technological shift is the use of distributed sensor networks. These networks consist of numerous low-power sensors strategically placed in fire-prone areas to monitor crucial environmental factors. These sensors continuously collect data on temperature, humidity, wind speed, and the presence of smoke or specific gases released during combustion. When a sensor detects anomalies that indicate a potential fire, it instantly transmits an alert to a central monitoring system. This allows for early detection, often before a fire has had the chance to grow and spread, which is critical for a swift and effective initial attack by firefighting crews (Verma et al., 2021; Kaur and Sood, 2019).

Drones equipped with thermal and infrared cameras are another key IoT application in wildfire management. They can be deployed to survey vast and often inaccessible terrains, providing a bird's-eye view of the landscape. This aerial surveillance helps in pinpointing the exact location of a fire, assessing its size and direction of spread, and identifying hotspots. The real-time video feeds from these drones provide invaluable situational awareness to incident commanders, enabling them to make informed decisions about resource allocation and firefighting strategies (Bushnaq, Chaaban and Al-Naffouri, 2021; Sethuraman et al., 2022).

Furthermore, IoT-based communication systems are crucial for coordinating a multiagency response. In remote areas where cellular coverage may be unreliable, IoT networks can establish a dedicated communication channel for firefighters and emergency responders. This ensures a constant flow of information between ground crews, aerial support, and the command center.

The data collected from these various IoT devices can also be fed into predictive models. By analyzing historical weather patterns, fuel conditions, and fire behavior, these models can help forecast the potential path and intensity of wildfires. This predictive capability allows authorities to issue timely evacuation warnings and proactively deploy resources to areas at high risk (Kaur and Sood, 2020; Zope et al., 2020).

To reduce the immense danger faced by human firefighters, robotics and automation are being developed for suppression tasks. One of the most critical and dangerous jobs in fighting a large wildfire is creating firebreaks. Traditionally, this involves crews of firefighters using chainsaws and bulldozers, often working perilously close to the flames, in thick smoke, and on unstable ground. This is where robotic bulldozers and mulchers shine. Operated from a safe distance, sometimes miles away, these unmanned ground vehicles (UGVs) can clear vast amounts of fuel like trees and brushes (Ghamry, Kamel and Zhang, 2016). This not only keeps firefighters out of the immediate danger zone but also allows containment lines to be built faster and more effectively, even overnight when visibility is low. This remote operation significantly reduces the risk of burnovers, smoke inhalation, and injuries from falling trees.

Beyond clearing vegetation, some UGVs are equipped for direct firefighting support. These robotic units can be outfitted with high-capacity water or fire-retardant hoses. They can be driven directly up to the fire's edge to cool hotspots or extinguish flames, all while the human operator remains at a safe command post (Gromek and Lowe, 2025). This is particularly valuable when dealing with volatile situations like vehicle fires within wildfires or protecting structures threatened by the advancing blaze. By sending in a robot, incident commanders avoid risking firefighter lives in situations with a high potential for sudden flare-ups or explosions (Ramasubramanian, Muthukumaraswamy and Sasikala, 2020).

20.4 The integration challenge: scalability, accessibility, and interoperability

While technology offers transformative potential, its deployment is fraught with systemic challenges that threaten to limit its impact. The journey from a promising pilot project to an effective tool is disabled by important barriers of scale, equity, and connectivity. These three challenges: scalability, accessibility, and interoperability; are not merely technical hurdles but are deeply intertwined with issues of finance, governance, and application.

20.4.1 Constraints on operational scalability

The operational scaling of advanced wildfire technologies from localized pilot studies to widespread, systemic deployment is impeded by several structural and systemic constraints. The transition from proof-of-concept to large-scale implementation faces significant financial, institutional, and human-capital barriers that collectively limit the effective reach of these innovations.

The first impediment is economic. The acquisition and maintenance of advanced technological systems (such as dense IoT sensor networks, AI-surveillance infrastructure, or autonomous aerial fleets) involve considerable expenditure of capital. These costs are superimposed upon public fire suppression budgets that are already escalating. Within such constrained fiscal environments, justifying large investments in novel technologies is challenging, particularly as comprehensive cost-benefit analyses are complicated by the difficulty of quantifying the long-term externalities of wildfires, such as public health impacts and the loss of ecosystem services. This financial pressure contributes to a significant research-to-operations gap, often termed the "valley of death" (Merceret et al., 2013), where promising technologies fail to secure the sustained funding required for operational deployment. Overcoming this requires innovative financing mechanisms, including public-private partnerships and strategic venture capital, to bridge the gap between initial validation and procurement.

Compounding these financial issues are institutional rigidity and human-capital deficits. Many fire management agencies operate within frameworks of established technological infrastructure and legacy systems, making the integration of novel, cloud-native platforms a complex technical and bureaucratic challenge (figure wildfire). Governmental procurement processes often exhibit risk aversion, which can stifle innovation by favoring incumbent technologies and disincentivizing private sector investment in next-generation systems. Furthermore, the effective utilization of these advanced tools is contingent upon a specialized workforce. A critical deficit exists not only in frontline responders but also in personnel skilled in data science, geographic information systems (GIS), and systems engineering. Without concurrent investment in targeted education and professional development to cultivate this expertise, the operational capacity of even the most sophisticated technological systems will remain fundamentally limited (McMahon, 2017).

20.4.2 Accessibility and equity in technology deployment

The operational deployment of advanced wildfire management technologies introduces significant challenges related to accessibility and equity. The high capital cost and selective implementation of these systems create a potential for the socio-economic stratification of technological protection, which can amplify preexisting social inequities.

A primary concern is that resource allocation for these technologies may be disproportionately correlated with property valuation rather than comprehensive social vulnerability indices. This can result in a de facto prioritization of affluent communities or high-value economic assets, while socially marginalized populations—often characterized by lower income, minority status, and language barriers—receive a lower standard of technological protection. Such a deployment strategy risks reinforcing systemic vulnerabilities, as these communities often possess fewer resources for private risk mitigation (e.g., home hardening) and may face greater barriers to evacuation and recovery.

Furthermore, the efficacy of technology-driven early warning and information systems is contingent upon equitable information access. The "digital divide"

Figure 20.5 Multiplatform. Postwildfire management multiplatform.

(disparities in internet availability, digital literacy, and linguistic accessibility) can create significant information asymmetries (Cullen, 2001). Emergency alerts and dynamic risk maps disseminated through digital platforms may fail to reach populations lacking reliable connectivity or proficiency in the dominant language, as observed in past fire events.

Addressing these accessibility challenges necessitates a fundamental shift in governance and resource allocation frameworks. Mitigation strategies require the formal integration of social vulnerability metrics into technology deployment models to ensure a more equitable distribution of protective resources. This must be coupled with the development of governance structures that actively involve representatives from marginalized communities in planning and decision-making processes. A truly accessible system must also employ a multichannel communication strategy, incorporating low-tech and multilingual dissemination methods to overcome the digital divide and ensure critical information reaches all segments of the at-risk population.

20.4.3 *Systemic interoperability deficits*

A critical impediment to effective, multiagency wildfire response is the systemic lack of interoperability across communication and data platforms. This deficit manifests as a heterogeneous technological ecosystem that degrades situational awareness and compromises operational coordination and safety.

At the tactical level, interoperability failures are evident in voice communications. Responding agencies from different jurisdictions (e.g., federal, state, local) frequently operate on disparate radio frequency protocols, precluding direct, real-time communication between field units. This necessitates the use of inefficient and often unreliable intermediaries, creating significant delays and increasing safety risks for personnel (Fig. 20.5).

At the strategic level, the problem is one of data fragmentation. The proliferation of proprietary technologies has resulted in vertically integrated, noninteroperable data architectures. Data streams from disparate sources (such as remote sensing platforms, IoT sensor networks, and predictive models) often use incompatible formats and semantics, preventing their aggregation into a unified common operational picture (Wolbers and Boersma, 2013). This fragmentation denies incident commanders a holistic view of the operational environment, forcing decisions based on incomplete information. The inherent heterogeneity of IoT devices and platforms further complicates data integration, requiring standardized data models and semantic frameworks to achieve true interoperability.

While technical solutions such as communication gateways and the adoption of open data standards are emerging to mitigate these issues, the root cause is often not technological but institutional. These interoperability deficits are frequently a direct consequence of fragmented governance structures, siloed procurement processes, and a lack of mandated cross-jurisdictional standards. Therefore, achieving a fully interoperable system requires not only technological innovation but also a fundamental commitment to collaborative governance and standardized operational protocols.

20.5 Conclusions

The global wildfire crisis represents a defining challenge of our time. The evidence is unequivocal: fires are becoming more frequent, more intense, and more destructive, driven by a warming climate and creating feedback loops that threaten to push critical ecosystems past points of no return. In response, a powerful and sophisticated suite of new technologies offers a beacon of hope, promising to revolutionize our ability to predict, detect, and combat these blazes.

Yet, this chapter has demonstrated that the very tools that could help save us can be constrained by systemic failures of our own making. The transformative potential of AI, satellites, and robotics is being fundamentally throttled by the interconnected challenges of scalability, accessibility, and interoperability. The path from a promising innovation to an effective, planet-wide solution is sometimes blocked by prohibitive costs, institutional inertia, deep-seated social inequities, and a fragmented governance landscape that promotes silos instead of synergy.

To navigate this new context in fire management, a paradigm shift is required. We must move beyond a purely reactive, suppression-focused mindset that has proven increasingly futile against the scale of the modern crisis. This new paradigm must:

- Embrace technology strategically, guiding its development and deployment with coherent, well-funded national policies that foster innovation while ensuring accountability.
- Prioritize equity relentlessly, ensuring that the benefits of these powerful new tools are used to protect the most vulnerable communities, not just the most valuable assets, thereby breaking the cycle of marginalization and disaster.
- Demand interoperability as a nonnegotiable standard, breaking down the institutional and technical silos that have long hampered effective response and building a truly collaborative, multiagency system where seamless communication is the norm, not the exception.

- Integrate knowledge holistically, weaving the precision and scale of modern science with the deep, place-based wisdom of Indigenous communities that have successfully managed fire-prone landscapes for millennia.
- The ultimate challenge is not simply to build better tools to fight fire. It is to build more resilient, equitable, and collaborative societies that can learn to live with it, using the full spectrum of human ingenuity—from the AI algorithm to the cultural burn—to steward our planet through the flames.

AI disclosure

During the preparation of this work, the author (s) used Gemini in order to improve English writing skills. After using this tool/service, the author (s) reviewed and edited the content as needed and take (s) full responsibility for the content of the publication.

References

Almeida, J.S., Huang, C., Nogueira, F.G., Bhatia, S., De Albuquerque, V.H.C., 2022. EdgeFireSmoke: a novel lightweight CNN Model for real-time video fire-smoke detection. IEEE Transactions on Industrial Informatics 18 (11), 7889–7898. Available at https://doi.org/10.1109/TII.2021.3138752, http://ieeexplore.ieee.org/xpl/RecentIssue.jsp?punumber=9424.

Andrianarivony, H.S., Akhloufi, M.A., 2024. Machine Learning and deep learning for wildfire spread prediction: a review. Fire 7 (12). Available at https://doi.org/10.3390/fire7120482, www.mdpi.com/journal/fire.

Bahhar, C., Ksibi, A., Ayadi, M., Jamjoom, M.M., Ullah, Z., Soufiene, B.O., et al., 2023. Wildfire and smoke detection using staged YOLO model and ensemble CNN. Electronics 12 (1), 228. Available at https://doi.org/10.3390/electronics12010228.

Barmpoutis, P., Dimitropoulos, K., Kaza, K., Grammalidis, N., 2019. Fire detection from images using faster R-CNN and multidimensional texture analysis ICASSP, IEEE International Conference on Acoustics, Speech and Signal Processing - Proceedings 8301 8305 2019. https://doi.org/10.1109/ICASSP.2019.8682647.

Bushnaq, O.M., Chaaban, A., Al-Naffouri, T.Y., 2021. The role of UAV-IoT networks in future wildfire detection. IEEE Internet of Things Journal 8 (23), 16984–16999. Available at https://doi.org/10.1109/JIOT.2021.3077593, http://ieeexplore.ieee.org/servlet/opac?punumber=6488907.

Casas, E., Ramos, L., Bendek, E., Rivas-Echeverría, F., 2023. Assessing the effectiveness of YOLO architectures for smoke and wildfire detection. IEEE Access 11, 96554–96583. Available at https://doi.org/10.1109/access.2023.3312217.

I.C. Change, 2014. Impacts, adaptation and vulnerability. Part A: global and sectoral aspects. Contribution of working group II to the fifth assessment report of the intergovernmental Panel on Climate Change, 1132.

Coughlan, M.R., Petty, A.M., 2012. Linking humans and fire: a proposal for a transdisciplinary fire ecology. International Journal of Wildland Fire 21 (5), 477–487. Available at https://doi.org/10.1071/WF11048.

Cullen, R., 2001. Addressing the digital divide. Online Information Review 25 (5), 311–320. Available at https://doi.org/10.1108/14684520110410517.

Davies, D., Ederer, G., Olsina, M., Wong, M., Cechini, R., 2019. Boller, NASA's fire information for resource management system (FIRMS): near real-time global fire monitoring using data from MODIS and VIIRS. EARSel Forest Fires SIG Workshop.

Delamere, J., Cable, J., Dierking, C., Larson, O., Peterson, M., Stream, B., et al., 2025. Engaging JPSS for wildland fire management in Alaska: past, present, and future. 105th Annual AMS Meeting 2025. 105, 453865.

Diaconu, B.M., 2023. Recent advances and emerging directions in fire detection systems based on machine learning algorithms. Romania Fire 6 (11). Available at https://doi.org/10.3390/fire6110441, www.mdpi.com/journal/fire.

Dosiou, A., Athinelis, I., Katris, E., Vassalou, M., Kyrkos, A., Krassakis, P., et al., 2024. Employing copernicus land service and sentinel-2 satellite mission data to assess the spatial dynamics and distribution of the extreme forest fires of 2023 in greece. Fire 7 (1), 20. Available at https://doi.org/10.3390/fire7010020.

Driscoll, D.A., Macdonald, K.J., Gibson, R.K., Doherty, T.S., Nimmo, D.G., Nolan, R.H., et al., 2024. Biodiversity impacts of the 2019–2020 Australian megafires. Nature 635 (8040), 898–905. Available at https://doi.org/10.1038/s41586-024-08174-6, https://www.nature.com/nature/.

Filizzola, C., Mazzeo, G., Marchese, F., Pietrapertosa, C., Pergola, N., 2025. The contribution of meteosat third generation–flexible combined imager (MTG-FCI) observations to the monitoring of thermal volcanic activity: the mount etna (Italy) february–march 2025 Eruption. Remote Sensing 17 (12), 2102. Available at https://doi.org/10.3390/rs17122102.

Ghamry, K.A., Kamel, M.A., Zhang, Y., 2016. International Conference on Unmanned Aircraft Systems, ICUAS Institute of Electrical and Electronics Engineers Inc. Canada Cooperative forest monitoring and fire detection using a team of UAVs-UGVs, pp. 1206–1211. Available at https://doi.org/10.1109/ICUAS.2016.7502585.

Govil, K., Welch, M.L., Ball, J.T., Pennypacker, C.R., 2020. Preliminary results from a wildfire detection system using deep learning on remote camera images. Remote Sensing 12 (1). Available at https://doi.org/10.3390/RS12010166, https://res.mdpi.com/d_attachment/remotesensing/remotesensing-12-00166/article_deploy/remotesensing-12-00166-v2.pdf.

Gromek, P., Lowe, T., 2025. Ground robot technologies in wildfire risk reduction. The viewpoint of the fire service. Progress in Disaster Science 26, 100435. Available at https://doi.org/10.1016/j.pdisas.2025.100435.

Gunay, O., Toreyin, B.U., Kose, K., Cetin, A.E., 2012. Entropy-functional-based online adaptive decision fusion framework with application to wildfire detection in video. IEEE Transactions on Image Processing 21 (5), 2853–2865. Available at https://doi.org/10.1109/tip.2012.2183141.

Heyns, A., Du Plessis, W., Kosch, M., Hough, G., 2019. Optimisation of tower site locations for camera-based wildfire detection systems. International Journal of Wildland Fire 28 (9), 651–665. Available at https://doi.org/10.1071/WF18196, http://www.publish.csiro.au/nid/126.htm.

Holmlund, K., Grandell, J., Schmetz, J., Stuhlmann, R., Bojkov, B., Munro, R., et al., 2021. Meteosat third generation (MTG): continuation and innovation of observations from geostationary orbit. Bulletin of the American Meteorological Society 102 (5), E990. Available at https://doi.org/10.1175/bams-d-19-0304.1.

Hu, Y., Yue, X., Tian, C., 2024. Climatic drivers of the Canadian wildfire episode in 2023. Atmospheric and Oceanic Science Letters 17 (4), 100483. Available at https://doi.org/10.1016/j.aosl.2024.100483.

Jain, P., Coogan, S.C.P., Subramanian, S.G., Crowley, M., Taylor, S., Flannigan, M.D., 2020. A review of machine learning applications in wildfire science and management. Environmental Reviews 28 (4), 478–505. Available at https://doi.org/10.1139/er-2020-0019, https://www.nrcresearchpress.com/loi/er.

Jain, P., Barber, Q.E., Taylor, E., Whitman, D.C., Acuna, Y., Boulanger, M.A., 2024. Parisien. Canada Under Fire–Drivers and Impacts of the Record-Breaking 2023 Wildfire Season.

Jones, M.W., Kelley, D.I., Burton, C.A., Di Giuseppe, F., Barbosa, M.L.F., Brambleby, E., et al., 2024. State of wildfires 2023-2024. Earth System Science Data 16 (8), 3601–3685. Available at https://doi.org/10.5194/essd-16-3601-2024, www.earth-system-science-data.net.

Kaur, H., Sood, S.K., 2019. Fog-assisted IoT-enabled scalable network infrastructure for wildfire surveillance. Journal of Network and Computer Applications 144, 171–183. Available at https://doi.org/10.1016/j.jnca.2019.07.005, http://www.elsevier.com/inca/publications/store/6/2/2/8/9/3/index.htt.

Kaur, H., Sood, S.K., 2020. Energy-efficient IoT-fog-cloud architectural paradigm for real-time wildfire prediction and forecasting. IEEE Systems Journal 14 (2), 2003–2011. Available at https://doi.org/10.1109/JSYST.2019.2923635, https://ieeexplore.ieee.org/xpl/mostRecentIssue.jsp?punumber=4267003.

Kelly, L.T., Giljohann, K.M., Duane, A., Aquilué, N., Archibald, S., Batllori, E., et al., 2020. Fire and biodiversity in the Anthropocene. Science (New York, N.Y.) 370 (6519). Available at https://doi.org/10.1126/science.abb0355.

Kinaneva, D., Hristov, G., Raychev, J., Zahariev, P., 2019. 42nd International Convention on Information and Communication Technology, Electronics and Microelectronics, MIPRO 2019 - Proceedings Institute of Electrical and Electronics Engineers Inc. Bulgaria Early forest fire detection using drones and artificial intelligence, pp. 1060–1065. Available at http://ieeexplore.ieee.org/xpl/mostRecentIssue.jsp?punumber=8747288, https://doi.org/10.23919/MIPRO.2019.8756696.

Kukuk, S.B., Kilimci, Z.H., 2021. Comprehensive analysis of forest fire detection using deep learning models and conventional machine learning algorithms. International Journal of Computational and Experimental Science and Engineering 7 (2), 84–94. Available at https://doi.org/10.22399/ijcesen.950045.

Lasslop, G., Coppola, A.I., Voulgarakis, A., Yue, C., Veraverbeke, S., 2019. Influence of fire on the carbon cycle and climate. Current Climate Change Reports 5 (2), 112–123. Available at https://doi.org/10.1007/s40641-019-00128-9, https://rd.springer.com/journal/40641.

Lattimer, B.Y., Hodges, J.L., Lattimer, A.M., 2020. Using machine learning in physics-based simulation of fire. Fire Safety Journal 114, 102991. Available at https://doi.org/10.1016/j.firesaf.2020.102991.

Li, X., Chen, Z., Wu, Q.J., Liu, C., 2018. 3D parallel fully convolutional networks for real-time video wildfire smoke detection. IEEE Transactions on Circuits and Systems for Video Technology 30 (1), 89–103.

Marcozzi, A.A., Johnson, J.V., Parsons, R.A., Flanary, S.J., Seielstad, C.A., Downs, J.Z., 2023. Application of LiDAR derived fuel cells to wildfire modeling at laboratory scale. Fire 6 (10). Available at https://doi.org/10.3390/fire6100394, www.mdpi.com/journal/fire.

Marengo, J.A., Espinoza, J.C., 2016. Extreme seasonal droughts and floods in Amazonia: causes, trends and impacts. International Journal of Climatology 36 (3), 1033–1050. Available at https://doi.org/10.1002/joc.4420.

Martin-Ducup, O., Dupuy, J.L., Soma, M., Guerra-Hernandez, J., Marino, E., Fernandes, P.M., et al., 2025. Unlocking the potential of airborne LiDAR for direct assessment of fuel bulk density and load distributions for wildfire hazard mapping. Agricultural and Forest

Meteorology 362. Available at https://doi.org/10.1016/j.agrformet.2024.110341, https://www.sciencedirect.com/science/journal/01681923.

McMahon, G.T., 2017. The leadership case for investing in continuing professional development. Academic Medicine 92 (8), 1075–1077. Available at https://doi.org/10.1097/ACM.0000000000001619, http://www.lww.com/product/?1040-2446.

Menezes, L.S., de Oliveira, A.M., Santos, F.L.M., Russo, A., de Souza, R.A.F., Roque, F.O., et al., 2022. Lightning patterns in the Pantanal: untangling natural and anthropogenic-induced wildfires. Science of the Total Environment 820. Available at https://doi.org/10.1016/j.scitotenv.2022.153021, www.elsevier.com/locate/scitotenv.

Merceret, F.J., O'Brien, T.P., Roeder, W.P., Huddleston, L.L., Bauman, W.H., Jedlovec, G.J., 2013. Transitioning research to operations: transforming the "valley of death" into a "valley of opportunity. Space Weather 11 (11), 637–640. Available at https://doi.org/10.1002/swe.20099.

Nasi, R., Dennis, R., Meijaard, E., Applegate, G., Moore, P., 2002. Forest fire and biological diversity. Unasylva 53 (209), 36–40.

Pausas, J.G., Keeley, J.E., 2021. Wildfires and global change. Frontiers in Ecology and the Environment 19 (7), 387–395. Available at https://doi.org/10.1002/fee.2359, http://onlinelibrary.wiley.com/journal/10.1002/(ISSN)1540-9309.

Post, E., Alley, R.B., Christensen, T.R., Macias-Fauria, M., Forbes, B.C., Gooseff, M.N., et al., 2019. The polar regions in a 2 C warmer world. Science advances 5 (12), eaaw9883.

Potapov, P., Tyukavina, A., Turubanova, S., Hansen, M.C., Giglio, L., Hernandez-Serna, A., et al., 2025. Unprecedentedly high global forest disturbance due to fire in 2023 and 2024. Proceedings of the National Academy of Sciences of the United States of America 122 (30). Available at https://doi.org/10.1073/pnas.2505418122.

Preeti, T., Kanakaraddi, S., Beelagi, A., Malagi, S., Sudi, A., 2021. Fire Prediction Using Machine Learning Techniques International Conference on Intelligent Technologies, CONIT Institute of Electrical and Electronics Engineers Inc. India Forest. Available at http://ieeexplore.ieee.org/xpl/mostRecentIssue.jsp?punumber=9497779, https://doi.org/10.1109/CONIT51480.2021.9498448.

Ramasubramanian, S., Muthukumaraswamy, S.A., Sasikala, A., 2020. Fire Detection using Artificial Intelligence for Fire-Fighting Robots Proceedings of the International Conference on Intelligent Computing and Control Systems, ICICCS 2020 Institute of Electrical and Electronics Engineers Inc. United Arab Emirates, pp. 180–185. Available at http://ieeexplore.ieee.org/xpl/mostRecentIssue.jsp?punumber=9118609, https://doi.org/10.1109/ICICCS48265.2020.9121017.

Saffre, F., Hildmann, H., Karvonen, H., Lind, T., 2022. Monitoring and cordoning wildfires with an autonomous swarm of unmanned aerial vehicles. Drones 6 (10), 301. Available at https://doi.org/10.3390/drones6100301.

Santín, C., Doerr, S.H., 2016. Fire effects on soils: the human dimension. Philosophical Transactions of the Royal Society B: Biological Sciences 371 (1696). Available at https://doi.org/10.1098/rstb.2015.0171, http://rstb.royalsocietypublishing.org/content/371/1696/20150171.full.pdf.

Sari, F., 2023. Identifying anthropogenic and natural causes of wildfires by maximum entropy method-based ignition susceptibility distribution models. Journal of Forestry Research 34 (2), 355–371. Available at https://doi.org/10.1007/s11676-022-01502-4.

Sethuraman, S.C., Tadkapally, G.R., Mohanty, S.P., Subramanian, A., 2022. iDrone: IoT-enabled unmanned aerial vehicles for detecting wildfires using convolutional neural networks. SN Computer Science 3 (3). Available at https://doi.org/10.1007/s42979-022-01160-7, https://www.springer.com/journal/42979.

Sullivan, A., Baker, E., Kurvits, T., Popescu, A., Paulson, A.K., Christianson, A.C., et al., 2022. Spreading like wildfire: The rising threat of extraordinary landscape fires.

Verma, S., Kaur, S., Rawat, D.B., Xi, C., Alex, L.T., Zaman Jhanjhi, N., 2021. Intelligent framework using IoT-based WSNs for wildfire detection. IEEE Access 9, 48185–48196. Available at https://doi.org/10.1109/ACCESS.2021.3060549, https://ieeexplore.ieee.org/xpl/RecentIssue.jsp?punumber=6287639.

Versini, P.-A., Velasco, M., Cabello, A., Sempere-Torres, D., 2013. Hydrological impact of forest fires and climate change in a Mediterranean basin. Natural Hazards 66 (2), 609–628. Available at https://doi.org/10.1007/s11069-012-0503-z.

Vieira, D.C.S., Fernández, C., Vega, J.A., Keizer, J.J., 2015. Does soil burn severity affect the post-fire runoff and interrill erosion response? A review based on meta-analysis of field rainfall simulation data. Journal of Hydrology 523, 452–464. Available at https://doi.org/10.1016/j.jhydrol.2015.01.071.

Wang, Z., Wang, Z., Zou, Z., Chen, X., Wu, H., Wang, W., et al., 2024. Severe global environmental issues caused by Canada's record-breaking wildfires in 2023. Advances in Atmospheric Sciences 41 (4), 565–571. Available at https://doi.org/10.1007/s00376-023-3241-0, https://www.springer.com/journal/376.

Wei, C., Xu, J., Li, Q., Jiang, S., 2022. An intelligent wildfire detection approach through cameras based on deep learning. Sustainability 14 (23), 15690. Available at https://doi.org/10.3390/su142315690.

Wells, E.M., Hatchett, B.J., Thiem, K., Tolby, Z., Hoekstra, S., Vickery, J., et al., 2025. NOAA fire weather testbed launches first in-person evaluation. Bulletin of the American Meteorological Society 106 (6), 359–363. Available at https://doi.org/10.1175/bams-d-24-0204.1.

Withey, K., Berenguer, E., Palmeira, A.F., Espírito-Santo, F.D.B., Lennox, G.D., Silva, C.V.J., et al., 2018. Quantifying immediate carbon emissions from El Niño-mediated wildfires in humid tropical forests. Philosophical Transactions of the Royal Society B: Biological Sciences 373 (1760). Available at https://doi.org/10.1098/rstb.2017.0312, http://rstb.royalsocietypublishing.org/.

Wolbers, J., Boersma, K., 2013. The common operational picture as collective sensemaking. Journal of Contingencies and Crisis Management 21 (4), 186–199. Available at https://doi.org/10.1111/1468-5973.12027.

Zope, V., Dadlani, T., Matai, A., Tembhurnikar, P., Kalani, R., 2020. IoT Sensor and Deep Neural Network based Wildfire Prediction System Proceedings of the International Conference on Intelligent Computing and Control Systems, ICICCS 2020 Institute of Electrical and Electronics Engineers Inc. India, pp. 205–208. Available at http://ieeexplore.ieee.org/xpl/mostRecentIssue.jsp?punumber=9118609 10.1109/ICICCS48265.2020.9120949.

A vision for the future of forest fire management: technology-enabled, nature-based solutions for adaptive and proactive strategies

21

Adrián Regos[1], Ana Novo[1], and João Gonçalves[2]
[1]Misión Biolóxica de Galicia, Consejo Superior de Investigaciones Científicas, Av. de Vigo, Santiago de Compostela, Spain, [2]BIOPOLIS Program in Genomics, Biodiversity and Land Planning, CIBIO, Centro de Investigação em Biodiversidade e Recursos Genéticos, InBIO Laboratório Associado, Vairão, Portugal

21.1 Introduction

Escalating wildfire activity in many regions of the world—driven by a changing climate, land-use legacies, rural abandonment, and the expansion of wildland–urban interfaces—has exposed the limits of predominantly reactive, suppression-centric approaches to fire management (Moreira et al., 2020). Longer fire seasons, more frequent extreme fire-weather events, and the increasing incidence of large, fast-moving wildfires have increased risks to people, ecosystems, infrastructure, and economies (Batllori et al., 2013). At the same time, there is a broad scientific consensus that fire is a natural ecological process in many biomes (McLauchlan et al., 2020). The central challenge, therefore, is not to eliminate fire, but to manage its regime—its frequency, intensity, severity, seasonality, and spatial pattern—so that it aligns with ecological functioning, ecosystem resilience (sensu [Johnstone et al., 2016]), and societal goals (Plumanns-Pouton et al., 2025; Kelly et al., 2020).

This chapter advances a vision for adaptive, proactive fire management, in which technology acts as an enabler rather than an end in itself, and where nature-based solutions (NbS)—including ecological restoration, agro-silvo-pastoral mosaics, green firebreaks, and culturally informed fire use—constitute the backbone of risk reduction and landscape resilience (Regos, 2022, 2026). We argue that the coming decade should prioritize the integration of scientific/biophysical observations and evidence, models, decision support, and governance into an iterative cycle that anticipates risk, acts before crises, and learns systematically from outcomes.

Technological capabilities have matured rapidly. Multisensor Earth observation (e.g., optical and thermal satellites, light detection and ranging [LiDAR], synthetic aperture radar [SAR]), constellations of small satellites, uncrewed aerial vehicles, in situ Internet of Things (IoT) sensor networks, and high-resolution fuel mapping now

Advances in Remote Sensing for Forest Fire Management. DOI: https://doi.org/10.1016/B978-0-443-41692-7.00022-6
© 2027 Elsevier Ltd. All rights reserved, including those for text and data mining, AI training, and similar technologies.

provide near-real-time situational awareness at unprecedented spatial and temporal scales (Fernández-Guisuraga et al., 2024; Chuvieco et al., 2020; Castro et al., 2025). Advances in machine learning and statistical modeling enable dynamic fire-danger forecasting, ignition likelihood mapping, early smoke detection, and spread simulations that integrate weather, topography, fuel structure, and human activity (Ziegler et al., 2019; Bisquert et al., 2014; Rothermel, 1972; Rossa and Fernandes, 2018; Vacchiano et al., 2018). Meanwhile, landscape-scale simulation platforms and digital twins are beginning to test "what-if" scenarios—exploring how alternative fuel treatments, land-use transitions, or restoration strategies would change future fire behavior and ecological outcomes (Sil et al., 2022; Pais et al., 2023; Millington et al., 2009; He et al., 2004; Duane et al., 2016; Duguy et al., 2007; Li et al., 2025). Importantly, these analytics are increasingly delivered through decision-support systems (DSS) usable by civil protection agencies, forest managers, municipalities, and communities (Martell, 2015).

Technology is a necessary driver, but it cannot resolve the structural drivers of wildfire risk. In fire-prone socio-ecological systems, fuels accumulate where agricultural use declines; continuity of flammable vegetation expands; and homogeneous plantations or dense shrublands connect headfires across large spatial extents. NbS directly address these root causes by reintroducing heterogeneity: restoring riparian corridors and wetlands that act as humid barriers; converting strategic zones to higher-moisture or lower-flammability vegetation; promoting silvopastoral systems and targeted grazing to reduce fine fuels; and supporting climate-smart agroforestry (e.g., chestnut groves, mixed orchards, vineyards) that combine economic value with fuel breaks and biodiversity benefits (Plumanns-Pouton et al., 2025; Regos et al., 2023; Ascoli et al., 2023). When planned with clear ecological objectives and equity considerations, NbS yield cobenefits in carbon storage, soil and water regulation, pollination, rural livelihoods, and cultural landscapes—strengthening social-ecological benefits for long-term fire risk governance (see e.g., Pais et al., 2020; Lecina-Diaz et al., 2023b).

We propose a simple organizing framework—the 4 A cycle—to articulate how technology and NbS interact across the full management continuum (Fig. 21.1):

1. **Assess**: Map fuels, vegetation structure, and ecosystem condition; identify values at risk; characterize social vulnerability and exposure.
2. **Anticipate**: Forecast fire weather and ignition likelihood; simulate fire spread and impacts under alternative scenarios; prioritize prevention (Tedim et al., 2016) and preparedness.
3. **Act**: Implement NbS and fuel management portfolios (restoration, grazing, prescribed fire, and green firebreaks).
4. **Adapt**: Monitor outcomes with remote sensing and field data; evaluate ecological responses and social acceptance; update policies and investments through adaptive management.

In this framework, technology supplies the data, models, and interfaces that make the cycle actionable and measurable. NbS provide the on-the-ground mechanisms and policy basis that foster measures and actions to reshape fuels and restore ecosystem function. Governance binds them together through coherent policy, sustainable funding, and participatory decision-making. The result is a "fire-smart" territory—one

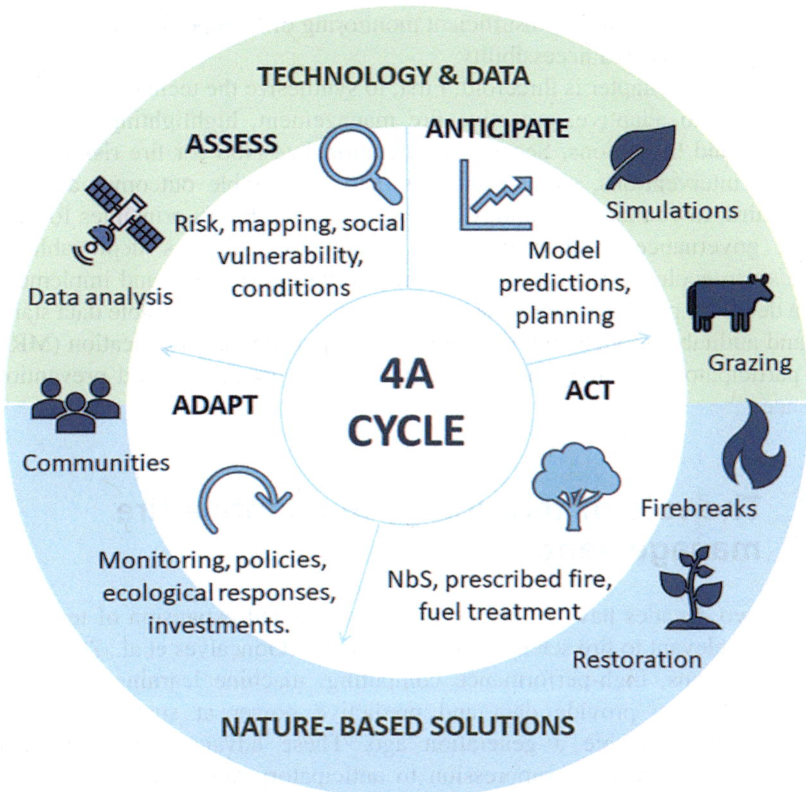

Figure 21.1 Conceptual framework illustrating the 4 A cycle (Anticipate, Assess, Act, Adapt) that integrates technology and nature-based solutions (NbS) across the full wildfire management continuum—from prevention and preparedness to response and recovery. The framework highlights how technological innovation supports each phase of adaptive, fire-smart landscape management, fostering continuous learning and resilience building.

whose structure, uses, and institutions dampen the probability of extreme events and reduce their consequences when they occur (Tedim et al., 2016).

This vision is not abstract. Emerging practices across Mediterranean and Atlantic–Mediterranean regions, Australia, North America, and Latin America show that combining cultural burning, ecological restoration, and strategic mosaics with risk analytics and DSS can reduce high-intensity fire spread, protect biodiversity, and sustain rural economies (Lecina-Diaz et al., 2023b; Mansuy et al., 2025; Mariani et al., 2022). European initiatives and national programs increasingly emphasize multi-sectoral policy coherence (e.g., aligning climate, agriculture, biodiversity, and civil protection) and place-based strategies that integrate local knowledge with scientific evidence (Ascoli et al., 2023; Plana et al., 2024). Nevertheless, significant barriers remain, including fragmented land tenure, short political cycles, limited societal acceptance of prescribed fire, insufficient monitoring of NbS performance, and gaps in data interoperability and accessibility.

The aim of this chapter is threefold. First, to synthesize the technological landscape most relevant to adaptive, proactive fire management, highlighting both practical capabilities and limitations. Second, to operationalize NbS for fire risk reduction—specifying interventions, siting principles, and measurable outcomes at landscape scales. Third, to connect technology and NbS through design principles for decision support, governance mechanisms, and investment pathways deployable across diverse socio-ecological contexts. We conclude with a research and implementation agenda detailing priorities for the next decade, including interoperable data standards, open and auditable models, robust monitoring, reporting, and verification (MRV) for NbS, participatory planning, and financing instruments that reward prevention and resilience.

21.2 The role of technology in adaptive fire management

The past two decades have witnessed an unprecedented expansion of technological capabilities relevant to fire science and management (Gonçalves et al., 2025). Remote sensing platforms, high-performance computing, machine learning, and decision-support tools now provide data and predictive power at spatial and temporal resolutions unimaginable a generation ago. These advances open the door to transitioning from reactive suppression to anticipatory, adaptive fire management, where managers can detect ignitions earlier, simulate alternative scenarios, optimize prevention and suppression resources, and monitor the effectiveness of interventions. However, the actual value of technology lies not in the tools themselves, but in how they are embedded within governance systems, combined with NbS, and translated into actionable strategies that shape fire regimes.

21.2.1 Earth observation and remote sensing

Earth observation technologies have transformed our capacity to monitor fire-prone landscapes across multiple spatial and temporal scales. Traditional coarse-resolution satellites such as the Moderate Resolution Imaging Spectroradiometer (MODIS) and the Visible Infrared Imaging Radiometer Suite (VIIRS) remain indispensable for global fire detection and burned-area mapping, providing consistent, near-daily coverage that supports fire-danger indices and emissions accounting (Bisquert et al., 2014; Giglio et al., 2003; Giglio et al., 2016; Fornacca et al., 2017; Hantson et al., 2013). However, the advent of higher-resolution constellations—including Sentinel-2, Landsat, and commercial CubeSats—now enables detailed assessments of fuel structure, vegetation stress, and postfire recovery at scales directly relevant for management (Silva-Cardoza et al., 2022; Chen et al., 2024). Thermal-infrared sensors allow the detection of active fire fronts and residual hotspots even under smoke plumes, supporting rapid response. SAR contributes valuable information under cloudy conditions, capturing changes in vegetation moisture and canopy structure.

Beyond "general-purpose" missions, there are ambitious plans to deploy "wildfire-centric" Earth observation constellations that will revolutionize the uptake of satellite data into actionable applications and DSS. FireSat is an example of such projects led by the Earth Fire Alliance (joining forces from public and private sectors). It is a planned constellation of over 50 small satellites dedicated to wildfire detection and monitoring, with full deployment targeted by 2030. Following the launch of the prototype in 2025, each satellite will carry multispectral infrared sensors (covering at least six thermal bands) and onboard AI-based processing to detect fires, along with GNSS-reflectometry to assess soil and vegetation moisture. This combined capability enables precise fire detection while also monitoring fuel conditions that drive fire behavior. When fully deployed, FireSat aims to provide near-real-time global coverage, revisiting high-risk regions every 15–20 minutes. Its ~5 m infrared imagery will be able to detect fires as small as 5–10 m, while onboard AI minimizes false alarms by analyzing changes across images. Fire characteristics are transmitted within minutes, giving fire managers rapid alerts and updated perimeter maps. OroraTech Wildfire Constellation, owned by a private company, is deploying a dedicated wildfire-monitoring constellation, beginning with eight nanosatellites in 2025. These carry thermal-infrared sensors capable of detecting fires as small as 4 × 4 m. Operating in a Sun-synchronous orbit, nanosats are optimized to capture afternoon and nighttime fire activity, addressing coverage gaps left by public satellites. Fire detections are processed onboard using AI algorithms and transmitted within minutes to support real-time firefighting decisions. The company plans to scale the system to ~100 satellites by 2028, targeting a global revisit time of around 30 minutes and enabling continuous tracking of ignitions and fire spread across both day and night. Finally, WildFireSat is a dedicated Canadian satellite mission scheduled for launch around 2029, consisting of a constellation of seven microsatellites specifically designed for wildfire monitoring. Each satellite will carry an infrared sensor optimized to measure Fire Radiative Power and track fire growth, enabling daily to twice-daily coverage over Canadian territories and contributing to global monitoring. The mission's focus is on real-time detection of new fires, quantifying their intensity, and aiding fire management decisions. The plans are to integrate with Canada's existing wildfire systems and contribute to international fire monitoring efforts.

At finer scales and very high spatial resolutions compared to satellites, airborne and unmanned aerial vehicle (UAV)–mounted LiDAR systems provide highly relevant data by mapping vertical fuel profiles, including canopy gaps and height, density, layering, and ground-to-crown fuel continuity with high precision, thereby improving fire behavior models and fuel treatment planning (González-Olabarria et al., 2012). Unlike optical sensors, LiDAR can penetrate through canopy layers, capturing understory vegetation and surface fuels often "invisible" to passive sensors (e.g., optical, thermal). When integrated with moisture and flammability data, LiDAR-derived fuel maps enable researchers, technicians, and fire managers to identify hazardous fuel accumulations.

Remote sensing is also a critical input for fuel moisture and vegetation stress indices, which are increasingly used in dynamic fire-danger forecasting (Yebra et al., 2018; Marino et al., 2020). For example, the Normalized Difference Water Index (NDWI) and shortwave infrared-based indices derived from Sentinel-2 can track spatial variations in live fuel moisture, while time-series analysis of vegetation indices reveals long-term trends in drought stress and flammability (Wang et al., 2008; Xu et al., 2024). Importantly, these capabilities are no longer confined to research. Civil protection agencies and land managers now have access to near-real-time data streams and dashboards that integrate satellite, meteorological, and in situ information for informed operational decision-making.

Finally, Earth observation supports the postfire phase, providing spatially explicit indicators of burn severity (e.g., differenced Normalized Burn Ratio [dNBR], Relativized Burn Ratio [RBR]) (Castro et al., 2025) as well as recovery that guides ecological restoration and monitoring of NbS (Gonçalves et al., 2025). When combined with field data and machine-learning approaches, satellite time series can evaluate the performance of restoration mosaics, grazing interventions, or green firebreaks over years to decades. This feedback loop makes remote sensing not only a tool for emergency response but also a foundation for adaptive and proactive fire management (Gonçalves et al., 2025).

21.2.2 Fire behavior and risk modeling

Modeling fire behavior and risk has become a cornerstone of proactive fire management, enabling practitioners to anticipate how ignitions might evolve under varying environmental conditions and to test the potential effectiveness of different interventions. Traditional mechanistic models such as Fire Behavior Simulator (FARSITE) and FlamMap simulate fire spread based on inputs of topography, fuels, and weather, generating spatially explicit projections of flame length, rate of spread, and crown fire potential (Ghodrat et al., 2023; Zigner et al., 2020). These tools remain widely used for operational planning and training, but uncertainties in input data and parameterization constrain their accuracy and reliability. Systematic validation with field observations, experimental prescribed fires, and long-term monitoring remains essential to ensure that model results are reliable and useful for decision-making (Hsu et al., 2025; Benali et al., 2023).

Complementing these, landscape-scale simulation platforms such as Landis-II, MEDFIRE (a fire-landscape dynamic model), and REMAINS (recently applied in the Mediterranean and Atlantic–Mediterranean regions) incorporate vegetation succession, disturbance dynamics, and climate scenarios to explore the long-term trajectories of fire regimes. These models enable the testing of "what if" scenarios, such as how introducing silvopastoral mosaics or prescribed burning might alter fire frequency, intensity, and ecological outcomes over decades. By coupling ecological processes with human management strategies, they provide insights into the trade-offs among biodiversity, carbon storage, and fire risk reduction (Pais et al., 2020; Mavsar et al., 2013; Cánibe et al., 2022).

The rapid growth of machine learning and artificial intelligence has further expanded modeling capacities. Data-driven approaches can integrate diverse predictors—including live fuel moisture indices, socio-economic data on ignitions, and weather reanalyses—to produce probabilistic maps of ignition likelihood or dynamic forecasts of fire danger. Some systems now incorporate ensemble approaches, where mechanistic spread models are combined with AI-derived ignition risk layers, offering more robust scenario planning (Castelli et al., 2015; Wu et al., 2022).

An emerging frontier is the development of digital twins of fire-prone landscapes, that is, high-resolution virtual replicas that integrate remote sensing, ecological models, and real-time meteorological data (Li et al., 2025; Huang et al., 2024). These platforms allow managers to interactively test the consequences of management interventions, evaluate alternative suppression strategies, and communicate risks to stakeholders through intuitive visualization. While still experimental, digital twins exemplify how modeling can bridge the gap between research and decision-making, turning complex dynamics into actionable intelligence for adaptive fire management.

21.2.3 Big data, artificial intelligence, and early detection

The proliferation of environmental sensors, coupled with advancements in AI and cloud computing, has opened up new frontiers for early fire detection and risk forecasting (Novo et al., 2020; Novo et al., 2025). IoT sensor networks measuring temperature, humidity, fuel moisture, and smoke particles are increasingly deployed in fire-prone landscapes, transmitting continuous data streams that feed into operational platforms (Chan et al., 2024; Li, 2025). When combined with meteorological forecasts, these inputs enable near-real-time assessment of ignition conditions at local scales.

At the same time, computer vision algorithms applied to high-resolution camera networks and UAV imagery allow automated detection of smoke plumes and flame signatures (see Bouguettaya et al., 2022; Moumgiakmas et al., 2021). These systems can reduce detection time from hours to minutes, particularly in the wildland–urban interface, where rapid suppression is critical. Similarly, satellite-based systems integrating thermal anomalies (e.g., MODIS, VIIRS, Sentinel-3) with AI-based filters have improved both sensitivity and specificity in identifying true ignitions versus false alarms (Talaat and ZainEldin, 2023; Liu et al., 2025). Moreover, the emergence of dedicated wildfire-monitoring constellations, such as FireSat and WildFireSat, is set to transform the role of satellites in fire management by enabling earlier detection of ignitions and continuous, near-real-time tracking of fire spread.

Beyond detection, AI is increasingly applied to predictive analytics and resource optimization. Machine learning models trained on historical ignition records, socio-economic variables, and environmental drivers can generate probabilistic ignition maps, highlighting priority areas for prevention and surveillance (see, e.g., Depicker et al., 2020; Bjånes et al., 2021; Zhang et al., 2021; Jain et al., 2020; Kondylatos et al., 2022). Reinforcement learning and optimization algorithms are also being explored for dynamic allocation of firefighting resources—deciding where to preposition aerial assets or when to trigger evacuations under uncertain conditions.

Crowdsourced and social media data provide an additional, often overlooked layer of intelligence. Platforms that mine geolocated tweets, mobile phone reports, or citizen science observations can enhance situational awareness, particularly in regions with sparse official monitoring (Tavra et al., 2021; Slavkovikj et al., 2014). While these sources raise issues of reliability and verification, their integration into multisource platforms offers valuable complements to traditional detection networks.

Overall, the convergence of big data and AI is shifting early detection and risk management from a reactive paradigm toward one of anticipation and rapid response, where information is continuously updated, analyzed, and operationalized for decision-makers and communities alike. AI frequently operates as a "black box," which can result in problems with data bias, validation, or explainability. Therefore, privacy, interoperability, and fair access to data must be considered when integrating into DSS.

21.2.4 Decision-support systems and operational tools

DSS are essential for transforming the vast data streams produced by satellites, sensors, and models into usable intelligence for proactive fire management. Rather than being designed solely for emergency response, modern DSS increasingly prioritize prevention, preparedness, and long-term planning, helping managers identify where and when to act before fires escalate into crises.

A key contribution of DSS is the capacity to integrate multicriteria spatial analyses that weigh diverse factors such as ignition probability, fuel loads, biodiversity values, water resources, and socio-economic vulnerability. This integration enables managers to identify priority zones for interventions such as prescribed burning, grazing, reforestation with less flammable species, or the establishment of green firebreaks. For example, platforms building on Copernicus Emergency Management Service (EMS) or the European Forest Fire Information System (EFFIS) already combine EO, meteorological forecasts, and fire danger indices to support strategic prevention at continental and national levels.

DSS also facilitates the optimization of resources for prevention. Spatial decision models can suggest where to allocate budgets for fuel reduction, where to target NbS interventions to maximize cobenefits, or how to design landscape mosaics that dampen fire spread (Pais et al., 2023). Importantly, these tools can incorporate stakeholder preferences and governance constraints (de Castro-Pardo et al., 2019), thereby moving beyond purely technical optimization toward socially acceptable solutions.

An emerging function of DSS is their role in monitoring and adaptive management. By linking remote sensing data with field monitoring, DSS can track the performance of prevention measures over time, evaluating, for instance, how a silvopastoral system or a restored riparian corridor modifies fire behavior under different conditions. This feedback enables learning by doing and helps secure long-term investments in NbS by demonstrating tangible risk reduction outcomes (Sakellariou et al., 2017).

Despite their promise, DSS face persistent challenges. Many platforms remain fragmented across agencies, with limited interoperability and inconsistent data standards. The "last mile" problem, which involves translating sophisticated analytics into formats and timelines that are usable by practitioners on the ground, remains a critical bottleneck. Overcoming these barriers requires not only technical innovation but also institutional coordination, participatory design, and training programs that ensure DSS support proactive prevention rather than reactive suppression.

21.3 Technology-enabling nature-based solutions

While technological advances are reshaping the landscape of fire management, their impact will remain limited if they are not coupled with structural changes in fuels, land use, and ecosystem functions. Satellites, models, and decision-support tools can indicate where risks are highest and which scenarios are most hazardous, but they cannot, by themselves, reduce fuel loads, restore degraded landscapes, or reintroduce the ecological processes that moderate fire regimes. This is where NbS become indispensable (Regos, 2022; Plana et al., 2024). By restructuring landscapes through ecological restoration, creating heterogeneous agro-silvo-pastoral mosaics, and implementing green firebreaks, NbS directly address the root causes of extreme wildfires (Regos, 2026). When combined with technology, they create a powerful feedback loop: data and models guide the strategic placement of NbS, while monitoring systems evaluate their effectiveness over time. The following section explores how NbS can operationalize proactive fire management, delivering both risk reduction and broader ecological and societal cobenefits.

NbS are increasingly recognized as essential to mitigating extreme wildfire risk by restructuring landscapes, reducing fuel continuity, and restoring ecosystem functions. However, their design and implementation have often been constrained by limited spatial data, inadequate monitoring, and difficulties in demonstrating long-term effectiveness. This has led to skepticism among policymakers and landowners, and has hindered large-scale investment.

Recent technological advances are beginning to remove these barriers. High-resolution remote sensing, ecological modeling, optimization algorithms, and participatory decision platforms now provide the tools to plan, implement, and evaluate NbS in unprecedented detail. Technology helps answer once-intractable questions—for instance, where should restoration or agro-silvo-pastoral systems be prioritized to maximize both fire risk reduction and biodiversity gains? How effective are prescribed burns or green firebreaks at reducing fire intensity under different conditions? How do NbS portfolios perform under future climate scenarios?

Crucially, technology does not replace NbS; it amplifies their potential by making them more strategic, transparent, and adaptive. Remote sensing guides the placement of interventions, models test their outcomes over decades, and monitoring systems generate evidence that strengthens political and social legitimacy. In this way, technology becomes the enabler that transforms NbS from isolated projects into the backbone of proactive, adaptive fire management strategies.

21.3.1 Spatial planning and prioritization

A fundamental challenge in implementing NbS lies in deciding where to act. Landscapes are heterogeneous mosaics of fuels, ecosystems, and human uses, and not every location offers the same potential for reducing wildfire risk or delivering cobenefits. Technology has radically improved the capacity to identify priority areas for intervention, transforming NbS from opportunistic projects into strategically targeted investments.

Remote sensing and geographic information systems (GIS) enable managers to map fine-scale fuel loads, vegetation types, land ownership, and values at risk, thereby creating the spatial foundation for effective prevention planning. When combined with climate and fire history data, these layers reveal hotspots where fuel continuity, ignition likelihood, and exposure converge. In such zones, NbS such as agro-silvo-pastoral mosaics, riparian restoration, or green firebreaks can provide disproportionate benefits in reducing extreme fire spread.

Advances in multi-criteria decision analysis (MCDA) and optimization algorithms further enhance spatial prioritization (Ball et al., 2009; Kukkala and Moilanen, 2013; Salgado-Rojas et al., 2025). These tools can integrate diverse objectives such as minimizing fire intensity, maximizing biodiversity connectivity, preserving carbon stocks, sustaining rural livelihoods, and producing spatially explicit recommendations on the placement of NbS portfolios (see, e.g., Cánibe et al., 2022; Iglesias et al., 2025). Widely used platforms, such as Zonation and Marxan, as well as newer R packages like "prioriactions," allow researchers and managers to explore trade-offs, set explicit targets, and identify cost-effective configurations of NbS. For example, optimization models can identify the optimal locations for silvopastoral systems that simultaneously reduce fine fuels, connect habitat patches, and generate economic returns.

Importantly, decision-support platforms are increasingly enabling participatory planning, where stakeholders can interact with maps, adjust the weights assigned to different criteria, and visualize trade-offs (de Castro-Pardo et al., 2019; Gonçalves et al., 2022; Lecina-Diaz et al., 2023a). This transparency not only improves the quality of prioritization but also increases social acceptance, ensuring that NbS align with local needs and landowner realities.

By making spatial prioritization data-driven and transparent, technology ensures that limited resources are directed to the areas of greatest impact, amplifying both the efficiency and legitimacy of NbS for wildfire prevention.

21.3.2 Monitoring and adaptive management

One of the main obstacles to scaling up NbS in fire management has been the lack of robust, long-term evidence of their effectiveness. Traditionally, interventions such as fuel breaks, grazing systems, or restoration projects were implemented without systematic monitoring, making it difficult to evaluate outcomes or justify sustained investment. Technology now provides the tools to transform NbS into evidence-based, adaptive strategies.

Remote sensing platforms (e.g., Sentinel-2, Landsat 8/9, PlanetScope) enable continuous monitoring of vegetation structure, fuel loads, and postfire recovery at high spatial and temporal resolutions. Spectral indices such as the Normalized Difference Vegetation Index (NDVI), NDWI, and shortwave infrared-based burn indices (dNBR, RBR) provide quantitative indicators of vegetation vigor, live fuel moisture, and burn severity (Silva-Cardoza et al., 2022; Picos et al., 2019; Fernández-Guisuraga et al., 2025). UAVs and drone-mounted LiDAR systems add precision at local scales, capturing fine-grained changes in canopy gaps, vertical structure, and surface fuels.

These data streams enable managers to track the impact of NbS interventions, such as silvopastoral mosaics, riparian restoration, or prescribed burning, and modify fire-relevant variables over time. In addition, combining short-term monitoring with long-term satellite time series would allow for observation of both immediate responses and cumulative impacts. By linking monitoring results to simulation models (e.g., Flammap, Landis-II, REMAINS), it becomes possible to test whether observed changes in vegetation structure translate into reduced fire intensity or altered regime dynamics. This feedback loop strengthens adaptive management: interventions are adjusted based on measured outcomes rather than assumptions.

Beyond biophysical monitoring, technology also enables social and economic evaluation of NbS. Mobile applications and participatory platforms can collect data on community perceptions, landowner engagement, and rural economic benefits, complementing ecological indicators. This integration ensures that NbS are judged not only on their ecological performance but also on their capacity to support resilient socio-ecological systems.

In short, technology provides the MRV backbone for NbS, transforming them into accountable and adaptable measures that can attract funding, inform policy, and maintain social legitimacy over the long term.

21.3.3 Modeling nature-based solutions scenarios under climate change

While monitoring provides evidence of how NbS perform today, modeling enables exploration of how they may function under future conditions of climate change, land-use transitions, and socio-economic dynamics. This ability is particularly relevant for fire-prone regions, where rising temperatures, prolonged droughts, and shifts in vegetation composition are expected to intensify wildfire regimes.

Landscape simulation models such as Landis-II (Scheller et al., 2007), MEDFIRE (Brotons et al., 2013) and REMAINS (Pais et al., 2023) integrate vegetation succession, disturbance regimes, and climate projections to assess how different NbS portfolios—ranging from silvopastoral systems to large-scale restoration corridors—may influence fire frequency, severity, and ecological outcomes over decades. These models can reveal trade-offs: for example, whether prioritizing carbon storage through dense reforestation might inadvertently increase future fire risk (Hermoso et al., 2021), or how creating heterogeneous agro-silvo-pastoral mosaics can simultaneously reduce fire intensity and support biodiversity (Regos et al., 2023).

By coupling these models with climate scenarios (e.g., RCP/SSP pathways) or people-nature-centric scenarios from the Nature Futures Framework, managers can anticipate which NbS strategies are robust across a range of possible futures. This foresight is crucial for avoiding "maladaptation"—investments that appear effective in the short term but fail under hotter, drier conditions (Cánibe et al., 2022; Sil et al., 2024; Fernandes, 2022).

An emerging frontier is the development of interactive modeling platforms and digital twins that allow stakeholders to visualize how landscapes might evolve under alternative NbS strategies. By presenting results as maps, animations, or even immersive environments, these tools can enhance risk communication and foster dialog among scientists, policymakers, and communities. Instead of abstract graphs, decision-makers can "see" how a landscape with restored riparian zones, expanded grazing, or prescribed burning patterns would behave during an extreme fire-weather event in 2050. The integration of Large Language Models, enabling natural language interaction with these platforms and digital twins, is expected to decrease barriers to adoption further while facilitating more sophisticated and comprehensive use.

In this way, modeling does not merely predict; it supports strategic foresight, helping managers design NbS portfolios that remain effective under a changing climates and shifting socio-ecological contexts (Ringland, 2010; Cook et al., 2014).

21.3.4 Challenges and research needs

Despite growing recognition of the potential of technology-enabled NbS, several challenges continue to limit their widespread adoption and effectiveness.

First, there are persistent data and interoperability gaps. Many fire-prone regions still lack high-resolution fuel maps, long-term monitoring networks, or standardized indicators for NbS performance. Even where data exist, they are often siloed across agencies, formats, and scales, making integration into DSS difficult. Developing open, interoperable standards for data and models remains a research priority.

Second, MRV frameworks for NbS are underdeveloped. While remote sensing offers powerful indicators, translating these into metrics that satisfy policymakers, insurers, or carbon markets is still a work in progress. Without robust MRV, scaling NbS through sustained funding and incentives remains challenging.

Third, there is the risk of techno-solutionism, that is, overreliance on advanced tools while neglecting governance, regional or local dynamics, as well as cultural, political, and socio-economic realities of communities. Algorithms may optimize NbS placement in theory, but fragmented land tenure, a lack of landowner incentives, and short political cycles often constrain their implementation on the ground. Future research should therefore focus not only on technical optimization, but also on designing policy instruments and governance arrangements that enable action.

Fourth, capacity building and equity remain critical. Many regions facing the most severe wildfire risks, such as Mediterranean landscapes, parts of Latin America, Africa, and Asia, have limited technical and institutional capacity to deploy advanced technologies. Without investment in training, accessibility, and coproduction, the risk of widening the gap between high-tech solutions and local realities exists.

Finally, the field requires long-term, transdisciplinary research agendas. This includes combining ecological modeling with social science methods to understand acceptance of NbS, developing financing mechanisms that reward prevention over suppression, and testing NbS portfolios across multiple climate and socio-economic futures. The integration of digital twins, participatory DSS, and MRV frameworks offers promising avenues, but their success will depend on embedding them in adaptive governance systems that learn from practice.

In short, the next decade must focus not only on technological innovation, but also on creating the institutional, financial, and social conditions that allow technology-enabled NbS to become the backbone of proactive fire management.

21.4 Conclusions and recommendations

Extreme wildfires are not only the product of adverse weather, but also of fuel accumulation, land-use legacies, and governance failures. As such, suppression-focused responses will remain insufficient. This chapter outlines a vision in which technology and NbS work in concert to reshape landscapes and institutions toward proactive, adaptive fire management (Fig. 21.2).

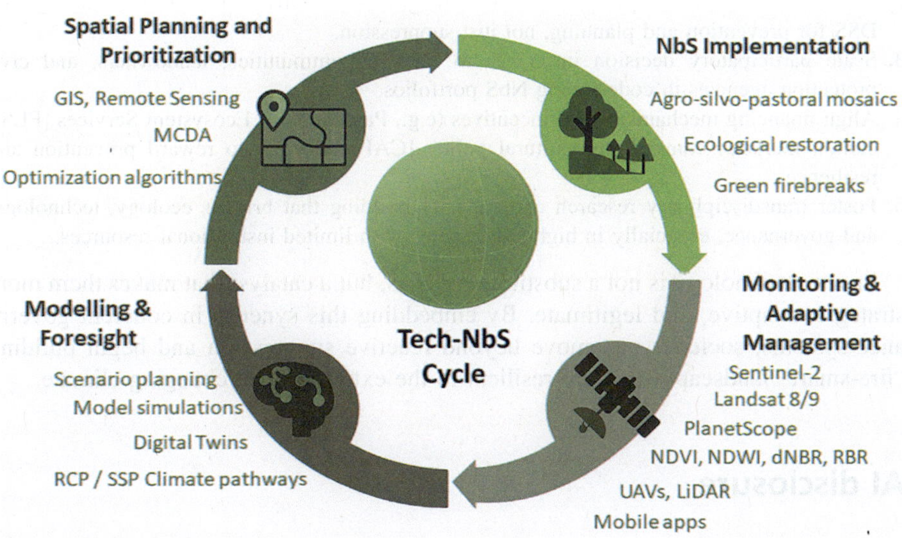

Figure 21.2 Circular framework illustrating the iterative cycle linking spatial planning, nature-based solutions (NbS) implementation, monitoring and adaptive management, and modeling and foresight.
This cycle emphasizes the continuous feedbacks between evidence generation, decision-making, and action, supporting the design of fire-smart, resilient landscapes through NbS and technological innovation.

Key insights include:

- **Technology provides foresight and accountability**. Remote sensing, modeling, artificial intelligence, and DSS enable managers to assess risks, anticipate future scenarios, and evaluate the performance of interventions. Without these tools, NbS remain difficult to target, justify, or scale.
- **NbS deliver the structural change needed to reduce extreme fire risk**. By restoring heterogeneity through silvopastoral mosaics, riparian corridors, prescribed burning, or rewilding, NbS alter fuel continuity while generating biodiversity, carbon, and social cobenefits.
- **The 4 A cycle—Assess, Anticipate, Act, Adapt—offers a practical framework** for integrating data, models, and NbS within adaptive governance. Technology enables each stage of the cycle, while NbS supply the on-the-ground mechanisms to shift fire regimes.
- **Policy coherence and social engagement are decisive**. No amount of data or optimization can overcome fragmented land tenure, short-term political horizons, or lack of legitimacy. Participatory platforms, transparent prioritization tools, and robust MRV systems are essential to ensure trust and long-term investment.

21.5 Recommendations for the next decade

1. Standardize monitoring and MRV frameworks to track NbS effectiveness in reducing fire intensity, emissions, and socio-ecological impacts.
2. Invest in interoperable, open platforms that integrate Earth Observation (EO), modeling, and DSS for prevention and planning, not just suppression.
3. Scale participatory decision interfaces to involve communities, landowners, and civil protection agencies in codesigning NbS portfolios.
4. Align financing mechanisms and incentives (e.g., Payments for Ecosystem Services [PES], carbon markets, common agricultural policy [CAP] reforms) to reward prevention and resilience.
5. Foster transdisciplinary research and capacity building that bridges ecology, technology, and governance, especially in high-risk regions with limited institutional resources.

In sum, technology is not a substitute for NbS, but a catalyst that makes them more strategic, adaptive, and legitimate. By embedding this synergy in coherent governance systems, societies can move beyond reactive suppression and begin building "fire-smart" landscapes that are resilient to the extremes of a changing climate.

AI disclosure

During the preparation of this work, the author(s) used ChatGPT (OpenAI GPT-5) to improve the readability and language quality of the text. The authors reviewed and approved all generated content and are fully responsible for the final version of the manuscript. After using this tool/service, the author(s) reviewed and edited the content as needed and take(s) full responsibility for the content of the publication.

Acknowledgement

A.R. was supported by the Ramón y Cajal fellowship programme funded by the Spanish Ministry of Science, Innovation and Universities (RYC2022-036822-I), and by the research projects RESFIRE (PID2023-152690OA-C22) and Paisajes resilientes (IN607D-2025-01). A. Novo would like to thank to the Spanish Ministry of Science and Innovation through the grant JDC2022–049758- I.

References

Ascoli, D., Plana, E., Oggioni, S.D., Tomao, A., Colonico, M., Corona, P., et al., 2023. Fire-smart solutions for sustainable wildfire risk prevention: bottom-up initiatives meet top-down policies under EU green deal. International Journal of Disaster Risk Reduction 92, 103715. Available at https://doi.org/10.1016/j.ijdrr.2023.103715.

Ball, I.R., Possingham, H.P., Watts, M.E., 2009. Marxan and Relatives: Software for Spatial Conservation Prioritization. Oxford University Press (OUP), pp. 185–195. Available at http://doi.org/10.1093/oso/9780199547760.003.0014.

Batllori, E., Parisien, M.A., Krawchuk, M.A., Moritz, M.A., 2013. Climate change-induced shifts in fire for Mediterranean ecosystems. Global Ecology and Biogeography 22 (10), 1118–1129. Available at https://doi.org/10.1111/geb.12065.

Benali, A., Guiomar, N., Gonçalves, H., Mota, B., Silva, F., Fernandes, P.M., et al., 2023. The portuguese large wildfire spread database (PT-FireSprd). Earth System Science Data 15 (8), 3791–3818. Available at https://doi.org/10.5194/essd-15-3791-2023, www.earth-system-science-data.net.

Bisquert, M., Sánchez, J., Caselles, V., 2014. Modeling fire danger in Galicia and Asturias (Spain) from MODIS images. Remote Sensing 6 (1), 540–554. Available at https://doi.org/10.3390/rs6010540.

Bjånes, A., De La Fuente, R., Mena, P., 2021. A deep learning ensemble model for wildfire susceptibility mapping. Ecological Informatics 65, 101397. Available at https://doi.org/10.1016/j.ecoinf.2021.101397.

Bouguettaya, A., Zarzour, H., Taberkit, A.M., Kechida, A., 2022. A review on early wildfire detection from unmanned aerial vehicles using deep learning-based computer vision algorithms. Signal Processing 190. Available at https://doi.org/10.1016/j.sigpro.2021.108309, https://www.journals.elsevier.com/signal-processing.

Brotons, L., Aquilué, N., de Cáceres, M., Fortin, M.-J., Fall, A., 2013. How Fire History, Fire Suppression Practices and Climate Change Affect Wildfire Regimes in Mediterranean Landscapes. PLoS ONE 8 (5), e62392. Available at https://doi.org/10.1371/journal.pone.0062392.

Cánibe, M., Hermoso, V., Campos, J., Carvalho-Santos, C., Fernandes, P.M., Freitas, T.R., et al., 2022. Climate and fire-smart landscape scenarios call for redesigning protection regimes to achieve multiple management goals. Journal of Environmental Management 322, 116045.

Castelli, M., Vanneschi, L., Popovič, A., 2015. Predicting burned areas of forest fires: an artificial intelligence approach. Fire Ecology 11 (1), 106–118. Available at https://doi.org/10.4996/fireecology.1101106.

Castro, P., Gonçalves, J., Mota, D., Marcos, B., Alves, C., Alonso, J., et al., 2025. Open access to burn severity data—a web-based portal for mainland portugal. Fire 8 (3), 95. Available at https://doi.org/10.3390/fire8030095.

Chan, C.C., Alvi, S.A., Zhou, X., Durrani, S., Wilson, N., Yebra, M., 2024. A survey on IoT ground sensing systems for early wildfire detection: technologies, challenges, and opportunities. IEEE Access 12, 172785–172819. Available at https://doi.org/10.1109/ACCESS.2024.3501336, http://ieeexplore.ieee.org/xpl/RecentIssue.jsp?punumber=6287639.

Chen, X., Zhang, Y., Wang, S., Zhao, Z., Liu, C., Wen, J., 2024. Comparative study of machine learning methods for mapping forest fire areas using Sentinel-1B and 2A imagery. Frontiers in Remote Sensing 5. Available at https://doi.org/10.3389/frsen.2024.1446641.

Chuvieco, E., Aguado, I., Salas, J., García, M., Yebra, M., Oliva, P., 2020. Satellite remote sensing contributions to wildland fire science and management. Current Forestry Reports, 6 (2), 81–96.

Cook, C.N., Inayatullah, S., Burgman, M.A., Sutherland, W.J., Wintle, B.A., 2014. Strategic foresight: how planning for the unpredictable can improve environmental decision-making. Trends in Ecology and Evolution 29 (9), 531–541. Available at https://doi.org/10.1016/j.tree.2014.07.005, www.elsevier.com/inca/publications/store/3/0/3/3/9.

de Castro-Pardo, M., Pérez-Rodríguez, F., Martín-Martín, J.M., Azevedo, J.C., 2019. Modelling stakeholders' preferences to pinpoint conflicts in the planning of transboundary protected areas. Land use Policy 89. Available at https://doi.org/10.1016/j.landusepol.2019.104233, www.elsevier.com/inca/publications/store/3/0/4/5/1/.

Depicker, A., De Baets, B., Baetens, J.M., 2020. Wildfire ignition probability in Belgium. Natural Hazards and Earth System Sciences 20 (2), 363–376. Available at https://doi.org/10.5194/nhess-20-363-2020, http://www.nat-hazards-earth-syst-sci.net/volumes_and_issues.html.

Duane, A., Aquilué, N., Gil-Tena, A., Brotons, L., 2016. Integrating fire spread patterns in fire modelling at landscape scale. Environmental Modelling & Software 86, 219–231. Available at https://doi.org/10.1016/j.envsoft.2016.10.001.

Duguy, B., Alloza, J.A., Röder, A., Vallejo, R., Pastor, F., 2007. Modelling the effects of landscape fuel treatments on fire growth and behaviour in a Mediterranean landscape (eastern Spain). International Journal of Wildland Fire 16 (5), 619–632. Available at https://doi.org/10.1071/WF06101.

Fernandes, P.M., 2022. Make Europe's forests climate-smart and fire-smart. Nature 609 (7925), 32. Available at https://doi.org/10.1038/d41586-022-02318-2, https://www.nature.com/nature/.

Fernández-Guisuraga, J.M., Monzón-González, A., Fernández-García, V., Peña-Pérez, S.A., Calvo, L., 2025. Estimation of Prometheus fuel types using physically based remote sensing techniques. Fire Ecology 21 (1). Available at https://doi.org/10.1186/s42408-025-00373-4, https://www.springer.com/journal/42408.

Fernández-Guisuraga, J.M., Fernández-Manso, A., Quintano, C., Fernández-García, V., Cerrillo, A., Marqués, G., et al., 2024. FIREMAP: cloud-based software to automate the estimation of wildfire-induced ecological impacts and recovery processes using remote sensing techniques. Ecological Informatics 81, 102591. Available at https://doi.org/10.1016/j.ecoinf.2024.102591.

Fornacca, D., Ren, G., Xiao, W., 2017. Performance of three MODIS fire products (MCD45A1, MCD64A1, MCD14ML), and ESA fire_CCI in a mountainous area of northwest Yunnan, China, characterized by frequent small fires. Remote Sensing 9 (11), 1131. Available at https://doi.org/10.3390/rs9111131.

Ghodrat, M., Shakeriaski, F., Fanaee, S.A., Simeoni, A., 2023. Software-based simulations of wildfire spread and wind-fire interaction. Fire 6 (1). Available at https://doi.org/10.3390/fire6010012, www.mdpi.com/journal/fire.

Giglio, L., Descloitres, J., Justice, C.O., Kaufman, Y.J., 2003. An enhanced contextual fire detection algorithm for MODIS. Remote Sensing of Environment 87 (2-3), 273–282. Available at https://doi.org/10.1016/S0034-4257(03)00184-6, www.elsevier.com/inca/publications/store/5/0/5/7/3/3.

Giglio, L., Schroeder, W., Justice, C.O., 2016. The collection 6 MODIS active fire detection algorithm and fire products. Remote Sensing of Environment 178, 31–41. Available at https://doi.org/10.1016/j.rse.2016.02.054, www.elsevier.com/inca/publications/store/5/0/5/7/3/3.

Gonçalves, C., Honrado, J.P., Cerejeira, J., Sousa, R., Fernandes, P.M., Vaz, A.S., et al., 2022. On the development of a regional climate change adaptation plan: integrating model-assisted projections and stakeholders' perceptions. Science of the Total Environment 805. Available at https://doi.org/10.1016/j.scitotenv.2021.150320, www.elsevier.com/locate/scitotenv.

Gonçalves, J., Portela, A.P., Regos, A., Sil, Â., Marcos, B., Alonso, J., et al., 2025. Fostering post-fire research towards a more balanced wildfire science agenda to navigate global environmental change. Fire 8 (2). Available at https://doi.org/10.3390/fire8020051, www.mdpi.com/journal/fire.

González-Olabarria, J.R., Rodríguez, F., Fernández-Landa, A., Mola-Yudego, B., 2012. Mapping fire risk in the model forest of Urbión (Spain) based on airborne LiDAR measurements. Forest Ecology and Management 282, 149–156. Available at https://doi.org/10.1016/j.foreco.2012.06.056.

Hantson, S., Padilla, M., Corti, D., Chuvieco, E., 2013. Strengths and weaknesses of MODIS hotspots to characterize global fire occurrence. Remote Sensing of Environment 131, 152–159. Available at https://doi.org/10.1016/j.rse.2012.12.004.

He, H.S., Shang, B.Z., Crow, T.R., Gustafson, E.J., Shifley, S.R., 2004. Simulating forest fuel and fire risk dynamics across landscapes - LANDIS fuel module design. Ecological Modelling 180 (1), 135–151. Available at https://doi.org/10.1016/j.ecolmodel.2004.07.003.

Hermoso, V., Regos, A., Morán-Ordóñez, A., Duane, A., Brotons, L., 2021. Tree planting: a double-edged sword to fight climate change in an era of megafires. Global Change Biology 27 (13), 3001–3003. Available at https://doi.org/10.1111/gcb.15625.

Hsu, A., Jones, M.W., Thurgood, J.R., Smith, A.J.P., Carmenta, R., Abatzoglou, J.T., et al., 2025. A global assemblage of regional prescribed burn recordss—GlobalRx. Scientific Data 12 (1). Available at https://doi.org/10.1038/s41597-025-04941-w, https://www.nature.com/sdata/.

Huang, Y., Li, J., Zheng, H., 2024. Modeling of wildfire digital twin: research progress in detection, simulation, and prediction techniques. Fire 7 (11). Available at https://doi.org/10.3390/fire7110412, www.mdpi.com/journal/fire.

Iglesias, M.C., Hermoso, V., Azevedo, J.C., Campos, J.C., Salgado-Rojas, J., Sil, Â., et al., 2025. Integrating multiple landscape management strategies to optimise conservation under climate and planning scenarios: a case study in the Iberian Peninsula. Ecosystem Services 74. Available at https://doi.org/10.1016/j.ecoser.2025.101742, https://www.sciencedirect.com/science/journal/22120416.

Jain, P., Coogan, S.C.P., Subramanian, S.G., Crowley, M., Taylor, S., Flannigan, M.D., 2020. A review of machine learning applications in wildfire science and management. Environmental Reviews 28 (4), 478–505. Available at https://doi.org/10.1139/er-2020-0019, https://www.nrcresearchpress.com/loi/er.

Johnstone, J.F., Allen, C.D., Franklin, J.F., Frelich, L.E., Harvey, B.J., Higuera, P.E., et al., 2016. Changing disturbance regimes, ecological memory, and forest resilience. Frontiers in Ecology and the Environment 14 (7), 369–378. Available at https://doi.org/10.1002/fee.1311, http://onlinelibrary.wiley.com/journal/10.1002/(ISSN)1540-9309.

Kelly, L.T., Giljohann, K.M., Duane, A., Aquilué, N., Archibald, S., Batllori, E., et al., 2020. Fire and biodiversity in the Anthropocene. Science (New York, N.Y.) 370 (6519). Available at https://doi.org/10.1126/science.abb0355, https://science.sciencemag.org/content/sci/370/6519/eabb0355.full.pdf.

Kondylatos, S., Prapas, I., Ronco, M., Papoutsis, I., Camps-Valls, G., Piles, M., et al., 2022. Wildfire danger prediction and understanding with deep learning. Geophysical Research Letters 49 (17). Available at https://doi.org/10.1029/2022gl099368.

Kukkala, A.S., Moilanen, A., 2013. Core concepts of spatial prioritisation in systematic conservation planning. Biological Reviews 88 (2), 443–464. Available at https://doi.org/10.1111/brv.12008.

Lecina-Diaz, J., Campos, J.C., Pais, S., Carvalho-Santos, C., Azevedo, J.C., Fernandes, P., et al., 2023a. Stakeholder perceptions of wildfire management strategies as nature-based solutions in two Iberian biosphere reserves. Ecology and Society 28 (1). Available at https://doi.org/10.5751/ES-13907-280139, https://ecologyandsociety.org/vol28/iss1/art39/ES-2022-13907.pdf.

Lecina-Diaz, J., Chas-Amil, M.-L., Aquilué, N., Sil, Â., Brotons, L., Regos, A., et al., 2023b. Incorporating fire-smartness into agricultural policies reduces suppression costs and ecosystem services damages from wildfires. Journal of Environmental Management 337, 117707. Available at https://doi.org/10.1016/j.jenvman.2023.117707.

Li, L., 2025. A comprehensive survey of fire weather index (FWI) systems and iot applications in peatland fire management. IEEE Access 13, 33579–33599. Available at https://doi.org/10.1109/access.2025.3543570.

Li, Y., Zhang, T., Ding, Y., Wadhwani, R., Huang, X., 2025. Review and perspectives of digital twin systems for wildland fire management. Journal of Forestry Research 36 (1). Available at https://doi.org/10.1007/s11676-024-01810-x, https://www.springer.com/journal/11676.

Liu, Q., Chen, H., Lin, D., 2025. Research and optimization of a multilevel fire detection framework based on deep learning and classical pattern recognition techniques. Scientific Reports 15 (1). Available at https://doi.org/10.1038/s41598-025-06721-3.

Mansuy, N., Buss, J., Hirsch, K., Guindon, L., Robinne, F.N., Boulanger, Y., et al., 2025. Integrating fire-smart fuels management with bioenergy benefits remote and Indigenous communities in Canada. Communications Earth and Environment 6 (1). Available at https://doi.org/10.1038/s43247-025-02313-1, https://www.nature.com/commsenv/.

Mariani, M., Connor, S.E., Theuerkauf, M., Herbert, A., Kuneš, P., Bowman, D., et al., 2022. Disruption of cultural burning promotes shrub encroachment and unprecedented wildfires. Frontiers in Ecology and the Environment 20 (5), 292–300. Available at https://doi.org/10.1002/fee.2395, http://onlinelibrary.wiley.com/journal/10.1002/(ISSN)1540-9309.

Marino, E., Yebra, M., Guillén-Climent, M., Algeet, N., Tomé, J.L., Madrigal, J., et al., 2020. Investigating live fuel moisture content estimation in fire-prone shrubland from remote sensing using empirical modelling and RTM simulations. Remote Sensing 12 (14). Available at https://doi.org/10.3390/rs12142251, https://res.mdpi.com/d_attachment/remotesensing/remotesensing-12-02251/article_deploy/remotesensing-12-02251-v2.pdf.

Martell, D.L., 2015. A review of recent forest and wildland fire management decision support systems research. Current Forestry Reports 1 (2), 128–137. Available at https://doi.org/10.1007/s40725-015-0011-y, https://www.springer.com/environment/sustainable+development/journal/40725.

Mavsar, R., Japelj, A., Kovač, M., 2013. Trade-offs between fire prevention and provision of ecosystem services in Slovenia. Forest Policy and Economics 29, 62–69. Available at https://doi.org/10.1016/j.forpol.2012.10.011.

McLauchlan, K.K., Higuera, P.E., Miesel, J., Rogers, B.M., Schweitzer, J., Shuman, J.K., et al., 2020. Fire as a fundamental ecological process: research advances and frontiers. Journal of Ecology 108 (5), 2047–2069. Available at https://doi.org/10.1111/1365-2745.13403, http://onlinelibrary.wiley.com/journal/10.1111/(ISSN)1365-2745.

Millington, J.D.A., Wainwright, J., Perry, G.L.W., Romero-Calcerrada, R., Malamud, B.D., 2009. Modelling Mediterranean landscape succession-disturbance dynamics: a landscape fire-succession model. Environmental Modelling and Software 24 (10), 1196–1208. Available at https://doi.org/10.1016/j.envsoft.2009.03.013.

Moreira, F., Ascoli, D., Safford, H., Adams, M.A., Moreno, J.M., Pereira, J.M.C., et al., 2020. Wildfire management in Mediterranean-type regions: paradigm change needed. Environmental Research Letters 15 (1). Available at https://doi.org/10.1088/1748-9326/ab541e, https://iopscience.iop.org/article/10.1088/1748-9326/ab541e/pdf.

Moumgiakmas, S.S., Samatas, G.G., Papakostas, G.A., 2021. Computer vision for fire detection on uavs—from software to hardware. Future Internet 13 (8). Available at https://doi.org/10.3390/fi13080200, http://www.mdpi.com/journal/futureinternet.

Novo, A., Fariñas-Álvarez, N., Martínez-Sánchez, J., González-Jorge, H., Fernández-Alonso, J.M., Lorenzo, H., 2020. Mapping forest fire risk—a case study in Galicia (Spain). Remote Sensing 12 (22), 3705. Available at https://doi.org/10.3390/rs12223705.

Novo, A., Regos, A., Fernández, C., 2025. Machine learning-based predictions of soil burn severity using spectral and LiDAR remote sensing data. Forest Ecology and Management 596, 123094. Available at https://doi.org/10.1016/j.foreco.2025.123094.

Pais, S., Aquilué, N., Honrado, J.P., Fernandes, P.M., Regos, A., 2023. Optimizing wildfire prevention through the integration of prescribed burning into 'fire-smart' land-use Policies. Fire 6 (12). Available at https://doi.org/10.3390/fire6120457, www.mdpi.com/journal/fire.

Pais, S., Aquilué, N., Brotons, L., Honrado, J.P., Fernandes, P.M., Regos, A., 2023. The REMAINS R-package: paving the way for fire-landscape modeling and management. Environmental Modelling and Software 168, 105801. Available at https://doi.org/10.1016/j.envsoft.2023.105801, www.elsevier.com/inca/publications/store/4/2/2/9/2/1. ISSN 1364-8152.

Pais, S., Aquilué, N., Campos, J., Sil, Â., Marcos, B., Martínez-Freiría, F., et al., 2020. Mountain farmland protection and fire-smart management jointly reduce fire hazard and enhance biodiversity and carbon sequestration. Ecosystem Services 44, 101143. Available at https://doi.org/10.1016/j.ecoser.2020.101143.

Picos, J., Alonso, L., Bastos, G., Armesto, J., 2019. Event-based integrated assessment of environmental variables and wildfire severity through sentinel-2 data. Forests 10 (11), 1021. Available at https://doi.org/10.3390/f10111021.

Plana, E., Serra, M., Smeenk, A., Regos, A., Berchtold, C., Huertas, M., et al., 2024. Framing coherence across EU policies towards integrated wildfire risk management and nature-based solutions. Fire 7 (11), 415. Available at https://doi.org/10.3390/fire7110415.

Plumanns-Pouton, E., Bakx, T.R.M., Buitenwerf, R., Espelta, J.M., Moreira, F., Regos, A., et al., 2025. Restoring fire regimes through rewilding. Current Biology 35 (13), R670. Available at https://doi.org/10.1016/j.cub.2025.04.026, https://www.sciencedirect.com/science/journal/09609822.

Regos, A., 2022. Nature-based solutions in an era of mega-fires. Nature 607 (7919), 449. Available at https://doi.org/10.1038/d41586-022-01955-x, https://www.nature.com/nature/.

Regos, A., 2026. A roadmap for nature-based solutions to extreme wildfires. iScience 29 (1), 114364. Available at https://doi.org/10.1016/j.isci.2025.114364.

Regos, A., Pais, S., Campos, J.C., Lecina-Diaz, J., 2023. Nature-based solutions to wildfires in rural landscapes of Southern Europe: let's be fire-smart!. International Journal of Wildland Fire 32 (6), 942–950. Available at https://doi.org/10.1071/WF22094, http://www.publish.csiro.au/nid/126.htm.

Ringland, G., 2010. The role of scenarios in strategic foresight. Technological Forecasting and Social Change 77 (9), 1493–1498. Available at https://doi.org/10.1016/j.techfore.2010.06.010.

Rossa, C.G., Fernandes, P.M., 2018. An empirical model for the effect of wind on fire spread rate. Fire 1 (2), 31. Available at https://doi.org/10.3390/fire1020031.

Rothermel, R., 1972. A Mathematical Model for Predicting Fire Spread in Wildland Fuels.

Sakellariou, S., Tampekis, S., Samara, F., Sfougaris, A., Christopoulou, O., 2017. Review of state-of-the-art decision support systems (DSSs) for prevention and suppression of forest fires. Journal of Forestry Research 28 (6), 1107–1117. Available at https://doi.org/10.1007/s11676-017-0452-1, https://www.springer.com/journal/11676.

Salgado-Rojas, J., Hermoso, V., Álvarez-Miranda, E., 2025. prioriactions: multi-action management planning in R. Methods in Ecology and Evolution 16 (6), 1028–1039. Available at https://doi.org/10.1111/2041-210X.14220, http://onlinelibrary.wiley.com/journal/10.1111/(ISSN)2041-210X.

Scheller, R.M., Domingo, J.B., Sturtevant, B.R., Williams, J.S., Rudy, A., Gustafson, E.J., Mladenoff, D.J., 2007. Design, development, and application of LANDIS-II, a spatial landscape simulation model with flexible temporal and spatial resolution. Ecological Modelling 201 (3–4), 409–419. Available at https://doi.org/10.1016/j.ecolmodel.2006.10.009.

Sil, Â., Azevedo, J.C., Fernandes, P.M., Alonso, J., Honrado, J.P., 2022. Fine-tuning the BFOLDS fire regime module to support the assessment of fire-related functions and services in a changing Mediterranean mountain landscape. Environmental Modelling and Software 155. Available at https://doi.org/10.1016/j.envsoft.2022.105464, www.elsevier.com/inca/publications/store/4/2/2/9/2/1.

Sil, Â., Azevedo, J.C., Fernandes, P.M., Honrado, J.P., 2024. Will fire-smart landscape management buffer the effects of climate and land-use changes on fire regimes? Ecological Processes 13 (1). Available at https://doi.org/10.1186/s13717-024-00535-3, https://ecologicalprocesses.springeropen.com.

Silva-Cardoza, A.I., Vega-Nieva, D.J., Briseño-Reyes, J., Briones-Herrera, C.I., López-Serrano, P.M., Corral-Rivas, J.J., et al., 2022. Evaluating a new relative phenological correction and the effect of sentinel-based earth engine compositing approaches to map fire severity and burned area. Remote Sensing 14 (13). Available at https://doi.org/10.3390/rs14133122, https://www.mdpi.com/2072-4292/14/13/3122/pdf?version=1656495196.

Slavkovikj, V., Verstockt, S., Van Hoecke, S., Van De Walle, R., 2014. Review of wildfire detection using social media. Fire Safety Journal 68, 109–118. Available at https://doi.org/10.1016/j.firesaf.2014.05.021, https://www.sciencedirect.com/science/journal/03797112.

Talaat, F.M., ZainEldin, H., 2023. An improved fire detection approach based on YOLO-v8 for smart cities. Neural Computing and Applications 35 (28), 20939–20954. Available at https://doi.org/10.1007/s00521-023-08809-1, https://www.springer.com/journal/521.

Tavra, M., Racetin, I., Peroš, J., 2021. The role of crowdsourcing and social media in crisis mapping: a case study of a wildfire reaching croatian city of split. Geoenvironmental Disasters 8 (1). Available at https://doi.org/10.1186/s40677-021-00181-3.

Tedim, F., Leone, V., Xanthopoulos, G., 2016. A wildfire risk management concept based on a social-ecological approach in the European Union: fire Smart Territory. International Journal of Disaster Risk Reduction 18, 138–153. Available at https://doi.org/10.1016/j.ijdrr.2016.06.005.

Vacchiano, G., Foderi, C., Berretti, R., Marchi, E., Motta, R., 2018. Modeling anthropogenic and natural fire ignitions in an inner-alpine valley. Natural Hazards and Earth System Sciences 18 (3), 935–948. Available at https://doi.org/10.5194/nhess-18-935-2018.

Wang, L., Qu, J.J., Hao, X., 2008. Forest fire detection using the normalized multi-band drought index (NMDI) with satellite measurements. Agricultural and Forest Meteorology 148 (11), 1767–1776. Available at https://doi.org/10.1016/j.agrformet.2008.06.005.

Wu, Z., Wang, B., Li, M., Tian, Y., Quan, Y., Liu, J., 2022. Simulation of forest fire spread based on artificial intelligence. Ecological Indicators 136, 108653. Available at https://doi.org/10.1016/j.ecolind.2022.108653.

Xu, H., Chen, J., He, G., Lin, Z., Bai, Y., Ren, M., et al., 2024. Immediate assessment of forest fire using a novel vegetation index and machine learning based on multi-platform, high temporal resolution remote sensing images. International Journal of Applied Earth Observation and Geoinformation 134, 104210. Available at https://doi.org/10.1016/j.jag.2024.104210.

Yebra, M., Quan, X., Riaño, D., Rozas Larraondo, P., van Dijk, A.I.J.M., Cary, G.J., 2018. A fuel moisture content and flammability monitoring methodology for continental Australia based on optical remote sensing. Remote Sensing of Environment 212, 260–272. Available at https://doi.org/10.1016/j.rse.2018.04.053, www.elsevier.com/inca/publications/store/5/0/5/7/3/3.

Zhang, G., Wang, M., Liu, K., 2021. Deep neural networks for global wildfire susceptibility modelling. Ecological Indicators 127 (2), 107735. Available at https://doi.org/10.1016/j.ecolind.2021.107735.

Ziegler, J.P., Ho, C.M., Mell, W., 2019. firebehavioR: An R Package for Fire Behavior and Danger Analysis.

Zigner, K., Carvalho, L.M.V., Peterson, S., Fujioka, F., Duine, G.J., Jones, C., et al., 2020. Evaluating the ability of farsite to simulate wildfires influenced by extreme, downslope winds in santa barbara, california. Fire 3 (3), 1–23. Available at https://doi.org/10.3390/fire3030029, https://www.mdpi.com/2571-6255/3/3/29/pdf.

Index

Note: Page numbers followed by "*f*" and "*t*" refer to figures and tables, respectively.

Q

R

S

T

9780443416927